T

U

V

W

Y

Z

CELL BIOLOGY RESEARCH PROGRESS

AUTOPHAGY

PRINCIPLES, REGULATION AND ROLES IN DISEASE

CELL BIOLOGY RESEARCH PROGRESS

Additional books in this series can be found on Nova's website
under the Series tab.

Additional E-books in this series can be found on Nova's website
under the E-book tab.

CELL BIOLOGY RESEARCH PROGRESS

AUTOPHAGY

PRINCIPLES, REGULATION AND ROLES IN DISEASE

NIKOLAI GORBUNOV
EDITOR

Nova Science Publishers, Inc.
New York

For permission to use material from this book please contact us:
Telephone 631-231-7269; Fax 631-231-8175
Web Site: http://www.novapublishers.com

Library of Congress Cataloging-in-Publication Data

Autophagy : principles, regulation, and roles in disease / [edited by] Nikolai Gorbunov.
p. cm.
Includes index.
ISBN 978-1-61942-266-7 (hardcover)
1. Autophagic vacuoles. I. Gorbunov, N. V. (Nikolai Vasil evich)
QH603.L9A98 2011
571.6'55--dc23

2011045801

Published by Nova Science Publishers, Inc. † New York

Contents

Preface

Tissue homeostatic responses to change of environmental conditions, genetically-programmed transformations, injury, and a variety of diseases require adjustment and/or reconstitution of the cellular content in order to support physiological balance, development and remodeling. Since the late 1950s and early 1960s the above events have been closely associated with a phenomenon named "autophagy" {a term derived from a Greek word meaning "self" (*auto*) and "eating" (*phagy*) [1]}. Current paradigm considers autophagy as an evolutionary conserved and strictly regulated lysosomal pathway employed by eukaryotic cells to degrade unnecessary or compromised organelles, cytoplasmic and endocytosed materials and invaded pathogens [2-4]. The morphology of autophagy was first identified in mammalian cells in the 1950s, and extensive morphological and pharmacological studies defined the basic steps of this process [5, 6]. Thus, it has been described several different types of autophagy based on (i) nature of cellular target (autophagosomes "cargo") of degradation, (ii) the mechanism of "cargo" delivery to lysosomes, and (iii) the pathways for delivery of "cargo" materials to autophagosome constructions [3-5, 7]. Thereby, formation of autophagosomes, double-membrane vacuoles lacking enzymes that are originated from the membrane of different organelles, is a major morphological feature of macroautophagy (often referred in the chapters of this book as autophagy), and autophagosomes *per se* serve for sequestration and delivery of intracellular components to lysosomal processing [2, 4-7].

The autophagy research has witnessed a remarkable growth since the initial description of autophagosomes that were seen by electron microscopy [1,2,4,6,7]. In particular, during past decade this topic has attracted widespread attention due to discovery of autophagy-related genes, autophagy-regulatory mechanisms, and invention of new autophagy research techniques [2,4,7]. Thus it has been determined that macroautophagy is controlled by the highly conserved *atg* –genes (*atg* for autophagy-related genes) [2-8]. Suppression of these genes and impairment or dysregulation of autophagy are associated with severe diseases and aging, and are shown to cause death in cell and animal models [2-4, 8, 9]. Therefore, overall autophagy is considered to be adaptive and protective cellular mechanisms [4,5,10]. Interestingly, as many other crucial pathways, autophagy plays dual role under normal and pathological conditions. In addition to the well-known role of autophagy in cell survival, autophagy-mediated type II programmed cell death has long been proposed [4,10,11]. The autophagic cell death was originally reported in tissues subjected active development and remodeling. That effect seems to be analogous to apoptosis in similar metamorphoses. However, the proposed function of autophagy in cell death is not restricted to the developmental programmed cell death; it also extends to cell death that occurs under various pathological conditions, e.g., immune disorders, tissue degeneration, gastrointestinal impairment, infections, *etc.* [2-4, 8-12]. Presumably, uncontrolled activation autophagy capable of massive removal of subcellular structures such as mitochondria, can be detrimental to the cell and tissue functions.

The objectives of the book "Autophagy: Principles, Regulation and Roles in Disease " are to give an up-to-date overview of biomedical aspects of autophagy research, a large interdisciplinary field emerging on intersection of (i) cell, developmental, and cancer biology, (ii) tissue remodeling and plasticity, aging and longevity; (iii) immunology, microbiology, toxicology, and pathobiology; (iv) regenerative and adaptive biomedicine; and (v) pharmacology. The provided analyses emphasize that in-depth understanding of the paradoxical role of autophagy in promoting cell death and survival is crucial for development of target therapy of autophagy-driving diseases. Giving a wide-angle perspective on biomedical aspects of autophagy we addressed this book to a broad audience of readers from students to practicing clinicians. Each chapter of this book is supplemented with extensive lists of references that include sources of the original data and concepts.

The field is expansive and the book is divided into three parts: Part 1 – Autophagy and immunity, immune disorders, inflammation and infection; Part 2 – Autophagy in cancer research and cancer therapy; Part 3 - Autophagy response to injury, metabolic and degenerative disorders, and exposure to nanomaterials.

Acknowledgment

The editor wishes to thank the Armed Forces Radiobiology Research Institute (AFRRI) and the Uniformed Services University of the Health Sciences (USU) for their encouragements and support in the preparation of this book.

References

[1] De Duve C, Wattiaux R. Annu Rev Physiol. 1966; 28:435-92.
[2] Eskelinen EL, Saftig P. Biochim Biophys Acta. 2009 1793(4):664-73.
[3] Todde V, Veenhuis M, van der Klei IJ. Biochim Biophys Acta. 2009 1792(1):3-13.
[4] Yang Z, Klionsky DJ. Nat Cell Biol. 2010 12(9):814-22.
[5] Yang Z, Klionsky DJ. *Curr Top Microbiol Immunol. 2009*; 335:1-32.
[6] CLARK SL Jr. *J Biophys Biochem Cytol.* 1957;3(3):349-62.
[7] Cuervo AM. *Nat Cell Biol.* 2010 12(8):735-7.
[8] Eskelinen EL. *Curr Opin Pharmacol.* 2011 11:294-300.
[9] Komatsu M, et al. *Nature.* 2006; 441:880–884.
[10] Das DK. *Antioxid Redox Signal.* 2011 14:2163-4.
[11] McPhee CK, Logan MA, Freeman MR, Baehrecke EH. *Nature.* 2010 465:1093-6.
[12] Kirkegaard K, Taylor MP, Jackson WT. *Nat Rev Microbiol.* 2004 2(4):301-14.

Part I.
Autophagy and Immunity, Immune Disorders, Inflammation and Infection

In: Autophagy: Principles, Regulation and Roles in Disease ISBN: 978-1-61942-266-7
Editor: Nikolai Gorbunov

Chapter I

The Diverse Roles of Autophagy in Immunity

Ian X. McLeod, Wei Jia and You-Wen He [*]
Department of Immunology,
Duke University Medical Center, Durham, NC, US

Abstract

Autophagy is a highly conserved process in all eukaryotic organisms that has been adapted for a multitude of divergent functions. The mammalian immune system is no exception, with various hematopoietic cells utilizing this intracellular remodeling process for immune homeostasis.

In this chapter, we will focus on the role of autophagy in three major areas of immune function: homeostasis of lymphocytes, antigen presentation and pathogen clearance, and thymic selection for immune tolerance.

First, autophagy has been shown to be crucial and indispensible for the survival and intracellular homeostasis of both T and B lymphocytes *in vivo*.

This has been demonstrated using multiple genetic models of autophagy ablation as well as pharmacological inhibition of various autophagic signaling pathways. Second, autophagy has been shown to enhance antigen presentation in several types of leukocytes of the innate immune system, including macrophages and dendritic cells, and to both promote and inhibit the clearance of several intracellular pathogens.

Lastly, in a very nascent field of autophagy in thymic epithelial cells, autophagy promotes the processing and surface display of many endogenous antigens in order to influence negative selection and tolerance of developing thymocytes. These examples will clearly demonstrate the complex applications that the panoply of immune cells has evolved to initiate and maintain immune function using an archaic process found in the most primal of eukaryotes.

[*] Correspondence to: You-Wen He, Department of Immunology, Box 3010, Duke University Medical Center, Durham, NC 27710. Phone: 9196137870; Fax: 91960848982. E-mail: he000004@mc.duke.edu.

1. Introduction-General Principles of Autophagy

Autophagy, an intracellular catabolic process that involves the generation of a double membrane structure, sequestration of various cytoplasmic components, and fusion with acidifying compartments, is conserved in all eukaryotic cell types. This process has been implicated in cellular survival during periods of starvation, cytokine withdrawal, removal of extraneous or damaged organelles, and turnover of long-lived proteins. Autophagy can be divided into three main classes, depending on the magnitude and nature of the import of substrates to be catabolized [1]. Macroautophagy, the best understood and most thoroughly studied form of autophagy, involves the *de novo* synthesis of autophagosomes around bulk substrates to be degraded and fusion of these autophagosomes with degradative vesicles to form autolysosomes. Microautophagy, while less studied, involves the small-scale blebbing and sequestration of soluble components within the immediate microenvironment of the previously formed autophagosome. Chaperone-mediated autophagy (CMA) involves the import of misfolded proteins, aggregates, ubiquitinated proteins, or even organelles into pre-formed autophagosomes via adaptors and receptors on the autophagosome. CMA is now considered to be of particular interest in the turnover of damaged organelles, and the clearance of pathogenic aggregates implicated in several human diseases, such as Parkinson's and Huntington's disease[2, 3]. Transporters in the autolysosome can convey nutrients and other recycled macromolecules back to the cytoplasm after lysosomal processing in all three forms of autophagy.

To date, at least 35 different gene products in yeast have so far been identified that play some role in the autophagic process [4], and the number of adapter and ancillary proteins associated with CMA are rapidly expanding. The autophagy machinery utilizes the coordinated activity of the class III PI3K, Vps34, to produce PI(3)P, and Beclin-1, which recruits a myriad of effectors with FYVE- or GLUE-domains, in order to flex and fuse membranes [5]. The serine/threonine kinase Ulk1 (Atg1l) is dependent on mTOR inhibition by TSC1/2 under periods of serum or nutrient starvation, and is essential for autophagy induction, though its direct targets are still being elucidated [6, 7]. The Ulk1 and PI3KIII complexes are brought together in active forms by the action of RalB on the nascent autophagosomal membrane during nutrient starvation [8]. Additionally required for membrane elongation are cascades of two ubiquitin-like conjugation systems, one of which involves the conjugation of Atg5-Atg12-Atg16, the function of which remains poorly understood. The other involves the cleavage, lipidation, and recycling of Atg8/LC3, the function of which may be related to the import of ubiquitinated aggregates into the autophagosome or other chaperone-mediated autophagic substrates [9].

While all eukaryotic cells studied thus far seem to have some form of autophagic capability, the levels and signals that induce them seem to vary widely. Many cell types seem to have a certain level of basal autophagy, which varies widely depending on cell type. For example, medullary thymic epithelial cells (mTECs), a major focus of this chapter, have extremely high levels of constitutive autophagy [10]. On the other end of the spectrum, mature naïve T lymphocytes have a very low level of constitutive autophagy (though still detectable), but can upregulate autophagy in response to several stimuli described below [11, 12]. Generally speaking, starvation of either serum or nutrients is a signal that induces a moderate level of autophagy, while stimulation of several immunity related receptors, such as the T cell receptor (TCR) or Toll-like receptors (TLRs), provide a very strong induction signal. Though the precise function of autophagy varies in specialized immune cells, as discussed below, autophagy in all cell types studied is an alternative source of nutrients scavenged from the cytoplasm, an alternative form of cellular degradation, augmenting the proteasomal system, and a major pathway for intracellular remodeling to reflect changes in the developmental stage of the cell. In this chapter, however, we will focus on specific roles of autophagy in highly specialized cells. First, how does autophagy support the homeostasis of lymphocytes and their precursors? Second, how is autophagy involved in antigen presentation, and innate immune sensing? Last, how does autophagy in non-hematopoietic cells support the development and education of thymocytes by combining strategies discussed in the first two sections?

2. Immune Homeostasis

Immune homeostasis requires the coordinate activity of two branches of the immune system, innate and adaptive. The two systems are highly intertwined, and intersect at several stages during a functional immune response. The adaptive immune system is dominated by T and B lymphocytes, which, due to the specificity of their cognate receptors, can form protective immune responses to diverse antigens including long-term immunologic memory. However, the adaptive system does not operate within a vacuum, and requires the innate immune system, with its pathogen sensing, uptake, and clearance duties to form a microenvironment in which lymphocytes can be primed, proliferate, and form a response leading to lasting protection. Specifically, the adaptive immune system requires constant contact with cells of the innate immune system, such as macrophages and dendritic cells (DCs), for survival cues and danger signals [13, 14]. Once these cues and signals have been deciphered by lymphocytes, they are still dependent upon the innate immune system to dictate the nature and magnitude of the immune response, through the induction of cytokines, and costimulatory ligands, both agonist and inhibitory. Finally, when the present danger has been eliminated, the adaptive immune system is dependent upon the signals received from the innate immune system to shut down operations and return to a homeostatic state, with antigen-specific lymphocyte clones ready to undertake the entire process once again.

Autophagy has been shown to play important roles in many of these stages of interaction between the innate and adaptive immune system. First, autophagy is essential for the homeostasis of immune effector cells, especially lymphocytes [12, 15, 16]. Second, autophagy also plays prime roles in both the recognition and clearance of pathogens (intracellular and extracellular) by the innate immune system [17], and if this is not sufficient to resolve the infection, antigen processing and presentation to cells of the adaptive immune system [17, 18]. Autophagy also helps to dictate the type of immune response by influencing cytokine production and secretion, key intercellular mediators in the immune system [19, 20]. Lastly, thymic epithelial cells (TECs), though not hematopoietic in lineage, have a role in developing immune tolerance that cannot be understated, and autophagy is indispensible for this function. By utilizing autophagy to present a diverse array of cell type specific potential antigens, under the control of Aire, TECs ensure that only functional but bridled T cells survive [21]. This chapter will highlight the important advances in our understanding of how autophagy contributes to a healthy immune system, capable of handling numerous specialized functions in the panoply of immune effector cells.

3. Autophagy in Lymphocytes

Lymphocytes are small (10-20uM), essential mediators of the adaptive immune system. Naïve, antigen-inexperienced lymphocytes have extremely limited cytoplasmic volume and are easily identified by a dominant nucleus. Thus it was surprising to note that autophagy genes were expressed in both T and B lymphocytes, including Atg5, Atg7, Beclin-1, LC3, and Atg3 [11, 12, 22]. Both fluorescence microscopy of LC3-GFP and TEM of highly characteristic double membrane structures have been utilized to identify autophagic structures in T lymphocytes, not only after starvation, stimulation by certain cytokines [23], or TCR and BCR engagement [12, 24], but even at basal levels [11]. In retrospect, the presence of autophagy in these lymphocytes with such limited cytoplasmic volume may be even more crucial for survival and homeostasis due to greater space limitations.

3.1. Autophagy Induction in Lymphocytes

As previously mentioned, autophagosomes are present in freshly isolated T and B lymphocytes. The number and size of autophagosomes are both increased slightly under serum or amino acid starvation, and this increase can be ablated by PI3K inhibitors, such as 3-MA and wortmannin, suggesting that the activity is regulated by mTOR and PI3K activity [11]. Stimulation of the TCR by anti-CD3 for 24-48 hours *in vitro* further increases levels of autophagy, as measured both by LC3 processing and LC3 punctae formation [11, 12]. Additionally, chronic restimulation of human $CD8^+$ T cells with a mitogen results in a large (3-4 fold) induction of autophagy by EM after long term culture in vitro [25]. This TCR-activation induced autophagy is also susceptible to PI3K inhibitors, such as 3-MA, as well as JNK inhibitors, and can be rapidly enhanced by administration of rapamycin (affecting mTOR) and zVAD (affecting caspase activity), suggesting a role for both these pathways, as well as PI3K activity and MAP kinase pathways [25]. The involvement of this diverse array of pathways should come as no surprise, as T cell homeostasis and activation require an immense cytoplasmic reorganization and mobilization of energy resources to affect a functional immune response [26]. Finally, autophagy is highly induced by the binding of the HIV envelope glycoprotein, gp120, to CXCR4 on the surface of CD4+ T cells [27]. This autophagy is induced on such a massive scale, that it has been implicated in the type II, caspase-independent programmed cell death pathway (PCDII), when apoptosis was inhibited in these bystander cells. Of note, apoptosis and autophagy were completely inhibited after CXCR4 engagement by knockdown of Atg7 or Beclin-1, or by addition of 3-MA or bafilomycin to inhibit autophagy [28]. The engagement of CXCR4 in autophagy induction strongly suggests a role for GPCRs in the induction of autophagy, if not more specifically, chemokine receptors, which could tune autophagy levels of lymphocytes entering specific microenvironments to affect function.

Autophagy induction in B lymphocytes follows some similar trends to T lymphocytes. Autophagy can be induced in B cells after prolonged BCR stimulation though IgM, leading to apoptosis [24]. This autophagy was inhibited through CD40 engagement, as was apoptosis, perhaps by the induction of Bcl-2 to inhibit both apoptosis and autophagy, as Bcl-2 can both prevent Bax activity and Beclin-1 association with Vps34 [29, 30]. This autophagy was also susceptible to wortmannin, once again suggesting roles for PI3K in autophagy induction.

Autophagy is also important for the development of B cells. Using a GFP-Beclin fusion, it was observed that Beclin-1 was expressed very highly in Pro-B cells, but downregulated into Pre-B transition [31], very analogous to high levels of Atg5, Beclin-1, and LC3 seen in DN thymocytes, but not DP thymocytes [12]. This correlates well with the observation that Atg5 is dispensable for later B cell development, using a CD19-cre transgene to induce deletion, in which the only phenotype observed was in self-renewing B1 cells being markedly decreased [32]. However, when Atg5 was knocked out using Atg5 null fetal liver chimeras, levels of Pre- and immature B cells were drastically reduced [12]. This suggests a stage-specific requirement for autophagy in B cells that affects development and survival of self-renewing populations but not mature, circulating B2 cells.

Even more recent evidence has confirmed that hematopoietic lymphocyte precursors require autophagy, not only for intracellular homeostasis and proper organelle quality control, but also to maintain a steady population not susceptible to leukemic transformations. When Atg7 was deleted in hematopoietic stem cells (HSCs), a rapid, myoproliferative disease developed, and yet paradoxically, Atg7-deficient HSC's could not reconstitute an irradiated mouse, nor support either lymphoid or myeloid development [33]. Therefore, autophagy is essential to maintain self-renewing populations of cells.

3.2. Autophagy in T Cell Survival and Death

Although the stage-specific requirements for autophagy in T and B cell development seem to be acute in early stages, autophagy plays a role in T cell homeostasis and survival as well. Autophagy is a crucial cell survival strategy in mature naïve T cells. Since T lymphocytes split time between cytokine and nutrient rich secondary lymphoid organs, nutrient rich but cytokine poor blood, and lymphatic ducts that are replete for both, it is essential that they are able to adjust to long and indefinite periods of starvation. T lymphocytes are acutely sensitive to apoptosis, and the levels of anti-apoptotic Bcl-2 family members, such as Bcl-2, Bcl-Xl, and Mcl-1 are highly regulated during development and cytokine signaling [34-36]. Therefore, it is of little surprise that cytokine signals, which are critical to maintain lymphocyte homeostasis, play substantial roles in the homeostatic autophagic process.

In the absence of core autophagy proteins Atg5 (using fetal liver chimeras and T cell-specific deletion) [12, 37], and Atg7 (utilizing a conditional knockout under an Lck driven cre transgene) [22], thymocyte development was essentially normal, with slight reductions in thymocyte number, but no increase in apoptosis ex vivo. However, there was a drastic reduction in the periphery of these autophagy-deficient cells, with reductions of up to 50% of mature $CD4^+$ and $CD8^+$ T lymphocytes, which rapidly succumb to apoptosis *ex vivo*, even in the presence of high levels of pro-survival homeostatic cytokines, and fail to proliferate in response to TCR stimulation [22]. Therefore, even though cytokines and constant TCR-MHC contact are required for the survival of naïve T lymphocytes, autophagy provides a non-redundant function for survival of these cells going beyond the requirements of other homeostatic signals.

A recent paper by *Kovacs et al.* demonstrated the importance of the PI3KIII complex protein, Beclin-1, in the survival and polarization of T cells during experimental autoimmune encephalitis (EAE). Using a CD4-cre mediated deletion of Beclin-1, the authors show that while peripheral T cell numbers are reduced to similar numbers as in an Atg7 knockout mouse, mitophagy remains intact. T cells rapidly die after TCR stimulation by apoptosis in the absence of Beclin-1, seemingly due to the inability to degrade the pro-apoptotic molecules pro-caspase-3, pro-caspase-8, and Bim. Additionally, cells polarized to the Th1 lineage have defects in survival and are resistant to EAE induction [38].

Autophagy has been shown to be critical for the removal of damaged or extraneous organelles in several systems. Parkin in neurons has been shown to be an E3 ubiquitin ligase for targeting mitochondria for mitophagy, especially when the outer mitochondrial membrane is damaged or stripped away [39, 40]. The BH3 only protein Nix has recently been shown to induce mitophagy in developing erythrocytes to leave them devoid of mitochondria [41], a process dependent on core autophagy machinery [42]. Autophagy in T lymphocytes has also been shown to prevent death by the removal of excess organelles *in vivo*. As mature SP thymocytes egress, the requirement or tolerance for mitochondria decreases, as they take on a mature, naïve phenotype. Autophagy deficient T cells are not able to complete this mitophagy, and can be found in the periphery containing increased numbers of mitochondria and ROS production [22], which are toxic and can induce apoptosis [43]. This is further confirmed by an enrichment in mitochondrial proteins present in autophagy deficient T lymphocytes, and the transcriptional upregulation of genes relating to mitochondrial maintenance [22, 37]. It remains to be determined what signal(s) induce this potent mitophagy in lymphocytes, and what developmental stage-specific requirement for mitophagy exists in T lymphocytes.

Autophagy can extend influence over other organelles besides mitochondria. Pexophagy, or the autophagy of peroxisomes has been described in cellular systems as a method by which cells can turn over excess peroxisomes, thereby reducing redox activity and eliminating cytotoxic ROS [44]. In T lymphocytes, the endoplasmic reticulum (ER) is carefully regulated in such a manner. In T cells inducibly lacking Atg7, ER membrane is expanded, and calcium storage is highly dysregulated [45]. As previously mentioned, these Atg7 deficient cells have a deficient response to mitogen in proliferative capacity, consistent with expanded ER and defective calcium flux [45]. The calcium storage is actually higher in the ER, leading to an imbalance between the intracellular and extracellular Ca^{2+}, in which calcium channels are open for a shorter period of time during TCR stimulation [45]. It remains to be determined whether ER turnover is regulated by autophagy during periods of cellular stress, or whether the ER membrane is naturally used in T lymphocytes as a membrane donor for autophagosomes, thus leading to a steady state of calcium storage over the life to a T cell. However, using an inducible-deletion system for the core autophagy protein Atg3, we have recently shown that ER expansion takes place over a period of weeks after deletion, suggesting a constant program of membrane trimming to counteract a constitutively active ER metabolic process [46].

Thus autophagy is essential for turning over excess organelles, or consistently trimming the levels of extraneous membranes. Both these processes can be viewed as stage specific in

T cells, but the possibility remains that T lineage cells simply have differential tolerances to the accumulation of defects that comes from autophagy inhibition. It remains to be seen whether lymphocytes have soluble targets or aggregates targeted to autophagosomes, other than the previously mentioned pro-death effector molecules [38], either in the context of TCR-signaling, or during naive homeostasis.

4. Autophagy in Innate Sensing and Antigen Presentation

The innate immune system is composed of a variety of immune sensing and effector cells. Of these, autophagy has been studied extensively in macrophages and DCs, as these cells have well defined roles in phagocytosis, antigen processing, and presentation. Since autophagy is most simply defined as a lysosomal targeting pathway, it is a logical transition to assume that it could affect outcome of these other processes, all of which involve late endosomal or lysosomal compartments. Macrophages and DCs have several major roles in immune protection, and autophagy influences three major roles in separate but related aspects: Pathogen clearance, antigen processing and presentation, and much more recently, pathogen sensing.

4.1. Autophagy in Pathogen Clearance

Though autophagy is responsible for intracellular remodeling, it can be centrally defined as a cytoplasmic to lysosomal targeting pathway. Likewise, phagocytosis can be defined as the formation of vesicles at the plasma membrane for lysosomal targeting, either for basic nutrition or for pathogen restriction. These vesicles undergo a functional maturation process, in which phagosomes are acidified and subjected to various proteases to cleave proteins and other macromolecules, ultimately fusing with a lysosome. Additionally, TLRs and other pattern recognition receptors (PRRs) present in the phagocytic pathway have been shown to play important roles in the induction of autophagy in innate immune cells [19, 47, 48]. Therefore, autophagy and phagocytosis have been shown to rely on similar molecules and intersect at several stages. DCs and Macrophages are especially active in phagocytic roles in order to sense and clear pathogens, and have employed autophagy as a mechanism by which to aid the phagocytic process in order to restrict both intracellular and extracellular pathogens.

Toll like receptors (TLRs) are a major class of PRRs, several of which are localized in endomembrane bound vesicles [49]. LPS, a ligand for TLR4, is a major cell wall component for many pathogens, including several intracellular species. When RAW 264.7 macrophages were stimulated with LPS, they rapidly induced LC3 processing and punctate formation [50]. This signaling was not mediated through the canonical MyD88 adaptor, but a MyD88-independent pathway, involving TRIF (Toll/IL-1 receptor domain-containing adaptor inducing IFN-β). This was shown by siRNA knockdown of TLR4, TRIF, and MyD88 [50]. Other TLR agonists, including ssRNA, PolyI:C, zymosan, and imiquimod, had similar effects on RAW 264.7 macrophages [51]. Many of the TLRs are themselves localized in phago/endosomal compartments, including TLR3, TLR7, and TLR9. In bone marrow derived

DCs, the NOD2 receptor was shown to induce autophagy when stimulated with muramyldipeptide in an Atg5, Atg7, and Atg16L-dependent manner, which resulted in a clearance of pathogen [52]. The previously mentioned LPS induced autophagy led to increased overlap between mycobacterium containing endosomes and autophagosomes [50]. Another study tested TLR3, 4, and 7 ligands for autophagy induction in RAW 264.7 cells using ssRNA and imiquimod, and observed LC3 punctae formation to be dependent on TLR-7 and MyD88 at early stages [53]. This TLR-7 dependent autophagy induction was found to target mycobacterium to autolysosomes and was able to overcome pathogen escape mechanisms for increased clearance [53]. Interestingly, the dependence on particular adaptors to induce autophagy seems to be related not only to the specific TLR stimulated, but also the time course of the stimulation, with MyD88 pathways dominating the early phase, and MyD88-independent pathways being necessary for later responses [50, 53].

Several studies using autophagy-related gene knockout mice and pharmacological inhibition of the autophagy pathway have shown that the outcome or polarization of the response to pathogen is dependent on autophagy. Inhibiting macroautophagy by administration of wortmannin or by knocking out Atg5 in plasmacytoid dendritic cells (pDCs) had an adverse effect on both the recognition and clearance of *Vesicular stomatitis virus* (VSV) in a TLR-7 dependent fashion [19]. This autophagic inhibition also led to decreased IFN-α secretion, a tell-tale cytokine in viral recognition that limits proliferation [19]. Group A *Streptococcus* proliferates with reckless abandon in the cytosol after endosomal evasion of MEFs in the absence of Atg5. However, they are rapidly sequestered by autophagic membranes and delivered to the lysosome in WT MEFs [54]. In other vertebrate infection models, macroautophagy can regulate the pathogenesis of the herpes simplex virus 1 in neurons, as virions lacking the beclin-1 interaction domain of ICP34.5 were unable to inhibit autophagy, through excluding Beclin-1 from the Vps34 complex, and replicated to lower viral loads and were deficient in intracerebral pathology [55]. Notably, the Levine group has also shown autophagy to be protective in neurons against *Sindbis* virus, not only for clearance of the viral load, but also for cytoprotective effects to counter apoptosis [56].

In a highly specialized manner, autophagy has been shown to protect APCs from adverse side effects and death associated with *Shigella* infection. During the usual lifespan of *Shigella*, phagosome rupture is induced, leading to fragmentation of phagosomes, and uncontrolled PRR signaling, leading to NFκB activation, dysregulated ROS production, and apoptosis. Chaperone mediated autophagy actually targets these signaling fragments for destruction through p62 and LC3 interactions, thus functioning in an anti-inflammatory pathway [57].

In sharp contrast to many of the previously mentioned studies, some viral sensing mechanisms are inhibited by autophagy, and some pathogens take advantage of autophagy related genes and intermediate structures as part of their normal life cycle. The classes of PRRs that fall into this category seem to exclude TLRs and have been studied more extensively in non-professional APC's. In MEFs, cytosolic based PRR's include the Rig-1-like receptor family (RLRs). In Atg5-deficient MEF's, type I interferon production was actually increased using dsRNA as a stimulus [48]. In agreement with the previous study, Atg5-deficient primary liver macrophages had higher production of type I interferon after RIG-1 stimulation [58]. When trying to build a consensus about the role of autophagy in response to PRR stimulation, it is important to remember that none of these systems exist in a vacuum, and many pathogen associated molecular patterns (PAMPs) can stimulate several PRR pathways. Most feasibly, it is the relative levels of induction and inhibition through orthogonal or even opposing pathways that determine the induction and contribution of autophagy in the ultimate progression of the immune response.

Several families of viruses use the autophagic membrane as a scaffold for replication. They are all single-stranded RNA viruses, including picornavirus (rhinovirus and poliovirus) [59, 60]. Perhaps uniquely, certain nidoviruses (coronaviridiae) such as MHV replicate normally in autophagy sufficient or deficient cells, regardless of whether autophagy is induced or not. However, when LC3 is depleted, replication is severely impacted. These viridiae actually bind to LC3 I, the non-lipidated, diffuse cytoplasmic form of the protein, and use it as a scaffold for replication [61]. Thus it is often difficult to predict how autophagy will affect the immune response to a given pathogen, as many have evolved mechanisms with which to subvert or utilize this process. However, a good rule of thumb seems to be that the completion of autophagy (fusion of amphisomes and phagosomes with lysosomes) is generally restrictive towards pathogen survival and escape, while incomplete autophagosomal maturation is beneficial towards pathogen replication.

4.2. Autophagy in Antigen Presentation

Antigen presentation is a process that most cells undergo in order to process and present peptides (in the context of MHC) or other macromolecules such as glycolipids (in the context of CD1d) to alert the adaptive immune system of the cytosolic health of the cell [62, 63]. Cells that have been transformed by mutagenesis or infected by intracellular pathogens can process these intracellular events and present these transformations to activated cells of the adaptive immune system, including T lymphocytes and NKT cells. In the case of T lymphocytes, $CD4^+$ T cells are restricted to peptides presented to them on MHC II molecules by professional APCs such as macrophages, DCs, and activated B cells. This is mainly accomplished through products derived from lysosomal degradation, and the peptides derived from phagocytosed remnants of other infected cells. As a result, most of the peptides generated from this process tend to be derived from membrane bound or secreted proteins, accessible to phagocytosis, and loaded onto MHC II molecules in MHC II bearing vesicles, structurally very similar to late endosomes. Cytotoxic $CD8^+$ lymphocytes are restricted to MHC I molecules present on most cell types. Peptides loaded onto MHC I molecules are thought to be mainly derived from proteasomal pathways, present in the cytoplasm and traditionally thought to be restricted to cytoplasmic proteins. The loading process is accomplished by import of processed peptides into the ER through the TAP transporter and direct loading onto *de novo* synthesized MHC I molecules. However, for a $CD8^+$ T cell to be primed in order to recognize transformed cells, it must first be activated by a professional APC, which may or may not be a target of the given pathogen. Therefore, in a process called cross presentation, phagocytosed material from infected cells is subjected to lysosomal or proteasomal treatment by APCs and loaded onto cognate MHCs for antigen presentation to T cell subsets. Macroautophagy is important in the context of many pathogens for accomplishing this task, in both lysosomal dominated MHC II peptide loading, and more recently described in MHC I cross presentation.

Much of the original work investigating the role of autophagy in antigen presentation came from studies looking at the nature of peptides presented on MHC I or II. Using acid extraction to isolate peptides from purified MHC molecules, these peptides can be sequenced [64]. Short-lived, cytosolic, and nuclear proteins have been identified as common ligands for MHC I molecules, highlighting the role of the proteasome in the degradative process. For example, on the short end of the half-life spectrum, ornithine decarboxylase, with a half-life of only a few minutes, gives rise to two peptides eluted off MHC I [64]. Even more striking was the elution of ten peptides off MHC I that were derived from the cyclin family of proteins [64], with only one being eluted off MHC II. Cyclin half-lives vary, but are generally degraded within minutes after translation [65]. On the other end of the spectrum, GAPDH, one of the most commonly held "Housekeeping" genes, has a half-life of several days and is highly resistant to proteasome degradation [66]. GAPDH is, however, a common endogenous ligand for MHC II, and has been eluted off five human MHC II molecules, but not once off an MHC I molecule [64, 67]. Hence, long-lived proteins which are proteasome resistant are common MHC II ligands. Consequentially, many of these cytoplasmic proteins are autophagy substrates, and in the case of GAPDH, are chaperone-mediated autophagy substrates [68]. However, the dogma of MHC I ligands being exclusively derived from cytoplasmic proteins processed through the proteasomal pathway is now under siege and will be discussed extensively.

Chaperone mediated autophagy (CMA) has also been shown to assist in the process of antigen processing/MHC II loading of endogenous cytoplasmic proteins. In B cell lines overexpressing the lysosomal integral protein Lamp-2 or the molecular chaperone Hsc70, which binds partially degraded proteins in the cytoplasm for import into the autolysosome, antigen presentation to $CD4^+$ T cells was aided immensely [69]. The relevance to this system *in vivo* remains to be determined, and it bears mentioning that the overexpression of these proteins might not necessarily be related to autophagy, itself, but to other lysosomal processes (such as formation and stabilization). However, this system provides a feasible pathway by which specific amino acid sequences (KFERQ, which binds Hsc70) can be targeted not only for degradation but antigen presentation [70].

4.3. Autophagy in MHC Class II Antigen Processing and Presentation

MHC II loading compartments have been described as multivesicular structures with similar morphologies to amphisomes [71]. These are now thought to be late endosomes. The Münz group has shown GFP-Atg8 autophagosomes to fuse with MHCII loading compartments (MHCIIC) with as high as an 80% overlap in professional APCs [72]. When the peptides for specific antigens were investigated by inducing or restricting autophagy in APCs, the results were consistently striking. Using M. tuberculosis as a model, infected macrophages or DCs in which autophagy was induced by starvation or rapamycin treatment, had greater presentation of the Ag85B antigen. Moreover, DCs infected with *M. tuberculosis* and treated with rapamycin when transferred to a naïve host could provide protection from further challenge [73]. Furthermore, Salmonella-infected murine macrophages stimulated through the NOD2 receptor induced macroautophagy in order to process and deliver antigen to MHCIIC. This processing and presentation was ablated in response to Atg5, -7, and -16L knockdown [52].

Macroautophagy has also been shown to be involved protection from *Y. pestis* infection when endosomal trafficking of MHC II is inhibited by the pathogen [74]. Finally, mice with a DC-specific deletion of Atg5 had deficient $CD4^+$ T cell priming when challenged with HSV and reduced MHCII presentation of viral peptides. However, in this case, inducing autophagy metabolically had no effect on antigen processing and presentation [75]. Taken together, the evidence for macroautophagy in providing an alternative processing and presentation route for bacteria is solid.

Autophagy has also been studied in its ability to act as an adjuvant in viral processing and presentation. Using the Epstein-Barr Virus (EBV) as a model, two major antigens have been shown to be processed by the autophagosome. Both nuclear antigen, EBNA1, and the oncogenic protein, LMP1, have been shown by Lee and Sugden to accumulate in autophagy deficient cells [76]. Using EBV-transformed B cells as antigen presenting cells, the Münz group demonstrated that antigen presentation to $CD4^+$ T cells was inhibited by silencing Atg12 or using 3-MA to inhibit macroautophagy induction, and that under normal pathogenesis, EBNA1 is translocated to MHCIIC via autophagosomal transport and processing [77].

Macroautophagy has also been shown to be involved in the MHCIIC processing and presentation of other Herpes viridiae, of which EBV is one type [78]. This is made especially apparent by the fact that HSV has evolved a protein, ICP34.5, with Beclin-1 binding capabilities for antagonizing PI3KIII [79].

Immunosurveillance for transformed cells and neoplasms is now recognized as another crucial function of the immune system. Numerous studies have been conducted highlighting the increase of malignancies in immunocompromised mice and patients. Although autophagy induction in cancer cells is, itself, still a controversial topic in cancer research, proper processing and presentation of tumor antigens is important for a robust immune response. Mucin-1 (MUC1), a tumor antigen, was overexpressed in transfected DCs. Using PI3K inhibitors, such as 3-MA and wortmannin, a substantially reduced $CD4^+$ T cell proliferation was observed, without much effect on $CD8^+$ T cell proliferation [80]. Consequentially, many studies have highlighted autophagy's role in tumor suppression. Beclin-1, initially identified as a tumor suppressor, is mutated in many kinds of cancer, which can directly relate to the ability transformed cells to process and present endogenous antigens. For example, in a recent study Beclin-1 was found to be downregulated at both the message and protein level in female patients with adenomyosis of the endometrium [81].

This research could help immensely with design of tumor vaccines, priming $CD4^+$ T cells to recognize tumor antigen bearing cells with only weak TCR affinities.

4.4. Autophagy in MHC Class I Antigen Processing and Presentation

There has been much data over the last decade concerning the role of macroautophagy in MHC II presentation. Models covering extracellular as well as intracellular bacteria, viruses, tumor antigens, and even endogenous proteins have all been explored. More recently, the role of macroautophagy in processing and presentation of extracellular and intracellular antigens for MHC I have been proposed and partially elucidated. As previously mentioned, the overwhelming majority of peptides presented on MHC I are endogenous, cytosolic proteins, processed by the proteasome, and transported into the ER lumen by TAP for presentation on newly synthesized MHC I [82]. However, in TAP-deficient cells, endogenously derived antigens were still able to be presented to $CD8^+$ cytotoxic T lymphocytes [83, 84].

The alternative pathway is now thought to be dependent upon autophagy, which would require other proteases to be able to efficiently generate peptides in a proteasome-independent manner. Vesicles in the endocytic pathway contain a huge diversity of proteases that can generate peptides of suitable length for MHC I and II loading [85, 86]. It is possible for endogenous antigens to enter the exogenous proteolytic pathway. Engineered epitopes fused to the MHC I heavy chain can be transported to the cell surface, endocytosed, and cleaved by late endosomal cathepsins, where they can be directly loaded onto the parental MHC I molecule in the same compartment as cleavage occurs [86].

This proteolytic activity, fused with MHC I loading in the same compartment simplifies proposed models, eliminating the need for as of yet unidentified vacuole to cytoplasmic transporters, as well as extra time the naked peptide would be exposed to the cytoplasmic and late endosomal environment [87, 88].

In a recent paper by *English et al.* the authors showed that macrophages infected by HSV1 could use autophagy to present peptides recognized by CD8+ T cells at later stages of infection [89]. Presentation of HSV1 peptides could be induced by pharmacological induction of autophagy, and was strongly inhibited by PI3K inhibitors and inhibition of lysosomal maturation. Moreover, in mice infected with ICP34.5 attenuated HSV1, the $CD8^+$ T cell proliferation was vastly increased [90]. Using another pathogen, *M. tuberculosis*, Jagannath demonstrated the efficacy of inducing autophagy in infected DCs prior to injection into mice towards generating a protective immunity [73].

Using models of endogenous tumor antigens presented to MHC I restricted $CD8^+$ T cells, a role for autophagy has also been shown. In B16 melanoma cells, IFNγ induces a potent autophagy, and its induction or pharmacological inhibition has drastic outcomes on the potency of MHC I presentation [91]. Additionally, using an *in vivo* model of tumor antigen presentation, a small peptide derived from HSV2 is able to induce autophagic degradation of aggregates in cancer cells for enhanced MHC I presentation [92]. These studies clearly demonstrate a role for autophagy in the MHC I presentation of endogenously-derived antigens which could potentially be used as adjuvants in the design of new vaccines.

In models of cross-presentation of exogenous antigens, autophagy has also been proposed to play an important role, and can be manipulated to the benefit of the APC. In these models, "donor cells" are phagocytosed by APCs, and peptides are presented on MHC I molecules of the APC. Using HEK 293T cells expressing ovalbumin as a model for antigen, autophagy was either induced by starvation or rapamycin, or inhibited by 3-MA. Cells were then irradiated to allow for efficient phagocytosis by DCs in coculture.

The autophagy induction in donor cells prior to coculture allowed for the robust activation of OT-1 $CD8^+$ T cells, where autophagy inhibition brought T cell proliferation back to nearly undetectable levels [93]. The same group also used gp100 as a model tumor antigen to test cross presentation from B16 F10 melanoma cells, and found strikingly similar results with tumor antigens as to a model endogenous protein [93].

Extending models of cross presentation in cells undergoing active autophagy to a viral infection model, Albert and colleagues took advantage of cells that died with a type II programmed cell death (PCDII) phenotype, characterized by the induction of large numbers of autophagosomes. Using WT or Bak/Bax deficient MEFs, in which caspase activity is blunted, but PCDII is induced upon cytotoxic stress, as a source of donor cells, the MEFs were exposed to influenza A. After infection, the MEFs were treated with etoposide or UV radiation to induce PCD, and injected into mice. Two weeks later, antigen specific $CD8^+$ T cells were isolated and tested for cytokine quality, and as was observed in models of tumor antigen presentation, autophagy effectively increased cross presentation *in vivo* [94]. Perhaps of the greatest interest to vaccine design, purified autophagosomes from dying cells can themselves be used to cross present antigen to APCs [94].

Thus autophagy can aid in the presentation of both intracellular and extracellular antigens for both classes of MHC restricted T cells. These strategies are a welcome addition to the arsenal of current therapies in chronic virally infected or cancer patients, since much of the priming can take place *in vitro* without having to worry about toxic side effects of autophagy-promoting agents.

5. Autophagy in Thymic Education

Thymic education involves the generation of T cell clones with TCRs that can react to peptide in the context of a cognate MHC molecule (positive selection), but have only mild affinity for self-peptides presented in the same manner (negative selection). Though non-hematopoietic in nature, thymic epithelial cells (TECs) are essential for selective purposes. DP thymocytes are especially susceptible to apoptosis as they express no IL-7Rα, and are wholly dependent on receiving pro-survival cues through TCR-mediated signals, and hence must recognize MHC or die by neglect in a process deemed positive selection. Positive selection takes place in the thymic cortex, under the auspices of cortical thymic epithelial cells (cTECs).

Upon completion, thymocytes migrate inward to the medulla and undergo a selective test for TCR affinity with self-antigens presented in the context of MHC on medullary thymic epithelial cells (mTECs). mTECs express *autoimmune regulator* (AIRE), which allows for the expression of a host of tissue-specific antigens that a mature T cell may encounter in the periphery, and therefore must be tolerized against in the thymic environment. Since many of these antigens are non-cytoplasmic in localization, autophagy plays a role in the processing and presentation of these antigens so that they may be properly displayed on MHC molecules.

In a series of studies, Nedjic, Aichinger, and Klein studied the role of thymic selection in the absence of the core autophagy protein, Atg5. As mentioned previously, TECs have extraordinarily high levels of constitutive (basal) autophagy. Atg5-deficient TECs, after engraftment into nude mice, supported the development of normal numbers of thymocytes. However, these mice had defects in both positive and negative selection, despite normal expression of AIRE, suggesting a role for macroautophagy in both selective processes [10]. When T cells generated from Atg5 null TECs were adoptively transferred, they recapitulated a massive wasting disease with colitis and lung infiltration.

The authors concluded that autophagy is essential for optimal expression of peptide-MHC complexes on the cell surface of TECs to foster responsive yet responsible T cells [10].

The high levels of autophagy, coupled with the diverse array of self-antigens expressed, and piled upon the unusual repertoire of proteases present in TECs, all point to the prominence of autophagy for generating peptides from endogenous proteins present in a wide range of subcellular locations, both for MHC I and II loading.

Since phagocytic activity is low in TECs, autophagy may play an even more prominent role in MHC II loading than in any other cell type [21]. It is tempting to ask whether increasing levels of autophagy even further in TECs caused an even more potent negative selection, or in other words, the generation of T cell clones that are resistant to autoimmunity in the context of an even greater range of tissue specific antigens.

Conclusion

Autophagy, at its core, remains a cytoplasmic to lysosomal targeting pathway. What is becoming apparent is that substrates and vesicles are able to become incorporated into the autophagosome/lysosome at many different stages of the pathway, from initiation to after fusion with the lysosome. Also, substrates incorporated into the autolysosome are not restricted to proteins that are cytoplasmic in nature.

From the inclusion of MHC I molecules internalized from the cell surface, phagocytosed antigens utilizing the autophagic machinery to fuse with late endosomal compartments, to the pro-apoptotic molecule Bim, perhaps from the mitochondrial membrane, in T lymphocytes, there can be no doubt that autophagosomal substrates can be derived from nearly any subcellular or extracellular location.

Studies of the innate immune system, with highly specialized cells for antigen processing and presentation have been instrumental in the elucidation of autophagy in cellular trafficking processes.

Alternatively, lymphocytes, with very low thresholds for cytoplasmic entropy, coinciding with low apoptotic thresholds, have been instrumental in the elucidation of the roles of autophagy as an alternative form of cell death, and in its ability to act in several pro-survival pathways.

Though autophagy is still enigmatic in its dual roles of cell death and survival, steady state basal autophagy must be considered a prosurvival pathway that only shows its adverse side under extreme pathogenic conditions, or when transformative events have occurred in the apoptotic pathway.

The role of autophagy in cell death must be considered beneficial for the organism as a whole, as autophagosomes from dying cells can be used for cross presentation to alert the immune system of danger, perhaps acting as a cue on par with any costimulatory molecule. Though much work remains to be done resolving specific substrates for autophagy in cells of the immune system, and how autophagy in dying cells can be used to aid in vaccine design, the ability to selectively induce autophagy in various cells of the immune system holds much promise.

Acknowledgments

This work is supported by NIH grants AI074944 and AI076814. We would like to thank Abigail Brunner for her thoughtful advice and careful reading of the manuscript.

Table 1. Autophagy in the control or escape of specific pathogens

Contributes to clearance/enhanced antigen presentation	Pathogen	Citations
	E. coli	[95]
	Group A. *Streptococcus*	[54]
	Rickettsia conorii	[96]
	Toxoplasma gondii	[97]
	Mycobacterium tuberculosis	[23]
	Salmonella	[52]
	HSV-1	[98]
		[89]
	Influenza	[94]
	Tobamovirus	[99]
	Togavirus	[100]
Specifically blocked by pathogen	HSV-1	[79]
	KHSV	[101]
	Picornavirus	[60]
	Rhinovirus	[102]
	Shigella flexneri	[103]
	Listeria monocytogenes	[104]
Induced by pathogen to replicate	Parvovirus	[105]
	Rotavirus	[106]
	BVDV	[107]
	Chlamydia	[108]
Induced by pathogen to kill host cells	HIV	[28]
	Salmonella	[109]

References

[1] Yorimitsu, T. and D.J. Klionsky, Autophagy: molecular machinery for self-eating. *Cell Death Differ,* 2005. 12 Suppl 2: p. 1542-52.

[2] Xilouri, M. and L. Stefanis, Autophagic pathways in Parkinson disease and related disorders. *Expert Rev. Mol. Med.* 2011. 13: p. e8.

[3] Nijholt, D.A., et al., Removing protein aggregates: the role of proteolysis in neurodegeneration. *Curr. Med. Chem.* 2011. 18(16): p. 2459-76.

[4] Yang, Z. and D.J. Klionsky, Eaten alive: a history of macroautophagy. *Nat. Cell Biol.* 2010. 12(9): p. 814-22.

[5] Matsunaga, K., et al., Two Beclin 1-binding proteins, Atg14L and Rubicon, reciprocally regulate autophagy at different stages. *Nat. Cell Biol.* 2009. 11(4): p. 385-96.

[6] Hara, T. and N. Mizushima, Role of ULK-FIP200 complex in mammalian autophagy: FIP200, a counterpart of yeast Atg17? *Autophagy.* 2009. 5(1): p. 85-7.

[7] Hara, T., et al., FIP200, a ULK-interacting protein, is required for autophagosome formation in mammalian cells. *J. Cell Biol.* 2008. 181(3): p. 497-510.

[8] Bodemann, B.O., et al., RalB and the exocyst mediate the cellular starvation response by direct activation of autophagosome assembly. *Cell.* 2011. 144(2): p. 253-67.

[9] Ohsumi, Y. and N. Mizushima, Two ubiquitin-like conjugation systems essential for autophagy. *Semin. Cell Dev. Biol.* 2004. 15(2): p. 231-6.

[10] Nedjic, J., et al., Autophagy in thymic epithelium shapes the T-cell repertoire and is essential for tolerance. *Nature*. 2008. 455(7211): p. 396-400.
[11] Li, C., et al., Autophagy is induced in CD4+ T cells and important for the growth factor-withdrawal cell death. *Journal of Immunology*. 2006. 177(8): p. 5163-8.
[12] Pua, H.H., et al., A critical role for the autophagy gene Atg5 in T cell survival and proliferation. *Journal of Experimental Medicine*. 2007. 204(1): p. 25-31.
[13] Cho, J.H., et al., T cell receptor-dependent regulation of lipid rafts controls naive CD8+ T cell homeostasis. *Immunity*. 2010. 32(2): p. 214-26.
[14] van Leeuwen, E.M., J. Sprent, and C.D. Surh, Generation and maintenance of memory CD4(+) T Cells. *Curr. Opin. Immunol.* 2009. 21(2): p. 167-72.
[15] McLeod, I.X. and Y. He, Roles of autophagy in lymphocytes: reflections and directions. *Cell. Mol. Immunol.* 2010. 7(2): p. 104-7.
[16] Pua, H.H. and Y.W. He, Autophagy and lymphocyte homeostasis. *Curr. Top Microbiol. Immunol.* 2009. 335: p. 85-105.
[17] Schmid, D. and C. Munz, Innate and adaptive immunity through autophagy. *Immunity*. 2007. 27(1): p. 11-21.
[18] Chemali, M., et al., Alternative pathways for MHC class I presentation: a new function for autophagy. *Cell Mol. Life Sci.* 2011. 68(9): p. 1533-41.
[19] Lee, H.K., et al., Autophagy-dependent viral recognition by plasmacytoid dendritic cells. *Science*. 2007. 315(5817): p. 1398-401.
[20] Harris, J., et al., T helper 2 cytokines inhibit autophagic control of intracellular Mycobacterium tuberculosis. *Immunity*. 2007. 27(3): p. 505-17.
[21] Nedjic, J., M. Aichinger, and L. Klein, Autophagy and T cell education in the thymus: eat yourself to know yourself. *Cell Cycle*. 2008. 7(23): p. 3625-8.
[22] Pua, H.H., et al., Autophagy is essential for mitochondrial clearance in mature T lymphocytes. *Journal of Immunology*. 2009. 182(7): p. 4046-55.
[23] Gutierrez, M.G., et al., Autophagy is a defense mechanism inhibiting BCG and Mycobacterium tuberculosis survival in infected macrophages. *Cell*. 2004. 119(6): p. 753-66.
[24] Watanabe, K., et al., Induction of autophagy by B cell antigen receptor stimulation and its inhibition by costimulation. *Biochemical & Biophysical Research Communications*. 2008. 374(2): p. 274-81.
[25] Gerland, L.M., et al., Autolysosomes accumulate during in vitro CD8+ T-lymphocyte aging and may participate in induced death sensitization of senescent cells. *Exp. Gerontol.* 2004. 39(5): p. 789-800.
[26] Rathmell, J.C., et al., IL-7 enhances the survival and maintains the size of naive T cells. *J. Immunol.* 2001. 167(12): p. 6869-76.
[27] Espert, L., P. Codogno, and M. Biard-Piechaczyk, What is the role of autophagy in HIV-1 infection? *Autophagy*. 2008. 4(3): p. 273-5.
[28] Espert, L., et al., Autophagy is involved in T cell death after binding of HIV-1 envelope proteins to CXCR4. *J. Clin. Invest.* 2006. 116(8): p. 2161-72.
[29] Watanabe, K. and T. Tsubata, Autophagy connects antigen receptor signaling to costimulatory signaling in B lymphocytes. *Autophagy*. 2009. 5(1): p. 108-10.
[30] Maiuri, M.C., et al., BH3-only proteins and BH3 mimetics induce autophagy by competitively disrupting the interaction between Beclin 1 and Bcl-2/Bcl-X(L). *Autophagy*. 2007. 3(4): p. 374-6.

[31] Arsov, I., et al., BAC-mediated transgenic expression of fluorescent autophagic protein Beclin 1 reveals a role for Beclin 1 in lymphocyte development. *Cell death and differentiation.* 2008. 15(9): p. 1385-95.

[32] Miller, B.C., et al., The autophagy gene ATG5 plays an essential role in B lymphocyte development. *Autophagy.* 2008. 4(3): p. 309-14.

[33] Mortensen, M., et al., The autophagy protein Atg7 is essential for hematopoietic stem cell maintenance. *J. Exp. Med.* 2011. 208(3): p. 455-67.

[34] Dzhagalov, I., A. Dunkle, and Y.W. He, The anti-apoptotic Bcl-2 family member Mcl-1 promotes T lymphocyte survival at multiple stages. *Journal of immunology.* (Baltimore, Md. : 1950), 2008. 181(1): p. 521-8.

[35] Hildeman, D., et al., Apoptosis and the homeostatic control of immune responses. *Curr. Opin. Immunol.* 2007. 19(5): p. 516-21.

[36] Zhang, N., et al., The role of apoptosis in the development and function of T lymphocytes. *Cell Res.* 2005. 15(10): p. 749-69.

[37] Stephenson, L.M., et al., Identification of Atg5-dependent transcriptional changes and increases in mitochondrial mass in Atg5-deficient T lymphocytes. *Autophagy.* 2009. 5(5): p. 625-35.

[38] Kovacs, J.R., et al., Autophagy promotes T-cell survival through degradation of proteins of the cell death machinery. *Cell Death Differ.* 2011.

[39] Narendra, D., et al., Parkin-induced mitophagy in the pathogenesis of Parkinson disease. *Autophagy.* 2009. 5(5): p. 706-8.

[40] Yoshii, S.R., et al., Parkin Mediates Proteasome-dependent Protein Degradation and Rupture of the Outer Mitochondrial Membrane. *J. Biol. Chem.* 2011. 286(22): p. 19630-40.

[41] Schweers, R.L., et al., NIX is required for programmed mitochondrial clearance during reticulocyte maturation. *Proc. Natl. Acad. Sci. U. S. A.* 2007. 104(49): p. 19500-5.

[42] Mortensen, M., et al., Loss of autophagy in erythroid cells leads to defective removal of mitochondria and severe anemia in vivo. *Proc. Natl. Acad. Sci. U. S. A.* 2010. 107(2): p. 832-7.

[43] Hildeman, D.A., et al., Reactive oxygen species regulate activation-induced T cell apoptosis. *Immunity.* 1999. 10(6): p. 735-44.

[44] Grunau, S., et al., The phosphoinositide 3-kinase Vps34p is required for pexophagy in Saccharomyces cerevisiae. *Biochem. J.* 2011. 434(1): p. 161-70.

[45] Jia, W., et al., Autophagy regulates endoplasmic reticulum homeostasis and calcium mobilization in T lymphocytes. *J. Immunol.* 2011. 186(3): p. 1564-74.

[46] Jia, W. and Y.W. He, Temporal regulation of intracellular organelle homeostasis in T lymphocytes by autophagy. *J. Immunol.* 2011. 186(9): p. 5313-22.

[47] Janeway, C.A., Jr., Approaching the asymptote? Evolution and revolution in immunology. *Cold Spring Harb. Symp. Quant. Biol.* 1989. 54 Pt 1: p. 1-13.

[48] Jounai, N., et al., The Atg5 Atg12 conjugate associates with innate antiviral immune responses. *Proc. Natl. Acad. Sci. U. S. A.* 2007. 104(35): p. 14050-5.

[49] Sanjuan, M.A., et al., Toll-like receptor signalling in macrophages links the autophagy pathway to phagocytosis. *Nature.* 2007. 450(7173): p. 1253-7.

[50] Xu, Y., et al., Toll-like receptor 4 is a sensor for autophagy associated with innate immunity. *Immunity.* 2007. 27(1): p. 135-44.

[51] Munz, C., Enhancing immunity through autophagy. *Annu. Rev. Immunol.* 2009. 27: p. 423-49.
[52] Cooney, R., et al., NOD2 stimulation induces autophagy in dendritic cells influencing bacterial handling and antigen presentation. *Nat. Med.* 2010. 16(1): p. 90-7.
[53] Delgado, M.A., et al., Toll-like receptors control autophagy. *EMBO J.* 2008. 27(7): p. 1110-21.
[54] Nakagawa, I., et al., Autophagy defends cells against invading group A Streptococcus. *Science.* 2004. 306(5698): p. 1037-40.
[55] Orvedahl, A., et al., HSV-1 ICP34.5 confers neurovirulence by targeting the Beclin 1 autophagy protein. *Cell Host Microbe*. 2007. 1(1): p. 23-35.
[56] Orvedahl, A., et al., Autophagy protects against Sindbis virus infection of the central nervous system. *Cell Host Microbe.* 2010. 7(2): p. 115-27.
[57] Dupont, N., et al., Shigella phagocytic vacuolar membrane remnants participate in the cellular response to pathogen invasion and are regulated by autophagy. *Cell Host Microbe.* 2009. 6(2): p. 137-49.
[58] Tal, M.C., et al., Absence of autophagy results in reactive oxygen species-dependent amplification of RLR signaling. *Proc. Natl. Acad. Sci. U. S. A.* 2009. 106(8): p. 2770-5.
[59] Dales, S., et al., Electron Microscopic Study of the Formation of Poliovirus. *Virology.* 1965. 26: p. 379-89.
[60] Jackson, W.T., et al., Subversion of cellular autophagosomal machinery by RNA viruses. *PLoS. Biol.* 2005. 3(5): p. e156.
[61] Reggiori, F., et al., Coronaviruses Hijack the LC3-I-positive EDEMosomes, ER-derived vesicles exporting short-lived ERAD regulators, for replication. *Cell Host Microbe.* 2010. 7(6): p. 500-8.
[62] Schiefner, A. and I.A. Wilson, Presentation of lipid antigens by CD1 glycoproteins. *Curr. Pharm. Des.* 2009. 15(28): p. 3311-7.
[63] Trombetta, E.S. and I. Mellman, Cell biology of antigen processing in vitro and in vivo. *Annu. Rev. Immunol.* 2005. 23: p. 975-1028.
[64] Rammensee, H., et al., SYFPEITHI: database for MHC ligands and peptide motifs. *Immunogenetics.* 1999. 50(3-4): p. 213-9.
[65] Barette, C., et al., Human cyclin C protein is stabilized by its associated kinase cdk8, independently of its catalytic activity. *Oncogene.* 2001. 20(5): p. 551-62.
[66] Dice, J.F. and A.L. Goldberg, Relationship between in vivo degradative rates and isoelectric points of proteins. *Proc. Natl. Acad. Sci. U. S. A.* 1975. 72(10): p. 3893-7.
[67] Dengjel, J., et al., Autophagy promotes MHC class II presentation of peptides from intracellular source proteins. *Proc. Natl. Acad. Sci. U. S. A.* 2005. 102(22): p. 7922-7.
[68] Aniento, F., et al., Uptake and degradation of glyceraldehyde-3-phosphate dehydrogenase by rat liver lysosomes. *J. Biol. Chem. 1*993. 268(14): p. 10463-70.
[69] Zhou, D., et al., Lamp-2a facilitates MHC class II presentation of cytoplasmic antigens. *Immunity.* 2005. 22(5): p. 571-81.
[70] Nedjic, J., et al., Macroautophagy, endogenous MHC II loading and T cell selection: the benefits of breaking the rules. *Curr. Opin. Immunol.* 2009. 21(1): p. 92-7.
[71] Zwart, W., et al., Spatial separation of HLA-DM/HLA-DR interactions within MIIC and phagosome-induced immune escape. *Immunity.* 2005. 22(2): p. 221-33.

[72] Schmid, D., M. Pypaert, and C. Munz, Antigen-loading compartments for major histocompatibility complex class II molecules continuously receive input from autophagosomes. *Immunity.* 2007. 26(1): p. 79-92.

[73] Jagannath, C., et al., Autophagy enhances the efficacy of BCG vaccine by increasing peptide presentation in mouse dendritic cells. *Nat. Med.* 2009. 15(3): p. 267-76.

[74] Russmann, H., et al., Alternative endogenous protein processing via an autophagy-dependent pathway compensates for Yersinia-mediated inhibition of endosomal major histocompatibility complex class II antigen presentation. *Infect. Immun.* 2010. 78(12): p. 5138-50.

[75] Lee, H.K., et al., In vivo requirement for Atg5 in antigen presentation by dendritic cells. *Immunity.* 2010. 32(2): p. 227-39.

[76] Lee, D.Y. and B. Sugden, The latent membrane protein 1 oncogene modifies B-cell physiology by regulating autophagy. *Oncogene.* 2008. 27(20): p. 2833-42.

[77] Paludan, C., et al., Endogenous MHC class II processing of a viral nuclear antigen after autophagy. *Science.* 2005. 307(5709): p. 593-6.

[78] Taylor, G.S., J. Mautner, and C. Munz, Autophagy in herpesvirus immune control and immune escape. *Herpesviridae.* 2011. 2(1): p. 2.

[79] Leib, D.A., et al., Interaction of ICP34.5 with Beclin 1 modulates herpes simplex virus type 1 pathogenesis through control of CD4+ T-cell responses. *J. Virol.* 2009. 83(23): p. 12164-71.

[80] Dorfel, D., et al., Processing and presentation of HLA class I and II epitopes by dendritic cells after transfection with in vitro-transcribed MUC1 RNA. *Blood.* 2005. 105(8): p. 3199-205.

[81] Ren, Y., et al., Decreased expression of Beclin 1 in eutopic endometrium of women with adenomyosis. *Arch. Gynecol. Obstet.* 2010. 282(4): p. 401-6.

[82] York, I.A. and K.L. Rock, Antigen processing and presentation by the class I major histocompatibility complex. *Annu. Rev. Immunol.* 1996. 14: p. 369-96.

[83] Esquivel, F., J. Yewdell, and J. Bennink, RMA/S cells present endogenously synthesized cytosolic proteins to class I-restricted cytotoxic T lymphocytes. *J. Exp. Med.* 1992. 175(1): p. 163-8.

[84] Hosken, N.A. and M.J. Bevan, An endogenous antigenic peptide bypasses the class I antigen presentation defect in RMA-S. *J. Exp. Med.* 1992. 175(3): p. 719-29.

[85] Shen, L., et al., Important role of cathepsin S in generating peptides for TAP-independent MHC class I crosspresentation in vivo. *Immunity.* 2004. 21(2): p. 155-65.

[86] Tiwari, N., et al., A transporter associated with antigen-processing independent vacuolar pathway for the MHC class I-mediated presentation of endogenous transmembrane proteins. *J. Immunol.* 2007. 178(12): p. 7932-42.

[87] Kovacsovics-Bankowski, M. and K.L. Rock, A phagosome-to-cytosol pathway for exogenous antigens presented on MHC class I molecules. *Science.* 1995. 267(5195): p. 243-6.

[88] Ackerman, A.L., et al., Early phagosomes in dendritic cells form a cellular compartment sufficient for cross presentation of exogenous antigens. *Proc. Natl. Acad. Sci. U. S. A.* 2003. 100(22): p. 12889-94.

[89] English, L., et al., Autophagy enhances the presentation of endogenous viral antigens on MHC class I molecules during HSV-1 infection. *Nat. Immunol.* 2009. 10(5): p. 480-7.

[90] Broberg, E.K., et al., Spread and replication of and immune response to gamma134.5-negative herpes simplex virus type 1 vectors in BALB/c mice. *J. Virol.* 2004. 78(23): p. 13139-52.

[91] Li, B., et al., Autophagy facilitates major histocompatibility complex class I expression induced by IFN-gamma in B16 melanoma cells. *Cancer Immunol. Immunother.* 2010. 59(2): p. 313-21.

[92] Fu, X., L. Tao, and X. Zhang, A short polypeptide from the herpes simplex virus type 2 ICP10 gene can induce antigen aggregation and autophagosomal degradation for enhanced immune presentation. *Hum. Gene Ther.* 2010. 21(12): p. 1687-96.

[93] Li, Y., et al., Efficient cross-presentation depends on autophagy in tumor cells. *Cancer Res.* 2008. 68(17): p. 6889-95.

[94] Uhl, M., et al., Autophagy within the antigen donor cell facilitates efficient antigen cross-priming of virus-specific CD8+ T cells. *Cell Death Differ.* 2009. 16(7): p. 991-1005.

[95] Pfeifer, J.D., et al., Phagocytic processing of bacterial antigens for class I MHC presentation to T cells. *Nature.* 1993. 361(6410): p. 359-62.

[96] Walker, D.H., et al., Cytokine-induced, nitric oxide-dependent, intracellular antirickettsial activity of mouse endothelial cells. *Lab. Invest.* 1997. 76(1): p. 129-38.

[97] Andrade, R.M., et al., CD40 induces macrophage anti-Toxoplasma gondii activity by triggering autophagy-dependent fusion of pathogen-containing vacuoles and lysosomes. *J. Clin. Invest.* 2006. 116(9): p. 2366-77.

[98] Leib, D.A., et al., Specific phenotypic restoration of an attenuated virus by knockout of a host resistance gene. *Proc. Natl. Acad. Sci. U. S. A.* 2000. 97(11): p. 6097-101.

[99] Liu, Y., et al., Autophagy regulates programmed cell death during the plant innate immune response. *Cell.* 2005. 121(4): p. 567-77.

[100] Liang, X.H., et al., Protection against fatal Sindbis virus encephalitis by beclin, a novel Bcl-2-interacting protein. *J. Virol.* 1998. 72(11): p. 8586-96.

[101] Pattingre, S., et al., Bcl-2 antiapoptotic proteins inhibit Beclin 1-dependent autophagy. *Cell. 2*005. 122(6): p. 927-39.

[102] Prentice, E., et al., Coronavirus replication complex formation utilizes components of cellular autophagy. *J. Biol. Chem.* 2004. 279(11): p. 10136-41.

[103] Ogawa, M., et al., Escape of intracellular Shigella from autophagy. *Science.* 2005. 307(5710): p. 727-31.

[104] Yoshikawa, Y., et al., Listeria monocytogenes ActA-mediated escape from autophagic recognition. *Nat. Cell Biol.* 2009. 11(10): p. 1233-40.

[105] Nakashima, A., et al., Survival of parvovirus B19-infected cells by cellular autophagy. *Virology.* 2006. 349(2): p. 254-63.

[106] Berkova, Z., et al., Rotavirus NSP4 induces a novel vesicular compartment regulated by calcium and associated with viroplasms. *J. Virol.* 2006. 80(12): p. 6061-71.

[107] Meyers, G., D. Stoll, and M. Gunn, Insertion of a sequence encoding light chain 3 of microtubule-associated proteins 1A and 1B in a pestivirus genome: connection with virus cytopathogenicity and induction of lethal disease in cattle. *J. Virol.* 1998. 72(5): p. 4139-48.

[108] Al-Younes, H.M., V. Brinkmann, and T.F. Meyer, Interaction of Chlamydia trachomatis serovar L2 with the host autophagic pathway. *Infect. Immun.* 2004. 72(8): p. 4751-62.

[109] Hernandez, L.D., et al., A Salmonella protein causes macrophage cell death by inducing autophagy. *J. Cell Biol.* 2003. 163(5): p. 1123-31.

In: Autophagy: Principles, Regulation and Roles in Disease ISBN: 978-1-61942-266-7
Editor: Nikolai Gorbunov

Chapter II

Cross-Presentation Induced by Autophagy towards Clinical Approaches

***Young-Jun Ju*[1]*, Jannatul Firdous*[1]*, Hyo-Shin Yoon*[1]*, Seung Hyun Han*[2,4] *and Cheol-Heui Yun*[1,3,4]**

[1] Department of Agricultural Biotechnology and
Research Institute for Agriculture and Life Sciences; Republic of Korea
[2] Department of Oral Microbiology & Immunology,
and Dental Research Institute, School of Dentistry, Seoul, Republic of Korea
[3] Center for Agricultural Biomaterials; Seoul, Republic of Korea
[4] Center for Food Safety and Toxicology,
Seoul National University, Gwanak-no, Gwanak-gu,
Seoul, Republic of Korea

Abstract

It is essential to maintain homeostasis in most cells, other than the cells undertaking death pathways, requiring sufficient nutrients from the environmental surroundings. However, when cells face undesirable conditions, the cells digest their own intracellular organelles. This metabolic sequence is defined as "autophagy", which has been recently in the spotlight due to its involvement into a variety of cell stress and defense responses including the host's immune responses. Autophagy is categorized based on (1) how cargo delivery is proceeding to lysosome such as macroautophagy, microautophagy and chaperone-mediated autophagy, and (2) its target, selective and non-selective autophagy, both of which can modulate antigen presentation and therefore, the cell-mediated immunity under the certain conditions.

Once autophagy begins, it undergoes specific steps for digesting the organelles through lysosomes and specific cytoplasmic components by forming a double membrane known as "autophagosomes". The autophagosomes are the best characterized components formed during the process of autophagy. Recently, immunologists have focused on "autophagy-related cross-presentation" due to the fact that it is a very efficient clinical tool for targeting numerous diseases including cancers and microbial infections by cross-priming of microbial and tumor specific antigens to $CD8^+$ T cells. Despite extensive efforts to discover the direct relationship between cross-presentation and autophagy, the precise mechanisms are yet not known. Therefore, we have traced a foot print of the autophagy and cross-presentation along with the clinical approaches to shed a new light to further studies and research strategies. In this chapter, the characteristics of autophagy, cross-presentation induced by autophagy and related research approaches are described in details. Furthermore, we provide a broad review of recent clinical data on implication of autophagy in immunopathophysiological conditions.

1. Introduction

The term **'autophagy'** was firstly coined in 1963 by Christian De Duve who received the Nobel Prize in Physiology or Medicine for the discovery of the lysosome in 1955. It was discovered by observing single- or double- membrane vesicles called, **'autophagosome'**, which contained self-organelles.[1] Since then, many researchers have been focusing on morphological studies of autophagy for the formation of autophagosome and their fusion with lysosomes. Studies at molecular level of autophagy have begun to expand dramatically by the late 1990s with the identification of the autophagy–related (Atg) genes in yeast, which are essential for autophagosome formation.[1, 2] Atg proteins are conjugated with each other forming multiple complexes at different levels during the process of autophagy (Figure 1).[3] Thus, it is not surprising that there have been many attempts to investigate specific functions of Atg genes in autophagy by modulating those Atg genes in many diseases models. Another aspect is that autophagy plays a role in antigen presentation. Classically, it has been reported that autophagy enhances antigen presentation by loading endogenous antigen to MHC class I on antigen presenting cells [APCs; dendritic cells (DCs), macrophages and B cells] that is presented to $CD8^+$ T cells and exogenous antigen on MHC class II that present to helper T (Th) cells. However, recent researches have suggested that autophagy may affect **"cross-presentation"**, especially to deliver the exogenous antigen to MHC class I on APCs which primes antigen-specific $CD8^+$ T cells to eliminate the abnormal and infected cells effectively. In this chapter, we describe cross-presentation induced by autophagy along with several therapeutic approaches.

2. Characteristics of Autophagy

2.1. Cell Survival Strategy

Maintaining homeostasis is the most important part of life cycle in most, if not all, cells. For this, organisms are taking nutrients as an energy source from their environment. Sometimes, however, organisms face unfavorable conditions including various stresses and inadequate levels of nutrients. When these occur, the cells digest their own intracellular organelles such as mitochondria, endoplasmic reticulum and some of their small proteins. It is called "autophagy" that stands for eating self-organelles and therefore, the cells synthesize the energy and struggle to survive against the stress conditions.[4, 5]

Autophagy can be classified based on (1) its mechanism of cargo delivery to lysosome (i.e., macroautophagy, microautophagy and chaperone-mediated autophagy (CMA)), or (2) its target (selective and non-selective autophagy). Macroautophagy is very rapid and radical phenomenon induced by various stress conditions such as a mal-nutrition, and cellular and DNA damage. The CMA is induced by prolonged and persisted cellular stresses via chaperone-mediated proteins, heat shock protein (HSP) 70 and HSP90, which are known to be involved in protein folding and stability. Furthermore, they are known to enhance the ability of cross-presentation in DCs.[6, 7] Microautophagy, originally found in bacteria, is extremely difficult to detect changes at the cellular and molecular level, at least with a current technology, and thus regarded as a relatively minor part of autophagy.[6-8]

It is worthwhile to note that, in immunology area, autophagy contributes to clearance of the cytosolic proteins derived from invading bacteria or viruses, called **xenophagy**. It is similar to autophagy in broad term, but there is a slight difference. Induction of a general autophagy (macroautophagy in this review) produced energy through specific autolysosomal pathway; however xenophagy occurs only for clearing non-self antigen and attempts to make the cell free from infected conditions via enveloping and degrading non-self antigens through autophagic process.[5, 9]

2.2. Genes and Proteins Involved in Autophagy

Key molecules and genes related to selective autophagy are discovered and grouped as **'Atg family'**, which participate in the formation of autophagosome composed of Atg complexes involving three modes of action; initiation, elongation and maturation. Briefly, during the process of autophagy maturation, three steps exist along with unique genes and proteins, respectively (Table 1). First, initiation step refers the forming of omegasome (described in detail below), and contributed by forming ULK/Atg complex, class III PI3K complex, and a few others. Second, elongation step refers to the formation of autophagosome, from omegasome, containing some unnecessary compartment in cytosol contributed by Atg12-conjugation system and LC3/Atg8-conjugation system before fusing with lysosome. Third, maturation step refers to autophagosome fusion with lysosome forming autolysosome that has degrading ability.[3, 4] The Atg7, E1 ligase, induces the conjugation of Atg5 to Atg12 and converts LC3-I to phosphatidylethanolamine conjugated LC3-II. The Atg5 and Atg12 complex, then, are conjugated with Atg16L and these complexes form a dimer for the formation of autophagosome.

The other protein, Atg14 (Barkor) is also conjugated with Beclin1 that complexes with Bcl2 and Vps family to induce PI3K signaling pathway and initiation of autophagosome. It is also known that autophagy requires PI3K/ERK/MAPK and mTOR/Akt signaling pathways.[3]

In the context of immunology, autophagy related genes and proteins are regulated by various stimulation including toll-like receptors (TLRs), immune-related GTPases and cytokines. Therefore, such key molecules (Table 1) could serve as markers to distinguish the initiation and elongation step during the process of autophagy. These autophagy-related genes are translated to proteins, and these proteins make complexes that are recruited in sequence and recruited protein complexes have a differential expression with distinguished function.

Thus, numerous studies have focused on the functions of combination of protein complex by analyzing the respective genes.

Table 1. Key molecules related to the formation of autophagosome and their functions[3, 4]

Initiation Step	Key molecules	Feature
ULK/Atg complex	ULK1, ULK2	Protein, phosphorylated by mTORC1
	Atg13	Phosphorylated by mTORC1
	FIP200	Scaffold for ULK1/2 and Atg13
	Atg101	Interacts with Atg13
Class IIIPI3-kinase complex	Vps34	PI3-kinase
	p150	Myristoylated
	Beclin1	BH3-only protein, interacts with Bcl-2
	Atg14	Autophagy-specific subunit
	Ambra1	Interacts with Beclin1
Others	Atg2	Interacts with Atg18 in yeast
	Atg9	Transmembrane protein
	WIPI1-4	PI(3)P-binding proteins
	DFCP1	PI(3)P-binding ER protein
	VMP1	Transmembrane protein
Elongation Step		
Atg12-conjugation system	Atg12	Ubiquitin-like, conjugates to Atg5
	Atg7	E1-like enzyme
	Atg10	E2-like enzyme
	Atg5	Conjugated by Atg12
	Atg16L1	Homodimer, interacts with Atg5
LC3/Atg8-conjugation system	LC3 (GATE-16, GABARAP)	Ubiquitin-like, conjugates to PE
	Atg4A-D	LC3/Atg8 C-terminal hydrolase, deconjugating enzyme
	Atg7	E1-like enzyme
	Atg3	E2-like enzyme

3. Action Mode of Autophagy

3.1. Initiation of Autophagy

Autophagy is initiated by various cellular stresses including nutrient deprivation, and DNA and ER damage. Once started, some molecules like mTOR substrate complex that is composed of ULK1/2, Atg13, FIP200, and Atg101 are translocated to ER from cytosol by mTOR stimulatory signals. Then ULK1 complex activates class III PI3K complex to produce phosphatidylinositol-3-phosphate (PIP3) which subsequently recruits double FYVE-containing protein 1 (DFCP1) and WD repeat domain phosphoinositide-interacting (WIPI) family proteins. These WIPIs and DFCP1 generate structure like "Ω", called "Omegasome" that it is an initial shape of autophagosome on the ER, nucleus and mitochondrial outer membrane. At the end of this step, LC3-PE is presented on inner membrane and it generates a perfect cargo shape structure, called autophagosome after the isolation of omegasome (*Step 1: Initiation* in Figure 1).[4, 10]

3.2. Elongation of Autophagosome

Omegasomes is then isolated from ER membrane with a circular closed double membrane structure to form autophagosome's envelope. When it is isolated from original sites (i.e., ER, mitochondria and nuclear membrane) two steps are found as follows. First, Atg12-Atg5 conjugation is performed by Atg7 (E1-like) and Atg10 (E2-like) enzymes and the complex (Atg12-Atg5) is recruiting Atg16L1 to form Atg12-Atg5-Atg16L1.[11] Second, the key molecules of autophagy, such as LC3, GATE16 and GABARAP (homology with Atg8, referred to Table 1)[12] are conjugated with PE (phosphatidylethanolamine) forming LC3-PE complex (LC3-II) through Atg7 and Atg3 (E2-like enzyme).[4, 10] Although recent studies suggest that autophagosome is originated from various source of membranes, there is a controversial issue because the detection of autophagosome is difficult and there are a few markers to define autophagosome uniquely except for LC3-I/II conversion (*Step 2 : Elongation* in Figure 1).[13]

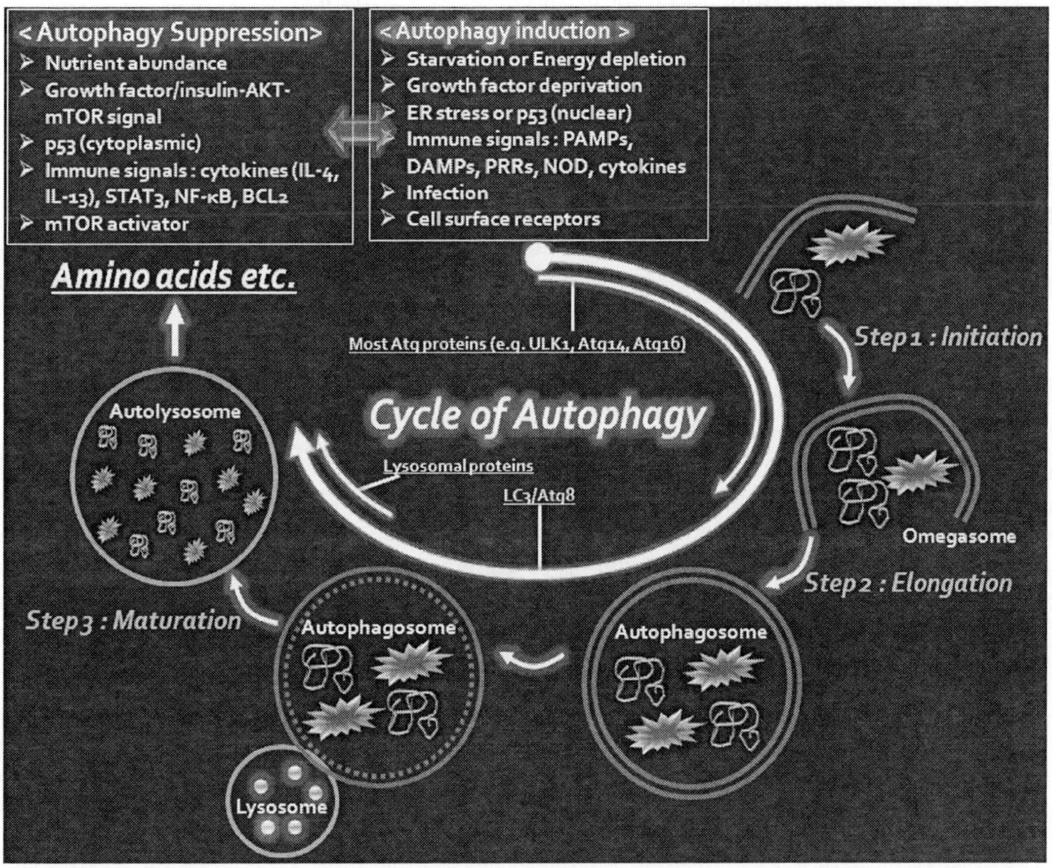

Figure 1. The progress of autophagy induction. *Step 1: Initiation* step is to start forming omegasome by stimulation with autophagy inducers from ER, mitochondrial or nuclear membranes; *Step 2: Elongation* step showed that omegasome elongates into autophagosome encapsulating cytosolic compartments; *Step 3: Maturation* step showed that autophagosome fuses with cytosolic lysosome forming into autolysosome for degrading encapsulated compartments.

3.3. Maturation Processing of Autophagosome

The autophagosome goes under maturation process by recruiting a novel binding protein of Atg8 homologues named OATL1 [a putative Rab guanosine triphosphatase–activating protein (GAP)]. This protein is targeting Rab33B, an Atg16L1-binding protein. Both of these proteins are involved in the fusion between autophagosomes and lysosomes through its GAP activity. After the fusion, lysosomal enzymes like cathepsin degrade inner membrane of autophagosome and encapsulate autophagosomal proteins. Now, it is called **"autolysosome"** that produces amino acids and ATP as a result of degradation process for cell survival (*Step 3: Maturation* in Figure 1).[4, 10, 14]

4. Autophagy and Antigen Presentation

4.1. Autophagy in the Classical Antigen Presentation

Antigen processing and presentation are the key concepts to bridge innate immunity to acquired immunity in APCs and therefore, are one of the most important phenomena in host protection. Numerous studies attempt to clarify the action mechanism of autophagy during the antigen processing and presentation and their related molecules. However, the precise mechanism is unclear because of limitations of current techniques: so far there are only few reliable tracking markers of autophagy, namely LC3-I/LC3-II conversion and expression of some Atg genes. Thus many scientists have devoted their work to find their own interpretation of the phenomena and to set the logical hypothesis.[15, 16]

When it comes to infection, innate immunity works as a first-line of defense in the host immune system through innate immune molecules and cells such as complement factors, neutrophils, macrophages, dendritic cells, and natural killer cells. But the innate immune system is not enough to clear all pathogens invading host and therefore, more powerful and pathogen-specific system is required via T and B lymphocytes. There are three major sequential steps for the activation of antigen-specific T and B lymphocytes. They are 1) Antigen uptake by APCs, 2) APCs homing to draining lymph nodes, 3) antigen presentation to and activation of T and B lymphocytes. Classically, exogenous antigens (some bacteria, parasites, and cancer cells, *etc.*) are required to be uptaken by APCs and presented through MHC class II pathway to $CD4^+$ T cell. Viruses are different from exogenous antigen where they insert their genome to host chromosome and utilize host replication system to (re)produce their own viral proteins, which is then recognized as endogenous antigens. Therefore, endogenous antigens are processed and presented via another mechanism in which ubiquitination is required for the detection and degradation of antigen peptides by proteasomes and these degraded peptides then enter into ER through TAP and are presented to $CD8^+$ T cell loading on MHC class I.[17]

The results of recent research suggested a potency of autophagy in antigen processing and presentation that it could enhance endogenous antigen presentation, simply, endogenous antigen on MHC class I or on MHC class II.[18] Intriguingly, it has been reported that autophagy can enhance delivery of cytosolic protein peptides uptaken exogenously leading MHC class I pathway via autophagosome.[19-21] When the omegasome forms into autophagosome, some cytosolic proteins, such as viral proteins produced in infected cells and phagocytosed bacterial peptide fragments, are encapsulated and become a part of autophagosomal compartment. Then lysosomes fuse with autophagosome, such a process called **"autophagosome maturation"**, and the lysosomal enzymes degrade the encapsulated proteins to peptide fragments to be loaded on MHC class II molecule after fused with autolysosome. And then the peptide loaded onto MHC class II molecule is transferred to plasma membrane and prime Th cells.[5] Thus, an exogenous antigen surrounded by phagosome escapes to cytosol and encapsulated by autophagosome. Then the antigen is degraded into little peptides after fusion with lysosome forming autolysosome. This processed peptide can be loaded on MHC class II molecules in exocytic vesicles and passage towards the cell surface. MHC class II molecules are synthesized in the ER and emerge from the Golgi complex to form Exocytic vesicles (green lines in Figure 2). In case of persistent viruses, endogenous antigen is cleaved by proteasome and the antigenic molecules enter the ER via TAP. They are loaded on MHC class I molecules and presented on the cell surface. However, endogenous viral antigens can be engulfed by autophagosome, which meets and fuses with lysosome to form autolysosomes. The processed peptides can be loaded on MHC class I molecules which are recycled by surrounded vacuole from the cell surfaces through a specific pathway known as autophagy-associated vacuolar pathway (red lines in Figure 3).

A recent study has suggested that autophagy can contribute to MHC class I processing pathway during HSV-1 infection via alternative pathway called 'vacuolar pathway'.[22] By using TAP deficient model, they have shown that viral peptides are still loaded on MHC class I molecules. In the vacuolar pathway, MHC class I molecules on the cell which are the parts of endovacuoles, and can be recycled, fuse with autophagolysosome that contains endogenous (viral) peptides degraded by lysosomal proteases, then these peptides can be loaded onto MHC class I molecules.[23-26] In fact, autophagosome works like carrier of antigen in both MHC class I and II pathways. Therefore, it is certain that a mode of autophagy induction is important during antigen processing and presentation. In addition, the fusion of autophagosome and lysosome is key points of autophagy-dependent antigen processing. Lysosomal enzymes degrade autophagosomal compartment to adequate size before the fusion with MHC class I molecule containing vacuole. After the formation of autolysosome, it fuses with recycled MHC class I containing vacuole and the degraded peptide are loaded on MHC class I molecule.

As a result, autophagy can contribute to antigen presenting efficiency via its specific production called autophagosome for using antigen carrier. Therefore, a detailed understanding of the autophagic mechanisms could provide with better management strategy for infectious and cancer diseases.

4.2. Cross-Presentation by Autophagy-Induced Cell

4.2.1. Cross-Presentation

Once APCs uptake antigen, they process and present peptides to antigen-specific $CD4^+$ T cells or $CD8^+$T cells; endogenous antigens such as viral antigens are loaded on MHC class I and activate $CD8^+$ T cells whereas exogenous antigens are loaded on MHC class II and activate naïve $CD4^+$ T cells. With antigen-specific exogenous antigens, $CD4^+$ T cells are eventually differentiated into antigen-specific Th cells. Then, the antigen-specific Th cells are expanding and subsequently helping differentiation of CD8$^+$ T cells into **cytotoxic T lymphocytes (CTLs)**. Thus, it is not easy to induce antigen-specific CTL response to eliminate the target cells at early time point after the microbial invasion. An alternative strategy for expanding antigen-specific CTLs directly and quickly could be "cross-presentation".[27]

In general, cross-presentation is defined as an alternative antigen presenting pathway: process and present exogenous antigens onto MHC class I molecules to activate CD8$^+$ T cells or endogenous antigens with MHC class II to activate CD4$^+$ T cells. When the antigen is cross-presented to naïve CD8$^+$ T cell, then it is differentiated into antigen-specific CTLs that could directly kill the tumor or infected cells, and more importantly generate antigen-specific CD8$^+$ memory T cells. For these reasons, cross-presentation has been thought as one of the most effective therapeutic strategies against tumor and bacterial infection.

4.2.2. Activation of T Cells via Autophagy-Dependent Cross-Presentation

A number of studies suggest that autophagy may affect cross-presentation through its special processing mechanism and may be dependent on proteasome which itself is a protein complex that degrades unnecessary or damaged proteins. However, there are controversial issues as following. Firstly, **proteasome-dependent hypothesis** suggests that phagocytosed antigens are escaping from phagosome into cytosol and then processed by ubiquitin-dependent proteasome degradation. At this point, there are two pathways again. One hand, it is known that degraded peptide is entered into ER and used the classical MHC class I pathway. According to this hypothesis, autophagy could affect escaping efficiency of phagocytosed antigen from phagosome. On the other hand, escaped peptide from phagosome is not entered into ER but may directly be processed through nuclear transport proteins, which has been postulated the involvement of ERAD family protein(s).[28-30] Therefore, the contrary is whether degraded peptides directly enter into ER or via autophagosome. Although the proteasome-dependent antigen presentation is the major pathway in normal cross-presentation, researches on autophagy-dependent cross-presentation do not seem to pay attention to the importance of proteasome (for further discussion see 5. Models to study autophagy-dependent cross-presentation). Secondly, **proteasome-independent hypothesis** suggest that exogenous antigen escapes from phagosome, then autophagosome surrounds and encapsulates the antigen. After forming autophagosome, it is fused with lysosome and the peptide is degraded by lysosomal enzymes.

Then recycled MHC class I containing vacuole is fused with autolysosome, and the peptide is loaded onto MHC class I in autophagosome. Finally, peptide-loading MHC class I molecule is exposed to cell surface, and it primes CD8$^+$ T cell directly (red lines in Figure 2.).[27, 31, 32] Although the exogenous antigen to MHC class I pathway is referred as cross-presentation, the contrast pathway that the endogenous antigen to MHC class II pathway is also called cross-presentation. In fact, autophagy is known to enhance MHC class II presentation during viral infection where the viral proteins are encapsulated by autophagosome. The autophagosome then fuses with lysosome forming autolysosome that degrades viral proteins to peptides. Consequently, they are loaded on MHC class II molecule and budding from ER, proceeding to exocytic vesicle, and presented to CD4$^+$ T cells (green lines in Figure 3).[32, 33]

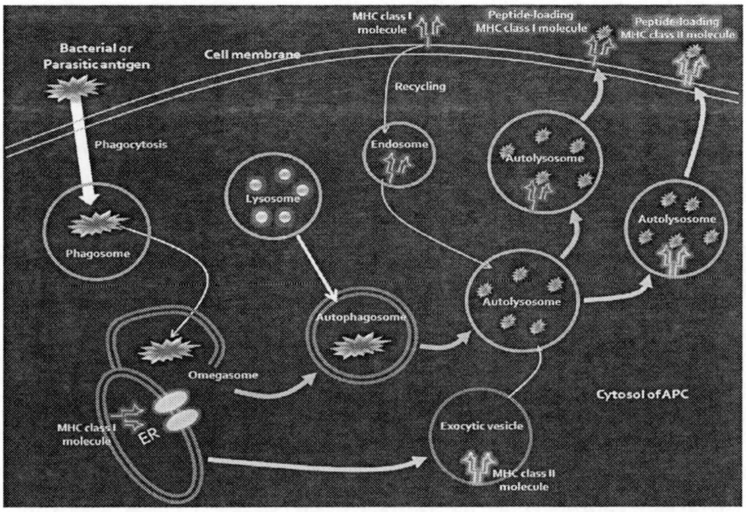

Figure 2. Exogenous antigen presentation via autophagy. Exogenous antigen escapes from phagosome. Then antigen is surrounded and encapsulated by autophagosome, which fused with lysosome forming autolysosome. In here, antigens are degraded to peptides. In MHC class I presentation pathway, autolysosome is fused with a vacuole containing recycled MHC class I, and the antigen peptides are loaded onto MHC class I molecule and presented to $CD8^+$ T cells (red lines); autophagy-dependent cross-presentation pathway. In MHC class II presentation pathway, autolysosome is fused with MHC class II molecule containing exocytic vesicle originated from ER, and the antigen peptides are loaded MHC class II molecule and presented to $CD4^+$ T cells (green lines); autophagy-dependent classical MHC class II pathway.

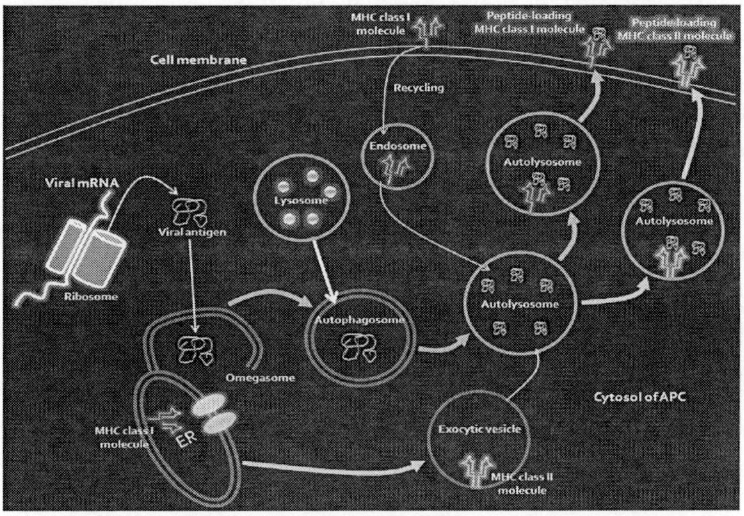

Figure 3. Endogenous antigen presentation via autophagy. Endogenous viral antigen synthesized from viral mRNA. Viral antigens are surrounded by omegasome and encapsulated by autophagosome, which fuses with lysosome to form autolysosome for degrading the viral antigens to peptides. In MHC class I presentation pathway, autolysosome is fused with a vacuole containing recycled MHC class I, and the antigenic peptides are loaded onto MHC class I molecule and presented to $CD8^+$ T cells (red lines); autophagy-dependent classical MHC class I pathway. In MHC class II presentation pathway, autolysosome is fused with MHC class II molecule containing exocytic vesicle originated from ER, and the antigen peptides are loaded MHC class II molecule and presented to $CD4^+$ T cells (green lines); autophagy-dependent MHC class II cross-presentation pathway.

5. Models to Study Autophagy-Dependent Cross-Presentation

It is to note that proteasome-dependent hypothesis for cross-presentation gets considerable attention compared to proteasome-independent pathway.[9, 13, 22, 31] In this chapter, we propose a concept of the proteasome-dependent but ER-independent pathway.. In our hypothesis, autophagosome works as carrier (or cargo) of MHC class I molecule and non-self antigen transport protein family. Therefore, the autophagosome carries the antigen peptide loading on MHC class I molecule after degrading the antigen by proteasome and transports it directly to cell surface, but not via ER. Then the transported peptide-loading MHC class I primes $CD8^+$ T cells. As we expect, cross-presentation during autophagy is faster than Th cell-dependent CTL response, this hypothesis has a novelty and a potentiality to new therapeutic approaches.

The details of this hypothesis are as follows:

1) Once autophagy-induced dendritic cells form phagocytosis and/or endocytosis containing the exogenous antigen, then it comes out to the cytoplasm from phagosome or endosome through yet to be defined function for cross-presentation through the involvement of transporter proteins such as TAP, tapasin, and Sec61 (***Step 1*** in Figure 4). Additionally these transporter proteins are suggested to contribute cross-presentation via two distinct pathways, "phagosome-to-cytosol pathway; major cross-presentation pathway" and "vacuolar pathway (different from autophagy-associated vacuolar pathway that does not utilize autophagosome for transporting the proteins), also called alternative cross-presentation pathway". In the "phagosome-to-cytosol pathway", the antigens are transported from phagosome to cytosol through Sec61 translocon, and back into phagosome through TAP after degradation by cytosolic proteasomes. Then, the peptides are presented after loading onto MHC class I molecule. In the "vacuolar pathway", phagosome fuses with ER directly, proteases in the ER degrade antigen to peptides, and the peptides are presented after loading onto MHC class I molecule.[34-40]
2) The antigen is ubiquitinated and proteasome recognizes the ubiquitinated-antigen that is degraded into small (about 10-20 amino acids) peptides (***Step 2*** in Figure 4).[41-43]
3) The peptides enter through TAP into autophagosome which contains MHC class I and its associated molecules (TAP, tapasin and β_2m, *etc*) together with transporter proteins those are derived from ER. Then these peptides are loaded on its MHC class I molecule (***Step 3*** in Figure 4).[34-40, 44, 45]
4) The peptide loaded autophagosome goes to cell surface membrane without autophagosome maturation when there is an inhibition of fusion between lysosome and autophagosome. Since optimal pH for the lysosomal enzyme is 4.8, H^+-ATPase proton pumps have to be utilized to regulate pH in lysosome for enzymatic activity. For instance lysosome inhibitors such as Bafilomycin A1 can prevent lysosome acidification by inhibiting H^+-ATPase pump on lysosomal membranes.[46-48] This step is a key to this hypothesis because if lysosome fuses with autophagosome, lysosomal enzymes degrade not only antigens but also potentially MHC class I

complex molecules (***Step 4*** in Figure 4).[46-48] This makes different from classical autolysosomal degrading mechanism as what degrades the protein to peptides, proteasome (in this hypothesis) or lysosome (generally referred one). Since it is the proteasome-dependent manner, there is no necessity to have lysosome to degrade the proteins and/or peptides again. In addition, peptide would be already loaded on MHC class I molecule before it reaches lysosome fusion step. Therefore, if lysosome is fused with autophagosome containing peptide-loading MHC class I molecule, lysosomal enzymes potentially degrade all of compartments in the autophagosome non-specifically. Thus, inhibiting lysosome with its inhibitors is considered to be a great therapeutic benefit to trigger higher MHC class I presentation.

5) Finally, the maturation-inhibited autophagosome would carry peptide loading MHC class I molecules onto the cell surface (***Step 5*** in Figure 4). It is worthwhile to mention that the destination of other proteins (TAP, MHC class I related molecule, and unloaded-peptides) is unknown not only in classical pathway but also with this hypothesis.

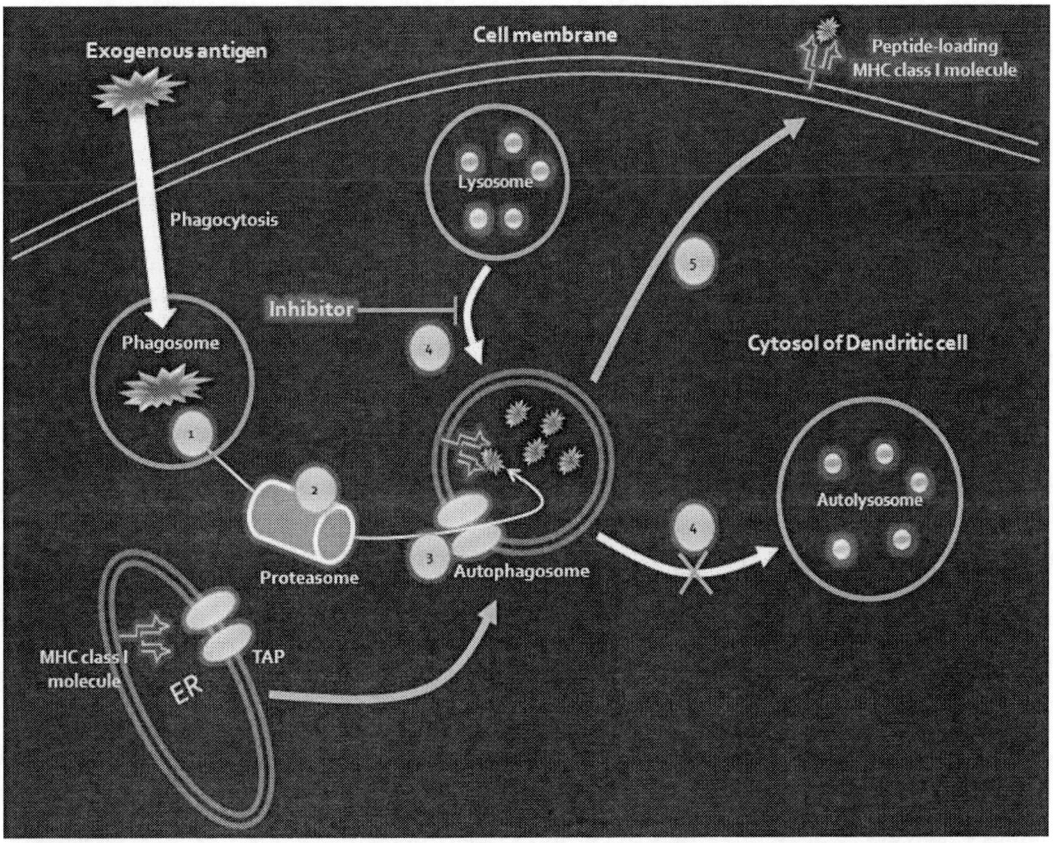

Figure 4. Schematic diagram for autophagy-dependent cross-presentation in dendritic cells. *Step 1*: Phagocytosed antigen is released from phagosome; *Step 2*: Antigen undergoes ubiquitin-dependent degradation by proteasome; *Step 3*: Peptides enter into autophagosome (not ER) and loaded onto MHC class I; *Step 4*: Lysosome inhibitor prevents the formation of autolysosome; *Step 5*: Autophagosome delivers peptide loaded MHC class I molecule to the surface of plasma membrane.

6. Strategic Approaches Utilizing Autophagy-Induced Cross-Presentation for Clinical Setting

6.1. Bacterial and Viral Infection

Autophagy has been reported to have potentials to enhance the ability of protection against bacterial and viral infection. As discussed, autophagy, especially xenophagy, is able to degrade numerous pathogens and deliver the microbial nucleic acids and viral proteins into endo/lysosomal compartments.[19, 21, 49]

It was noticed that studies on bacterial and viral infection have been focused mostly on autophagy, not on cross-presentation, due to the ambiguous mechanism of cross-presentation which has less efficiency on priming than classical priming of acquired immunity. In this review, we describe the close relationship between autophagy and cross-presentation. Although the efficiency of activating T cells is not impressed by inducing cross-presentation in clinical approaches yet, there is a great possibility to enhance the degree of cross-presentation by using induction of autophagy as aforementioned. Therefore, we would like to discuss the models to study autophagy-related diseases and the potential role of cross-presentation by autophagy for clinical necessity.

It is well known that immune cells can sense bacterial invasion through not only TLRs but also NOD1/2 molecules. Interestingly NOD1/2, the intracellular pattern recognition receptors, is found to recruit autophagy protein, ATG16L1. NOD1/2 introduces this ATG16L1 to bacterial entry sites and regulates autophagy ultimately for clearing the bacteria. However, when NOD2 mutation occurs, it causes Crohn's disease due to the impaired formation of antigen targeting autophagosomes.[50-53]

Autophagy plays a vital role in immune responses against intracellular mycobacteria not only by killing them but also enhancing antigen presentation and modulating the secretion of pro-inflammatory cytokines. It has been found that autophagy is stimulated by Th1 cytokines such as TNF-α and suppressed by Th2 cytokines such as IL-10.[8] In addition to cytokines, vitamin D and TLR stimulation also shown to promote autophagy.[54] Therefore, systemic or local administration of autophagy inducing agents to the lung could have a clinical application as adjunctive treatment against tuberculosis. Moreover, latent tuberculosis could be successfully removed through autophagy inducing therapeutic vaccines and/or adjuvants.[54]

Some bacteria such as group A *Streptococcus[55], Rickettsia conorii[56], Listeria monocytogenes[57], Porphyromonas gingivalis[58], Brucella abortus[59], and Legionella pneumophila[60]* are known to have ability to invade host cell that autophagy has not been induced yet. For example, group A *Streptococcus* is typically considered as an extracellular pathogen but also invades into the cytoplasm of epithelial cells. Then the cell modulates itself to induce autophagy for clearing the invading bacteria. When it happens, the bacteria are encapsulated by autophagosome, which is fused with lysosome forming autolysosome, and then cleared efficiently and rapidly from the cells.[55, 61]

During herpes simplex virus type-1 (HSV-1) infection, autophagy enhances the ability of MHC class I presentation and plays a role in clearing HSV-1 infected macrophages. HSV-1 has the most convincing evidence for describing the importance of autophagy in progression of disease. One research work showed that a vacuolar response that has the ability to increase the presentation of a peptide of HSV-1 glycoprotein B (gB) on MHC class I molecules to $CD8^+$ T cells, can be triggered in HSV-1-infected macrophages. Various cytokines and stress conditions could modulate this vacuolar response which has close relation to both rapid macroautophagy and prolonged CMA (chaperone mediated autophagy). Macroautophagy enhances virus clearing effect at early time point (*e.g.*, 8 hours post-infection) and CMA is conducting its effects for prolonged time after the infection and clear the virus permanently.[22]

Autophagy may prevent inflammation and autoimmunity, for example in an autoimmune disease like systemic lupus erythematosis (SLE). Once autoantibody binds to the cell surface, inflammation occurs and cellular components are released due to cell destruction thus, soluble immune complexes are formed. Deposition of immune complexes causes thickening of the blood vessels, kidneys, joints and other tissues. Under healthy conditions, autophagy can remove the dead and dying cells as well as cellular debris rapidly from the cells without eliciting immune responses.

However, defective autophagy cannot clear the autoantigens and could break the peripheral self-tolerance.[62] It is therefore possible that defective autophagy may contribute to the pathogenesis of SLE or other autoimmune diseases. An example of defective autophagy mediated disease is Crohn's disease as mentioned earlier. The pathogenic mechanisms are postulated to be involved with a dysregulated immune response to commensal intestinal bacteria, altered mucosal barrier function, and/or defective bacterial clearance because Crohn's disease is defective in NOD2 genes that could induce autophagy.[63]

Although we would like to suggest that induction of autophagy in bacteria- or virus-infected cells can be new clinical approaches to clear the infectious components.[64, 65] Apparent paradox is that side effects on the induction of autophagy (usually induced by rapamycin, or energy deprivation) showing non-negligible abnormality and cell death. So when patients are under the clinical trial, selection of the molecule(s) together with their action mode of signaling pathway involved in the induction of autophagy is extremely important for therapeutic approaches.

6.2. Cancer

With no doubt, cross-presentation is one of the most important concepts in the fields of tumor immunology. During the tumorigenesis, tumor cells secrete suppressive molecules which induce regulatory T cells and inhibit the activation of Th cells and CTLs, allowing the tumor cells to survive without being recognized by host immune system. Therefore, an alternative way to expand $CD8^+$ T cells without the involvement of Th cells would be very attractive to many research scientists, when and if it is ever possible.

For this, cross-presentation is the most effective strategy to expand tumor antigen-specific $CD8^+$ T cells to protect tumor progression since its effective role has been shown not only in priming but also in retention of fully differentiated CTLs augmenting protective responses.[66, 67]

Table 2. Autophagy inducing drug/therapy in trials or clinics and their mechanism of action

Drug/ Therapy	Target	Action mechanism	Effects on autophagy	Diseases
HDAC inhibitors (e.g. Butyrate and suberoylanilide hydroxamic acid)	Hisone deacetylase	Blocking DNA supercoiling in cancer cell by targeting histone deacetylase	Increased due to blockage of caspase activity[68]	Chronic myelogenous leukemia
Tamoxifen	Oestrogen receptor (DNA-binding transcription factor to regulate gene expression)	Binding estrogen receptors on tumors to produce a nuclear complex that decreases DNA synthesis	Increased since active cell death is induced[69]	Breast cancer
EB1089	Vitamin D receptor (regulate gene expression or repression)	Binding the target to repress the expression of several markers associated with cancer progression	Increased due to dramatic lysosomal changes and involvement of Beclin 1[70]	Sporadic breast cancer, prostate and lung cancer (in clinical trials)
Radiation	DNA	Radiation therapy works by damaging the DNA of cancerous cells by either photon energy or charged particle	Increased due to developing acidic vesicular organelles (autophagome feature)[71]	Various cancers
Angiogenesis inhibitors (e.g. Kringle5, ADH-1, AG013736 and anti-VEGF antibody)	Angiogenesis activations	Block the target and inactivate the process of the growth of new blood vessels from pre-existing vessels	Increased by up-regulating Beclin 1[72]	Numerous cancers (in clinical trials)
Tyrosine kinase inhibitor (Imatinib)	Several tyrosine kinases	Inhibits the enzyme responsible for the activation of signal transduction cascades	Increased by activating the cellular autophagy machinery[73]	Chronic myelogenous leukemia
Alkylating agent (Temozolomide, N-methyl-N'-nitro-N'-nitrosoguanidine: MNNG)	DNA	Extensively damaging DNA	Increased by recruiting the autophagy machinery[74]	Primary or recurrent high-grade gliomas
Akt inhibitors (e.g. IL-6-hydroxymethyl-chiro-inositol 2(R)-2-O-methyl-3-O-octadecylcarbonate)	Akt (Anti-apoptotic protein; inhibiting autophagy)	Downregulating the anti-apoptotic protein & lowering the sensitivity threshold of cancer cells(75)	Increased by mediating caspase-independent death & reducing resistance to chemotherapeutic drugs, TRAIL, ATRA, and ionizing radiation[76]	Various cancers
mTOR inhibitors (Rapamycin, CCi-779, RAD-001, AP23573)	mTORC1	Prevents cell growth, proliferation, motility, survival, protein synthesis, and transcription	Increased since the site of action of mTOR is upstream of Apg proteins[77]	Numerous cancers (in clinical trials)

Recent reports indicate that inducing autophagy in tumor cells can facilitate the formation of autophagosome, where it is utilized to trap small cytosolic compartments including mitochondria, nucleus, and other part of compartment from tumor cells. DCs then engulf the tumor antigens-autophagosome and degrade peptides into the compartments. Other reports show that defective autophagy is linked to tumorigenesis and persistence of viral infection. These studies imply that autophagy plays an important role in the degradation of intracellular pathogens simultaneously with the activation of innate immune cells, and efficient cross-presentation allows to activate adaptive immunity to prevent infections and tumor formation.[78] Due to various obstacles in the quest to define the degradation mechanism of autophagosome compartment containing tumor antigens, this has been considered one of the most interesting but challenging issue in the scientific community. Previous reports suggested the potential for autophagy-mediated cross-presentation, where it demonstrated promising therapeutic approaches to inhibit tumor progression and introduced the possibility of creating a new paradigm for vaccine development.[78, 79] It was also suggested that the isolated autophagosomes could be used as potent vaccines for immunotherapeutic treatment against cancer and infectious diseases.[69]

Some anti-cancerous drugs, cytokines, therapeutic hormones and genes are represented as 'tumor cell death inducers' that activate the formation of autophagy. Numerous drugs and treatments are known to up-regulate autophagy formation and some are used therapeutically and others are currently undergoing clinical trials (Table 2).[79]

On the other side of the coin, one research group has shown that the formation of autophagy is related to promote tumor growth. According to this study, the depletion of Atg5 suppresses the growth of tumors suggesting that its growth can be suppressed if treated with anti-cancer drug along with autophagy inhibitors. In addition, it has been shown that tumor cells induce the formation of autophagy by increasing glycogenesis suggesting that co-treatment of glucose-uptake inhibitors and autophagy inhibitor would be a potential anti-cancer therapy.[80] Based on these studies, autophagy can be not only a potential therapeutic strategy to remove tumors, but also dangerous arm to promote tumor progression, thus it is rather obvious to conduct a patient-tailored therapy. However, in cancer treatment, clinical approaches based on autophagy targeting tumor cells slightly differ from bacterial and viral infection in a sense that the latter is targeting immune cells. Thus, it is possible in certain cases that induction of autophagy in both tumor cells and immune cells would be a good strategy compared to a single therapy.

Conclusion

Autophagy is a process for recycling unnecessary intracellular molecules in the cytoplasm by disposing the defective organelles and invasive pathogens, and also acts as a cytoprotective mechanism during starvation for maintaining homeostasis. However, there is a notion that harmful effects are also observed after active autophagy formation. Since morphological studies have shown that autophagy is linked to disease progression, the strategies to utilize autophagy for clinical setting has been under the spotlight. More importantly, autophagy has been reported to cross-present non-self antigens to T cells. Therefore, defining the precise roles of autophagy in context of specific disease and finding a way to control autophagy are certainly the endeavors of the research. The mechanism for sequestering vesicle formation and determining the complex signals which are responsible for the different specificities and magnitudes for autophagy along with the regulatory pathways are important subjects of future investigations. It is likely that these studies will eventually lead to the utilization of autophagy to fight a variety of diseases. Moreover; these studies will build up a bridge between basic research and clinical approach and show a blueprint to therapeutic application in near future.

Acknowledgment

This work was supported by the National Research Foundation (NRF-2010-616-E00006, 2010-0027222 and 2010-0003291), and the Agriculture Research Center program of the Ministry for Food, Agriculture, Forestry and Fisheries, Republic of Korea.

References

[1] Klionsky DJ. Autophagy: from phenomenology to molecular understanding in less than a decade. *Nat Rev Mol Cell Biol.* 2007;8(11):931-7.

[2] Yang Z, Klionsky DJ. Eaten alive: a history of macroautophagy. *Nat Cell Biol.* 2010;12(9):814-22.

[3] Virgin HW, Levine B. Autophagy genes in immunity. *Nat Immunol.* 2009;10(5):461-70.

[4] Levine B, Mizushima N, Virgin HW. Autophagy in immunity and inflammation. *Nature.* 2011;469(7330):323-35.

[5] Vyas JM, Van der Veen AG, Ploegh HL. The known unknowns of antigen processing and presentation. *Nat Rev Immunol.* 2008;8(8):607-18.

[6] Ichiyanagi T, Imai T, Kajiwara C, Mizukami S, Nakai A, Nakayama T, et al. Essential role of endogenous heat shock protein 90 of dendritic cells in antigen cross-presentation. *J Immunol.* 2010;185(5):2693-700.

[7] Murshid A, Gong J, Calderwood SK. Heat shock protein 90 mediates efficient antigen cross presentation through the scavenger receptor expressed by endothelial cells-I. *J Immunol.* 2010;185(5):2903-17.

[8] Kurotaki T, Tamura Y, Ueda G, Oura J, Kutomi G, Hirohashi Y, et al. Efficient cross-presentation by heat shock protein 90-peptide complex-loaded dendritic cells via an endosomal pathway. *J Immunol.* 2007;179(3):1803-13.

[9] Chemali M, Radtke K, Desjardins M, English L. Alternative pathways for MHC class I presentation: a new function for autophagy. *Cell Mol Life Sci.* 2011;68(9):1533-41.

[10] Maiuri MC, Zalckvar E, Kimchi A, Kroemer G. Self-eating and self-killing: crosstalk between autophagy and apoptosis. *Nat Rev Mol Cell Biol.* 2007;8(9):741-52.

[11] Fujita N, Itoh T, Omori H, Fukuda M, Noda T, Yoshimori T. The Atg16L complex specifies the site of LC3 lipidation for membrane biogenesis in autophagy. *Mol Biol Cell.* 2008;19(5):2092-100.

[12] Weidberg H, Shvets E, Shpilka T, Shimron F, Shinder V, Elazar Z. LC3 and GATE-16/GABARAP subfamilies are both essential yet act differently in autophagosome biogenesis. *EMBO J.* 2010;29(11):1792-802.

[13] Mizushima N, Yoshimori T, Levine B. Methods in mammalian autophagy research. *Cell.* 2010;140(3):313-26.

[14] Rabinowitz JD, White E. Autophagy and metabolism. *Science.* 2010;330(6009):1344-8.

[15] Munz C. Autophagy and antigen presentation. *Cell Microbiol.* 2006;8(6):891-8.

[16] Crotzer VL, Blum JS. Autophagy and its role in MHC-mediated antigen presentation. *J Immunol.* 2009;182(6):3335-41.

[17] Jensen PE. Recent advances in antigen processing and presentation. *Nat Immunol.* 2007;8(10):1041-8.

[18] Crotzer VL, Blum JS. Autophagy and adaptive immunity. *Immunology.* 2010;131(1):9-17.

[19] Schmid D, Munz C. Innate and adaptive immunity through autophagy. *Immunity.* 2007;27(1):11-21.

[20] Huang J, Klionsky DJ. Autophagy and human disease. *Cell Cycle.* 2007;6(15):1837-49.

[21] Levine B, Deretic V. Unveiling the roles of autophagy in innate and adaptive immunity. *Nat Rev Immunol.* 2007;7(10):767-77.

[22] English L, Chemali M, Duron J, Rondeau C, Laplante A, Gingras D, et al. Autophagy enhances the presentation of endogenous viral antigens on MHC class I molecules during HSV-1 infection. *Nat Immunol.* 2009;10(5):480-7.

[23] Esquivel F, Yewdell J, Bennink J. RMA/S cells present endogenously synthesized cytosolic proteins to class I-restricted cytotoxic T lymphocytes. *J Exp Med.*

[24] Hosken NA, Bevan MJ. An endogenous antigenic peptide bypasses the class I antigen presentation defect in RMA-S. *J Exp Med.* 1992;175(3):719-29.

[25] Sijts AJ, De Bruijn ML, Nieland JD, Kast WM, Melief CJ. Cytotoxic T lymphocytes against the antigen-processing-defective RMA-S tumor cell line. *Eur J Immunol.* 1992;22(6):1639-42.

[26] Tiwari N, Garbi N, Reinheckel T, Moldenhauer G, Hammerling GJ, Momburg F. A transporter associated with antigen-processing independent vacuolar pathway for the MHC class I-mediated presentation of endogenous transmembrane proteins. *J Immunol.* 2007;178(12):7932-42.

[27] Rock KL, Shen L. Cross-presentation: underlying mechanisms and role in immune surveillance. *Immunol Rev.* 2005;207:166-83.

[28] Kario E, Amar N, Elazar Z, Navon A. A new autophagy-related checkpoint in the degradation of an ERAD-M target. *J Biol Chem.* 2011;286(13):11479-91.

[29] Kroeger H, Miranda E, MacLeod I, Perez J, Crowther DC, Marciniak SJ, et al. Endoplasmic reticulum-associated degradation (ERAD) and autophagy cooperate to degrade polymerogenic mutant serpins. *J Biol Chem.* 2009;284(34):22793-802.

[30] Willer M, Forte GM, Stirling CJ. Sec61p is required for ERAD-L: genetic dissection of the translocation and ERAD-L functions of Sec61P using novel derivatives of CPY. *J Biol Chem.* 2008;283(49):33883-8.

[31] Rock KL, Farfan-Arribas DJ, Shen L. Proteases in MHC class I presentation and cross-presentation. J Immunol. 2010;184(1):9-15.

[32] Munz C. Antigen processing via autophagy--not only for MHC class II presentation anymore? *Curr Opin Immunol.* 2010;22(1):89-93.

[33] Russmann H, Panthel K, Kohn B, Jellbauer S, Winter SE, Garbom S, et al. Alternative endogenous protein processing via an autophagy-dependent pathway compensates for Yersinia-mediated inhibition of endosomal major histocompatibility complex class II antigen presentation. *Infect Immun.* 2010;78(12):5138-50.

[34] Desjardins M. ER-mediated phagocytosis: a new membrane for new functions. *Nat Rev Immunol.* 2003;3(4):280-91.

[35] Rock KL. The ins and outs of cross-presentation. *Nat Immunol.* 2003;4(10):941-3.

[36] Oh YK, Harding CV, Swanson JA. The efficiency of antigen delivery from macrophage phagosomes into cytoplasm for MHC class I-restricted antigen presentation. *Vaccine.* 1997;15(5):511-8.

[37] York IA, Rock KL. Antigen processing and presentation by the class I major histocompatibility complex. *Annu Rev Immunol.* 1996;14:369-96.

[38] Kovacsovics-Bankowski M, Rock KL. A phagosome-to-cytosol pathway for exogenous antigens presented on MHC class I molecules. *Science.* 1995;267(5195):243-6.

[39] Rodriguez A, Regnault A, Kleijmeer M, Ricciardi-Castagnoli P, Amigorena S. Selective transport of internalized antigens to the cytosol for MHC class I presentation in dendritic cells. *Nat Cell Biol.* 1999;1(6):362-8.

[40] Roy CR. Immunology: professional secrets. *Nature.* 2003;425(6956):351-2.

[41] Hershko A, Ciechanover A. The ubiquitin system for protein degradation. *Annu Rev Biochem.* 1992;61:761-807.

[42] Rock KL, Goldberg AL. Degradation of cell proteins and the generation of MHC class I-presented peptides. *Annu Rev Immunol.* 1999;17:739-79.

[43] Kisselev AF, Akopian TN, Woo KM, Goldberg AL. The sizes of peptides generated from protein by mammalian 26 and 20 S proteasomes. Implications for understanding the degradative mechanism and antigen presentation. *J Biol Chem.* 1999;274(6):3363-71.

[44] Houde M, Bertholet S, Gagnon E, Brunet S, Goyette G, Laplante A, et al. Phagosomes are competent organelles for antigen cross-presentation. *Nature.* 2003;425(6956):402-6.

[45] Rock KL. A new foreign policy: MHC class I molecules monitor the outside world. *Immunol Today.* 1996;17(3):131-7.

[46] van Schalkwyk DA, Chan XW, Misiano P, Gagliardi S, Farina C, Saliba KJ. Inhibition of Plasmodium falciparum pH regulation by small molecule indole derivatives results in rapid parasite death. *Biochem Pharmacol.* 2010;79(9):1291-9.

[47] Klionsky DJ, Abeliovich H, Agostinis P, Agrawal DK, Aliev G, Askew DS, et al. Guidelines for the use and interpretation of assays for monitoring autophagy in higher eukaryotes. *Autophagy.* 2008;4(2):151-75.

[48] Yamamoto A, Tagawa Y, Yoshimori T, Moriyama Y, Masaki R, Tashiro Y. Bafilomycin A1 prevents maturation of autophagic vacuoles by inhibiting fusion between autophagosomes and lysosomes in rat hepatoma cell line, H-4-II-E cells. *Cell Struct Funct.* 1998;23(1):33-42.

[49] Serna A, Ramirez MC, Soukhanova A, Sigal LJ. Cutting edge: efficient MHC class I cross-presentation during early vaccinia infection requires the transfer of proteasomal intermediates between antigen donor and presenting cells. *J Immunol.* 2003;171(11):5668-72.

[50] Travassos LH, Carneiro LA, Ramjeet M, Hussey S, Kim YG, Magalhaes JG, et al. Nod1 and Nod2 direct autophagy by recruiting ATG16L1 to the plasma membrane at the site of bacterial entry. *Nat Immunol.* 2010;11(1):55-62.

[51] Cooney R, Baker J, Brain O, Danis B, Pichulik T, Allan P, et al. NOD2 stimulation induces autophagy in dendritic cells influencing bacterial handling and antigen presentation. *Nat Med.* 2010;16(1):90-7.

[52] Shin JN, Eissa NT. Autophagy gets the 'NOD' to enhance bacterial handling and antigen presentation. *Immunol Cell Biol.* 2010;88(4):343-5.

[53] Brain O, Allan P, Simmons A. NOD2-mediated autophagy and Crohn disease. *Autophagy.* 2010;6(3):412-4.

[54] Ni Cheallaigh C, Keane J, Lavelle EC, Hope JC, Harris J. Autophagy in the immune response to tuberculosis: clinical perspectives. *Clin Exp Immunol.* 2011;164(3):291-300.

[55] Nakagawa I, Amano A, Mizushima N, Yamamoto A, Yamaguchi H, Kamimoto T, et al. Autophagy defends cells against invading group A Streptococcus. *Science.* 2004;306(5698):1037-40.

[56] Walker DH, Popov VL, Crocquet-Valdes PA, Welsh CJ, Feng HM. Cytokine-induced, nitric oxide-dependent, intracellular antirickettsial activity of mouse endothelial cells. *Lab Invest.* 1997;76(1):129-38.

[57] Rich KA, Burkett C, Webster P. Cytoplasmic bacteria can be targets for autophagy. *Cell Microbiol.* 2003;5(7):455-68.

[58] Dorn BR, Dunn WA, Jr., Progulske-Fox A. Porphyromonas gingivalis traffics to autophagosomes in human coronary artery endothelial cells. *Infect Immun.* 2001;69(9):5698-708.

[59] Pizarro-Cerda J, Moreno E, Sanguedolce V, Mege JL, Gorvel JP. Virulent Brucella abortus prevents lysosome fusion and is distributed within autophagosome-like compartments. *Infect Immun.* 1998;66(5):2387-92.

[60] Sturgill-Koszycki S, Swanson MS. Legionella pneumophila replication vacuoles mature into acidic, endocytic organelles. *J Exp Med.* 2000;192(9):1261-72.

[61] Kirkegaard K, Taylor MP, Jackson WT. Cellular autophagy: surrender, avoidance and subversion by microorganisms. *Nat Rev Microbiol.* 2004;2(4):301-14.

[62] Grossmayer GE, Munoz LE, Gaipl US, Franz S, Sheriff A, Voll RE, et al. Removal of dying cells and systemic lupus erythematosus. *Mod Rheumatol.* 2005;15(6):383-90.

[63] Baumgart DC, Carding SR. Inflammatory bowel disease: cause and immunobiology. *Lancet.* 2007;369(9573):1627-40.

[64] Espert L, Denizot M, Grimaldi M, Robert-Hebmann V, Gay B, Varbanov M, et al. Autophagy is involved in T cell death after binding of HIV-1 envelope proteins to CXCR4. *J Clin Invest.* 2006;116(8):2161-72.

[65] Levine B, Kroemer G. Autophagy in the pathogenesis of disease. *Cell.* 2008;132(1):27-42.

[66] Calzascia T, Di Berardino-Besson W, Wilmotte R, Masson F, de Tribolet N, Dietrich PY, et al. Cutting edge: cross-presentation as a mechanism for efficient recruitment of tumor-specific CTL to the brain. *J Immunol.* 2003;171(5):2187-91.

[67] Hayashi A, Wakita H, Yoshikawa T, Nakanishi T, Tsutsumi Y, Mayumi T, et al. A strategy for efficient cross-presentation of CTL-epitope peptides leading to enhanced induction of in vivo tumor immunity. *J Control Release.* 2007;117(1):11-9.

[68] Shao Y, Gao Z, Marks PA, Jiang X. Apoptotic and autophagic cell death induced by histone deacetylase inhibitors. *Proc Natl Acad Sci U S A.* 2004;101(52):18030-5.

[69] Bursch W, Ellinger A, Kienzl H, Torok L, Pandey S, Sikorska M, et al. Active cell death induced by the anti-estrogens tamoxifen and ICI 164 384 in human mammary carcinoma cells (MCF-7) in culture: the role of autophagy. *Carcinogenesis.* 1996;17(8):1595-607.

[70] Hoyer-Hansen M, Bastholm L, Mathiasen IS, Elling F, Jaattela M. Vitamin D analog EB1089 triggers dramatic lysosomal changes and Beclin 1-mediated autophagic cell death. *Cell Death Differ.* 2005;12(10):1297-309.

[71] Paglin S, Hollister T, Delohery T, Hackett N, McMahill M, Sphicas E, et al. A novel response of cancer cells to radiation involves autophagy and formation of acidic vesicles. *Cancer Res.* 2001;61(2):439-44.

[72] Nguyen TM, Subramanian IV, Kelekar A, Ramakrishnan S. Kringle 5 of human plasminogen, an angiogenesis inhibitor, induces both autophagy and apoptotic death in endothelial cells. *Blood.* 2007;109(11):4793-802.

[73] Ertmer A, Huber V, Gilch S, Yoshimori T, Erfle V, Duyster J, et al. The anticancer drug imatinib induces cellular autophagy. *Leukemia.* 2007;21(5):936-42.

[74] Kanzawa T, Germano IM, Komata T, Ito H, Kondo Y, Kondo S. Role of autophagy in temozolomide-induced cytotoxicity for malignant glioma cells. *Cell Death Differ.* 2004;11(4):448-57.

[75] Martelli AM, Tazzari PL, Tabellini G, Bortul R, Billi AM, Manzoli L, et al. A new selective AKT pharmacological inhibitor reduces resistance to chemotherapeutic drugs, TRAIL, all-trans-retinoic acid, and ionizing radiation of human leukemia cells. *Leukemia.* 2003;17(9):1794-805.

[76] 7Takashina K, Brun M, Ota T, Maude DK, Fujiwara A, Ono Y, et al. Anomalous resistance ridges along filling factor nu=4i. *Phys Rev Lett.* 2007;99(3):036803.

[77] Noda T, Ohsumi Y. Tor, a phosphatidylinositol kinase homologue, controls autophagy in yeast. *J Biol Chem.* 1998;273(7):3963-6.

[78] Li Y, Wang LX, Yang G, Hao F, Urba WJ, Hu HM. Efficient cross-presentation depends on autophagy in tumor cells. *Cancer Res.* 2008;68(17):6889-95.

[79] Li Y, Wang LX, Pang P, Twitty C, Fox BA, Aung S, et al. Cross-presentation of tumor associated antigens through tumor-derived autophagosomes. *Autophagy.* 2009;5(4):576-7.

[80] Kelekar A. Introduction to the review series Autophagy in Higher Eukaryotes--a matter of survival or death. *Autophagy.* 2008;4(5):555-6.

In: Autophagy: Principles, Regulation and Roles in Disease ISBN: 978-1-61942-266-7
Editor: Nikolai Gorbunov

Chapter III

Autophagy as a Hallmark of Hemophagocytic Diseases

E. Marion Schneider,[1,*,†] Myriam Lorenz[1] and Paul Walther[2]
[1] Sektion Experimentelle Anaesthesiologie, University Hospital Ulm, Ulm, Germany
[2] Zentrale Einrichtung Elektronenmikroskopie, University of Ulm, Ulm, Germany

Abstract

Background: Hemophagocytosis occurs in states of genetic and transient immune deficiencies including virus infections, toxicities, malignancies, autoimmune diseases and severe bacterial infections such as sepsis. The documentation of hemophagocytosis in blood and bone marrow smears or organ biopsies is often difficult.

Procedure: It is demonstrated that hemophagocytes can be successfully enriched from Ficoll-separated blood cells by cell culture methods. Isolates from classical hemophagocytic lymphohistiocytosis (HLH patients) and patients with sepsis were compared. High-pressure freezing techniques were applied to study morphological characteristics on the ultrastructural level, and surface markers were studied by flow cytometry.

Results: Hemophagocytosis of autologous blood cells, long-term survival, and the apparently retarded degradation of phagocytosed cells were found in both HLH and sepsis derived cultures. Macroautophagy of cytoplasm and mitochondria was prominent in early cultures and progressed during prolonged cell culture. The structure of the nuclei in both diseases was different, however, HLH hemophagocyte nuclei were polymorph in shape and highly flexible, sepsis-associated hemophagocyte nuclei were round and rigid. Phenotypically, both types of hemophagocytes expressed a similar pattern of largely immature dendritic cell markers. However, the HLH hemophagocytes more often lacked TLR3, whereas the sepsis-derived hemophagocytes were TLR3 positive.

---

* Correspondence to E. Marion Schneider, Sektion Experimentelle Anaesthesiologie, Universitaetsklinikum Ulm, Steinhoevelstr. 9, 89075 Ulm, Germany, ph: +49-731-500-60080/81, Fax: +49-731-500-60082, e-mail: marion.schneider@uni-ulm.de.

† This work is dedicated to Prof. Dr. Dr. hc Michael Georgieff on the occasion of his 60th birthday.

Conclusion: Ultrastructural studies substantiate phagocytosis of autologous erythrocytes and morphologically intact leukocytes in enriched hemophagocyte cultures from patients with HLH and defined cases of sepsis-associated hemophagocytosis. Autophagy is a characteristic feature of these hemophagocytes, which may explain apoptosis resistance in vitro and in vivo.

Keywords: HLH, sepsis, autophagy, hemophagocytosis, dendritic cells, ultrastructural analysis by high-pressure freezing and electron tomography

Introduction

Hemophagocytic lymphohistiocytosis (HLH) is a complex disorder of the immune system characterized by high fever, hypercytokinemia, hepatosplenomegaly, and tissue-/organ-infiltrating histiocytes (Fisman, 2000; Henter et al., 2004). Infections and defective NK- and cytotoxic function lead to a cytokine storm, which may cause a manifestation of hemophagocytosis. The hemophagocytic syndrome was first described in 1939 (Scott R, 1939), and 13 years later a familial (syn. primary or genetic) form was discovered (Farquhar and Claireaux, 1952). Primary HLH affects young children as well as newborns and is inherited in an autosomal recessive manner, and secondary HLH is an infection-associated, transient manifestation of hemophagocytosis that may occur in patients of all ages. In addition to viral infections, secondary hemophagocytic syndromes may also emerge during advanced systemic inflammatory response syndrome (SIRS) and severe sepsis intrinsically coherent with bacterial infections (Kuwata et al., 2006; Risdall et al., 1984) which may be complicated by virus infections (Buyse et al., 2010 ; Creput et al., 2008).

Our newly established cell cultures are the first to identify the presence of hemophagocytic cells in the circulation and propose an answer to the difficulty of proving hemophagocytosis in blood and bone marrow smears (Janka and Zur Stadt, 2005). These cultures allow the comparison of hemophagocytosis of different patients in vitro. Accordingly, cultures of very young patients with classical HLH have been studied in parallel to cultures of adult individuals suffering from hemophagocytic syndromes associated with severe sepsis. We were able to observe phagocytosis of autologous blood cells as well as leukocytes. Using high-pressure freezing and electron tomography, we also found massive macroautophagy in both types of patients. Along these lines, morphological differences in the structure of the nuclei were observed but the phenotypic marker profile of the hemophagocytes (resembling immature myeloid dendritic cells) was similar in both diseases.

Materials and Methods

Patients

Patients with acute hemophagocytosis were diagnosed following the criteria of the HLH2004 study group (Henter et al., 2007). None of the pediatric patients had a homozygous HLH-specific mutation in the following genes: Perforin, Munc13-4, Syntaxin11, or Rab27a. All suffered from acute viral infections (EBV or CMV) and were <5 years old. The disease state was resolved upon HLH-specific therapy. Adult patients with sepsis were 42 to 64 years old and the hemophagocytes were isolated shortly before the manifestation of septic shock, which is often associated with virus reactivation (EBV, CMV, or HSV1).

Following a major operative trauma, the episodes of patients with severe thrombocytopenia, anemia, low leukocyte counts, increased ferritin, and fever are studied for hemophagocytosis by screening blunt edge blood smears, as suggested for patients with advanced systemic inflammatory response syndrome (SIRS) (Kuwata et al., 2006).

Cell Culture

Ficoll-isolated peripheral blood- or bone marrow-derived mononuclear cells are depleted from non-adherent cells following 2h of plastic adherence. During four weeks of culture in RPMI1640 supplemented with 10% fetal calf serum (selected by very low endotoxin content, <2 endotoxin units (EU)/ml), 25 mM HEPES and antibiotics, the hemophagocytes are enriched and cultures are free from lymphocytes, erythrocytes and granulocytes. The hemophagocytes are amoeboid, floating or loosely adherent and can be easily harvested by gentle agitation.

Electron Microscopy

For electron microscopy, the classical chemical fixation high-pressure freezing method was applied. Chemical fixation was performed with 2% glutar-dialdehyde (grade I, Sigma 5882) for 20 min *in situ*, followed by three washes in PBS and two washes in RPMI1640. Next, fixed cells were treated with 2% osmium tetroxide, dehydrated in a graded series of isopropanol, and block stained with 2% uranyl acetate. High-pressure freezing and freeze substitution were performed (Buser and Walther, 2008), with minor modifications.

The cells were mounted onto sapphire discs and frozen via high pressure directly from the living state using a Compact-01 (Wohlwend GmbH, Sennwald, Switzerland). The freeze substitution medium consisted of acetone with 0.2% osmium tetroxide, 0.1% uranyl acetate and 5% water. Both chemically fixed and high-pressure frozen samples were embedded in EPON. Ultra thin sections (80 nm) were cut on a Leica microtome E, stained with 0.2% lead citrate, and imaged using an EM 10 (Zeiss, Oberkochen, Germany) at an accelerating voltage of 80 kV.

For electron tomography, samples were high-pressure frozen, freeze substituted, and embedded as described above. Semi-thin sections (500 nm) were imaged with a Titan 300 (FEI, Eindhoven, The Netherlands) at an accelerating voltage of 300 kV using the scanning transmission (STEM) mode and by recording the dark field signal with a high angle annular dark field detector (HAADF). For a single tomogram, a total of 73 images were recorded by tilting the specimen from -72° to +72°. The tomogram was reconstructed by weighted back projection using the IMOD software (Kremer et al., 1996) version 3.13.5.

A total of seven patients were studied, four with HLH and three with severe sepsis and hemophagocytic syndromes.

Results in Morphology

Hemophagocytic Cells Enriched by in vitro Culture

Figure 1 summarizes different states and targets of hemophagocytosis identified in representative cultures after chemical fixation. Figures 1A to 1D represent hemophagocytes as well as leukocytes of HLH patients. Figures 1E and G show ingested cells from hemophagocytes of sepsis patients. Degradation of cells in Figures 1E and F appears to be more advanced than in HLH hemophagocytes. Figure 1G shows even further degraded material, such that only a dark electron-dense matrix and additional vacuoles can be identified as content of the phagosome. The prominent signs of autophagy are not restricted to 28-day-old hemophagocyte cultures, since autophagy can be observed by studying freshly prepared bone marrow smears of end-stage-patients (Beutel et al., 2009; Fisman, 2000; Schaer et al., 2002).

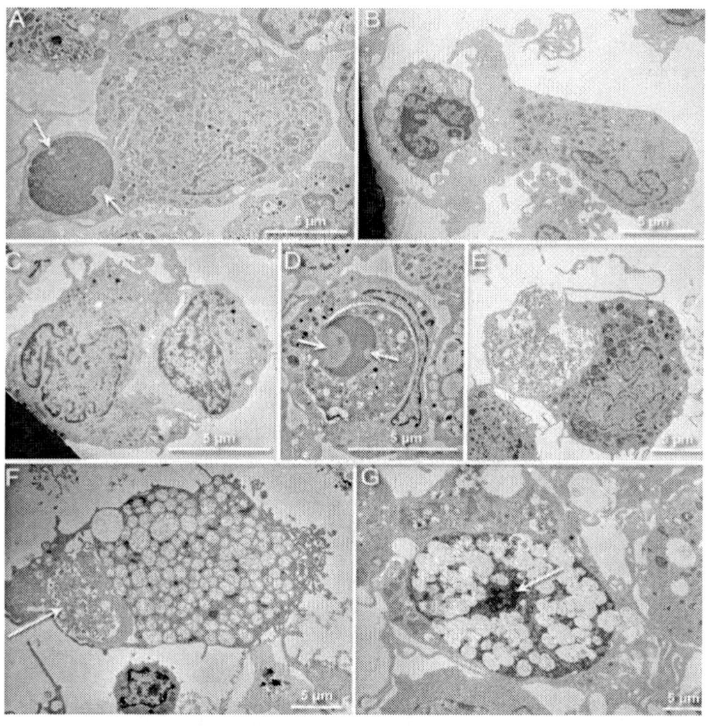

Figure 1. Structural features of hemophagocytes. Hemophagocytes with ingested and partially digested erythrocytes (arrows) located in a distal phagosome (HLH, chemical fixation). (A). Early hemophagocytosis of another autologous hemophagocyte with remarkable autophagous vacuoles (HLH, chemical fixation) (B). Hemophagocytosis of a hemophagocyte (HLH, chemical fixation) (C). More advanced state of erythrocyte digestion in a phagosome being engulfed by another autophagosome (HLH, chemical fixation) (D). As in C, a more advanced state of phagocytosis and digestion (sepsis, chemical fixation) (E). As in E, a hemophagocyte shows advanced signs of autophagy (sepsis, chemical fixation) (F). Autophagic vacuole located in another autophagosome, similar to D but at a more advanced state of digestion. The center bears electron dense material (sepsis, chemical fixation) (G).

Vesicles, Membrane Contact, and Fusion Zones

In the chemically fixed specimens, the cytoplasm of most cultured hemophagocytes contains numerous empty vacuolar structures (Fig. 1F). However, when applying high-pressure freezing techniques, these vacuolar structures represent vesicles containing a material with different staining intensity (Fig. 2A-F). Close contact and some signs of vesicle fusion events such as in Figures 2G-I are characteristic for hemophagocytes cultured for 28 days and beyond.

Autophagy and Cytoplasmic Inclusions

Membrane structures were studied in greater detail using electron tomography. Figure 3 shows four computed sections of a tomogram (HLH). The caged area of section 3A documents two individual membranes lying in parallel stretches (Fig. 3B); however, in Figure 3C only a single membrane bilayer is detectable. The structure vanishes completely in the uppermost section of the tomogram (Fig. 3D). These images also imply phagophore formation initiated by an isolation membrane, which is depicted just below the dashed area (Fig. 3B-D), where cytoplasm is enclosed by a rounded bilayer. A structurally completed autophagosome can be seen in the lower left of the tomogram, and its content can be distinguished easily from the surrounding cytoplasm.

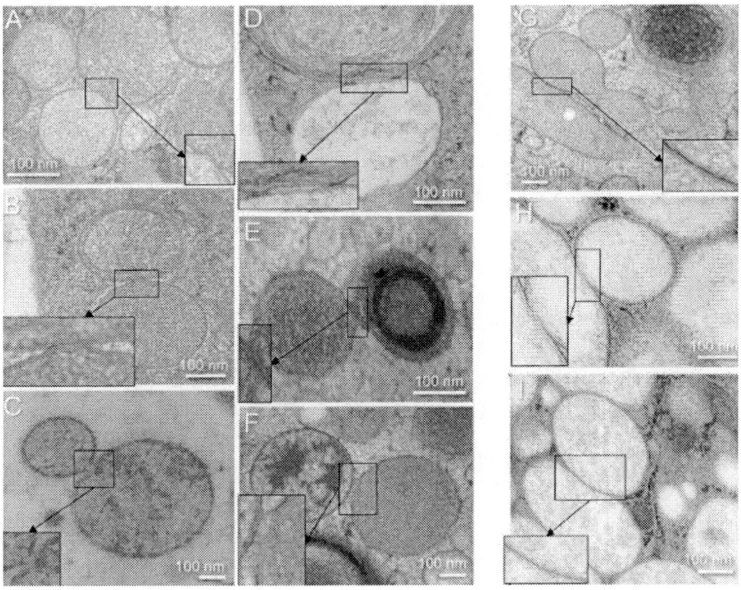

Figure 2. Vesicle contact areas in high-pressure frozen hemophagocytes. A-C (top to bottom): Structurally similar autophagic vesicles in early states of autophagy with cytoplasmic content (A, B) and partially digested material (C). Multi-lamellar double membrane-coated vesicle and a mono-layer membrane vesicle with more electron transparent-staining content (D). Bilayered membrane-surrounded vesicles with protein-rich material (E). Vesicle with cytoplasmic contents ín contact with a vesicle with an electron dense precipitate (F). G-I (top to bottom): Single membrane-surrounded vesicles frequently observed in cultures of >28 days (G-I).

Mitochondria

Mitochondria differ in their grade of lamellar structures and cristae. Figure 4A shows both parallel (M*) and tubular (M) cristae derived from a patient with sepsis. Figure 4B shows an HLH-derived hemophagocyte at lower magnification: both types of mitochondria (M, M*), inclusion bodies (I) and electron dense vesicles (L) can be observed. In Figure 4C (HLH), a multi-layered membrane surrounds a mitochondrion with fuzzy tubular cristae (M). Figure 4D is derived from a tomogram showing phagophore formation around two mitochondria (M*, M, arrows). The outer structure of an adjacent object of low electron density (on the right) resembles the form of a mitochondrion, but no cristae can be identified. Numerous membranes are in close vicinity.

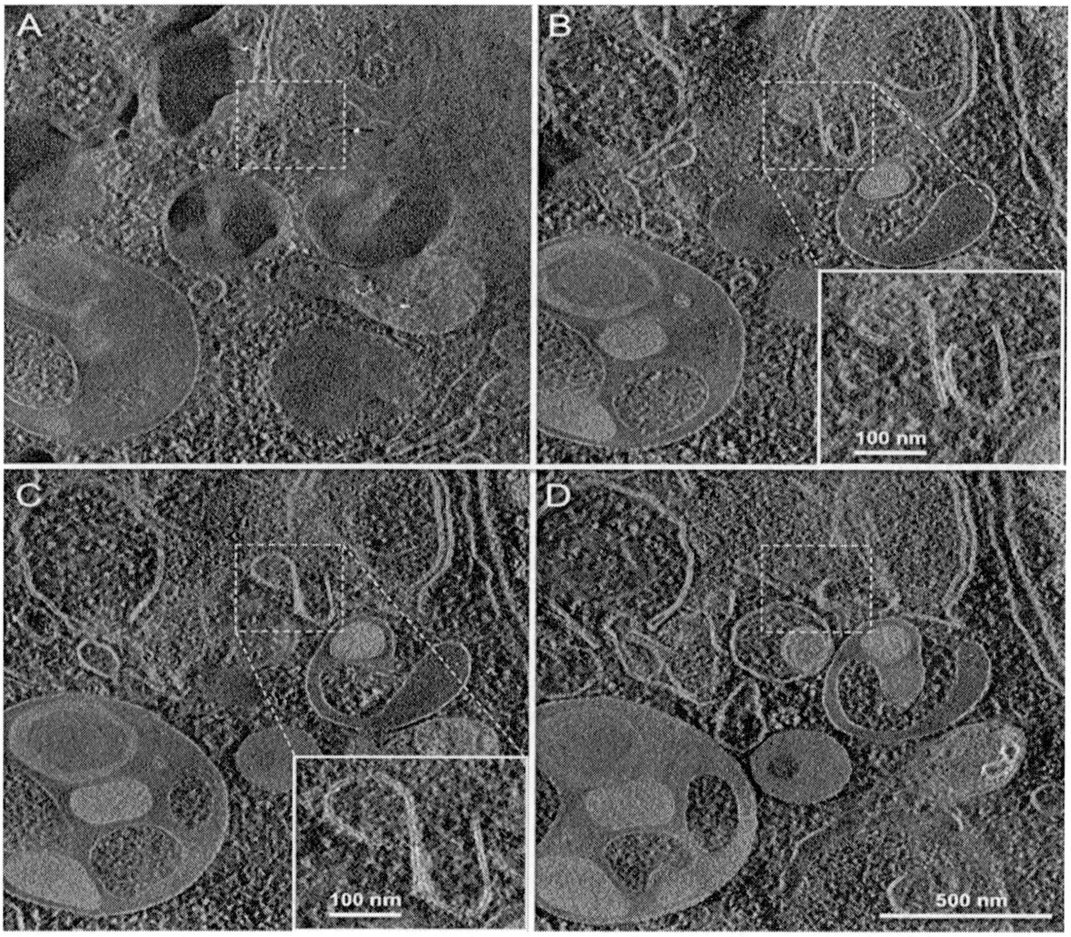

Figure 3. Four computed sections of an electron tomogram of the autophagic vesicle rich cytoplasm in a hemophagocyte of a semi-thin section after high-pressure freezing preparation (HLH). The dashed area is membrane-free (3A) but tomography reveals newly formed or degrading membranes with a stretch of parallelized location (3B). In another computed section (3C) of the same tomogram, the parallel stretch of the two membranes shows only one membrane bilayer and this membrane composition has disappeared in the computed section 3D.

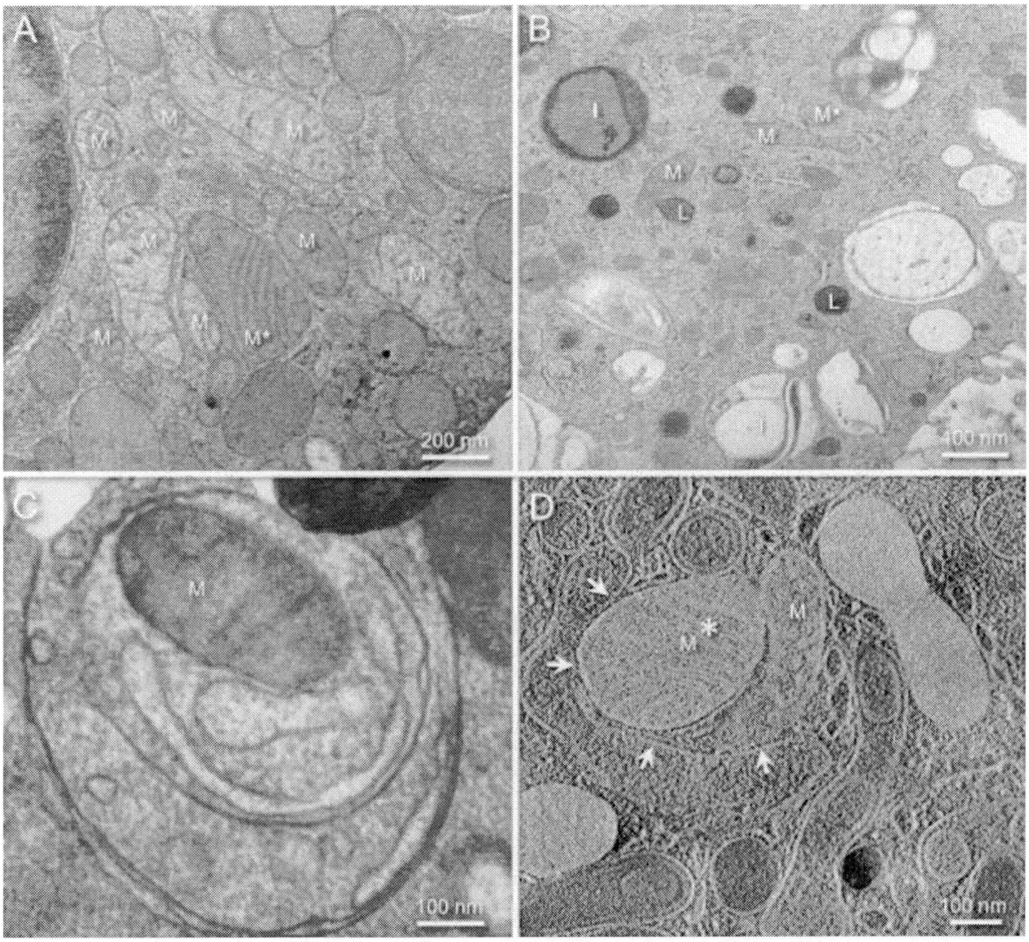

Figure 4. High-pressure frozen specimen of hemophagocytes with structurally different mitochondria. Sections of a hemophagocyte from a patient with sepsis containing mitochondria with dense and parallel cristae as well as various mitochondria with more tubular cristae (A), and a section of an HLH patient's hemophagocytes with tubular structured cristae in mitochondria, few lysosomes (L), vesicles with inclusion bodies (I) (B). A mitochondrion with fuzzy cristae derived from a patient with HLH surrounded by more than two layers of folded double membranes resembling an isolation membrane (C). A similar aspect in a computed section of a tomogram highlighting early phagophore formation around two mitochondria.

Phenotypes of Hemophagocytes

Figure 5 summarizes the surface marker profiles of hemophagocytes cultured from patients with pediatric HLH and adult sepsis, respectively. In cells cultured from patients with sepsis, CD14 expression varies over a broad range, whereas the majority of pediatric HLH hemophagocytes are more homogenously CD14 positive. The antigen presentation marker CD86 is expressed in variable antigen density of both sepsis and HLH hemophagocytes, and ß-integrin CD11c is found in all isolates. Finally, CCR7 is found only in a lower percentage of sepsis as well as pediatric HLH hemophagocytes.

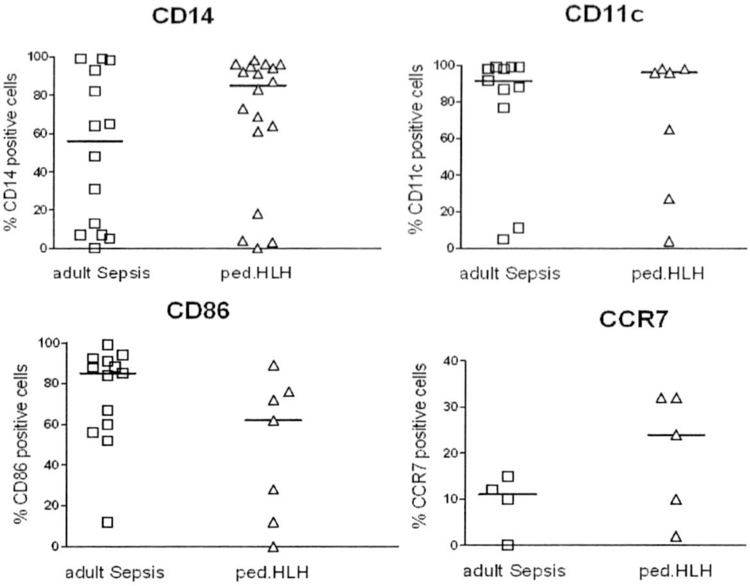

Figure 5. Phenotype of hemophagocytes cultured for 28 days derived from patients with HLH and severe sepsis. The amount of positive cells stained for CD14, CD86, CD11c and CCR7 has been determined by flow cytometry.

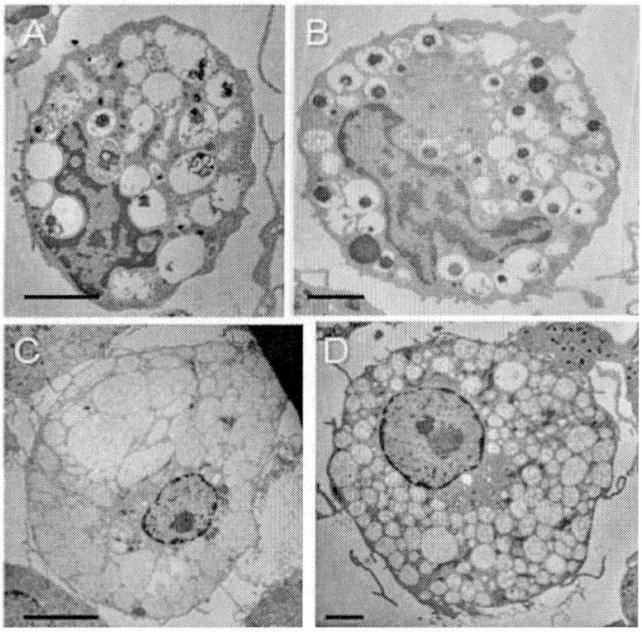

Figure 6. Hemophagocytes of patients with HLH and sepsis after 28 days of in vitro culture chemical fixation. Whole cell sections of HLH hemophagocytes show numerous vacuoles with degraded contents. Vacuoles indent the nucleus at various sites (A, B). Whole cell sections from patients with sepsis-associated hemophagocytosis show more amorphous vacuole contents. The nuclei are located eccentrically and are not mechanically indented by vacuoles (C, D).

Differences of Nuclear Shape in Hemophagocytes from Patients with HLH and Sepsis

Figure 6 compares hemophagocyte structures of both groups of patients at low magnification to create an overview. Chemical fixation was used, and all isolates display vacuolar cytoplasm. These structures appear to be either empty or filled with denatured proteineous material. The nuclei of both HLH and sepsis cultures are located peripherally but differ in shape. For example, HLH nuclei are irregular in shape and indented by vacuolar structures. Furthermore, they contain patchy homo and heterochromatin (Figs. 6A, B). In contrast, sepsis-derived hemophagocytes show round nuclei with distally condensed heterochromatin (Figs. 6C, D).

Discussion

Phagocytosis

Hemophagocytosis is defined as pathology of primary and secondary hemophagocytic diseases. The engulfment of hematopoietic cells is generally difficult to prove using blood and bone marrow smears or even biopsies of lymph nodes and affected organs (Janka and Zur Stadt, 2005). Furthermore, no direct evidence has shown that hemophagocytosis is responsible for cytopenia and anemia in HLH. Nevertheless, hemophagocytosis has been suggested to play a major role in the pathogenesis of cytopenia (Yarali et al., 2009). The current study demonstrates that i) active blood cell phagocytosis occurs in cultures of HLH hemophagocytes, and ii) a remarkably similar process occurs in cultures from severely septic patients suffering from hemophagocytosis. Earlier research showed similarities between selected sepsis patients and HLH such as virus reactivation (von Muller et al., 2007), occurrence of hemophagocytes in blood smears (Kuwata et al., 2006) and/or elevated plasma concentrations of soluble CD25, a reliable diagnostic marker for HLH (Schneider et al., 2002; Henter et al., 2007;). Recently, severe cases with virus-associated HLH have been described in H1N1 (Beutel et al., 2011).

Hemophagocytosis continues during culturing; therefore, hemophagocytes can be easily enriched in vitro. Individual cultures of hemophagocytes of patients survive for longer than 8 months in culture which strongly supports the cytoprotective process of autophagy impairing spontaneous necrosis or apoptotic cell death. To guarantee a homogenous and hemophagocyte-rich population, our ultrastructural studies have been standardized to a culture period of 28 days.

By contrast, hemophagocytic cells can only be enriched from healthy donors, when stored buffy coat preparations were used for culture intitiation (Schneider, 2011). A phenotypic and morphological comparison of individual preparations of healthy donors emphazise the importance of HLA-class II expression density in patients' derived hemophagocytes (Schneider et al., 2011).

The complication of severe sepsis with hemophagocytic syndromes has been previously described (Buyse et al., 2010; Creput et al., 2008; Francois et al., 1997; Ito et al., 2006; Schaer et al., 2006) and appears to occur in 15% of all patients with severe sepsis and advanced SIRS (Kuwata et al., 2006). Selecting clinically defined patients, we prepared 28-day-old hemophagocyte cultures for ultrastructural analysis and identified phagocytosis of erythrocytes and their incomplete degradation. Remarkably, the ultrastructural analysis also identified phagocytosis of autologous hemophagocytes (Fig. 1 B, C, and E). Although phagocytic cells are very similar in patients with sepsis and pediatric HLH, the phagocytes of septic patients appear to digest the phagocytosed cells more rapidly than HLH phagocytes, which may be a result of more efficient phagosome-lysosome fusion events.

This observation suggests that hemophagocytosis contributes to the manifestation of anemia and cytopenia of different hematopoietic lineages in vivo.

Under physiological conditions, phagocytes recognize and ingest apoptotic and necrotic cells via the recognition of numerous cell surface antigens including scavenger receptors, Fc receptors, tetraspanins, and others (see (Kinchen and Ravichandran, 2007) for a review). After the induction of apoptosis, the first phagocytic signal is provided by phosphatidyl serine positivity. Generally, "eat me" signals involve a complex series of cell signaling events (Kinchen and Ravichandran, 2007) involving Rho family GTPases. RhoA attenuates phagocytosis, whereas other GTPases such as Cdc42 and the early endosome marker Rab5 stimulate phagocytosis (Nakaya et al., 2006). Regarding hemophagocytic syndromes, one may speculate that either recognition receptors or Rho-GTPases are disorganized or that hemophagocytes induce phosphatidyl serine positivity in target cells. A more detailed investigation of signaling events is now perceivable using in vitro enriched hemophagocytes. Moreover, a study by Zhang and colleagues showed that the primary HLH associated syntaxin 11 directly influences phagocyte activity in macrophages (Zhang et al., 2008); however, delayed degradation of phagocytosed cells as suggested by the structural conservation of ingested erythrocytes in 28-day cultured hemophagocytes has not been proven (Fig. 1). The importance of lysosomal function for hemophagocytic syndromes is evident in cases of primary HLH manifesting in patients with Rab27a mutations (known as Griscelli syndrome) because Rab27a is responsible for lysosome acidification (Jancic et al., 2007).

Autophagy

Hemophagocytes contain vacuole-like structures, which have been observed by light microscopy. Examining these morphological features by electron microscopy revealed that they are indicators of autophagy. When cryofixation is applied, the vacuoles contain material of various electron densities (Fig. 2). Moreover, possible interactions occur between structurally similar and structurally different vesicles in hemophagocytes isolated from patients with HLH and sepsis. Structures that may represent heterolysosomes were also found (cf. Fig. 2E and F), but unequivocal lysosome-endosome fusion events are rare to detect. Such characteristics have been described as biosignatures of autophagy (Uchiyama et al., 2008). Interestingly, mutations described in lysosomal storage diseases have been shown to promote autophagy (Uchiyama et al., 2008), and these diseases have a high risk to manifest reactive histiocytosis in bone marrow (Chang et al., 2003; Kvasnicka and Thiele, 2002).

The first step of autophagy is the formation of an autophagosome, which occurs when a portion of the cytoplasm is engulfed by a double membrane vacuole. The current study provides evidence for de-novo membrane- and phagophore formation in cultured hemophagocytes (Fig. 3, Hohn et al. 2011). Such phagophores, which sequester cytoplasm and eventually seal to vesicles, which are then termed autophagosomes, can also be fenestrated as demonstrated by the tomography in Figure 3. Theoretically, these images suggest extensive membrane remodeling but would not distinguish degradation from membrane synthesis. According to Itoh and co-workers, *de novo* synthesis of a double membrane sac, the isolation membrane, and its elongation involves LC3 (Itoh et al., 2008). Indeed, LC3 expression can be detected in all hemophagocytes studied here. The insertion of LC3 into membrane vesicles has been demonstrated as a key process in autophagy as exemplified by Jackson et al. who studied dendritic cell autophagy promoted by virus infections (Jackson et al., 2005). The LC3 staining pattern is associated with membranes forming a foamy pattern of the cytoplasm. Therefore, these foamy hemophagocytes appear to be in an advanced phase of autophagy.. Recently, LC3-independent autophagy has been described as well (Nishida et al., 2009).

Mitophagy in HLH and Sepsis-Associated Hemophagocytosis

Another criterion of autophagy is the presence of isolation membranes around mitochondria followed by mitochondrial dysfunction, called mitophagy (Kim et al., 2007). High-pressure freezing optimizes the structural preservation of mitochondria in our cultures. We found mitochondria with highly parallel cristae [M*] as well as mitochondria with tubular and less organized structures [M]. The tubular structure may indicate remodeling, which has been reported to be associated with autophagy and cell death (Kim et al., 2007; Perkins et al., 2009; Tinari et al., 2007). Unequivocal mitophagy is characterized by the presence of an isolation membrane around a mitochondrion, which has been observed in Figures 4 C and D (Scarlatti et al., 2009).

Autophagy and Immune Function

Autophagy plays a role as an immune defense mechanism against bacterial (Alonso et al., 2007; Deretic, 2006; Nakagawa et al., 2004; Ogawa and Sasakawa, 2006) as well as viral infections (Lee et al., 2008; Orvedahl and Levine, 2008; Talloczy et al., 2002), both of which play a major role in the manifestation of hemophagocytic syndromes in HLH and sepsis. Nevertheless, autophagy has also been described to promote virus replication (Jackson et al., 2005). Therefore, the documentation of macroautophagy in dendritic cells of patients is not unexpected. Vyas and colleagues report that pathogen-derived TLR ligands stimulate the maturation of autophagic dendritic cells, a process participating in the generation of self tolerant T-cells (Vyas et al., 2008). The authors further argue that autophagy improves the antigen presentation capacity of dendritic cells (Vyas et al., 2008); however, patients with hemophagocytosis suffer from anergy and non-responsiveness against infectious agents (Castillo and Carcillo, 2009; Hotchkiss et al., 2009). Therefore, hemophagocytes do not appear to be good antigen presenters, as indicated by low expression of CD80 and CD86 (Fig. 5). Another possible explanation for autophagy is the starvation of cultured cells by inadequate conditions, which may be caused by an artifact. However, autophagy is already implied by the foamy cytoplasm in hemophagocytes of blood and bone marrow smears of end-stage sepsis patients with hemophagocytosis as well as in Giemsa-stained specimen (see refs. (Beutel et al., 2009; Fisman, 2000; Grom, 2004) for more examples). More profound evidence for early autophagy is provided by electron micrographs of 14-day-old hemophagocytes in culture (unpublished results).

Cell Lineage of Hemophagocytes

According to our phenotypic analysis, the cell type displaying hemophagocytosis upon in vitro culturing is an antigen-presenting cell of either macrophage or myeloid dendritic cell origin. In some isolates of both classical HLH and sepsis-associated hemophagocytosis, macrophage antigens such as CD14 (Fig. 5) and CD68 (Schneider et al. 2011) may be prominent, whereas in others, dendritic cell-like properties such as CCR7 (Fig. 5) and CD178 (Schneider et al. 2011) are expressed at high density on all cells. In all of our cultured hemophagocytes, CD11c identifies myeloid dendritic cells. Variable phenotypes of monocyte-like and dendritic cell-like markers in hemophagocytic syndromes have been described before (Ito et al., 2006; Schaer et al., 2006). In vivo elevated M-CSF concentrations have been proposed to guide the differentiation of monocytic cells into the macrophage or dendritic cell lineage (Francois et al., 1997). Additionally, these cytokines may influence in vitro differentiation.

Similarities and Differences of Hemophagocytes in HLH and Severe Sepsis

It has been demonstrated that hemophagocytic syndromes of patients with HLH and sepsis share clinical characteristics (Creput et al., 2008; Kuwata et al., 2006; Risdall et al., 1984; Schaer et al., 2006). The current study emphasizes that hemophagocytes of both types of patients display similar morphological and phenotypical properties. Nevertheless, ultrastructural analysis also emphasizes the following differences: first, hemophagocytes of sepsis patients appear to digest ingested blood cells faster than HLH hemophagocytes; second, the eccentric nuclei of sepsis-derived hemophagocytes are round and present with distally concentrated heterochromatin, whereas the nuclei of hemophagocytes from HLH patients are irregularly indented by autophagic vacuoles; third, hetero- and homo-chromatin is more patchy in HLH hemophagocytes.

Conclusion

The current study illustrates macroautophagy in antigen presenting cells associated with the infection associated immune deficiency syndromes, HLH and sepsis. In addition to autophagy, these dendritic cells display prominent hemophagocytosis of autologous blood cells in vitro, which is likely to mirror hemophagocytosis in vivo. The ultrastructural research performed suggests that the morphological characteristics of foamy cytoplasm of hemophagocytes observed in Giemsa-stained blood and bone marrow smears is due to macroautophagy and impaired apoptosis. Autophagy may explain the prolonged life span in vitro as well as the persistence of hemophagocytes in vivo. Despite the inflammatory environment in which these hemophagocytes are activated, we observed impaired expression of antigen presentation markers and an inferior function to degrade ingested material. Consecutively, this cell type is sensitive to danger-assoiated molecular pattern (DAMP) signaling and is likely to condition states of chronic inflammation. A major signaling mechanism is expected to occur from ATP-stimulated P2RX7 responses, or undigestable material in phagocytes leading to IL-1βß processing by the inflammsome machinery (Latz, 2011). So far, IL-1β is a leading biomarker for sepsis associated immune dysfunction but does not play a major role in HLH. We therefore believe that autophagy associated hemophagocytosis is a so far neglected mechanism for immune dysfunction not only by diminishing myeloid progenitors, but also for targeting defined natural killer cell subpopulations which are rare in sepsis as well as HLH. The molecular basis of hemophagocytosis remains to be an intriguing issue of molecular research in the future.

Acknowledgments

We thank Manfred Weiss for diagnosis of sepsis patients, Harald Hohmann and Fengguang Liu for phenotypic analysis of hemophagocytic cells, Eberhard Schmid for electron microscopical specimen preparation, Simone Gehrer, Maria Baier and Julian M. Schneider for immunohistochemical work and image processing. The support by the Histiocytosis Association of America, the Deutsche Histiozytosehilfe and the Erwachsenenhistiozytose (EHX) is greatly acknowledged.

Conflict of Interest Disclosure

The authors declare no competing financial interests.

References

Alonso, S., K. Pethe, D.G. Russell, and G.E. Purdy. 2007. Lysosomal killing of Mycobacterium mediated by ubiquitin-derived peptides is enhanced by autophagy. *Proc. Natl. Acad. Sci. U. S. A.* 104:6031-6. Epub 2007 Mar 26.

Beutel, G., O. Wiesner, M. Eder, C. Hafer, A.S. Schneider, J.T. Kielstein, C. Kuhn, A. Heim, T. Ganzenmuller, H.H. Kreipe, A. Haverich, A. Tecklenburg, A. Ganser, T. Welte, and M.M. Hoeper. 2011. Virus-associated hemophagocytic syndrome as a major contributor to death in patients with 2009 influenza A (H1N1) infection. *Crit. Care*. 15:R80.

Beutel, K., U. Gross-Wieltsch, T. Wiesel, U.Z. Stadt, G. Janka, and H.J. Wagner. 2009. Infection of T lymphocytes in Epstein-Barr virus-associated hemophagocytic lymphohistiocytosis in children of non-Asian origin. *Pediatr. Blood Cancer*. 53:184-90.

Buser, C., and P. Walther. 2008. Freeze-substitution: the addition of water to polar solvents enhances the retention of structure and acts at temperatures around -60 degrees C. *J. Microsc.* 230:268-77.

Buyse, S., L. Teixeira, L. Galicier, E. Mariotte, V. Lemiale, A. Seguin, P. Bertheau, E. Canet, A. de Labarthe, M. Darmon, M. Rybojad, B. Schlemmer, and E. Azoulay. Critical care management of patients with hemophagocytic lymphohistiocytosis. 2010. *Intensive Care Med.* 36:1695-702.

Castillo, L., and J. Carcillo. 2009. Secondary hemophagocytic lymphohistiocytosis and severe sepsis/ systemic inflammatory response syndrome/multiorgan dysfunction syndrome/macrophage activation syndrome share common intermediate phenotypes on a spectrum of inflammation. *Pediatr. Crit. Care Med.* 10:387-92.

Chang, K.L., K.K. Gaal, Q. Huang, and L.M. Weiss. 2003. Histiocytic lesions involving the bone marrow. *Semin. Diagn. Pathol.* 20:226-36.

Creput, C., L. Galicier, S. Buyse, and E. Azoulay. 2008. Understanding organ dysfunction in hemophagocytic lymphohistiocytosis. *Intensive Care Med.* 34:1177-87.

Deretic, V. 2006. Autophagy as an immune defense mechanism. *Curr. Opin. Immunol.* 18:375-82. Epub 2006 Jun 19.

Farquhar, J.W., and A.E. Claireaux. 1952. Familial haemophagocytic reticulosis. *Arch. Dis. Child.* 27:519-25.

Fisman, D.N. 2000. Hemophagocytic syndromes and infection. *Emerg. Infect. Dis.* 6:601-8.

Francois, B., F. Trimoreau, P. Vignon, P. Fixe, V. Praloran, and H. Gastinne. 1997. Thrombocytopenia in the sepsis syndrome: role of hemophagocytosis and macrophage colony-stimulating factor. *Am. J. Med.* 103:114-20.

Grom, A.A. 2004. Natural killer cell dysfunction: A common pathway in systemic-onset juvenile rheumatoid arthritis, macrophage activation syndrome, and hemophagocytic lymphohistiocytosis? *Arthritis Rheum.* 50:689-98.

Henter, J.I., A. Horne, M. Arico, R.M. Egeler, A.H. Filipovich, S. Imashuku, S. Ladisch, K. McClain, D. Webb, J. Winiarski, and G. Janka. 2007. HLH-2004: Diagnostic and therapeutic guidelines for hemophagocytic lymphohistiocytosis. *Pediatr. Blood Cancer.* 48:124-31.

Henter, J.I., C. Tondini, and J. Pritchard. 2004. Histiocyte disorders. *Crit. Rev. Oncol. Hematol.* 50:157-74.

Hohn, K., M. Sailer, L. Wang, M. Lorenz, M.E. Schneider, and P. Walther. 2011. Preparation of cryofixed cells for improved 3D ultrastructure with scanning transmission electron tomography. *Histochem Cell Biol.* 135:1-9.

Hotchkiss, R.S., C.M. Coopersmith, J.E. McDunn, and T.A. Ferguson. 2009. The sepsis seesaw: tilting toward immunosuppression. *Nat. Med.* 15:496-7.

Ito, S., N. Takada, A. Ozasa, M. Hanada, M. Sugiyama, K. Suzuki, Y. Nagae, T. Inagaki, Y. Suzuki, and H. Komatsu. 2006. Secondary hemophagocytic syndrome in a patient with methicillin-sensitive Staphylococcus Aureus bacteremia due to severe decubitus ulcer. *Intern. Med.* 45:303-7.

Itoh, T., N. Fujita, E. Kanno, A. Yamamoto, T. Yoshimori, and M. Fukuda. 2008. Golgi-resident small GTPase Rab33B interacts with Atg16L and modulates autophagosome formation. *Mol. Biol. Cell.* 19:2916-25.

Jackson, W.T., T.H. Giddings, Jr., M.P. Taylor, S. Mulinyawe, M. Rabinovitch, R.R. Kopito, and K. Kirkegaard. 2005. Subversion of cellular autophagosomal machinery by RNA viruses. *PLoS Biol.* 3:e156.

Jancic, C., A. Savina, C. Wasmeier, T. Tolmachova, J. El-Benna, P.M. Dang, S. Pascolo, M.A. Gougerot-Pocidalo, G. Raposo, M.C. Seabra, and S. Amigorena. 2007. Rab27a regulates phagosomal pH and NADPH oxidase recruitment to dendritic cell phagosomes. *Nat. Cell Biol.* 9:367-78.

Janka, G., and U. Zur Stadt. 2005. Familial and acquired hemophagocytic lymphohistiocytosis. *Hematology Am. Soc. Hematol. Educ. Program*:82-8.

Kim, I., S. Rodriguez-Enriquez, and J.J. Lemasters. 2007. Selective degradation of mitochondria by mitophagy. *Arch. Biochem. Biophys*. 462:245-53.

Kinchen, J.M., and K.S. Ravichandran. 2007. Journey to the grave: signaling events regulating removal of apoptotic cells. *J. Cell Sci.* 120:2143-9.

Kremer, J.R., D.N. Mastronarde, and J.R. McIntosh. 1996. Computer visualization of three-dimensional image data using IMOD. *J. Struct. Biol.* 116:71-6.

Kuwata, K., S. Yamada, E. Kinuwaki, M. Naito, and H. Mitsuya. 2006. Peripheral hemophagocytosis: An early indicator of advanced systemic inflammatory response syndrome/hemophagocytic syndrome. *Shock*. 25:344-50.

Kvasnicka, H.M., and J. Thiele. 2002. [Bone marrow manifestation of Niemann-Pick disease. A review of histiocytic proliferations of the bone marrow]. *Pathologe*. 23:486-9.

Latz, E. The inflammasomes: mechanisms of activation and function. 2010. *Curr. Opin. Immunol.* 22:28-33.

Lee, D.Y., J. Lee, and B. Sugden. 2008. The Unfolded Protein Response and Autophagy: Herpesviruses Rule! *J. Virol.* 10:10.

Nakagawa, I., A. Amano, N. Mizushima, A. Yamamoto, H. Yamaguchi, T. Kamimoto, A. Nara, J. Funao, M. Nakata, K. Tsuda, S. Hamada, and T. Yoshimori. 2004. Autophagy defends cells against invading group A Streptococcus. *Science*. 306:1037-40.

Nakaya, M., M. Tanaka, Y. Okabe, R. Hanayama, and S. Nagata. 2006. Opposite Effects of Rho Family GTPases on Engulfment of Apoptotic Cells by Macrophages. Vol. 281. *J. Biol. Chem.* 8836-8842.

Nishida, Y., S. Arakawa, K. Fujitani, H. Yamaguchi, T. Mizuta, T. Kanaseki, M. Komatsu, K. Otsu, Y. Tsujimoto, and S. Shimizu. 2009. Discovery of Atg5/Atg7-independent alternative macroautophagy. *Nature*. 461:654-8.

Ogawa, M., and C. Sasakawa. 2006. Shigella and autophagy. *Autophagy*. 2:171-4.

Orvedahl, A., and B. Levine. 2008. Eating the enemy within: autophagy in infectious diseases. *Cell Death Differ*. 5:5.

Perkins, G., E. Bossy-Wetzel, and M.H. Ellisman. 2009. New insights into mitochondrial structure during cell death. *Exp. Neurol.* 218:183-92.

Risdall, R.J., R.D. Brunning, J.I. Hernandez, and D.H. Gordon. 1984. Bacteria-associated hemophagocytic syndrome. *Cancer*. 54:2968-72.

Scarlatti, F., R. Granata, A.J. Meijer, and P. Codogno. 2009. Does autophagy have a license to kill mammalian cells? *Cell Death Differ*. 16:12-20.

Schaer, D.J., F.S. Boretti, G. Schoedon, and A. Schaffner. 2002. Induction of the CD163-dependent haemoglobin uptake by macrophages as a novel anti-inflammatory action of glucocorticoids. *Br. J. Haematol.* 119:239-43.

Schaer, D.J., C.A. Schaer, G. Schoedon, A. Imhof, and M.O. Kurrer. 2006. Hemophagocytic macrophages constitute a major compartment of heme oxygenase expression in sepsis. *Eur. J. Haematol.* 77:432-6.

Schneider, E.M., Flacke, S., Liu, F., Lorenz, M. R, Schilling, P., Nass, M. E., Foehr, K. J, Huber-Lang, M., Weiss, M. E. 2011. Autophagy and ATP-induced anti-apoptosis in antigen presenting cells (APC) follows the cytokine storm in patients after major trauma. *Journal of Cell Communication and Signaling.*

Schneider, E.M., I. Lorenz, M. Muller-Rosenberger, G. Steinbach, M. Kron, and G.E. Janka-Schaub. 2002. Hemophagocytic lymphohistiocytosis is associated with deficiencies of cellular cytolysis but normal expression of transcripts relevant to killer-cell-induced apoptosis. *Blood*. 100:2891-8.

Scott R, R.-S.A. 1939. Histiocytic medullary reticulosis. *Lancet*. 2:194-8.

Talloczy, Z., W. Jiang, H.W.t. Virgin, D.A. Leib, D. Scheuner, R.J. Kaufman, E.L. Eskelinen, and B. Levine. 2002. Regulation of starvation- and virus-induced autophagy by the eIF2alpha kinase signaling pathway. *Proc. Natl. Acad. Sci. U .S. A.* 99:190-5. Epub 2001 Dec 26.

Tinari, A., T. Garofalo, M. Sorice, M.D. Esposti, and W. Malorni. 2007. Mitoptosis: different pathways for mitochondrial execution. *Autophagy*. 3:282-4.

Uchiyama, Y., M. Shibata, M. Koike, K. Yoshimura, and M. Sasaki. 2008. Autophagy-physiology and pathophysiology. *Histochem. Cell Biol*. 129:407-20.

von Muller, L., A. Klemm, N. Durmus, M. Weiss, H. Suger-Wiedeck, M. Schneider, W. Hampl, and T. Mertens. 2007. Cellular immunity and active human cytomegalovirus infection in patients with septic shock. *J. Infect. Dis*. 196:1288-95.

Vyas, J.M., A.G. Van der Veen, and H.L. Ploegh. 2008. The known unknowns of antigen processing and presentation. *Nat. Rev. Immunol*. 8:607-18.

Yarali, N., I. Balaban, N. Akyurek, S. Ucar, and P. Zorlu. 2009. Hemophagocytosis: the cause of anemia and thrombocytopenia in congenital syphilis. *Pediatr Hematol Oncol*. 26:461-6.

Zhang, S., D. Ma, X. Wang, T. Celkan, M. Nordenskjold, J.I. Henter, B. Fadeel, and C. Zheng. 2008. Syntaxin-11 is expressed in primary human monocytes/macrophages and acts as a negative regulator of macrophage engulfment of apoptotic cells and IgG-opsonized target cells. *Br. J. Haematol*. 142:469-79.

In: Autophagy: Principles, Regulation and Roles in Disease ISBN: 978-1-61942-266-7
Editor: Nikolai Gorbunov

Chapter IV

Autophagy-Mediated Innate Defense Mechanism in Crypt Paneth Cells Responding to Impairment of Small Intestine Barrier after Total-Body Gamma-Photon Irradiation

***Nikolai V. Gorbunov**[*1] **and Juliann G. Kiang**[*2]*
[1] The Henry M. Jackson Foundation for the Advancement of Military Medicine, Bethesda, MD, US
[2] Radiation Combined Injury Program; Armed Forces Radiobiology Research Institute; Bethesda, MD, US

Abstract

Macroautophagy (mATG) is a major lysosomal mechanism for the bulk degradation of cytoplasmic constituents, including proteins, damaged organelles, and penetrated pathogens. It is characterized by sequestration of entire portions of the cytoplasm by a double-membrane bounded vacuole called autophagosome. Despite its action as a self-digestion process, autophagy is mainly considered to be a well-regulated pathway to provide cell survival and remodeling. A line of evidence indicates that mATG plays a crucial role in both innate and adaptive immunity. It has been shown that mATG can suppress viral infections, inactivate intracellular bacteria and parasites, and deliver cytoplasmic antigens for MHC class II presentation to the adaptive immune system. Thus, the immune system utilizes mATG for degradation of cytoplasmic materials, to both restrict intracellular pathogens and regulate adaptive immunity. Exposure to cytotoxic agents often interferes with this cross-talk, impairs immune system, and compromises host barrier functions that can ultimately lead to pathogen translocation and

[*] Dr. Nikolai V. Gorbunov: Telephone: 301-295-1352; Fax: 301-562-8835; Gorbunov@afrri.usuhs.mil.
[*] Dr. Juliann G. Kiang: Telephone: 301-295-1076; Fax: 301-295-1731; kiang@afrri.usuhs.mil.

sepsis. Very little is known on the role of mATG in the innate mechanisms emerging to compensate immune homeostasis under the immunosuppressive conditions. This chapter is focused on role of mATG in the innate defense mechanism activated in intestinal crypt Paneth cells following acute ionizing irradiation. Total-body acute ionizing irradiation either due to accidental exposure or radiotherapy of *malignant diseases* is associated with life-threatening effects where severity of outcomes is predominantly determined by impairment of dose-limiting immune tissue and small intestine. Thus, radiation exposure can induce death of radio-sensitive proliferating crypt epithelial precursor cells and cells of the gut-associated lymphoid tissue (GALT) that compromise integrity of the intestinal mucosal barrier, increase probability of bacterial translocation from the gut lumen and sepsis. Recent data demonstrate important role of Paneth cells in innate defense mechanisms of intestinal mucosa. Paneth cells are located at the base of the crypts of Lieberkühn adjacent to and/or surrounding the multi-potent crypt stem cells. They represent a relatively stable population of long-lived, radio-resistant, and differentiated cells. Paneth cells secrete several bacteriolytic enzymes and antimicrobial peptides into the intestinal lumen to suppress bacterial translocation. Among the secreted peptides are α-defensins, cationic peptides with a broad spectrum of antimicrobial activity and the ability to rapidly galvanize antigen-presenting cells, activate adaptive immune responses, and interfere with pro-inflammatory cytokines. Herein, we show that radiation-induced systemic immuno-suppression and damage to intestinal mucosa can promote bacterial translocation from gut lumen to crypt lumen. That is correlated with mATG up-regulation and mATG-mediated secretary activity of crypt Paneth cells. We suggest that effective countermeasures against radiation injury shall include therapeutic modulation of mATG activity in intestinal Paneth cells.

1. Introduction

Ionizing irradiation impairs epithelial tissue, damages lymphatic and reticulo-endothelial systems and produces immunosupression. These detrimental effects lead to alterations in interaction between the host and symbiotic microbiota [1-3]. While microbes outnumber the host cells by 1-2 orders of magnitude, they are normally spatially separated from the host in a manner that enables nutrient metabolism as well as evolvement of physiologic features such as innate and adaptive immunity, immune tolerance, and intestinal epithelial structure and barrier function [6-9]. The major interface barrier between the microbiota and host tissue is constituted by epithelium and mucosa-associated lymphoid tissue (MALT or GALT when it is referred to gut associated lymphoid tissue). Considering a relative resistance of gut microorganisms to ionizing radiation (compared to the host cells) , radiation-induced impairment of tissue barriers can result in breakdown of this symbiotic homeostasis, bacterial translocation, and lead to pathologies such as tissue inflammation and sepsis [6-9]. Deciphering the mechanisms of these events, in particular the role of autophagy in sustaining barrier functions, can provide new insights for developing effective countermeasures against acute radiotoxicity and associated sepsis [2-5]. Indeed, although there are reports indicating that particular types of infections utilize the host autophagy mechanism for propagation [10-12], the majority of literature data suggest that autophagy per se is a crucial cytoprotective mechanism against a variety of stress factors, pathogens, and parasites [10, 13, 14]. In gut, the mucosal layer is exposed to resident microflora colonizing gut lumen and to a dietary influx of microorganisms. The gut bacterial population is gradually distributed from small intestine to colon with lowest bacterial density in the duodenum and highest in colon [15]. In this spatial distribution the resident bacteria are usually found only in the intestinal lumen while the dense mucus covering the mucosal surface and the crypts of Lieberkuhn containing enteric epithelium stem cells and host defense Paneth cells represent rather a sterile environment [15-17]. Recent findings indicate that both innate and adaptive intestinal immune systems monitor bacterial interaction with the mucosal interface and activate immune responses in order to balance bacterial density and to limit bacterial contact with the epithelial surface [9, 15]. This host-bacteria homeostasis can be dramatically changed due to ionizing irradiation [2, 3, 18]. Indeed, exposure to ionizing irradiation or ingestion of radionuclides results in the death of the radio-sensitive, proliferating crypt epithelial and GALT cells that can breach integrity the mucosal barrier [19-20]. Compensatory response to this damage may involve radioresistant innate defense mechanisms in order to restore the barrier homeostasis [21]. Recent data indicate that Paneth cells play important role in innate defense mechanisms of intestinal mucosa [15, 22, 23]. Paneth cells are highly granulated pyramidally shaped epithelial-type of cells which are located at the base of the crypts of Lieberkühn within clusters of approximately 15 cells adjacent to and/or surrounding the multi-potent crypt stem cells [4, 15, 24]. They represent a relatively stable population of long-lived (20 days), radio-resistant, differentiated secretory type of epithelial cells [22, 25]. Paneth cells secrete several bacteriolytic enzymes and antimicrobial peptides into the intestinal lumen upon stimulation with Gram-negative bacteria, Gram-positive bacteria and bacterial products, and can promote pro-inflammatory effects [15, 22, 23]. Among the secreted peptides are α-defensins, cationic peptides with a broad spectrum of antimicrobial activity and the ability to rapidly galvanize antigen-presenting cells, activate adaptive immune responses, and interfere with pro-inflammatory cytokines [15, 26-28]. While the exocytosis of these antimicrobial factors is produced by characteristic cytoplasmic granules present in Paneth cells, the formation of the secretory granules involves autophagy mechanism, and is down-regulated by suppression of Atg16L1, Atg5 and Atg7 autophagy genes [29, 30]. Thus, it has been demonstrated recently that autophagy-deficient Paneth cells exhibit a striking loss of function in the granule exocytosis pathway whereby the gene expression associated with inflammatory responses is increased [29, 30]. The objective of the current chapter is to demonstrate that autophagy pathway plays an essential role in compensatory defense response of Paneth cells to impairment of intestinal barrier due to total-body gamma-photon irradiation. Briefly, using an animal model of the hematopoietic acute radiation syndrome produced by total-body gamma-irradiation we demonstrated that radiation-induced systemic immuno-suppression and damage to intestinal mucosa promoted bacterial translocation from gut lumen to crypt lumen. That was correlated with mATG up-regulation and mATG-mediated secretory activity of crypt Paneth cells.

2. Hypothesis Test: Experimental Procedures and Technical Approach

2.1. Animal Model

A model of acute hematopoietic syndrome was established using CD2F1 male mice (Harlan Laboratories, Indianpolis, IN) and described previously [5]. Animals were subjected to 9.25 Gy total-body ^{60}Co γ-photon radiation (0.6 Gy/min). Whole blood (0.7 – 1 ml) was collected by cardiac puncture from anesthetized mice followed by euthanasia. Small intestine (ileum) tissue samples were collected for histology, immunofluorescence, the transmission electron microscopy (TEM), and immunoblotting analyses as indicated below. White blood cell profile in the peripheral blood was assessed using the ADVIA 2120 Hematology System (Siemens, Deerfield, IL). The experiments were performed in a facility accredited by the Association for the Assessment and Accreditation of Laboratory Animal Care-International (AAALAC-I). The animals used in this study received humane care in compliance with the Animal Welfare Act and other federal statutes and regulations relating to animals and experiments involving animals. This study adheres to principles stated in the Guide for the Care and Use of Laboratory Animals, NRC Publication, 1996 edition.

2.2. Analysis of the Small Intestine Proteins

Proteins from small intestine (SI) specimens were extracted in accordance with the protocol described previously [5, 18]. Protein aliquots in tris buffer (pH=6.8) containing 1% sodium dodecyl sulfate (SDS) and 1% 2-mercaptoethanol were resolved on SDS-polyacrylamide slab gels (NuPAGE 4-12% Bis-Tris; Invitrogen, Carlsbad, CA). After electrophoresis, proteins were blotted onto a PDVF membrane and the blots were incubated with antibodies (1 μg/ml) raised against microtubule-associated protein 1 light chain 3 (LC3), BAG-1, LAMP1, caspase-3 (casp-3), and α-defensin-4 (AD4) (Santa Cruz Biotechnology Inc., LifeSpan Biosciences, Inc.) followed by incubation with species-specific IgG peroxidase conjugate. IgG levels were not altered by radiation. Therefore, IgG was used as a control for protein loading. Protein bands of interest were quantitated densitometrically and normalized to IgG.

Interaction of chaperone heat shock protein 70 (HSP-70) and Bcl-2 associated athanogene (BAG-1) and the autophagy LC3 protein was analyzed with immunoprecipitation techniques as described previously [5, 18].

2.3. Fluorescent *In Situ* Hybridization Analysis of α-Defensin-4 mRNA

Fluorescent *in situ* hybridization was performed using a protocol published previously [18]. Sections were incubated overnight at 50°C with a biotin-dual-end-labeled oligonucleotide probe (5'- TCA GCG GCG GGG GCA GCA GTA CAA AAA TCG TAT TCC ACA AGT CCC ACG AA -3', purchased from Integrated DNA Technologies), complementary to the divergent 3' region of mouse AD4 cDNA (GenBank Accession NM_010039). The denatured specimens were incubated with streptavidin-AlexaFluor 488 conjugate and counterstained with Hoechst 33342. The procedure of positive staining for CD15 (a specific marker of Paneth cells) in small intestine and fluorescence imaging techniques are indicated below in the following paragraph. The fluorescence signals from CD15 protein and the FISH conjugates were collected in the red and green channels, respectively. The granularity of crypt Paneth cells was determined with DIC imaging techniques.

2.4. Immunofluorescent Staining and Image Analysis

Small intestine specimens (5 per each animal group) collected at necropsy were processed for the immunofluorescence analysis and analyzed with fluorescence confocal microscopy [18]. Donkey normal serum and antibody were diluted in phosphate buffered saline (PBS) containing 0.5% BSA and 0.15% glycine. Any nonspecific binding was blocked by incubating the samples with purified donkey normal serum (Santa Cruz Biotechnology, Inc., Santa Cruz, CA) diluted 1:20. The primary antibodies were raised against CD15 (mouse monoclonal biotin conjugated IgM from eBioscience), LC3, and AD4 (vendors indicated above). This was followed by incubation with secondary fluorochrome-conjugated antibody and/or streptavidin-AlexaFluor 610 conjugate (Molecular Probes, Inc., Eugene OR), and with Heochst 33342 (Molecular Probes, Inc., Eugene OR) diluted 1:3000. The secondary antibodies were AlexaFluor 488 and AlexaFluor 594 conjugated donkey IgG (Molecular Probes Inc., Eugene OR). Negative controls for nonspecific binding included normal goat serum without primary antibody or with secondary antibody alone. Five confocal fluorescence and DIC images of crypts (per specimen) were captured with a Zeiss LSM 7100 microscope. The immunofluorescence image analysis was conducted as described previously [18]. The index of spatial correlation (*r*) of proteins was determined by multiple pixel analysis for pairwise signal interaction of green and red channels. Paneth cell identification was conducted by i) their spatial localization in crypts, ii) presence of immunoreactivity to CD15 and AD4, which are specific to this epithelial phenotype, and iii) spatial appearance of the immunoreactivity to CD15, AD4, and the FISH reactivity to AD4 mRNA (see above) in morphologically identified Paneth cells.

2.5. Transmission Electron Microscopy

Two- to three-millimeter slices of tissue were fixed in 4% formaldehyde, 4% glutaraldehyde in PBS overnight, post-fixed in 2% osmium tetroxide in PBS, dehydrated in a graduated series of ethanol, and embedded in Spurr's epoxy resin. Blocks were processed as described previously [5, 18]. The sections of embedded specimens were analyzed with a Philips CM100 electron microscope.

3. Effects of Whole-Body Ionizing Irradiation

3.1. Acute Hematopoietic Effect

Mice exposed to 9.25 Gy ^{60}Co γ-photon radiation died within 15 days post-exposure [5]. The mortality rate was 85-100%. The conducted hematological, histological, and bio-analytical observations covered the first seven days post-irradiation. Irradiation resulted in complete depletion of lymphocytes and neutrophils in the peripheral blood within the first day, while the lymphocyte count in sham-treated mice did not change and was 73% ± 5 % (n=6) of total WBC. These effects were accompanied by depletion of hematopoietic cells in bone marrow. The obtained results suggested the presence of severe damage to lymphoid tissue caused by irradiation.

3.2. Alterations in the Crypt/Villus Structures Induced by Ionizing Radiation

It is well documented that response of the crypt proliferating cells to radiation-induced damage is relatively fast across a wide range of radiation doses [19]. In animal models apoptotic alterations due to direct radiation damage have shown to have transient character and return to basal levels during the first 24 h post-radiation period [19]. Our observations covered a period from the 1st through the 7th day post-exposure. As indicated in Fig. 1, histological assessment did not show any significant structural alterations along the crypt-villus axis of the irradiated intestine within this time period.

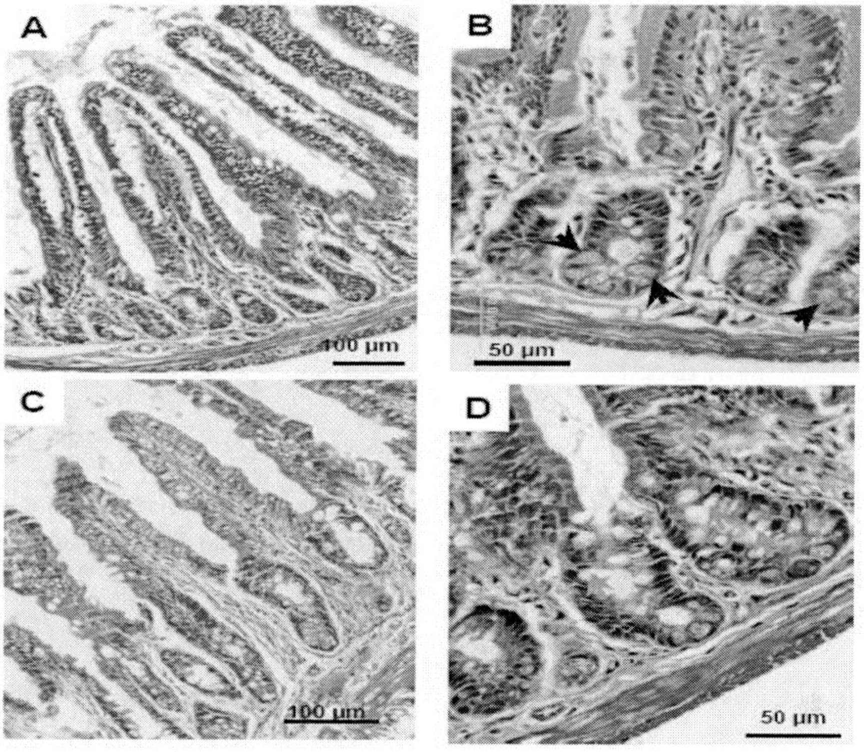

Figure 1. Histology of crypt-villus structures in the mouse small intestine.

Panels A-B are specimens obtained from a sham treated animal

Panels C-D are specimens obtained from an irradiated animal at the 7th day post-irradiation.

Paneth cells are indicated with black arrows in the panels "B" and "D". The Paneth cells were identified by a characteristic pyramid-like shape and abundance with granules (appear in red). Increase in degranulation in Paneth cells was observed at the 7th day post-irradiation and is indicated with green arrows in the panel "D".

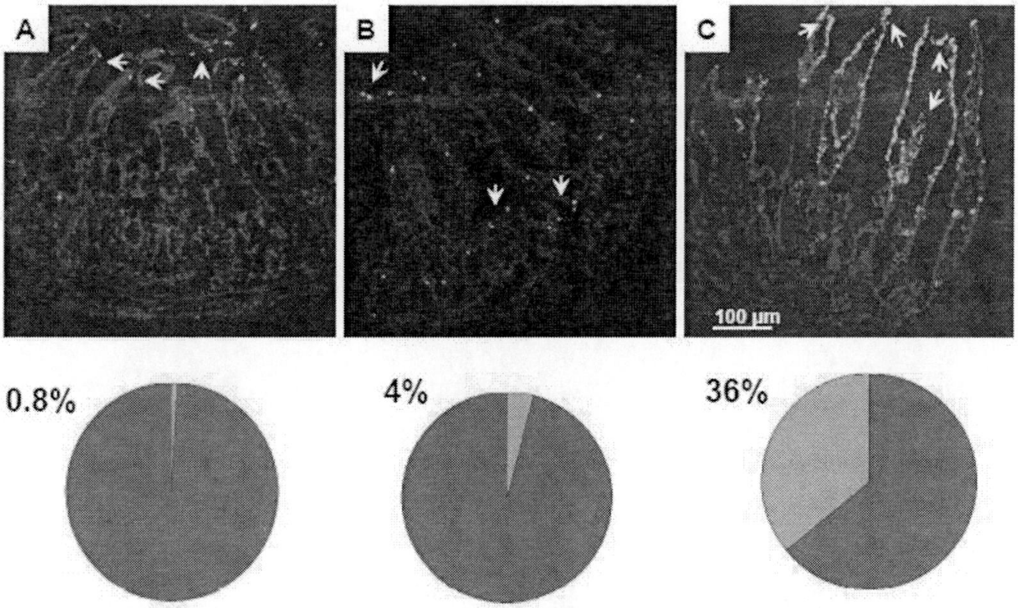

Apoptotic cells (%) are in green.

Figure 2. TUNEL Assay of Apoptosis in Small Intestine. (From: Gorbunov NV, Kiang JG. J Pathol. 2009;219(2):242-52; with copyright permission).

Panel A is projection of apoptotic cells in a sham treatment specimen.

Panel B is same as "A" except image was taken at the day post-irradiation.

Panel C is same as "A" except image was taken at the 7th day post-irradiation.

The presence of apoptotic cells (in green) is indicated with white arrows. Counterstaining of nuclei with Hoechst 33342 appears in blue color.

The apoptosis effect was expressed as percent of apoptotic events in nuclei of the selected ROI which was calculated from spatial correlation of signals obtained in the respective blue and green channels.

A, B and C are representative images of analyses conducted using 5 specimens in each groups.

However, the fluorescence TUNEL analysis revealed a positive staining for apoptosis in enterocytes at the top of villi (toward the edge of mucosal interface) and a decrease in proliferative activity in crypt cells that occurred between the 5th and 7th day observations. This was accompanied by a dramatic rise of casp-3 amounts (another marker of apoptosis) as determined in lysates of intestinal tissue.

All together these observations indicated increases in apoptotic events in the irradiated small intestine prior to appearance of symptoms of the radiation-induced illness. We hypothesized that these delayed transformations of villus cells facilitated mucosal permeability for gut bacteria that in turn, could up-regulate innate defense mechanisms mediated by Paneth cells resistant to the given irradiation dose. Indeed, appearance of bacteria in the crypt lumen was observed at the 7th day post-irradiation that was accompanied by an increase in the granule density at the apical membrane of Paneth cell.

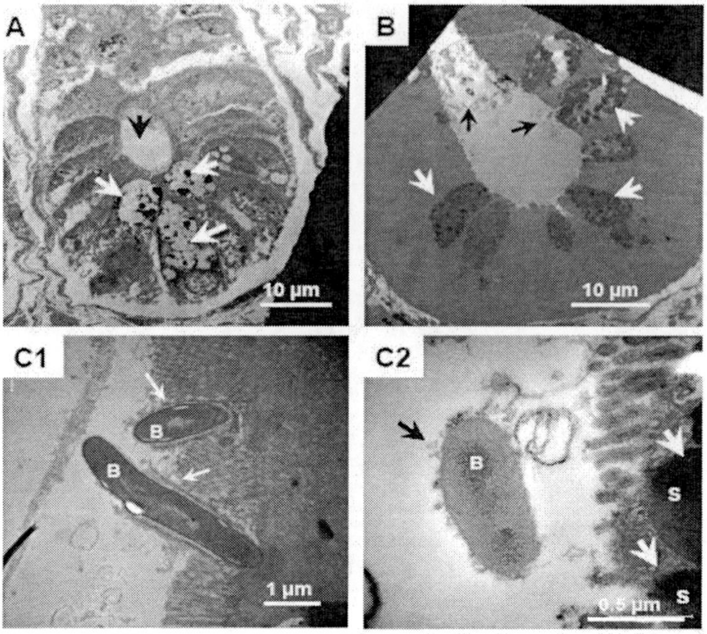

Figure 3. Transmission electron micrographs of bacteria translocation into crypt lumen of the irradiated small intestine. (Adapted from: Gorbunov NV, Garrison BR, Kiang JG. Int J Immunopathol Pharmacol. 2010;23(4):1111-23; with copyright permission).

Panel A: TEM image of intestinal crypt (sham, the 7th day post-treatment); crypt lumen is indicated with black arrow, Paneth cells containing granules are indicated with white arrows.

Panel B: TEM image of intestine crypt (irradiation, the 7th day post-exposure); Paneth cells containing granules are indicated with white arrows, appearance of bacteria in the crypt lumen is indicated with black arrows.

Panel C1: Bacterial intrusion into the crypt cells. Panel C2: Appearance of bacteria and release of granules in crypt lumen are indicated with white arrows. Note the secretory granules (S) contain homogenous electron-dense material; bacteria are indicated "B".

The specimens were collected at the 7th day post-irradiation.

The ultrastructure of mouse Paneth cells has been well discussed and been characterized by the presence of secretory granules [31]. It has been suggested that Paneth cells play a significant role in controlling the intestinal microbiota by releasing antibacterial substances, and that the changes of bacterial milieu affect morphology and secretory activity of Paneth cells; overall, these features make Paneth cells a part of intestinal innate defense mechanism [22, 23, 26-28, 31, 32]. There is a line of evidence that increase in apical vesicles of Paneth cells is accompanied by degranulation of the cells indicating that those may be exocytotic vesicles [31].

As shown in Fig. 4, the secretory-type granules are situated in the supranuclear region of Paneth cells and represent a bipartite substructure: a large round central core of high electron density is surrounded by a peripheral halo of lower density. The histochemical data published recently indicate that the core contains basic protein (including lysozyme, phospholipase A2 and α-defensins) and the peripheral halo is composed of acid mucopolysaccharides [22, 27,

31, 32]. Another line of reports suggests that secretory activity of Paneth cells involves autophagy; however, the detailed aspects of this mechanism in morphogenesis and/or secretion of granules are not clear [29, 30]. The images presented in panels C-D (Fig. 4) show different stages of development of the secretory granules and their interaction with autolysosome-like structures in sham-treated animals. These data suggest that the "basal autolysosomal activity" most likely controls dynamic and maturation of the secretory granules via crinophagy mechanisms, i.e., direct fusion of secretory vesicles with lysosomes. Note the structure of autophagosomes, lysosomes, and late autophagic vacuoles identified with TEM were identical to those communicated recently by Dr. Eskelinen, (2008) [33].

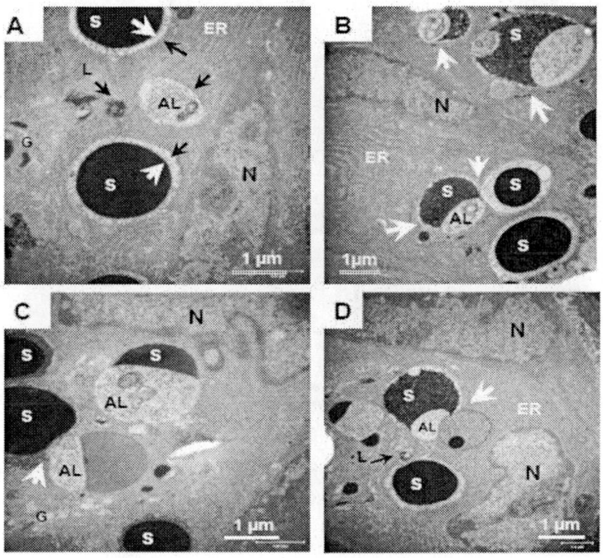

Figure 4. Transmission electron micrographs of morphogenesis of secretory granule in crypt Paneth cells.

Panels A-C: TEM images of Paneth cells in sham-treated intestine (the 7th day post-treatment). Components indicated in the panels are: the secretory granules (S) containing bipartite structure; L: lysosome-like bodies; ER: endoplasmic reticulum; AL: autolysosome-like structure; N: nucleus. Formation of secretory lysosomes and crinophagy events, i.e., fusion of the secretory granules with lysosomes and partial degradation of the granules, are indicated with white arrow heads.

Evidently, the electron density and fusion of secretory granules as well as their quantity in Paneth cells and secretory release to the crypt lumen were elevated after irradiation (Fig. 5). That was accompanied by formation of large autolysosomes containing massive amounts of membrane components. These observations suggest that besides the essential regulatory role in the Paneth cell secretory function [29, 30], the autophagy/lysosomal mechanism is likely important for elimination of vacuolar debris resulted from degranulation.

Along with the above alterations we observed fusion and elongation of mitochondria in Paneth cells (Fig. 5). The mitochondria were characterized by presence of a dense cristate (Fig. 5). Numerous communications indicate that the mitochondrial shape determines the fate of the cell during autophagy, and that elongated mitochondria can be easily spared from autophagic degradation [34]. Therefore, we suggest that observed mitochondrial remodeling in Paneth cells of irradiated tissue (Fig. 5E) was likely a part of mechanisms protecting cells from self-destruction by activated autophagy.

The alterations in Paneth cells determined in irradiated tissue with TEM were corroborated by immunochemical observations of autophagy markers and AD4, an antibacterial peptide specific for mouse Paneth cells.

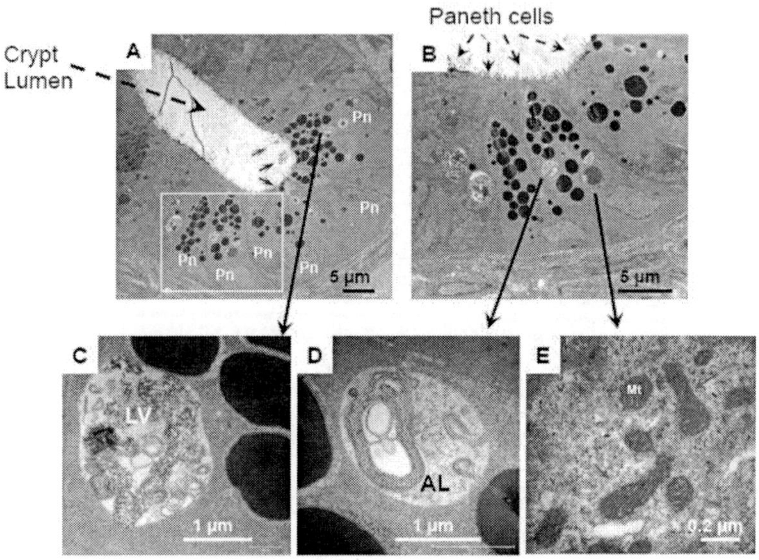

Figure 5. Transmission electron microscopy assessment of formation of autolysosome vacuoles (AL) and secretory granules in crypt Paneth cells of irradiated animals. (Adapted from: Gorbunov NV, Kiang JG. J Pathol. 2009;219(2):242-52; with copyright permission).

Panel A is a low magnification image of a small intestine crypt. The pyramidal shape Paneth cells (Pn) are observed at the bottom of the crypt. Appearance of bacteria and Paneth cells degranulation are indicated with red and blue arrows respectively.

Panel B is a magnified image of the white rectangular box indicated in the Pannel A; Paneth cells are indicated with black arrows. Secretory granule fusion and morphogenesis are indicated with white arrows. Elimination of massive amounts of membrane fragments in autolysosomes is indicated with a green arrow.

Formation of lysosomal vacuoles (LV) is shown in Panel C. Membrane re-arrangements in an autophagosome vacuole are shown in Panel D. Mitochondria elongation and fusion is shown in Panel E with red arrows.

3.3. Upregulation of Autophagy in Small Intestine of Irradiated Mice

A key step in the autophagosome biogenesis is the conversion of light-chain protein 3 type I (LC3-I, also known as ubiqitin-like protein, Atg8) to type II (LC3-II). The conversion occurs via the cleavage of the LC3-I carboxyl terminus by the Atg4 cysteine protease. The subsequent binding of the modified LC3-I to phosphatidylethanolamine, i.e., process of lipidation of LC3-1, on the isolation membrane, as it forms, is mediated by E-1- and E-2-like enzymes Atg7 and Atg3 (14). Therefore, conversion of LC3-1 to LC3-II and formation of LC3-positive vesicles are considered to be a marker of activation of autophagy (14).

As shown in Fig. 6A, a substantial increase in LC3-II isoform was observed in immunoblots of small intestine specimens at the 7th day post-exposure. We did not see alterations in the LC3-II level at the 2nd day post-irradiation; the elevation in the LC3-II isoform at the 3rd day post-IIr was not statistically significant, whereas the increase in the LC3-II levels observed at the 7th day post-irradiation was approximately 7.5-fold higher compared to sham treatment (Fig. 6A and B). Based on our observations (Figs. 6 A, C) the increased immunoreactivity of LC3 protein in the irradiated tissue was predominantly due to accumulation of type II LC3 isoform bound to autophagosome vesicles and thus, indicated activation of mAG pathway. The accumulation of LC3-II in the damaged tissue was correlated with up-regulation of Atg5, another autophagy protein, and the lysosomal associated membrane protein-1 (LAMP1), a marker of lysosomes. LAMP1 is primarily located in the lysosomes of the cell. It is one of the most abundant lysosomal proteins that constitutes up to 50% of the total membrane proteins of the late endosomes and lysosomes and plays a crucial role in lysosome biogenesis and autophagy [35, 36]. A predicted amino acid sequence of the LAMP1 core protein is of approximately 400 residues and the molecular weight is about 40 kD [35, 36]. LAMP1 is extensively glycosylated with asparagine-linked oligosaccharides in a post-translational manner; the molecular weight of the glycosylated isoforms varies in the range of 100 kD to 130 kD [www.cellsignal.com/products/3243.html]. It is suggested that while the C-terminus of the protein is located in the cytoplasm of the cell, most of the core protein is located in the lysosomal space and these glycans protect the underlying polypeptides from the proteolytic environment of the lysosome [35]. Our analysis demonstrated that radiation-induced alterations were characterized by accumulation of the both nonglycosylated and glycosylated LAMP1 isoforms, thus indicating increase in lysosomal activity. These observations were farther confirmed by confocal immunofluorescence image analysis of LAMP1 in crypt Paneth cells that showed increase in spatial co-localization of LAMP1and LC3 proteins after irradiation.

A growing body of evidence suggests involvement of chaperone heat shock protein 70 (HSP-70) and Bcl-2 - associated athanogene (BAG-1) in regulation of LC3-translocation [37]. BAG-1 is a multifunctional protein, acts as a co-chaperone, and exerts many of its actions (including suppression of apoptosis) via binding HSP-70 and Bcl-2 proteins [38]. Expression of HSP-70 can suppress LC3-dependent vacuolization in vitro [39]. The tissue lysates obtained from sham-treated and irradiated animals were immunoprecipitated with HSP-70 - specific antibody and the immunoprecipitates were analyzed for autophagosomal membrane protein LC3-II and BAG-1. The results of analysis (Fig. 6 D and E) showed that irradiated small intestine had lower amount of LC3 capable of interacting with HSP-70 (compared to sham treatment). This effect could result in a decrease in HSP-70-dependent inhibition of LC3-dependent vacuolization in irradiated tissue. We found a small significant reduction in BAG-1 co-precipitation with HSP-70 between irradiated and sham-treated specimens (Fig. 6 D and E).

The radiation-associated up-regulation of the autophagy pathway determined by immunochemical analysis of LC3 protein was confirmed with confocal immunofluorescence imaging of LC3 protein in small intestine specimens. The projections of LC3 protein in the specimens and relative intensity of immunofluorescence of LC3 are shown in Fig. 7. The intensity of LC3 immunofluorescence was significantly higher after irradiation with its prevalent expression at the region of the small intestine crypts (Fig. 7B and D) as compared to that in sham-treated tissue (Fig. 7A and C). The increase in the LC3 immunofluorescence was accompanied by appearance of LC3-abundant vesicular formations in the crypt cells (Fig. 7B inset), which were further identified as Paneth cells by their characteristic morphology (see below Figs. 10 and 11) and the presence of CD15 immunoreactivity (Fig. 8). Interestingly, we did not observe a mitophagy effect in the irradiated tissue when mitochondria were stained for prohibitin, and autophagosomes - for LC3 and/or autophagic vacuoles - for LAMP1 (Gorbunov & Kiang, unpublished data).

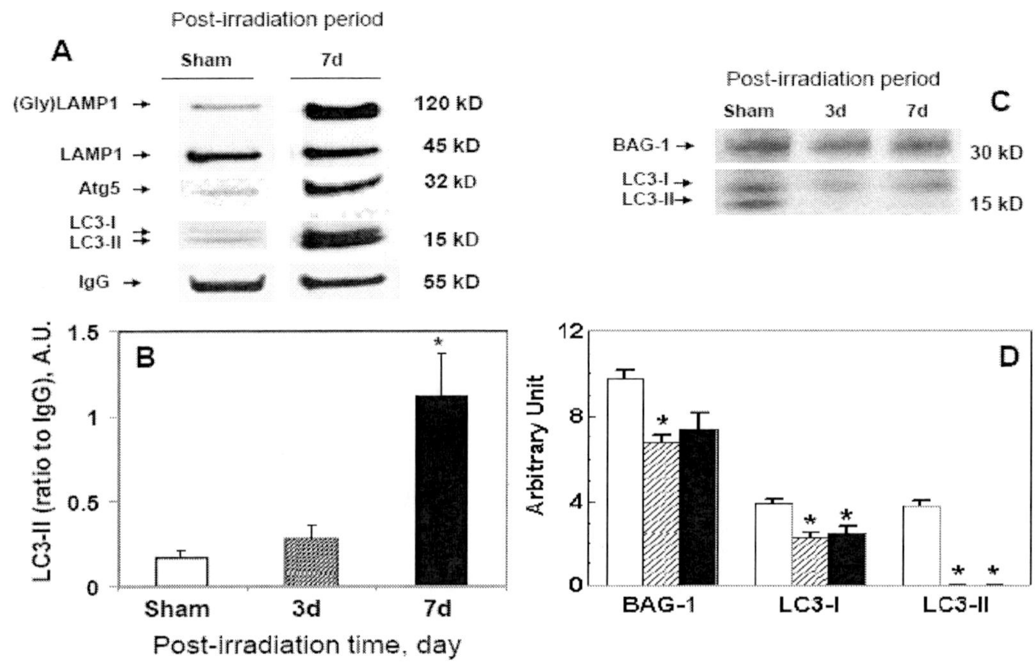

Figure 6. Immunoblot analysis LC3, Atg5, and LAMP1 autolysosomal proteins in the irradiated small intestine. (Adapted from: Gorbunov NV, Kiang JG. J Pathol. 2009 219(2):242-52; with copyright permission).

Representative immunoblot bands of LC3-I / LC3-II, Atg5 and LAMP1 proteins is shown in panel (A) and densitometrical analysis of the bands of the LC3-II protein is presented in panel (B); *$p < 0.05$ vs. sham treatment, the 3rd and the 7th day post-irradiation, $n = 5$, determined by one way ANOVA with Tukey post hoc correction. Interactions of LC3 and BAG-1 with HSP-70 was conducted using co-immunoprecepitation of the LC3 and BAG-1 proteins with HSP-70 protein. The results are presented in panel C and panel D (densitometry histograms); where open bars are “sham treatments”, hatched bars are “the 3rd post-irradiation”, and the solid bars are “the 7th day post-irradiation” animal groups.

*$p < 0.05$ *vs.* sham treatment, $n = 3$, determined by Student’s t-test.

The results of immunofluorescence imaging of autophagy were corroborated with TEM data presented in Fig. 5 and discussed above supporting the hypothesis of involvement of autophagy in the response of Paneth cells to the radiation-induced increase in permeability of mucosa barrier to gut bacteria.

Figure 7. Immunofluorescence confocal imaging of the LC3 protein in the small intestine. (Gorbunov NV, Kiang JG. J Pathol. 2009 219(2):242-52; with copyright permission).

Panels A and B are representative projections of the LC3 protein (green channel) in specimens obtained from a sham-treated animal and an irradiated animal, respectively. Panels C and D are assessment of relative immunofluorescence of the LC3 protein in the projections presented respectively in the panels A and B. Counterstaining of nuclei with Hoechst 33342 appears in blue color.

The confocal images were taken with pinhole setup to obtain 0.5 µm Z-sections.

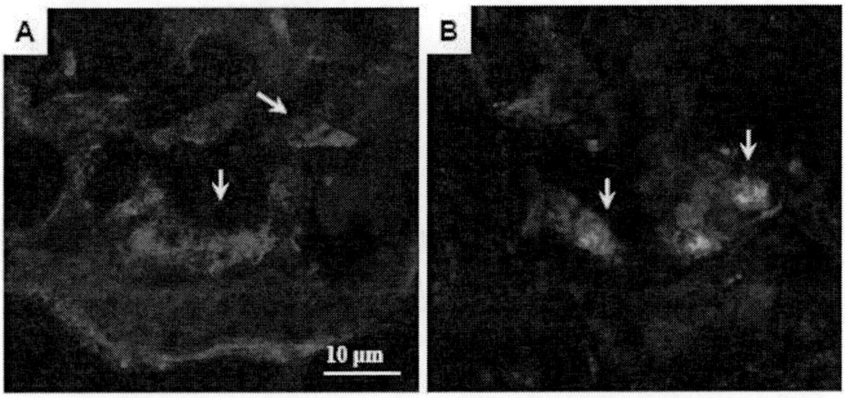

Figure 8. Assessment of LC3 immunoreactivity in the CD15-positive (Paneth) crypt cells.

Panel A is projections of CD15 (red channel) and LC3 (green channel) in a sham treatment specimen. (From: Gorbunov NV, Kiang JG. J Pathol. 2009;219(2):242-52; with copyright permission).

Panel B is same as "A" except image was taken at the 7th day post-irradiation.

Co-localization of CD15 and LC3 is appeared in yellow color as result of interference of green and red colors (Panel B). CD15-positive cells are indicated with arrows.

Vacuole vesicles with 200 – 900 nm diameters were identified as autophagosomes.

The confocal images were taken with pinhole setup to obtain 0.5 µm Z-sections.

A and B are representative images of analyses conducted using 5 specimens in each groups.

3.4. Expression of α-Defensin-4 in Small Intestine of Irradiated Mice

Assessment of AD4 expression in small intestine was conducted using Western immunoblotting of proteins. To prove increase in expression of AD4 in Paneth cells, we used two independent techniques:(i) FISH analysis of AD4 mRNA and (ii) immunofluorescence imaging of AD4 peptide. The results of the Western blot analysis presented in Fig. 9 showed that the 8 kD form of AD4 in the ileum of irradiated mice was dramatically increased (7-fold compared to sham treatment, $p<0.01$) at the 7th day. The confocal image analysis of AD4 immunofluorescence in ileum specimens presented in Fig. 10 A-B revealed predominant localization of AD4 at the bottom of crypt villi in both the sham-treated and irradiated animals at the 7th day post-irradiation.

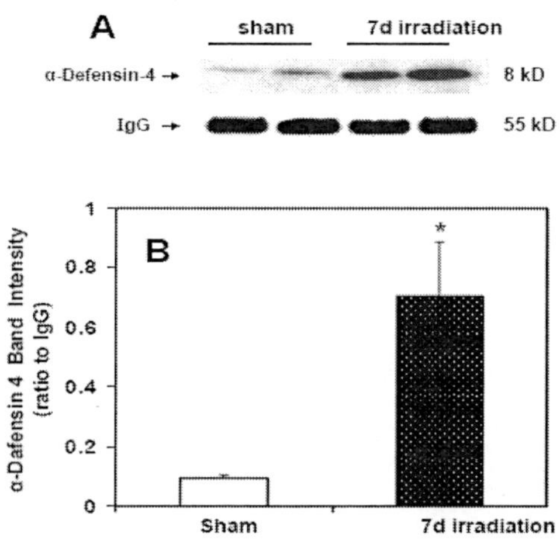

Figure 9. Immunoblot analysis of α-defensin-4 in the small intestine of irradiated mice. (From: Gorbunov NV, Garrison BR, Kiang JG. Int J Immunopathol Pharmacol. 2010;23(4):1111-23; with copyright permission).

Panel A is a representative immunoblot of α-Defensin-4. The protein bands were quantified densitometrically, normalized to IgG, and presented in panel B.

The tissue specimens were collected at the 7th day post-treatment (indicated as "7d").

*$p<0.005$ *vs*. sham treatment, n = 5.

However, intensity of AD4 immunofluorescence was approximately 4-fold higher in the irradiated animals *vs*. sham-treated animals (Figs. 10B2 *vs*. 10A2 and 10B5 *vs*. 10A5). Spatial analysis of AD4 in the crypt cells showed exclusive expression of AD4 in the pyramidally shaped cells identified as Paneth cells by positive expression of CD15 adhesion molecules (Fig. 10 A-B). It should be noted that the relative increase in AD4 immunoreactivity was observed in the mucosa at the base of villi of the irradiated animals that likely resulted from degranulation of Paneth cells and secretion of AD4 into the crypt lumen (Fig. 10B5, indicated with arrows). The observed up-regulation of AD4 in Paneth cells of irradiated animals was corroborated by FISH analysis of AD4 mRNA (Fig. 11). Indeed, increased levels of AD4 mRNA were observed in granulated crypt cells at the 7th day after irradiation (Fig. 11A, indicated with arrows). These cells were identified as CD15-positive Paneth cells (Fig. 11B).

A similar finding was also observed in another strain of mice [40]. It should be noted that under circumstances of irradiation followed by wound trauma, wound trauma exacerbates the radiation-induced autophagy in small intestine tissue [40].

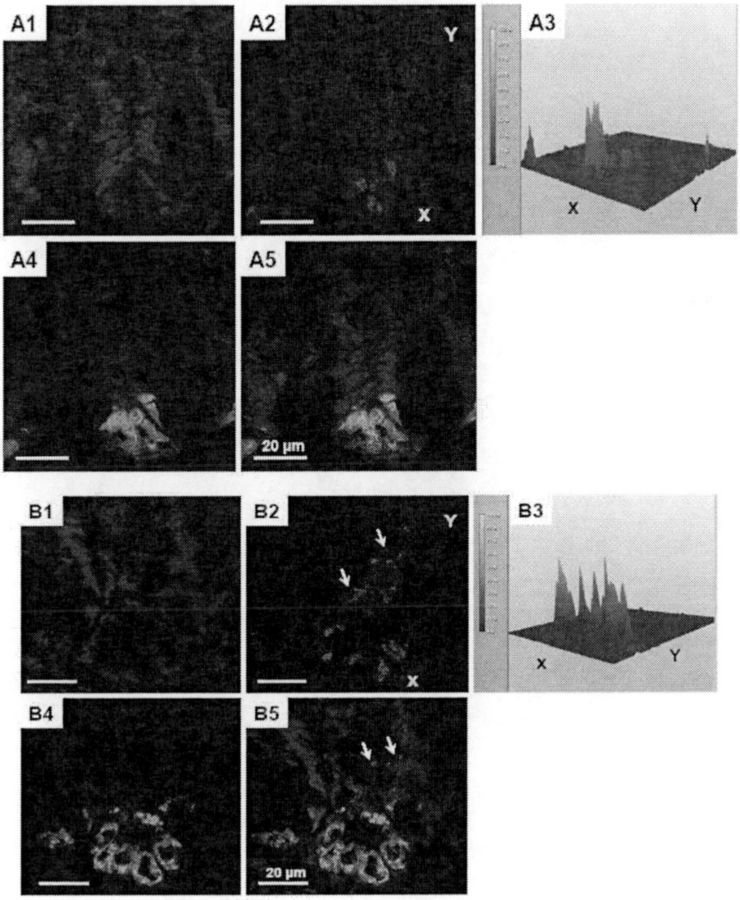

Figure 10. Spatial localization of α-defensin-4 in the CD15-positive (Paneth) crypt cells.

(From: Gorbunov NV, Garrison BR, Kiang JG. Int J Immunopathol Pharmacol. 2010;23(4):1111-23; with copyright permission).

A – Sham treatment; B – Post-irradiation at the 7th day.

1 - Counterstaining of nuclei with Hoechst 33342 (blue channel).

2 - Immunoreactivity of α-defensin-4 (red channel).

3 - Relative immunofluorescence of α-defensin-4 in the projections presented in “2”.

4- Immunoreactivity of CD-15 (green channel).

Note co-localization of α-defensin-4 and CD15 projections is shown in panels A5 and B5. Appearance of immunoreactivity of α-defensin-4 (red channel) in crypt mucosa of the exposed animals is indicated with arrows. The confocal fluorescence images were taken as described in *Methods*. Images shown in panels A and B are representative of analyses conducted using 5 specimens in each group.

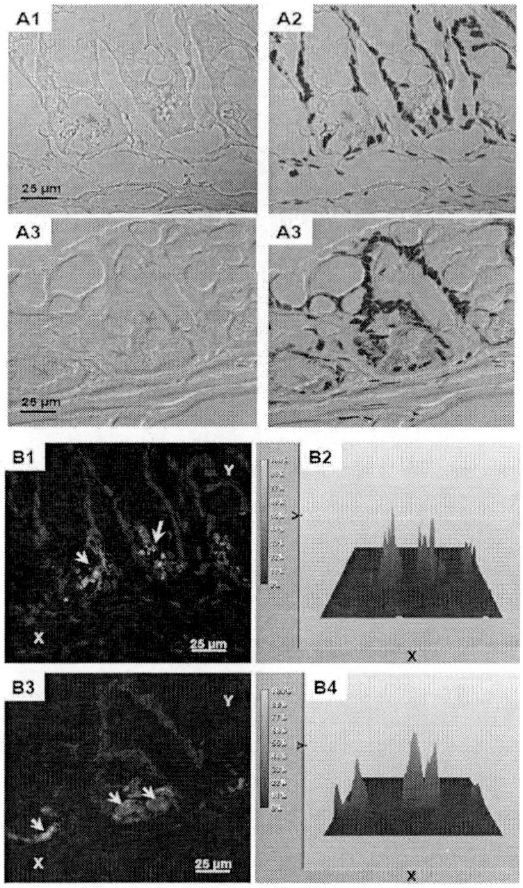

Figure 11. FISH analysis of α-defensin-4 mRNA in the small intestine of irradiated mice.

(From: Gorbunov NV, Garrison BR, Kiang JG. Int J Immunopathol Pharmacol. 2010;23(4):1111-23; with copyright permission).

A – Overlay of fluorescence images of the AD4 oligonucleotide probe in crypt cells with DIC images used for visualization of granular crypt cells. Panels A1 and A2 are images of specimens obtained from sham-treated animals. Panels A3 and A4 are images of specimens obtained at the 7th day following Irradiation. Note increased levels of the AD4 mRNA were observed in granulated crypt cells (indicated with red arrows) at the 7th day after irradiation.

B1 – Fluorescence image of spatial localization of the AD4 mRNA-compatible probe (green channel, indicated with white arrows) and CD15 (red channel) in a sham specimen. Blue channel - Counterstaining of nuclei with Hoechst 33342.

B2 - Relative fluorescence of the AD4 oligonucleotide probe in the projections presented in “B1”.

B3 – Fluorescence image of spatial localization of the AD4 oligonucleotide probe (green channel, indicated with white arrows) and CD15 (red channel) in a specimen obtained from an irradiated animal.

B4 - Relative fluorescence of the AD4 oligonucleotide probe in the projections presented in “B3”.

The confocal images were taken with pinhole setup to obtain 0.5 µm Z-sections.

Sets A and B are representative images of analyses conducted using 5 specimens in each group.

Conclusion

A large body of evidence indicates that irradiation-induced response of the small intestine, one of the most complex and dynamic tissues of the organism, is driven by multi-cellular systems which constitute the mucosal epithelium, endothelium, and the lymphoepithelial, lymphoreticular, and peripheral neuronal autonomic structures [3, 4, 19-21, 24, 25]. The predictive analysis of the small intestine radiation syndrome is complicated by the presence of a complex hierarchy of cellular structures, a sophisticated interactive function of the multi-cellular systems (including host-bacterial interaction), and the differential lifespan and radio-sensitivity [3, 4, 19-21, 24, 25, 41]. Therefore, we expected activation of different driving molecular and cellular mechanisms at different stages of the pathogenesis of the intestine radiation-related injury and inflammation in order to support the nutritional and barrier homeostasis [5, 18]. In the normal gut, the integrity and homeostasis of the intestine barrier are sustained by the mucus layer, the mucosal epithelium, and GALT [20, 42], and is mediated and regulated by molecular components comprising extracellular and intracellular matrix, and the cell communication systems, i.e., cell adhesion molecules, pattern-recognition molecules, antibacterial peptides, immunoglobulins, cytokines, etc. [43-48]. It is known that acute irradiation results in a rapid depletion of GALT cells and antibacterial IgA due to the high radiosensitivity of the intestine lymphocytes and plasma cells [20, 49, 50]. The response of the mucosal epithelium to radiation has a more complicated character due to the presence of multiple cell phenotypes with variations in proliferating and differentiating activity [20, 24]. Indeed, the radiation response of the crypt proliferating cells is relatively fast across a wide range of radiation doses. Subsequently, they are subjected to apoptotic elimination from the crypt cell population [19, 51]. The completely differentiated epithelial cells constitute the apex of villi. The status and dynamic of these short-living cells are determined by different intrinsic and extrinsic factors (including cell-cell and cell-matrix interactions); they are subjected to repopulation each 3-5 days (in mouse) [43-46, 50]. Currently, there is a large body of information and descriptive models addressing the dynamics of radiation-induced injury in the intestine. However, no data cover the existing gaps in understanding the molecular mechanisms that sustain crypt antibacterial barrier during the post-irradiation period. In this light, an attractive subject for investigation is the role of autophagy in crypt Paneth cells responding to radiation. Crypt Paneth cells play a key role in innate host defense [22-32]. Our recent reports demonstrated that Paneth cells respond to intestinal alterations induced in the hematopoietic radiation syndrome [5, 18]. This response was associated with appearance of resident bacteria in crypt lumens and involved up-regulation of AD4, mAG, and secretory activity of Paneth cells [5, 18]. We postulated that up-regulation of these pathways in Paneth cells could be a part of innate compensatory response to translocation of gut resident bacteria following the radiation-induced immunosuppression, and damage to GALT and gut epithelium. Importance of mAG in recovery from injury and secretory function has been documented recently [29, 30]. This view supports our findings. Indeed, based on our observations up-regulation of mAG in Paneth cells were correlated with increase in the cell granularity, lysosomal activity, and secretion of AD4 (a Paneth cell-specific antibacterial peptide). This radiation-induced elevation of the both mAG and AD4 in crypt Paneth cells occurred during neutropenia but before appearance of symptoms of radiation-induced morbidity, i.e., at the 7th day post-irradiation. The mechanism of this delayed effect is not clear but it was likely associated with increase in mucus layer permeability to resident bacteria in the pathogenesis of the radiation-induced disease. This could be crucial for the outcome of radiation injury and needs to be discussed. It has been recently shown that activation of mAG pathway due to direct effects of irradiation can occur at transcriptional, translational, and post-translational levels and can involve endoplasmic reticulum (ER) stress response [52-54]. Although it has been reported that activation of these events requires minutes to hours [52-54]. In contrast we did not find a significant difference in the LC3-II and Atg5 (either AD4) levels between sham and irradiated mice within the first two days post-irradiation. Therefore, it is highly likely that the "direct" effect of radiation exposure was not a major cause of activation of the mAG pathway and of Paneth cell secretory activity in intestine at the 7th day post-irradiation. Evidently, delayed activation of mAG and AD4 was a part of innate defense response to the irradiation-induced GALT impairment and loss of integrity of the mucosal layer as well as to the systemic immune suppression characterized by lymphopenia and neutropenia. However, while the damage to GALT in our model can be explained by high radiosensitivity of lymphoid cells, the 9.25 Gy-radiation effect on epithelial barrier needs to be clarified. Epithelial lineages are rapidly and continuously renewed through a spatially well-organized sequence of proliferation, commitment, migration-associated differentiation, and exfoliation. This renewal is driven by a stem cell hierarchy maintained in the intestinal crypts [24, 41]. The descendants of the multipotent stem cells undergo several rounds of cell divisions in the midpoint of the crypt which appears of high radiosensitivity [19, 41]. Cells are exfoliated into the intestinsl lumen at the apical extrusion zone of the villus that involves the anoikis mechanisms. This sequence of cell divisions, migration-associated differentiation and exfoliation is completed within 3-5 days. Our observations indicated that irradiation of CD2F1 male mice with doses of 9.25 Gy did not, per se, causes the gastrointestinal radiation subsyndrome and associated collapse of villi, while B6D2F1/J mice display shortened and widened villi of ileum after irradiation [48]. Taking into consideration the concept of dynamic response of entorocytes to irradiation discussed above, we postulated that enteritis associated with the development of acute radiation syndrome reported in this and previous communications is not likely due to the radiation-produced apoptosis of the proliferating crypt cells, but rather due to (i) temporal arrest of epithelial cell proliferation and renewal along the crypt/villus axis, and (ii) increase in bacterial-enterocyte interaction followed by pro-inflammatory alterations and decrease in integrity of enterocyte monolayer. Our hypothesis was supported by several events observed at the same time-point in the irradiated small intestine. These delayed events were (i) appearance of TUNEL-positive enterocytes at mucosal interface; (ii) appearance of bacteria in crypt lumens; (iii) accumulation of AD4, a Paneth cell-specific antibacterial secretory factor; and (iv) activation of mAG in Paneth cells accompanied by increase in autophagy-related vacuolization and secretory activity. Finally, it should keep in mind that under circumstances of irradiation followed by wound trauma, wound trauma exacerbates the radiation-induced autophagy [40].

Perspective

The role of Paneth cells in mediating innate defense response has been intensively discussed in the literature [15, 22, 23, 27, 32]. Paneth cells are shown to be capable of directly sensing bacteria invaded into crypt lumen and regulating microbiota across intestine [15, 22, 23]. Our analysis of LC3-II and AD4 in CD15-positive Paneth cells suggests that observed increase in immunoreactivity of these proteins was associated with autophagy and secretory responses and was likely due to impairment of mucosal barrier function.

We propose that up-regulation of Paneth cells in the investigated model is connected to the consequent compensatory response to complex alterations in intestinal mucosa that are temporally and functionally associated with further development of symptoms of the acute radiation illness.

Paneth cells are pivotal players in crypt function.

They provide immunochemical homeostasis in intestine by secretion of antibacterial peptides, epithelial growth factors, alarmins, and chemoattractants [15, 26-28, 32]. Therefore, activation of autophagy in Paneth cells can be a part of local regulatory mechanisms and/or innate defense response in the irradiated intestine that is essential during the post-exposure period.

Overall, understanding the role of autophagy in responses of specialized innate defense cells to ionizing irradiation can open new horizons for engineering of radioprotectors, radiomitigators, and radiotherapeutics agents. In particular, regulation of autophagy and secretory activity of Paneth cells can be a part of required countermeasures against gastrointestinal impairment in military and civilian personnel suffering from radiation mass casualties [55].

Acknowledgments

The authors thank Dr. Thomas B. Elliott for his comprehensive discussion of this paper; Dr. Dennis P. McDaniel for his help with TEM analysis, HM1 Neil Agravante, Ms. Joan Smith, Ms. Sarah Leggate, Mr. Rex Hartzorger and Mr. Bradley Garrison for their technical support.

Grants

This work was supported by AFRRI Intramural RAB2CF (to JGK) and NIAID YI-AI-5045-04 (To JGK).

There are no ethical and financial conflicts in the presented work.

Disclaimer

The opinions or assertions contained herein are the author's private views and are not to be construed as official or reflecting the views of the Uniformed Services University of the Health Sciences, the United States Department of Defense, or the United Stated National Institute of Health.

References

[1] Lawrence JH, Tennant R. The comparative effects of neutrons and X-rays on the whole body. *J Exp Med.* 1937;66(6):667-88.

[2] Elliott TB, Ledney GD, Harding RA, Henderson PL, Gerstenberg HM, Rotruck JR, Verdolin MH, Stille CM, Krieger AG. Mixed-field neutrons and gamma photons induce different changes in ileal bacteria and correlated sepsis in mice. *Int J Radiat Biol.* 1995;68(3):311-20.

[3] Packey CD, Ciorba MA. Microbial influences on the small intestinal response to radiation injury. *Curr Opin Gastroenterol.* 2010;26(2):88-4.

[4] Crawford PA, Gordon JI Microbial regulation of intestinal radiosensitivity. *Proc Natl Acad Sci USA.* 2005;102(37):13254-59.

[5] Gorbunov NV, Kiang JG. Up-regulation of autophagy in small intestine Paneth cells in response to total-body gamma-irradiation. *J Pathol.* 2009;219(2):242-52.

[6] Ismail AS, Behrendt CL, Hooper LV. Reciprocal interactions between commensal bacteria and gamma delta intraepithelial lymphocytes during mucosal injury. *J Immunol.* 2009; 182(5):3047-54.

[7] Bäckhed F, Ley RE, Sonnenburg JL, Peterson DA, Gordon JI. Host-bacterial mutualism in the human intestine. *Science.* 2005;307(5717):1915-20.

[8] Preidis GA, Versalovic J. Targeting the human microbiome with antibiotics, probiotics, and prebiotics: gastroenterology enters the metagenomics era. *Gastroenterology.* 2009;136(6):2015-31.

[9] Duerkop BA, Vaishnava S, Hooper LV. Immune responses to the microbiota at the intestinal mucosal surface. *Immunity.* 2009;31(3):368-76.

[10] Kirkegaard K, Taylor MP, Jackson WT. Cellular autophagy: surrender, avoidance and subversion by microorganisms. *Nat Rev Microbiol.* 2004;2(4):301-14.

[11] Heiseke A, Aguib Y, Schatzl HM. Autophagy, prion infection and their mutual interactions. *Curr Issues Mol Biol.* 2010;12(2):87-97.

[12] Shrivastava S, Raychoudhuri A, Steele R, Ray R, Ray RB. Knockdown of autophagy enhances the innate immune response in hepatitis C virus-infected hepatocytes. *Hepatology.* 2011;53(2):406-14.

[13] Meiffren G, Joubert PE, Grégoire IP, Codogno P, Rabourdin-Combe C, Faure M. Pathogen recognition by the cell surface receptor CD46 induces autophagy. *Autophagy.* 2010;6(2):299-300.

[14] Yang Z, Klionsky DJ. Eaten alive: a history of macroautophagy. *Nat Cell Biol.* 2010;12(9):814-22.

[15] Ouellette AJ. Paneth cells and innate mucosal immunity. *Curr Opin Gastroenterol.* 2010;6:547-53.

[16] Gusils C, Morata V, and Gonzales S. Determination of bacterial adhesion to intestinal mucus. *Methods Mol Biol.* 2004;268: 411-15.

[17] Müller CA, Autenrieth IB, Peschel A. Innate defenses of the intestinal epithelial barrier. *Cell Mol Life Sci.* 2005;62(12):1297-07.

[18] Gorbunov NV, Garrison BR, Kiang JG. Response of crypt paneth cells in the small intestine following total-body gamma-irradiation. *Int J Immunopathol Pharmacol.* 2010;23(4):1111-23.

[19] Potten CS, Merritt A, Hickman J, Hall P, Faranda A. Characterization of radiation-induced apoptosis in the small intestine and its biological implications. *Int J Radiat Biol.* 1994;65:71-8.

[20] Hauer-Jensen M, Wang J, Denham JW. Mechanisms and Modification of the radiation response of gastrointestinal organs. In *Modification of radiation response.* C. Nieder, L. Milas, K.K. Ang, ed. Springer. Berlin, 2003; p. 49.

[21] Janot L, Sirard JC, Secher T, Noulin N, Fick L, Akira S, Uematsu S, Didierlaurent A, Hussell T, Ryffel B, Erard F. Radioresistant cells expressing TLR5 control the respiratory epithelium's innate immune responses to flagellin. *Eur J Immunol.* 2009;39(6):1587-96.

[22] Elphick DA, Mahida YR. Paneth cells: their role in innate immunity and inflammatory disease. *Gut.* 2005;54:1802-09.

[23] Vaishnava S, Behrendt CL, Ismail AS, Eckmann L, Hooper LV. Paneth cells directly sense gut commensals and maintain homeostasis at the intestinal host-microbial interface. *Proc Natl Acad Sci USA.* 2008;105(52):20858-63.

[24] Barker N, van de Wetering M, Clevers H. The intestinal stem cells. *Genes Dev.* 2008; 2:1856-64.

[25] Hampton J.C. Effects of Ionizing Radiation on the Fine Structure of Paneth Cells of the Mouse. *Radiat Res.* 1966;28:71-83.

[26] Yang D, Liu ZH, Tewary P, Chen Q, de la Rosa G, Oppenheim JJ. Defensin participation in innate and adaptive immunity. *Curr Pharm Des.* 2007; 13:3131-39.

[27] Salzman NH, Hung K, Haribhai D, Chu H, Karlsson-Sjöberg J, Amir E, Teggatz P, Barman M, Hayward M, Eastwood D, Stoel M, Zhou Y, Sodergren E, Weinstock GM, Bevins CL, Williams CB, Bos NA. Enteric defensins are essential regulators of intestinal microbial ecology. *Nat Immunol.* 2010;11(1):76-83.

[28] Liévin-Le Moal V, Servin AL The front line of enteric host defense against unwelcome intrusion of harmful microorganisms: mucins, antimicrobial peptides, and microbiota. *Clin Microbiol Rev.* 2006;19(2):315-37.

[29] Cadwell K, Patel KK, Komatsu M, Virgin HW 4th, Stappenbeck TS. A common role for Atg16L1, Atg5 and Atg7 in small intestinal Paneth cells and Crohn's disease. *Autophagy.* 2009;5(2):250-52.

[30] Cadwell K, Stappenbeck TS, Virgin HW. Role of autophagy and autophagy genes in inflammatory bowel disease. *Curr Top Microbiol Immunol.* 2009;335:141-67.

[31] Satoh Y, Yamano M, Matsuda M, Ono K. Ultrastructure of Paneth cells in the intestine of various mammals. *J Electron Microsc Tech.* 1990; 16(1):69-80.

[32] Keshav S. Paneth cells: leukocyte-like mediators of innate immunity in the intestine. *J Leukoc Biol.* 2006;80(3):500-08.

[33] Eskelinen EL. Fine structure of the autophagosome. *Methods Mol Biol.* 2008;445:11-28.

[34] Gomes LC, Benedetto GD, Scorrano L. During autophagy mitochondria elongate, are spared from degradation and sustain cell viability. *Nat Cell Biol.* 2011;13(5):589-98.

[35] Kundra R, Kornfeld S. Asparagine-linked oligosaccharides protect Lamp-1 and Lamp-2 from intracellular proteolysis. *J Biol Chem.* 1999;274(43):31039-46.

[36] Eskelinen EL. Roles of LAMP-1 and LAMP-2 in lysosome biogenesis and autophagy. *Mol Aspects Med.* 2006;27(5-6):495-502.

[37] Bursch W. The autophagosomal-lysosomal compartment in programmed cell death. *Cell Death Differ.* 2001;8:569-81.

[38] Takayama S, Reed JC. Molecular chaperone targeting and regulation by BAG family proteins. *Nat Cell Biol.* 2001;3:E237-41.

[39] Yacoub A, Park MA, Gupta P, Rahmani M, Zhang G, Hamed H *et al.* Caspase-, cathepsin-, and PERK-dependent regulation of MDA-7/IL-24-induced cell killing in primary human glioma cells. *Mol. Cancer Ther.* 2008;7:297-313.

[40] Kiang JG, Garrison BR, Gorbunov NV. Radiation combined injury: DNA damage, apoptosis, and autophagy. *Adapt Med.* 2010;2:1-10.

[41] Hermiston ML, Gordon JI. Organization of the crypt-villus axis and evolution of its stem cell hierarchy during intestinal development. *Am J Physiol.* 1995;268:G813-22.

[42] Garrett WS, Gordon JI, Glimcher LH. Homeostasis and inflammation in the intestine. *Cell.* 2010;140(6):859-70.

[43] Bouchard V, Harnois C, Demers MJ, Thibodeau S, Laquerre V, Gauthier R, Vézina A, Noël D, Fujita N, Tsuruo T, Arguin M, Vachon PH. B1 integrin/Fak/Src signaling in intestinal epithelial crypt cell survival: integration of complex regulatory mechanisms. *Apoptosis.* 2008;13(4):531-42.

[44] Zhan M, Zhao H, Han ZC. Signalling mechanisms of anoikis. *Histol Histopathol.* 2004;19(3):973-83.

[45] Fasano A. Zonulin and its regulation of intestinal barrier function: the biological door to inflammation, autoimmunity, and cancer. *Physiol Rev.* 2011 91(1):151-75.

[46] Louis NA and Lin PW. The Intestinal Immune Barrier. *NeoReviews.* 2009;10(4): e180-e190.

[47] Han X , Fink MP, Delude R.L, Proinflammatory cytokines cause NO*-dependent and -independent changes in expression and localization of tight junction proteins in intestinal epithelial cells. *Shock.* 2003;19:229-37.

[48] Kiang JG, Jiao W, Cary LH, Mog SR, Elliott TB, Pellmar TC, Ledney GD. Wound trauma increases radiation-induced mortality by activation of iNOS pathway and elevation of cytokine concentrations and bacterial infection. *Radiat Res.* 2010;173:339-32.

[49] Somosy Z, Horváth G, Telbisz A, Réz G, Pálfia Z. Morphological aspects of ionizing radiation response of small intestine. *Micron.* 2002;33:167-78.

[50] Cerveny TJ, MacVittie, Young RW. Acute radiation Syndrome in Humans. In: *Medical consequence of nuclear warfare.* Ed.: Zajtchuk R. Office of the Surgeon General. Department of the Army, USA.1998. p. 15- 36.

[51] Matsuu-Matsuyama M, Nakashima M, Shichijo K, Okaichi K, Nakayama T, Sekine I. Basic fibroblast growth factor suppresses radiation-induced apoptosis and TP53 pathway in rat small intestine. *Rad Res.* 2010;174:52-61.

[52] Ito H, Daido S, Kanzawa T, Kondo S, Kondo Y. Radiation-induced autophagy is associated with LC3 and its inhibition sensitizes malignant glioma cells. *Int J Oncol.* 2005;26(5):1401-10.

[53] Lomonaco SL, Finniss S, Xiang C, Decarvalho A, Umansky F, Kalkanis SN, Mikkelsen T, Brodie C. The induction of autophagy by gamma-radiation contributes to the radioresistance of glioma stem cells. *Int J Cancer.* 2009;125(3):717-22.

[54] Kim KW, Moretti L, Mitchell LR, Jung DK, Lu B. Endoplasmic reticulum stress mediates radiation-induced autophagy by perk-eIF2alpha in caspase-3/7-deficient cells. *Oncogene.* 2010;29(22):3241-51.

[55] Gastrointestinal Kinetics. In: Medical Management of Radiological Casualties. Third Ed. Military Medical Operations. *AFRRI, Bethesda,* Maryland 2009.

In: Autophagy: Principles, Regulation and Roles in Disease
Editor: Nikolai Gorbunov
ISBN: 978-1-61942-266-7

Chapter V

Autophagy in Periodontal Disease

***Hiromasa Tsuda*[1,3*] *and Yoshikazu Mikami*[2,3]**
[1]Department of Biochemistry,
Nihon University School of Dentistry, Tokyo, Japan
[2] Department of Biochemistry Anatomy,
Nihon University School of Dentistry, Tokyo, Japan
[3] Division of Functional Morphology, Dental Research Center,
Nihon University School of Dentistry, Tokyo, Japan

Abstract

Periodontal diseases are featured by chronic infection, inflammation and degeneration of the periodontal tissue that is induced by anaerobic gram-negative bacteria (e.g., *Porphyromonas gingivalis*). We have recently reported that butyrate, a metabolite produced by periodontopathic bacteria, induced death of the gingival epithelial cells mainly *via* autophagy. Another study showed that the release of high-mobility group box-1 (HMGB1), a molecule that acts as a group of proinflammatory cytokine when released extracellularly, was detected in the culture medium of gingival epithelial cells after butyrate-induced cell death. These results suggest a possible connection between autophagy and periodontal diseases.

P. gingivalis, a causative bacterium of periodontal diseases and butyrate producer, can persist and proliferate in host epithelial cells. The bacterium is internalized in phagosomes of the host cells and then transferred to autophagosomes, where it persists and proliferates. Inhibition of autophagy by 3-methyladenine or wortmannin caused lysosomal fusion with phagosomes containing *P. gingivalis*, and the lysosomal enzymes digest the bacterium. This suggests that *P. gingivalis* can utilize a part of the host autophagy mechanism for their own protection from being damaged by host anti-bacterial system.

* Corresponding author's name and mailing address: Hiromasa Tsuda, Department of Biochemistry, Nihon University School of Dentistry, 1-8-13 Kanda Surugadai, Chiyoda-ku, Tokyo 101-8310, Japan; Phone: +81-3-3219-8123; Fax: +81-3-3219-8334; E-mail: tsuda-h@dent.nihon-u.ac.jp.

In this chapter, we provide an overview of the relationship between autophagy and the processes of periodontal disease and propose a hypothesis for the mechanism of periodontal disease via autophagy.

1. Introduction

Periodontal disease is a common chronic inflammation of tooth-supporting tissues, such as gingival epithelium, periodontal ligaments, cementum, and alveolar bone. Alveolar bone destruction is one of the most important clinical symptoms of the disease, because destruction of the bone does not stop without proper treatment and it is very hard to recover the destroyed bone. Gingival crevice is a groove between the surface of tooth and the epithelium lining free gingiva (Fig. 1). The epithelium lining inside of the groove is attaching tooth surface using hemidesmosomes at the bottom of the gingival crevice, which is termed junctional epithelium (Fig. 1). The junctional epithelium is permeable and serum derived exudate from subgingival connective tissue pass through the junctional epithelium because of the presence of wide intercellular space of epithelium. The serum exudate is known as gingival cervical fluid (GCF) and other oral epithelial surfaces do not show such exudation of tissue fluid. GCFs contain polymorphonuclrear leukocytes (PMNs), gamma globulins, complements. Therefore, it has been suggested that the phagocytic and immunological properties of the GCFs are important in the host-defense mechanisms of gingival grooves, in addition to barriers of the gingival crevice wall epithelium. However, it is thought by some that the GCFs only exude into the gingival crevices in response to pathogenic stimuli and is absent in perfectly healthy gingival grooves. Once bacteria accumulate at a rim of gingival crevice as dental plaques, the GCFs pass through from subgingival connective tissue and an inflammatory response is triggered. Defensins are also play an important role for host defense around gingival crevices. Human defensins are small cationic peptides that comprise an important component of innate immune defense. They have broad spectrum antimicrobial activity against gram-negative, gram-positive bacteria, fungi, and viruses, and were classified as α- and β-defensin [20]. The α-defensins are secreted mainly by PMNs and Paneth cells in the small intestine and the β-defensins are ubiquitously expressed throughout epithelium. In the gingival crevice, α-defensins are secreted by PMNs in response to bacterial infection, whereas β-defensins are produced by gingival epithelial cells [21]. Indeed, these antimicrobial peptides were detected in GCFs [21] and may have important roles in response to mucosal infection. Yet, how the host-defense breakdown occurs around gingival crevice in periodontal disease remains unclear.

Dental plaque consists of numerous species of microorganisms and their metabolic products. The participation of anaerobic gram-negative bacteria (e.g., *Porphyromonas*, *Prevotella*, *Treponema*, and *Fusobacterium* species) that colonize subgingival crevices as dental plaque is strongly implicated in the pathogenesis of periodontal disease. These bacteria produce a variety of virulence factors (e.g., lipopolysaccharides, proteases, fimbriae, and capsular polysaccharides), and the bacteria-derived components induce the production of effectors of inflammation. Therefore, these factors have been considered to be the main candidates that cause periodontal diseases. However, there is no sufficient explanation of the mechanisms by which the bacterial products trigger the initiation of periodontal disease.

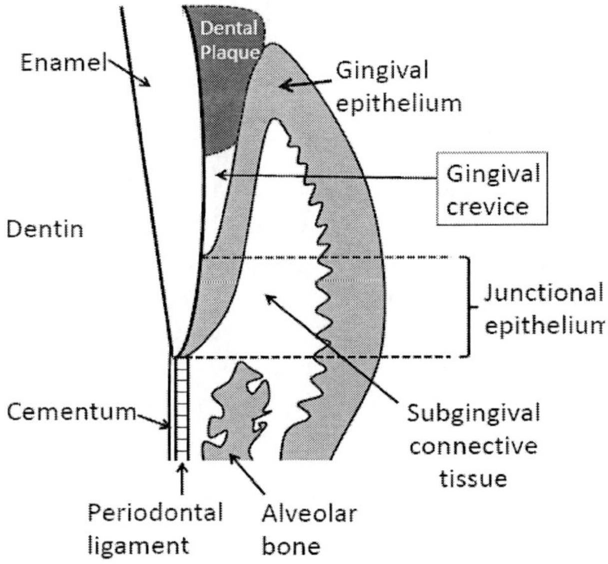

Figure 1. Diagrammatic representation of gingival tissues.

High-mobility group box-1 (HMGB1), also known as amphoterin, is a highly conserved, widely expressed, and abundant cellular protein [7, 8]. HMGB1 protein works as a DNA chaperone in the nucleus, and facilitates the assembly of site-specific DNA-binding proteins at their target promoters [7-10]. HMGB1 protein also plays a significant role in extracellular signaling once it is released into the extracellular environment [11-14]. The released HMGB1 is reported to function as an activator molecule during inflammation [14-16] and as a chemoattractant for inflammatory leukocytes [17] as well as participate in cell migration [18, 19]. In addition, in vivo experiments suggested that an antibody to HMGB1 can reduce endotoxin-induced lethality [23] and collagen-induced arthritis [24]. Therefore, HMGB1 plays a role of proinflammatory cytokines. HMGB1 is passively released from cell-membrane damaged cells such as necrotic cells or actively secreted from activated immune cells [22]. Interestingly, high concentrations of HMGB1 are found in the GCFs of patients with periodontal disease [25]. Recently, we reported that butyrate, one of the metabolite produced by periodontopathic bacteria, induces the death of gingival epithelial cells mainly through autophagy [1]. In addition, another group reported that butyrate-induced cell death resulted in the passive release of HMGB1 into the extracellular milieu [2]. The first aim of this chapter is to describe the possible relationship between HMGB1 release by cell death through autophagy and periodontitis.

Many examples of relationship between autophagy pathway and bacterial infection have been reported and they are well reviewed by Kirkegaard et al [60]. In the *Rickettsia conorii*-infected mouse endothelial cells, which were treated with tumor-necrosis factor-α (TNF-α) or interferon, the bacterial cells were found in double-membrane-bound structures and the some bacteria seemed to be damaged. In the *Listeria monocytogenesis*-infected mouse macrophage J774A.1 cells, the bacteria were found in autophagic vacuoles after chloramphenicol treatment. *Salmonella typhimurium* inject SipB protein into host macrophages using its type-III-secretion systems. The injected SipB protein binds to caspase-1 and induces apoptosis by increased-caspase-1 activities. In the caspase-1 deficient condition, host cells showed characteristics of autophagy [61]. In these examples written above, bacterial cells were found in possible autophagosome structures. In contrast, it is reported that there are several types of bacteria which can subvert the autophagic pathway and persist and proliferate in the host cells. *Burucella abortus* localize in the double membranes structure that resemble autophagosomes after endosomal uptake by mammalian cells. A treatment with 3-methyladenine or wortmannin can reduce infection and cause the bacteria digestion in the lysosomal pathway. This is one of the examples that the bacteria, which were endocytosed in the host cells, subvert autophagy process for degradation of autophagosomal contents and survive in the host cells. *Porphyromonas gingivalis* (*P. gingivalis*), a butyrate-producing periodontopathic bacterium, also take same strategy to persist in host cells [3, 4]. Therefore, we show the process of the subversion of autophagy by *P. gingivalis*.

In this chapter, we discuss these two autophagy-related events and propose an autophagy-mediated mechanism implicated in periodontal disease.

2. Butyrate Induces Gingival Cell Death through Autophagy

Dental plaque consists of microorganism and the extracellular matrix. During plaque maturation, the microbial constitution changes from aerobic gram-positive to anaerobic gram-negative bacteria. The latter types of bacteria produce high concentrations of butyrate (2.6-14 mM) in mature subgingival dental plaques [5]. Our previous data indicated that high concentrations (0.5-10 mM) of butyrate induce gingival epithelial cell death [1]. In addition, the amount of the butyrate-induced cell death in the presence of 1% fetal bovine serum (FBS) was greater than that of 10% FBS (unpublished data). During butyrate challenge, some features of apoptosis, such as a phosphatidylserine redistribution, increased caspase-3 activity and decreased anti-apoptotic *bcl-2* expression, were observed. However, only a small amount of the cells was rescued by pan-caspase inhibition. During cell death, the conversion of microtubule-associated protein 1 light chain 3 (LC3)-I to LC3-II and LC3 molecule accumulation were observed. In addition, 3-methyladenine, an inhibitor of autophagy, rescued cells from butyrate-induced cell death. These results suggest that butyrate induces gingival epithelial cell death mainly through autophagy and partially through apoptosis. Supporting our results, butyrate-induced autophagic cell death in HeLa cervical cancer cells was also reported [6]. Currently, the mechanism underlying butyrate-induced epithelial cell death remains unclear.

Autophagy is also induced by nutrition shortage conditions. Thus, the conditions at the border of the gingival epithelial layer and mature dental plaque may synergistically promote the autophagy of epithelial cells. As mentioned above, mature dental plaques contain a high concentration of butyrate, which can induce autophagy. Moreover, bacteria in mature dental plaques consume nutrients that are present in the serum-derived GCFs. Nutrients in the GCFs are also considered to be one of the sources of nutrients for the epithelial cells lying along the gingival crevices. Therefore, the gingival epithelial cells must compete against the abundant live bacteria present in the plaques for the nutrients, and the subsequent energy shortage in epithelial cells may induce autophagy. Indeed, in our preliminary experiments, butyrate-induced cell death via autophagy decreased on increasing serum concentration in the media (unpublished data). In the next paragraph, we describe the results of our recent study conducted to clarify the role of autophagy and periodontal inflammation in the mediation of butyrate-induced death of epithelial cells.

3. HMGB1 Is Released by Autophagy: Its Possible Relationship of Autophagy with Periodontal Disease

High concentrations of HMGB1 are found in the gingival crevicular fluid of patients with periodontal disease [25]. HMGB1 is observed mainly in the nuclei of healthy gingival epithelial cells [2]. In contrast, translocation of HMGB1 to the cytoplasm is observed in patients with periodontal disease [2]. Considering that released HMGB1 recruits PMNs and stimulates inflammation and the immune response, HMGB1 may play an important role in the initiation and progression of periodontal inflammation. Therefore, the role of HMGB1 in the mechanism of periodontal disease formation is being investigated.

As mentioned above, we demonstrated that high concentrations of butyrate (0.5-10 mM) induced gingival epithelial cell death through autophagy [1]. Most of the dead cell displayed the loss of membrane integrity (SYTOX-Green permeable) that was accompanied by cell swelling; evidently, the cell death is supposed to be necrotic or pyroptotic (unpublished observation). It is considered that loss of cellular membrane integrity may cause leakage of cytoplasmic HMGB1 to outside of the cells. In fact, Ebe and co-authors recently reported that 10 mM butyrate induced necrotic death of Ca9-22 gingival epithelial cells, thereby releasing HMGB1 into the extracellular environment [2]. Taken together with these reports, it is possible that autophagy closely related to the butyrate-induced HMGB1 release. Further investigations are needed to elucidate the exact importance of the relationships between autophagy and HMGB1 release in periodontal disease.

In the cell, HMGB1 is located in the nucleus and cytoplasm [9, 26-28]. The nucleo-cytoplasmic shuttling of HMGB1 is controlled by post-transcriptional modifications that include acetylation [9, 26]. Histone acetyltransferases (HATs) and histone deacetylases (HDACs) mediate the acetylation status of HMGB1; HATs acetylate HMGB1 and HDACs remove the acetyl groups from the molecule (Fig.2). Acetylated HMGB1 is shuttled from the nucleus to the cytoplasm, where it is released. Butyrate is also known as an HDAC inhibitor [29]. Butyrate-inhibited HDAC activity may increase the acetylated HMGB1, capable of translocation to the cytoplasm (Fig. 2). This seems to correspond to the observation that most of the HMGB1 population is found in the cytoplasm of periodontal-pocket gingival cells, which may be exposed to a high concentration of butyrate in the mature dental plaque [2]. In such situations, the acetylated HMGB1 in the cytoplasm can be easily released from the gingival cells by butyrate-induced cell death through autophagy.

Butyrate reportedly induced intracellular adhesion molecule-1 (ICAM-1) expressions on the gingival epithelial cells [62]. Vinolo et al reported that subcutaneous injection of butyrate with air (air pouch assay) on rat induced the migration of PMNs, the cytokine-induced neutrophil chemoattractants-2 (CINC-2) production, and L-selectin expression on neutrophil surface [63]. Because HMGB1 is reported to work as activator of inflammation [14-16], these reports imply the possibility of the relationship between the butyrate-induced HMGB1 release and the PMN migration into gingival crevices. Therefore, HMGB1 may play an important role in the initiation or enhancement of inflammation of periodontal disease (Fig. 2).

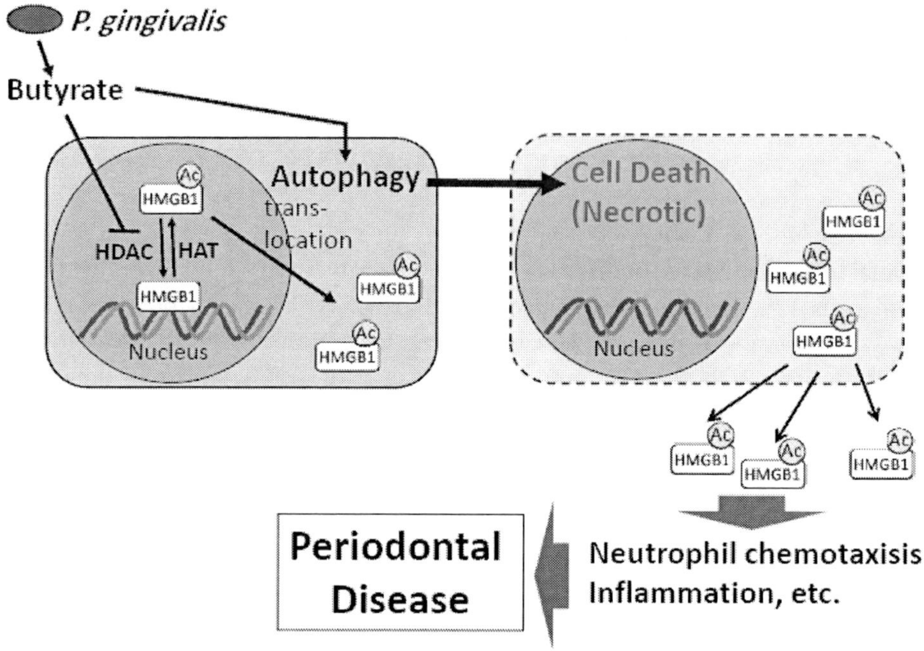

Figure 2. Proposed mechanisms of periodontal disease initiation. Histone acetyltransferases (HATs) and histone deacetylases (HDACs) control the acetylation status of HMGB1. Acetylated HMGB1 is shuttled from the nucleus to the cytoplasm. *P. gingivalis*, a pathogen which cause periodontal disease, produces butyrate, which is a bacterial metabolite known as an HDAC inhibitor. Butyrate deactivates HDACs and increases the amount of acetylated HMGB1, which localizes to the cytoplasm. Butyrate also induces necrotic cell death via autophagy. Passive release of HMGB1 from dying gingival epithelial cells triggers inflammation.

4. Autophagy Subversion Strategy of *P. Gingivalis* to Persist in the Host Cells

P. gingivalis is considered to be one of the most important pathogenic causes of periodontal disease. The bacterium has been found not only in gingival and buccal epithelial cells [30-33] but also in atherosclerotic plaques [34, 35] and coronary artery cells [36, 37]. Transient bacteremia caused by frequent hemorrhages from periodontal tissues during periodontal disease can provide a direct route for *P. gingivalis* into the blood circulation [38-41]. Epidemiological studies have shown the presence of a strong relationship between periodontal disease and coronary heart disease [42-47]. Thus, the mechanisms promoting the internalization of *P. gingivalis* into both oral epithelial and cardiovascular cells have been investigated in biomedical and clinical studies. The accumulated evidence indicates that autophagy plays an important role in internalization, persistence, and proliferation of *P. gingivalis* in host cells [4, 30, 31, 33, 37].

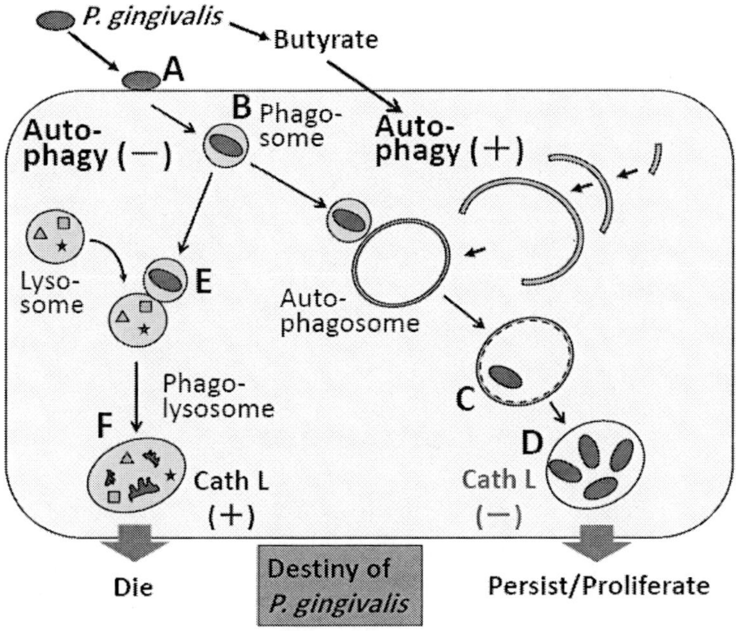

Figure 3. Model of *P. gingivalis* trafficking in epithelial and endothelial cells. *P. gingivalis* adheres to the cell surface (A) and is internalized into early phagosomes (B). The invading bacteria are transferred to autophagosomes in the absence of lysosome fusion (C). *P. gingivalis* persists and proliferates in the autophagosomes (D). Inhibition of autophagy by 3-methyladenine or wortmannin causes lysosomes to fuse with early phagosomes (E) and the internalized bacteria are digested by lysosomal enzymes (F). Butyrate, which is produced by *P. gingivalis*, induces autophagy and triggers uptake of *P. gingivalis* by host cells. Abbreviation: Cath L, Cathepsin L.

Bélanger et al. have proposed a model of *P. gingivalis* incorporation and trafficking within endothelial cells on the basis of their transmission electron microscopy observation and immunofluorescence analysis of interactions between *P. gingivalis* microorganisms and autophagy-related proteins [3, 4]. The initial adherence of *P. gingivalis* to human coronary artery endothelial cells (HCAECs) mediated by multiple adhesive molecules, such as *fimA* and *hagB* gene products, on the surface of *P. gingivalis* cells occurs within 15 min after the inoculation of live *P. gingivalis* into HCAEC culture [Fig. 3A, Refs. 4, 36, 48-51]. They also reported that the binding of *P. gingivalis* on HCAEC surfaces occurs via lipid rafts [3, 52-55]. In addition, ICAM-1 is reportedly required for the invasion of *P. gingivalis* into human oral-epithelial KB cells [56]. The second step is internalization of the bacterium into early phagosomes via lipid rafts (Fig. 3B). This step occurs 15-60 min after challenge of *P. gingivalis* to host cells. The invading bacteria in the phagosomes are next transferred to autophagosomes without fusing the phagosomes with lysosomes (60-240 min after challenge of P. gingivalis to host cells), and they persist in these autophagosomes (Fig. 3C). Thus, bacteria survive, persist, and proliferate in the host cell while being incorporated into the autophagosomes where they are protected from hydrolytic enzymes such as lysosomal cathepsin L [Fig. 3 D, Refs. 3, 4, 36, 48, 57-59]. Interestingly, inhibition of early autophagosome formation by 3-methyladenine or wortmannin causes lysosomes to fuse with early phagosomes (Fig. 3E), and the internalized bacteria are digested by lysosomal enzymes [Fig. 3F, Refs. 3, 4]. Therefore, autophagy is a key event in *P. gingivalis* invasion and residence in host cells, although it remains unclear how *P. gingivalis* activates autophagy. Our previous study demonstrated that butyrate produced by *P. gingivalis* induces autophagy [1]; this indicates that butyrate-induced autophagy may have an important role in the persistence of *P. gingivalis* in host cells.

Takigawa et al reported that butyrate induced ICAM-1 expressions on the surface of gingival epithelial cells [62]. As described above, *P. gingivalis* uses ICAM-1 to invade into the host cells. As butyrate also induces autophagy, the bacterial metabolite may help *P. gingivalis* to subvert the mechanism to digest autophagosomal contents and to persist in the host. Therefore, *P. gingivalis* might use the butyrate as an effector molecule to escape from host defense mechanism around or in the gingival crevices.

Conclusion

P. gingivalis, which is the main causative bacteria of periodontal disease, produces butyrate as a metabolite. This metabolic product induces autophagy in gingival epithelial cells and subsequent cell death, followed by the release of HMGB1 which in turn can elicit inflammation and host immune response. Additionally, *P. gingivalis* invades into epithelial or endothelial cells, escapes from lysosomal enzymes, persists, and proliferate in autophagosomes using the machinery of autophagy. These two processes are the possible factors promoting periodontal disease, and autophagy may play an important role in the establishment of periodontal disease. Therefore, autophagy-targeting therapy is a potential approach for preventing periodontal disease.

References

[1] Tsuda H, Ochiai K, Suzuki N, Otsuka K. Butyrate, a bacterial metabolite, induces apoptosis and autophagic cell death in gingival epithelial cells. *J. Periodontal. Res.* 2010;45(5):626-634.

[2] Ebe N, Hara-Yokoyama M, Iwasaki K, Iseki S, Okuhara S, Podyma-Inoue KA, et al. Pocket epithelium in the pathological setting for HMGB1 release. *J. Dent. Res.* 2011;90(2):235-240.

[3] Belanger M, Rodrigues PH, Dunn WA, Jr., Progulske-Fox A. Autophagy: a highway for *Porphyromonas gingivalis* in endothelial cells. *Autophagy.* 2006;2(3):165-170.

[4] Dorn BR, Dunn WA, Jr., Progulske-Fox A. *Porphyromonas gingivalis* traffics to autophagosomes in human coronary artery endothelial cells. *Infect. Immun.* 2001;69(9):5698-5708.

[5] Pollanen MT, Salonen JI, Uitto VJ. Structure and function of the tooth-epithelial interface in health and disease. *Periodontol. 2000.* 2003;31:12-31.

[6] Shao Y, Gao Z, Marks PA, Jiang X. Apoptotic and autophagic cell death induced by histone deacetylase inhibitors. *Proc. Natl. Acad. Sci. U. S. A.* 2004;101(52):18030-18035.

[7] Muller S, Ronfani L, Bianchi ME. Regulated expression and subcellular localization of HMGB1, a chromatin protein with a cytokine function. *J. Intern. Med.* 2004 ;255(3):332-343.
[8] Thomas JO, Travers AA. HMG1 and 2, and related 'architectural' DNA-binding proteins. *Trends Biochem. Sci.* 2001;26(3):167-174.
[9] Bonaldi T, Langst G, Strohner R, Becker PB, Bianchi ME. The DNA chaperone HMGB1 facilitates ACF/CHRAC-dependent nucleosome sliding. *EMBO J.* 2002;21(24):6865-6873.
[10] Lotze MT, Tracey KJ. High-mobility group box 1 protein (HMGB1): nuclear weapon in the immune arsenal. *Nat. Rev. Immunol.* 2005;5(4):331-342.
[11] Bianchi ME, Beltrame M. Flexing DNA: HMG-box proteins and their partners. *Am. J. Hum. Genet.* 1998;63(6):1573-1577.
[12] Campana L, Bosurgi L, Rovere-Querini P. HMGB1: a two-headed signal regulating tumor progression and immunity. *Curr. Opin. Immunol.* 2008;20(5):518-523.
[13] Klune JR, Dhupar R, Cardinal J, Billiar TR, Tsung A. HMGB1: endogenous danger signaling. *Mol. Med.* 2008;14(7-8):476-484.
[14] Scaffidi P, Misteli T, Bianchi ME. Release of chromatin protein HMGB1 by necrotic cells triggers inflammation. *Nature.* 2002;418(6894):191-195.
[15] Messmer D, Yang H, Telusma G, Knoll F, Li J, Messmer B, et al. High mobility group box protein 1: an endogenous signal for dendritic cell maturation and Th1 polarization. *J. Immunol.* 2004;173(1):307-313.
[16] Yang D, Chen Q, Yang H, Tracey KJ, Bustin M, Oppenheim JJ. High mobility group box-1 protein induces the migration and activation of human dendritic cells and acts as an alarmin. *J. Leukoc. Biol.* 2007;81(1):59-66.
[17] Orlova VV, Choi EY, Xie C, Chavakis E, Bierhaus A, Ihanus E, et al. A novel pathway of HMGB1-mediated inflammatory cell recruitment that requires Mac-1-integrin. *EMBO J.* 2007;26(4):1129-1139.
[18] Limana F, Germani A, Zacheo A, Kajstura J, Di Carlo A, Borsellino G, et al. Exogenous high-mobility group box 1 protein induces myocardial regeneration after infarction via enhanced cardiac C-kit$^+$ cell proliferation and differentiation. *Circ. Res.* 2005;97(8):e73-83.
[19] Palumbo R, Sampaolesi M, De Marchis F, Tonlorenzi R, Colombetti S, Mondino A, et al. Extracellular HMGB1, a signal of tissue damage, induces mesoangioblast migration and proliferation. *J. Cell Biol.* 2004;164(3):441-449.
[20] Salzman NH. Paneth cell defensins and the regulation of the microbiome: détente at mucosal surfaces. *Gut. Microbes.* 2010;1(6):401-406.
[21] Diamond DL, Kimball JR, Krisanaprakornkit S, Ganz T, Dale BA. Detection of beta-defensins secreted by human oral epithelial cells. *J. Immunol. Methods.* 2001;256(1-2):65-76.
[22] Campana L, Bosurgi L, Rovere-Querini P. HMGB1: a two-headed signal regulating tumor progression and immunity. *Curr. Opin. Immunol.* 2008;20(5):518-523.
[23] Wang H, Bloom O, Zhang M, Vishnubhakat JM, Ombrellino M, Che J, et al. HMG-1 as a late mediator of endotoxin lethality in mice. *Science.* 1999;285(5425):248-251.
[24] Andersson U, Erlandsson-Harris H. HMGB1 is a potent trigger of arthritis. *J. Intern. Med.* 2004;255(3):344-350.

[25] Morimoto Y, Kawahara KI, Tancharoen S, Kikuchi K, Matsuyama T, Hashiguchi T, et al. Tumor necrosis factor-alpha stimulates gingival epithelial cells to release high mobility-group box 1. *J. Periodontal. Res*. 2008;43(1):76-83.
[26] Evankovich J, Cho SW, Zhang R, Cardinal J, Dhupar R, Zhang L, et al. High mobility group box 1 release from hepatocytes during ischemia and reperfusion injury is mediated by decreased histone deacetylase activity. *J. Biol. Chem*. 2010;285 (51):39888-39897.
[27] Tang D, Kang R, Livesey KM, Cheh CW, Farkas A, Loughran P, et al. Endogenous HMGB1 regulates autophagy. *J Cell Biol*. 2010;190(5):881-892.
[28] Youn JH, Shin JS. Nucleocytoplasmic shuttling of HMGB1 is regulated by phosphorylation that redirects it toward secretion. *J. Immunol*. 2006;177(11):7889-7897.
[29] Vidali G, Boffa LC, Bradbury EM, Allfrey VG. Butyrate suppression of histone deacetylation leads to accumulation of multiacetylated forms of histones H3 and H4 and increased DNase I sensitivity of the associated DNA sequences. *Proc. Natl. Acad. Sci. U. S. A*. 1978;75(5):2239-2243.
[30] Duncan MJ, Nakao S, Skobe Z, Xie H. Interactions of *Porphyromonas gingivalis* with epithelial cells. *Infect. Immun*. 1993;61(5):2260-2265.
[31] Lamont RJ, Oda D, Persson RE, Persson GR. Interaction of *Porphyromonas gingivalis* with gingival epithelial cells maintained in culture. *Oral Microbiol. Immunol*. 1992;7(6):364-367.
[32] Rudney JD, Chen R, Sedgewick GJ. Intracellular *Actinobacillus actinomycetem-comitans* and *Porphyromonas gingivalis* in buccal epithelial cells collected from human subjects. *Infect. Immun*. 2001;69(4):2700-2707.
[33] Sandros J, Papapanou P, Dahlen G. *Porphyromonas gingivalis* invades oral epithelial cells in vitro. *J. Periodontal. Res*. 1993;28(3):219-226.
[34] Chiu B. Multiple infections in carotid atherosclerotic plaques. *Am. Heart J*. 1999;138(5 Pt 2):S534-536.
[35] Haraszthy VI, Zambon JJ, Trevisan M, Zeid M, Genco RJ. Identification of periodontal pathogens in atheromatous plaques. *J. Periodontol*. 2000;71(10):1554-1560.
[36] Deshpande RG, Khan MB, Genco CA. Invasion of aortic and heart endothelial cells by *Porphyromonas gingivalis*. *Infect. Immun*. 1998;66(11):5337-5343.
[37] Dorn BR, Dunn WA, Jr., Progulske-Fox A. Invasion of human coronary artery cells by periodontal pathogens. *Infect. Immun*. 1999;67(11):5792-5798.
[38] Carroll GC, Sebor RJ. Dental flossing and its relationship to transient bacteremia. *J. Periodontol*. 1980;51(12):691-692.
[39] Iwai T. Periodontal bacteremia and various vascular diseases. *J. Periodontal. Res*. 2009;44(6):689-694.
[40] Sconyers JR, Crawford JJ, Moriarty JD. Relationship of bacteremia to toothbrushing in patients with periodontitis. *J. Am. Dent. Assoc*. 1973;87(3):616-622.
[41] Silver JG, Martin AW, McBride BC. Experimental transient bacteraemias in human subjects with varying degrees of plaque accumulation and gingival inflammation. *J. Clin. Periodontol*. 1977;4(2):92-99.
[42] Beck J, Garcia R, Heiss G, Vokonas PS, Offenbacher S. Periodontal disease and cardiovascular disease. *J. Periodontol*. 1996;67(10 Suppl):1123-1137.

[43] Beck JD, Eke P, Heiss G, Madianos P, Couper D, Lin D, et al. Periodontal disease and coronary heart disease: a reappraisal of the exposure. *Circulation*. 2005;112(1):19-24.
[44] Beck JD, Offenbacher S. Systemic effects of periodontitis: epidemiology of periodontal disease and cardiovascular disease. *J. Periodontol*. 2005;76(11 Suppl):2089-2100.
[45] DeStefano F, Anda RF, Kahn HS, Williamson DF, Russell CM. Dental disease and risk of coronary heart disease and mortality. *BMJ*. 1993;306(6879):688-691.
[46] Mattila KJ, Nieminen MS, Valtonen VV, Rasi VP, Kesaniemi YA, Syrjala SL, et al. Association between dental health and acute myocardial infarction. *BMJ*. 1989;298(6676):779-781.
[47] Mattila KJ, Valtonen VV, Nieminen M, Huttunen JK. Dental infection and the risk of new coronary events: prospective study of patients with documented coronary artery disease. *Clin. Infect. Dis*. 1995;20(3):588-592.
[48] Dorn BR, Burks JN, Seifert KN, Progulske-Fox A. Invasion of endothelial and epithelial cells by strains of *Porphyromonas gingivalis*. *FEMS Microbiol. Lett*. 2000;187(2):139-144.
[49] Njoroge T, Genco RJ, Sojar HT, Hamada N, Genco CA. A role for fimbriae in *Porphyromonas gingivalis* invasion of oral epithelial cells. *Infect. Immun*. 1997;65(5):1980-1984.
[50] Song H, Belanger M, Whitlock J, Kozarov E, Progulske-Fox A. Hemagglutinin B is involved in the adherence of *Porphyromonas gingivalis* to human coronary artery endothelial cells. *Infect. Immun*. 2005;73(11):7267-7273.
[51] Xie H, Cai S, Lamont RJ. Environmental regulation of fimbrial gene expression in *Porphyromonas gingivalis*. *Infect. Immun*. 1997;65(6):2265-2271.
[52] Duncan MJ, Li G, Shin JS, Carson JL, Abraham SN. Bacterial penetration of bladder epithelium through lipid rafts. *J. Biol. Chem*. 2004;279(18):18944-18951.
[53] Duncan MJ, Shin JS, Abraham SN. Microbial entry through caveolae: variations on a theme. *Cell Microbiol*. 2002;4(12):783-791.
[54] Tsuda K, Amano A, Umebayashi K, Inaba H, Nakagawa I, Nakanishi Y, et al. Molecular dissection of internalization of *Porphyromonas gingivalis* by cells using fluorescent beads coated with bacterial membrane vesicle. *Cell Struct. Funct*. 2005;30(2):81-91.
[55] Watarai M. Interaction between Brucella abortus and cellular prion protein in lipid raft microdomains. *Microbes Infect*. 2004;6(1):93-100.
[56] Tamai R, Asai Y, Ogawa T. Requirement for intercellular adhesion molecule 1 and caveolae in invasion of human oral epithelial cells by *Porphyromonas gingivalis*. *Infect. Immun*. 2005;73(10):6290-6298.
[57] Deshpande RG, Khan M, Genco CA. Invasion strategies of the oral pathogen *Porphyromonas gingivalis*: implications for cardiovascular disease. *Invasion Metastasis*. 1998;18(2):57-69.
[58] Dorn BR, Dunn WA, Jr., Progulske-Fox A. Bacterial interactions with the autophagic pathway. *Cell Microbiol*. 2002;4(1):1-10.
[59] Nakhjiri SF, Park Y, Yilmaz O, Chung WO, Watanabe K, El-Sabaeny A, et al. Inhibition of epithelial cell apoptosis by *Porphyromonas gingivalis*. *FEMS Microbiol. Lett*. 2001;200(2):145-149.
[60] Kirkegaard K, Taylor MP, Jackson WT. Cellular autophagy: surrender, avoidance and subversion by microorganisms. *Nat. Rev. Microbiol*. 2004;2(4):301-314.

[61] Wehkamp J, Wang G, Kübler I, Nuding S, Gregorieff A, Schnabel A, et al. The Paneth cell alpha-defensin deficiency of ileal Crohn's disease is linked to Wnt/Tcf-4. *J. Immunol.* 2007;179(5):3109-3118.

[62] Takigawa S, Sugano N, Nishihara R, Koshi R, Murai M, Yoshinuma N, et al. The effect of butyric acid on adhesion molecule expression by human gingival epithelial cells. *J. Periodontal. Res.* 2008;43(4):386-390.

[63] Vinolo MA, Rodrigues HG, Hatanaka E, Hebeda CB, Farsky SH, Curi R. Short-chain fatty acids stimulate the migration of neutrophils to inflammatory sites. *Clin. Sci.* (Lond). 2009;117(9):331-338.

Part II.
Autophagy in Cancer Research and Cancer Therapy

In: Autophagy: Principles, Regulation and Roles in Disease ISBN: 978-1-61942-266-7
Editor: Nikolai Gorbunov

Chapter VI

The Induction of Autophagy in Gliomas Treated by Temozolomide

Manabu Natsumeda*, Hiroshi Aoki and Yukihiko Fujii
Department of Neurosurgery, Brain Research Institute,
Niigata University, Niigata, Japan

Abstract

Autophagy, rather than apoptosis, is induced in gliomas by various treatments, such as radiation and temozolomide (TMZ). The mechanisms of autophagy are gradually being clarified, however, the role of autophagy in cancer therapy remains controversial. Our goal is to elucidate the role of autophagy for application to clinical treatment. We set out to establish methods of monitoring autophagy in *in vitro* studies and in glioma samples. There are only a few previous reports discussing the detection of autophagy in tissue samples.

In vitro studies employing electron microscopy, analysis of LC3I/II conversion by western blotting, and GFP-LC3B immunofluorescence assays were undertaken to detect the induction of autophagy. None of these methods alone are sufficient for monitoring autophagy. Therefore, multiple assays to verify an autophagic response were incorporated.

The development of a reliable method to monitor autophagy in surgically obtained tissues is vital in elucidating the role of autophagy in the clinical setting. To achieve this goal, immunohistochemistry was utilized in surgical tissue. Macroautophagy was monitored by immunohistochemistry employing anti-LC3B and anti-lysosome associated membrane protein 1 (LAMP1) antibodies. Likewise, chaperone-mediated autophagy was monitored using anti-LAMP2A antibodies. An increase of autophagy after TMZ treatment was noted. LC3B expression by western blotting in frozen tissue samples showed increases in both LC3-I and LC3-II expression after TMZ treatment.

* Address correspondence to: Manabu Natsumeda, MD; Department of Neurosurgery, Brain Research Institute, Niigata University ;1-757 Asahimachidori, Chuo-ku, Niigata 951-8585, Japan; Tel: 81-25-227-0653; Fax: 81-25-227-0819; E-mail: mnatsumeda@bri.niigata-u.ac.jp.

By employing immunohistochemical approaches mentioned, it is possible to monitor autophagy in gliomas, opening the door for monitoring therapeutic effects and/or resistance to treatments in gliomas in the clinical setting.

1. Introduction

Malignant gliomas are locally invasive primary brain tumors that have poor prognosis despite treatment with a combination of surgery, radiotherapy, and chemotherapy. Temozolomide has been shown to increase overall survival in glioblastoma by 2 months with concomitant adjuvant radiochemotherapy followed by adjuvant temozolomide compared to radiotherapy alone.[1,2]

Autophagy has been induced by rapamycin, arsenic trioxide, various chemotherapeutics drugs[3,4] including temozolomide (TMZ)[5], and ionizing radiation[6,7] in glioma cell lines. TMZ is an alkylating agent that has been shown to be effective against glioblastoma.[2] Analysis of prognostic factors in TMZ-treated malignant gliomas has focused on factors pertaining to the apoptotic effects of TMZ. However, in vitro glioma studies show that autophagy, and not apoptosis, is induced by TMZ.[5,8]

The role of autophagy in cancer treatment remains controversial. There has long been debate over whether autophagy is cell killing or cell protective in cancer.[3,9-11] The concept of "autophagic cell death" has recently been recognized as not being the essential role of autophagy.[12,13] From this perspective, the inhibition of autophagy can be a "new" anticancer treatment strategy.

In this chapter, we focus on *in vitro* data and analysis of surgical samples regarding autophagy induced by TMZ in gliomas.

2.0. Induction to Autophagy by Temozolomide *In Vitro*

2.1. Materials and Methods

2.1.1. Tumor Cell Lines

Human malignant glioma U87-MG, U373-MG cells were purchased from the American Tissue Culture Collection, (Manassas, VA, USA). Cells were cultured in Dulbecco's modified Eagle's medium (DMEM, GIBCO BRL, Grand Island, NY, USA) supplemented with 10% fetal bovine serum (GIBCO BRL), 4 mM glutamine, 100 U/ml penicillin and 100 μg/ml streptomycin.

2.1.2. Cell Proliferation Assay

The cytotoxic effect of TMZ on cultured U87-MG and U373-MG cells was determined by employing cell proliferation reagent WST-1 (Roche Applied Science, Indianapolis, IN) as described previously.[8,14,15]

2.1.3. Cell Cycle Analysis

U87-MG and U373-MG cells detached with trypsin-EDTA were collected, fixed with 70% ethanol, and stained with a 10% propidium iodide solution (cellular DNA flow cytometric analysis reagent set; Roche) according to the manufacturer's instructions. DNA content was analyzed with a FACScan flow cytometer (Becton Dickinson, San Jose, CA, USA), and data were analyzed with the manufacturer's software (CellQuest; Becton Dickinson). Three replicates were performed for each experiment.[6,16]

2.1.4. Apoptosis Detection Assay

Tumor cells were seeded on Lab- Tek chamber slides (Nunc, Rochester, NY) and incubated overnight and then were treated with 200 μg TMZ for 72 hours and stained with the terminal deoxynucleotidyl transferase-mediated dUTP nick-end labeling (TUNEL) technique using an ApopTag apoptosis detection kit (Chemicon, Temecula, CA) as described previously.[16] Two hundred cells were counted and scored for the incidence of apoptotic chromatin changes under a fluorescence microscope. Paclitaxel (5 nM) was used as a positive control to induce apoptosis.[15]

2.1.5. Electron Microscopy

To morphologically demonstrate the induction of autophagy in TMZ treated tumor cells, the cells were treated with or without TMZ (200 μM) for 72 hours were harvested by trypsinization, washed twice with PBS and fixed with ice-cold glutaraldehyde (3% in 0.1 M cacodylate buffer, pH 7.4) for 30 minutes. After washing in PBS, the cells were postfixed in osmium tetroxide and embedded in Spurr's low-viscosity medium; 0.1 mm thin sections were stained with uranyl acetate / lead citrate (Fluka) and viewed with a Philips EM 400 electron microscope.[5,15]

2.1.6. Detection of Autophagic Cells with GFP-LC3 Dots

Using green fluorescent (GFP) tagged-LC3 expression vector kindly provided by Dr. Noboru Mizushima (Tokyo Medical and Dental University, Tokyo, Japan), the incidence of treated tumor cells expressing GFP-LC3 dots were assessed. Transfection was performed on 100 mm plates with 10 μg plasmid/DNA plate with GenePorter (Gene Therapy System INC., San Diego CA, USA) according to the manufacturer's instructions. After overnight exposure, cells were washed three times with PBS and cultured with complete medium. GFP-LC3-transfected U373-MG cells were cultured on the chamber slide dish (Fisher Inc.). After exposure to TMZ for 72 hours, cells were fixed in 1% paraformaldehyde and analyzed by fluorescence microscopy. To quantify the induction of cells expressing GFP-LC3 dots, 200 GFP-positive cells were counted for each treatment and the percentage of autophagic cells were calculated. 3-MA and bafilomycin A1 were used to inhibit autophagy.[5,15]

2.1.7. Statistical Analysis

Statistical analyses were performed by using two-tailed Student's t-tests. Data are expressed as means and 95% confidence intervals (CI). The criterion for statistical significance was set at $p<0.05$.

2.2. Results

2.2.1. Effect of TMZ on Cell Viability of Malignant Glioma Cell Lines

TMZ inhibited the cell viability of glioma cells in a dose-dependent manner. IC_{50} (concentration resulting in cell viability of 50% compared to control) of TMZ was between 100 and 200 μM.[8,14,15] This concentration was used for further experiments.

2.2.2. Induction of Apoptosis by Paclitaxel, but Not TMZ

No significant number of apoptotic cells was detected in U87-MG and U373-MG cell treated with or without 200 μM of TMZ. Paclitaxel induced apoptosis in 12.7% of U87-MG and 20.9% of U373-MG cells (data not shown). Similarly, analysis of apoptosis by DAPI staining revealed that there was no increase in the number of cells with apoptotic nuclei by 200 μM of TMZ compared to control, whereas 5 nM of paclitaxel significantly increased the percentage of tumor cells with apoptotic nuclei (data not shown). These results indicated that the principle cytotoxicity of TMZ is not due to induction of apoptosis.

2.2.3. Induction of Autophagy by TMZ

Ultrastructural analysis of TMZ-treated U373-MG cells revealed that numerous autophagic vacuoles were observed, while untreated tumor cells exhibited few such features.[5,8]

GFP-LC3 cells presented diffuse distribution under control, while a punctate pattern of GFP-LC3 was increased in number and fluorescence intensity by autophagy in TMZ-treated cells.[15] The number of tumor cells expressing GFP-LC3 dots was significantly greater in U87-MG and U373-MG cells treated with 200 μM of TMZ than in those treated with vehicle alone ($p < 0.05$).[15] These results indicate that autophagy is induced by TMZ.

2.2.4. The Protective Nature of TMZ-Induced Autophagy

The increased percentage of cells with GFP-LC3 dots in TMZ-treated cells were decreased in the presence of 10 mM of 3-MA. Cell viability of TMZ treated cells were significantly decreased in the presence of 10 mM of 3-MA.[15] These results suggest the inhibition of autophagy by 3-MA led to an induction of cell killing. Thus, TMZ-induced autophagy is protective.

3.0. Induction of Autophagy after Temozolomide Treatment in Glioma Surgical Samples

From our *in vitro* results, we conclude that:

A. TMZ induces autophagy, but not apoptosis.
B. TMZ-induced autophagy is cell protective autophagy.

To confirm induction of autophagy by TMZ in surgically obtained glioma samples, immunohistochemical analysis and western blotting were employed. This has been discussed in detail in a previously published report.[17] A brief synopsis is provided below.

3.1. Materials and Methods

3.1.1. Patients

Fourteen glioma patients undergoing surgery before and after TMZ treatment at Niigata University from 1995 to 2009 were included in this study. Written consent for the analysis of specimens after approval from the institutional review board. Histological diagnoses were made by qualified neuropathologists in accordance with the 2007 World Health Organization (WHO) classification.[18]

These patients underwent a total of 41 surgeries. Of the 41 specimens obtained, the initial surgical specimen and first surgical specimen obtained after initialization of TMZ treatment were analyzed. Thus, a total of 28 specimens from 14 patients were analyzed. There were no exclusions from this study.

3.1.2. Immunohistochemical Analysis

Twenty percent buffered formalin-fixed, paraffin-embedded surgical tissue was obtained from all 28 surgical specimens. Histological examination of 4.5-μm-thick consecutive sections was performed by staining with hematoxylin-eosin (HE) and 4 different primary antibodies: anti-LC3B antibody (Abcam, Cambridge, UK; 1:12,000), anti-LAMP1 antibody (Abcam; 1:1600), anti-LAMP2A antibody (Abcam; 1:8000), and anti-Caspase-3 antibody (Cell Signaling Techonology, Danvers, MA, USA 1:200). Antigen retrieval was performed by autoclaving at 121°C for 10 minutes in 10mM citric acid (pH 6.0).

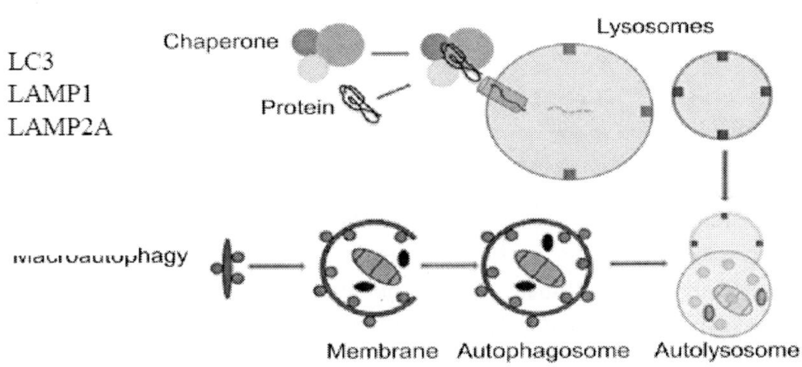

	Macroautophagy				Chaperone-mediated autophagy
	Membrane	Autophagosome Early	Autophagosome Late	Auto-lysosome	
LC3B	+	+	+	±	−
LAMP1	−	−	+	+	−
LAMP2A	−	−	−	−	+

Figure 1. Schematic representation of the processes for chaperone-mediated autophagy and macroautophagy. LC3B is located on the surface of the transmembrane and autophagosomes, and is present during most stages of the macroautophagy cycle. LAMP1 and LAMP2A are membrane proteins located on the surface of lysosomes and play a role in the transport of proteins to the lysosomes in macroautophagy and chaperone-mediated autophagy, respectively.

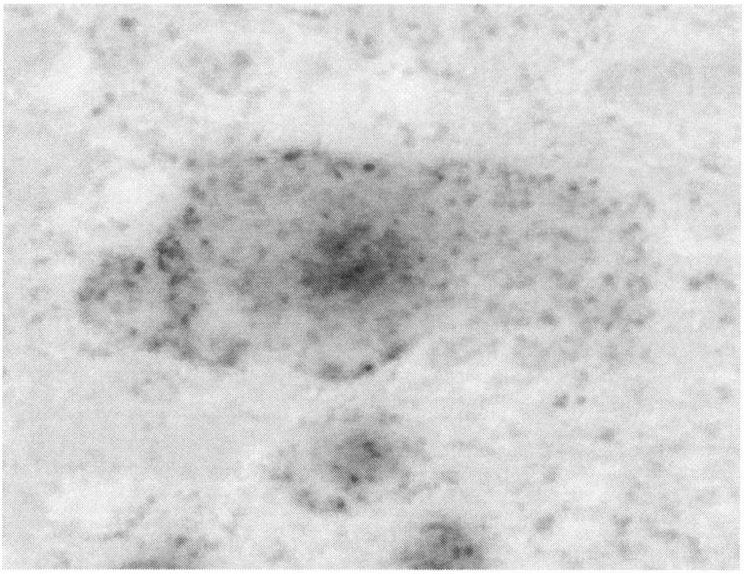

Figure 2. Dot-like staining in the cytosol of tumor cells for the presence of LAMP1 in anaplastic astrocytoma. Bar = 15 μm.

Table 1. Cutoffs for semi-quantitative analysis of LC3B, LAMP1 and LAMP2A immunostains

	(+)	(++)	(+++)
LC3BLC3B	0 20%	21-80%	81-100%
LAMP1	0-20%	21-80%	81-100%
LAMP2A	0-20%	21-50%	51-100%

Induction of macroautophagy was monitored by semiquantitative analysis of LC3B and LAMP1 immunohistochemical staining. Similarly, induction of chaperone-mediated autophagy (CMA) was monitored by analysis of LAMP2A immunostaining. The proteins are known to show immunopositivity during different stages of autophagy (Fig. 1). Punctate staining of the cytoplasm of neoplastic cells (Fig. 2: LAMP1) was considered significant. The area of highest positivity was selected for analysis, and more than 1,000 neoplastic cells were counted. The percentage of positive neoplastic cells was calculated and semiquantitatively subdivided into three groups: slightly positive (1 point), positive (2 points), and highly positive (3 points). The cutoffs for semiquantitative assessment of each stain are shown in Table 1.

3.1.3. Western Blotting

Western blotting for LC3B was performed on 6 frozen sections obtained from 3 patients in whom frozen sections were taken before and after initialization of TMZ treatment. These sections were preserved at −80°C. Equal amounts of proteins isolated from the frozen sections were separated by 15% SDS-PAGE (ATTO, Tokyo, Japan) as previously described,[7,15] and LC3B-I (cytosolic) and LC3B-II (autophagosomal) bands were analyzed. β-actin (1:160000; Clone AC-15, Sigma, St. Louis, MO, USA) was used as the loading control.

3.1.4. Statistical Analysis

Statistical analyses of the mean semiquantitative scores for the pre- and post-TMZ groups were performed using paired t-tests. The level for statistical significance was set at $p < 0.05$.

3.3. Results

3.3.1. Patient Demographics

Initial WHO grading for the 14 newly diagnosed tumors ranged from grades 2 to 4, consisting of both astrocytic and oligodendroglial tumors. The median age at first surgery (33.5 years; range 1–74 years) was considerably lower than that for gliomas in the average population at our institution, indicating the inclusion of WHO grade 2 gliomas and a tendency towards performing multiple surgeries in the younger age group.

3.3.2. Immunohistochemical Analysis

All 28 specimens showed punctate staining for LC3B, LAMP1, and LAMP2A in the cytoplasm of tumor cells. Punctate staining for LC3B and LAMP1 indicated the presence of autophagosomes. The size of the largest intracytoplasmic dots was about 1 μm (Fig. 2), approximating the size of autophagosomes.

The results of paired t-tests for semiquantitative analysis of the 3 stains are shown in the previous report.[16] A combination of LC3B, LAMP1 and LAMP2A stain scores, as well as the additive score of LC3B and LAMP1 stains were significantly increased in the post-TMZ group. However, analysis by single stains alone did not show a significant difference.

Caspase-3 staining was performed on all 28 specimens to monitor apoptosis. However, all specimens obtained before or after TMZ treatment showed very low positivity (< 2%).

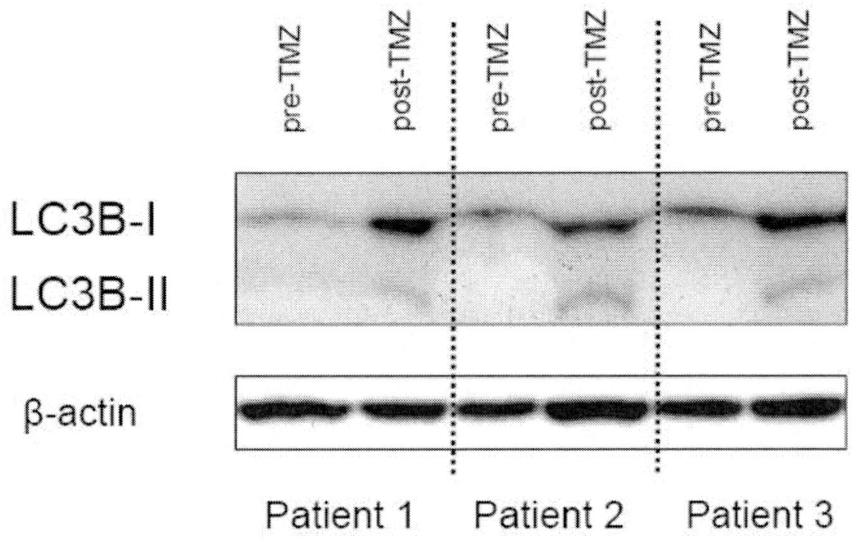

Figure 3. LC3B protein analysis by Western blotting was performed on six specimens obtained from three patients before and after temozolomide (TMZ) treatment. LC3B-I and LC3B-II protein increased in post-TMZ tissue samples. LC3B-I expression was higher than LC3B-II expression.

3.3.3. Western Blotting

LC3B expression was analyzed by western blotting performed on 6 frozen sections obtained from 3 patients, before and after TMZ treatment. Increased LC3B-I (cytosolic) and LC3B-II (autophagosomal) expression was observed in the post-TMZ group (Fig. 3). LC3B-I expression was higher than LC3B-II expression in all samples.

4. Discussion

4.1. The Role of Autophagy in Cancer

There has long been debate over whether autophagy is cell killing or cell protective in cancer.[3,9-11] The answer seems to be that autophagy can be both cell killing and cell protective under different circumstances.[9,10]

Autophagic pathways share molecules, such as Akt and mTOR, with the oncogenic pathways that are regulated receptor tyrosine kinases (RTKs). RTKs regulate two main pathways, the PI3K/Akt/mTOR pathway and Ras/RAF1/MEK1/2/ERK1/2 pathways (Fig 4). The PI3K/Akt/mTOR pathway inhibits autophagy, whereas the Ras/RAF1/MEK1/2/ERK1/2 pathway can both inhibit and promote autophagy. The oncogene *Ras* inhibits autophagy through activation of the class I PI3K pathway and activates autophagy through the selective activation of the Ras/RAF1/MEK1/2/ERK1/2 pathway. The PI3K/Akt/mTOR pathway activates cancer growth but inhibits autophagy, and the Ras/RAF1/MEK1/2/ERK1/2 pathway promotes cancer growth and autophagy. Autophagy induced by the inhibition of PI3K/Akt/mTOR pathway may cause growth inhibition and death of cancer cells, whereas autophagy induced by activation of the Ras/RAF1/MEK1/2/ERK1/2 pathway protects cancer cells from death.[3]

Tumor suppressors such as beclin-1, phosphatase and tensin homologue on chromosome 10 (PTEN), p53, tuberous sclerosis complex (TSC) 1 and 2 are known to stimulate autophagy.[1] These pathways and molecules are summarized in figure 4.

There are two potential therapeutic strategies for autophagy in anticancer treatment. First, to induce autophagy and enhance its antitumor effect, and second, to inihibit autophagy and induce apoptosis.[3] Chloroquine, an autophagy inhibitor, in combination with conventional anticancer therapies has produced encouraging results for glioblastoma multiforme treatment.[19] Suppression of autophagy using autophagy inhibitors is known to promote cell death by imatinib in chronic myelogenous leukemia[20,21] and by tamoxifen in lymphoma.[22]

4.2. Establishment of Battery of Tests to Precisely Assess Induction of Autophagy

The development of a reliable method to monitor autophagy in surgically obtained tissues is vital in elucidating the role of autophagy in cancer. Accepted methods for analyzing macroautophagy *in vitro* include electron microscopy, western blotting for LC3B, and GFP-

LC3B immunoflourescence.[23] Methods to characterize autophagy in glioma and other cells and tissues have been discussed.[24]

More than 40 proteins are known to be directly or indirectly associated with macroautophagy, e.g. proteins of the autophagy-related gene (Atg) family (i.e. Atg5-Atg12 complex, Atg8, Atg16) and LAMP family (i.e. LAMP1, LAMP2). These proteins enable immunohistochemical monitoring of autophagy in formalin-fixed, paraffin-embedded tumor tissues. As mentioned above, several pathways that regulate autophagy, such as the mTOR[25], ERK1/2[3,15] and p62[26,27] pathways, and several molecules associated with autophagy, such as PTEN, and p53,[3,4,10,28-30] have been identified, and their components are targets to evaluate autophagic activity. Precise methods to measure the induction of macroautophagy and CMA should be determined. At the present time, multiple assays are needed because no single assay satisfies all the criteria for assessing autophagy.[23]

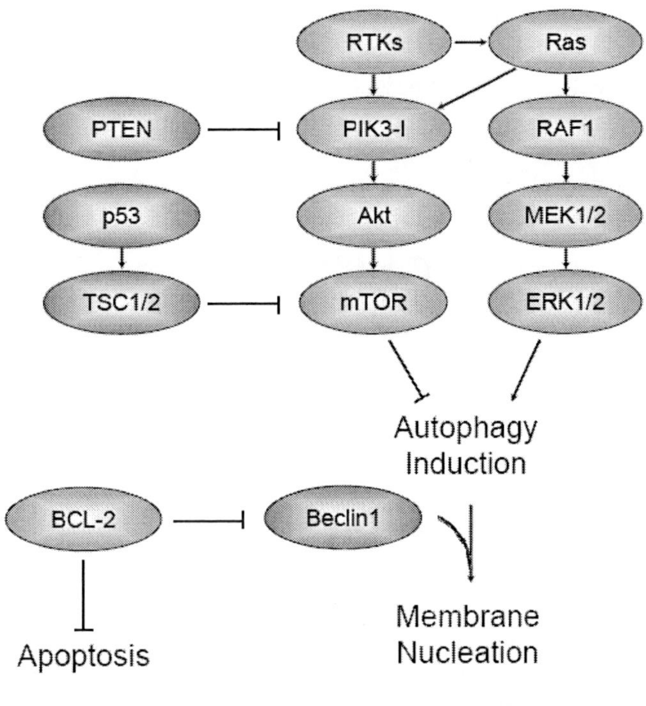

Figure 4. Schematic representation of pathways and molecules associated with autophagy. Tumor suppressors are shown in red. The oncogene *Ras* inhibits autophagy through activation of the class I PI3K pathway and activates autophagy through the selective activation of the Ras/RAF1/MEK1/2/ERK1/2 pathway.

Assessment of Induction of Autophagy in Surgical Tissues

As research for the manipulation of autophagy for the treatment of glioma and other types of cancer continue, the detection of autophagy in surgical tissues becomes vital. Immunostaining of Atg5[31] and LC3B,[8] as well as LAMP1 and LAMP2A,[17] are promising markers in evaluating glioma treatment. Immunohistochemistry of other proteins mentioned above, including p62, are promising candidates and should be tested in glioma samples.

Caution is necessary when interpreting the results of autophagy induction in surgical tissues. Other causes for induction other than TMZ can be postulated, such as the effects of radiation and starvation of tumor cells or regional ischemia in tumors with a high mitotic rate, for example glioblastoma.

4.3. Determining the Role of Chaperone-Mediated Autophagy (CMA) in Glioma Treatment

CMA is a uniquely selective form of autophagy by which specific cytosolic proteins are transported one-by-one across the lysosomal membrane for degradation.[32,33] LAMP2A is a protein expressed on the lysosomal membrane during chaperone-mediated autophagy.[34] The role of CMA in cancer treatment has not been extensively investigated,[35] although a recent report investigates its role in pancreatic cancer.[36] LAMP2A expression was analyzed by western blotting in 11 frozen specimens obtained from 8 patients before and after TMZ treatment. LAMP2A expression was noted in specimen obtained both before and after TMZ treatment (data not shown). We are presently conducting *in vitro* experiments to help understand the role of CMA in glioma.

Conclusion

The role of autophagy in cancer is multidimensional. To fully understand its role, the further establishment of feasible and effective methods to monitor autophagy *in vitro* and in surgical samples is vital.

References

[1] Stummer W, Pichlmeier U, Meinel T, Wiestler OD, Zanella F, Reulen HJ. Fluorescence-guided surgery with 5-aminolevulinic acid for resection of malignant glioma: a randomised controlled phase III trial. *Lancet Oncol.* 2006;7(5):392-401.

[2] Stupp R, Mason WP, van den Bent MJ, Weller M, Fisher B, Taphoorn MJ *et al.* Radiotherapy plus concomitant and adjuvant temozolomide for glioblastoma. *N. Eng. J. Med.* 2005;352(10):987-96.

[3] Kondo Y and Kondo S. Autophagy and cancer therapy. *Autophagy.* 2006;2(2):85-90.

[4] Kondo Y, Kanzawa T, Sawaya R and Kondo S. The role of autophagy in cancer development and response to therapy. *Nat. Rev. Cancer.* 2005;5(9):726-34.

[5] Kanzawa T, Germano IM, Komata T, Ito H, Kondo Y, Kondo S. Role of autophagy in temozolomide-induced cytotoxicity for malignant glioma cells. *Cell Death Differ.* 2004;11(4):448-57.

[6] Yao KC, Komata T, Kondo Y, Kanzawa T, Kondo S, Germano IM. Molecular response of human glioblastoma multiforme cells to ionizing radiation: Cell cycle arrest, modulation of the expression of cyclin-dependent kinase inhibitors, and autophagy. *J. Neurosurg.* 2003;98(2):378-84.

[7] Ito H, Daido S, Kanzawa T, Kondo S, Kondo Y. Radiation-induced autophagy is associated with LC3 and its inhibition sensitizes malignant glioma cells. *Int. J. Oncol.* 2005;26(5):1401-10.

[8] Aoki H, Kondo Y, Aldape K, Yamamoto A, Iwado E, Yokoyama T *et al.* Monitoring autophagy in glioblastoma with antibody against isoform B of human microtubule-associated protein 1 light chain 3. *Autophagy.* 2008;4(4):1–9.

[9] Marx J. Autophagy: Is it cancer's friend or foe? *Science.* 2006;312(5777):1160-1.

[10] Levine B. Cell biology: autophagy and cancer. *Nature.* 2007;446(7137):745-47.

[11] Morselli E, Galluzzi L, Kepp O, Vicencio JM, Criollo A, Mauri MC *et al.* Anti- and pro-tumor functions of autophagy. *Biochimica et Biophysica Acta.* 2009;1793(9):1524-32.

[12] Kroemer G and Levine B. Autophagic cell death: The story of a misnomer. *Nat. Rev. Mol. Cell Biol.* 2008;9(12):1004-10.

[13] Kroemer G, Galluzzi L, Vendenabeele P, Abrams J, Alnemri ES, Baehrecke EH *et al.* Classification of cell death: recommendations of the Nomenclature Committee on Cell Death 2009. *Cell Death Differ.* 2009;16(1):3-11.

[14] Daido S, Kanzawa T, Yamamoto A, Takeuchi H, Kondo Y, Kondo S. Pivotal role of the cell death factor BNIP3 in ceramide-induced autophagic cell death in malignant glioma cells. *Cancer Res.* 2004;64(12):4286-93.

[15] Aoki H, Takada Y, Kondo S, Sawaya R, Aggarwal BB, Kondo Y. Evidence that curcumin suppresses the growth of malignant gliomas in vitro and in vivo through induction of autophagy: role of Akt and extracellular signal-related kinase signaling pathways. *Mol. Pharmacol.* 2007;72(1):29–39.

[16] Takeuchi H, Kondo Y, Fujiwara K, Kanzawa T, Aoki H, Mills GB *et al.* Synergistic augmentation of rapamycin-induced autophagy in malignant glioma cells by phosphatidylinositol 3-kinase/ protein B inhibitors. *Cancer Res.* 2005;65(8):3336-46.

[17] Natsumeda M, Aoki H, Miyahara H, Yajima N, Uzuka T, Toyoshima Y *et al.* Induction of autophagy in temozolomide treated malignant gliomas. *Neuropathology.* 2011;31(5):486-93.

[18] Louis DN, Ohgaki H, Wiestler OD, Ed(s). In *WHO classification of tumours of the central nervous system.* 2007; International Agency for Research on Cancer (IARC), Lyon.

[19] Sotelo J, Briceno E, Lopez-Gonzalez MA. Adding chloroquine to conventional treatment for glioblastoma multiforme: a randomized, double-blind, placebo-controlled trial. *Ann. Intern. Med.* 2006;144(5):337-43.

[20] Bellodi C, Lidonnici MR, Hamilton A, Vignir Helgason G, Soliera AR, Ronchetti M *et al.* Targeting autophagy potentiates tyrosine kinase inhibitor-induced cell death in Philadelphia chromosome-positive cells including primary CML stem cells. *J. Clin. Invest.* 2009:119(5):1109-23.

[21] Salomoni P, Calabretta B. Targeted therapies and autophagy: new insights from chronic myeloid leukemia. *Autophagy.* 2009;5(7):1050-1.

[22] Amaravadi RK, Yu D, Lum JJ, Bui T, Christophorou MA, Evan GI *et al.* Autophagy inhibition enhances therapy-induced apoptosis in a Myc-induced model of lymphoma. *J. Clin. Invest.* 2007;117(2):326-36.

[23] Klionsky DJ, Abeliovich H, Agostinis P, Agrawal DK, Aliev G, Askew DS *et al.* Guidelines for the use and interpretation of assays for monitoring autophagy in higher eukaryotes. *Autophagy.* 2008;4(2):151-75.

[24] Jiang H, White EJ, Conrad C, Gomez-Manzano C, Fueyo J. Autophagy pathways in glioblastoma. In *Methods in Enzymology*: Autophagy in Disease and Clinical Applications. 453, Part C. Ed(s): 273-87. PMID: 9780123743107.

[25] Mathew R, Karantza-Wadsworth V, White E. Role of autophagy in cancer. *Nat. Rev. Cancer.* 2007;7(12):961-67.

[26] Komatsu M, Kurokawa H, Waguri S, Taguchi K, Kobayashi A, Ichimura Y *et al.* The selective autophagy substrate p62 activates the stress responsive transcription factor Nrf2 through inactivation of Keap1. *Nat. Cell Biol.* 2010;12(3):213-23.

[27] Moscat J and Diaz-Meco MT. p62 at the crossroads of autophagy, apoptosis, and cancer. *Cell.* 2009;137(6):1001-4.

[28] Ito H, Aoki H, Kuhnel F, Kondo Y, Kubicka S, Wirth T *et al.* Autophagic cell death of malignant glioma cells by a conditionally replicating adenovirus. *J. Natl. Cancer Inst.* 2006;98(9):625-36.

[29] Tasdemir E, Maiuri MC, Galluzzi L, Vitale I, Djavaheri-Mergny M, D'Amelio M *et al.* Regulation of autophagy by cytoplasmic p53. *Nat. Cell Biol.* 2008;10(6):676-87.

[30] Levine B. Unraveling the role of autophagy in cancer. *Autophagy.* 2006;2(2):65-6.

[31] Jiang H, Gomez-Manzano C, Aoki H, Alonso MM, Kondo S, McCormick F *et al.* Examination of the therapeutic potential of Delta-24-R GD in brain tumor stem cell: role of autophagic cell death. *J. Natl. Cancer Inst.* 2007;99(18):1410-4.

[32] Cuervo AM. Chaperone-mediated autophagy: selectivity pays off. *Trends Endocrinol. Metab.* 2009;21(3):142-50.

[33] Orenstein SJ, Cuervo AM. Chaperone-mediated autophagy: molecular mechanisms and physiological relevance. *Sem. Cell Dev. Biol.* 2010;21(7):719-26.

[34] Cuervo AM, Dice JF. Unique properties of lamp2a compared to other lamp2 isoforms. *J. Cell Sci.* 2000;113:4441-50.

[35] Kon M, Cuervo AM. Chaperone-mediated autophagy in health and disease. *FEBS Lett.* 2010;584(7):1399-404.

[36] Welsch T, Younsi A, Disanza A, Rodriguez JA, Cuervo AM, Scita G *et al.* Eps8 is recruited to lysosomes and subjected to chaperone- mediated autophagy in cancer cells. *Exp. Cell Res.* 2010;316(12):1914-24.

In: Autophagy: Principles, Regulation and Roles in Disease
Editor: Nikolai Gorbunov
ISBN: 978-1-61942-266-7

Chapter VII

Autophagy Pathways and Drug Targets for Cancer Therapy

Huai-long Xu[1*], Lei-lei Fu[1*], Yu Yang[2], Bo Liu[3], Xu Zhao[1], Wen-wen Li[1] and Jin-ku Bao[1]

[1] School of Life Sciences, Sichuan University, Chengdu 610064, China
[2] School of Life Science and Biopharmaceutics,
Shenyang Pharmaceutical University, Shenyang 110016, China
[3] State Key Laboratory of Biotherapy and Cancer Center,
West China Hospital, West China Medical School,
Sichuan University, Chengdu 610041, China

Abstract

Autophagy is evolutionarily conserved lysosomal mechanism mediating a variety of vital homeostatic processes, such as stress responses, remodeling, differentiation, secretion, innate and adaptive immune responses, as well as cell death. Therefore, it is not surprising the presence of a growing scientific interest in the role of autophagy in pathogenesis of a wide range of diseases, particularly of cancer. The autophagy is a multistage complex process and many of its pathways remain obscured. However, key regulatory elements such as autophagy-related genes (ATGs), class I and III PI3kinases, mTOR and p53 are suggested to be crucial in control of regulation of autophagic process in cancer. Often autophagy is portrayed as two-face Janus, facing opposite directions: past and future, death and life. This dual role of autophagy in determining the destiny of cancer cells makes it an attractive target for treatment with anti-cancer drugs such as a former one, rapamycin, and more recent tamoxifen, imatinib, gefitinib and chloroquine. In this light, a predictive analysis of potential targets for regulation of autophagy in cancer cells may give a fundamental basis for anticancer drug development.

Keywords: Autophagy; Cancer; ATGs; PI3Kinase; mTOR; p53; Rapamycin; Target; Drug

---

* These authors contributed equally to this work. Correspondence to: Tel./Fax: 86-28-85415171. E-mail addresses: liubo2400@163.com (Bo Liu); jinkubao@126.com (Jin-ku Bao).

1. Introduction

Autophagy, a term from Greek “auto” (self) and “phagy” (to eat), refers to an evolutionarily conserved catabolic process in which cell degrades misfolded and compromised proteins and damaged organelles, such as the endoplasmic reticulum, Golgi apparatus, and mitochondria. By these means autophagy serves as a critical adaptive response to starvation (amino acid and nutrient deprivation) and metabolic stress that results in nutrient recycling and energy boosting. Based on the mode of delivery to lysosome, three types of autophagy have been identified, namely macroautophagy, microautophagy and chaperone-mediated autophagy [1]. The main focus of the current chapter is macroautophagy, a comprehensively investigated pathway that will further referred as autophagy.

Autophagy is a multi-step process highly regulated by a limited number of conserved autophagy-related genes (ATGs). Originally discovered in yeast, ATG genes were reported to play vital roles in autophagosome formation and autophagy regulation, and so far they have been delineated the physiological mechanisms of autophagy and revealed the relations with several important pathological process, such as cancer initiation and progression [2]. Thus, a numerous data have shown that autophagy can play a multi-diverse role in cancer development and that under some conditions, autophagy can function as a tumor suppressor by eliminating damaged cells via so-called type-II programmed cell death (type-II PCD) [2]. In addition, autophagy can indirectly protect cells from carcinogenesis by maintaining genomic stability and homeostasis [3]. Paradoxically, the cytoprotective effects of autophagy can also contribute to tumor development under stress and can play the key roles in tumor cell resistance to chemotherapy while inhibition of autophagy can enhance the cytotoxicity of cancer chemotherapeutic agents [4, 5]. The findings from all these studies indicate the presence of intricate relationship between autophagy and carcinogenesis. The objective of this chapter is to provide crucial insights into development of new anticancer drugs aiming at autophagic molecular mechanisms.

2. Molecular Basis of Autophagy

The delineation of molecular mechanisms of autophagy originated from the year of 1993 when the autophagy-related genes (ATGs) were first identified in the yeast *Saccharomyces cerevisiae* [6]. It is well known that the molecular basis of autophagy is conserved from yeast to mammals, and the orthologs of most of the yeast ATGs have been discovered in mammalian cells. The complete autophagic flow is a highly regulated multi-step process which, in general, can be divided into several stages, including induction, vesicle nucleation, vesicle elongation and completion, docking and fusion, degradation and recycling [7, 8] (Figure.1).

Furthermore, autophagy can be induced by a variety of stimuli (e.g., nutrient deprivation, hypoxia, cytokines, hormones and DNA damage), and the induction of autophagy is, in most cases, associated with the mammalian target of rapamycin complex 1 (mTORC1) which is a central controller of cell growth [9]. In yeast, Tor signaling can negatively regulate Atg1 and Atg13 while the association of which is required for the early activation of autophagy [10].

The vesicle nucleation is mediated by activation of Class III phosphatidylinositol 3-kinase (PI3KCIII) which is an ortholog of Vps34 [11]. The activity of Vps34 is regulated by a complex consisting of Beclin-1 (a mammalian ortholog of Atg6) and the myristylated serine kinase Vps15/p150 [12]. Two PI3KCIII positive regulators, ultraviolet irradiation resistance-associated gene (UVRAG) and Bif-1, can enhance PI3KIII activity by interacting with Beclin-1 [13, 14]. In yeast, Atg14 can convert Vps34 complex to pre-autophagosomal structure (PAS) [15]. Although the mammalian ortholog of Atg14 has been discovered, its intricate mechanisms have not been clarified yet [16]. In the vesicle nucleation step, PI3P, produced by PI3KIII, plays an important role by binding and recruiting PX and FYVE domain-containing proteins such as Atg18-Atg2 complex [17, 18].

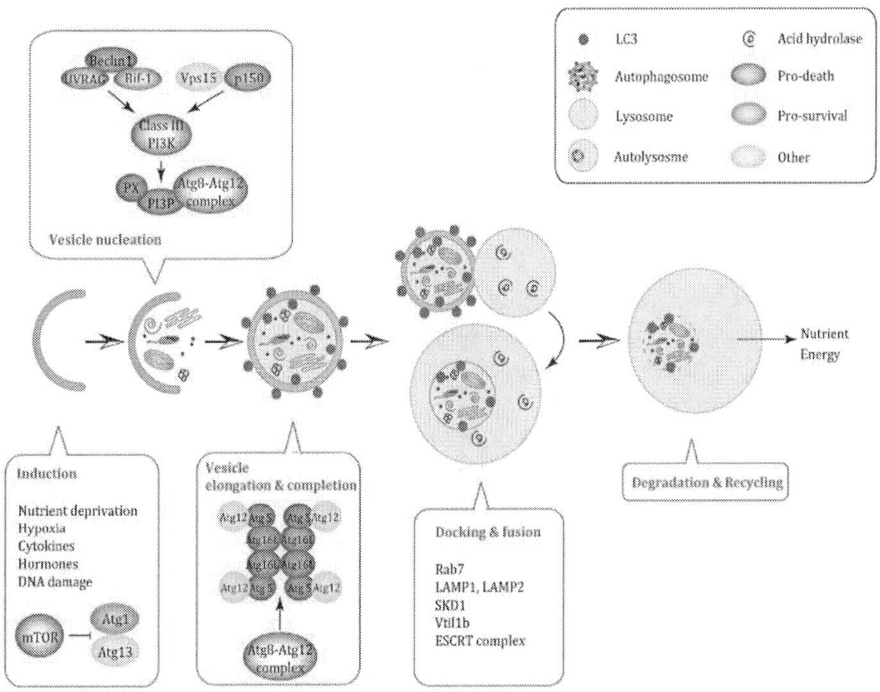

Figure 1. The stages of autophagy.

Autophagy is stimulated by nutrient deprivation, hypoxia, cytokines, hormones and DNA damage. The early stages of activation require Atg 1 and Atg13 which in turn can be inhibited by mTOR. Vesicle nucleation depends on Beclin-1-class III PI3K- Vps15 core complexes and other proteins. Vesicle elongation and completion mediated by the Atg16L complex and LC3. Docking and fusion refer to the maturation of autolysosome promoted by Rab7, LAMP1, LAMP2, SKD1, Vtil 1b and ESCRT complex. In the last stage, autophagosomal cargoes are digested, and then nutrient and energy are recycled.

The vesicle elongation and completion process require two ubiquitin-like cascades involving two ubiquitin-like proteins, Atg12 and Atg8 proteins, respectively [19, 20]. The Atg12-Atg5 conjugate can interact with a small coiled-coil protein, Atg16L (an ortholog of Atg 16 in yeast), to form an Atg12-Atg5-Atg16L complex. In conjunction with Atg12-Atg5, Atg16L directs the complex to autophagic isolation membrane that is essential for the second pathway and autophagosome formation [21]. The second ubiquitin-like pathway involves

LC3 (mammalian ortholog of Atg8) lipidation which plays an essential role in membrane dynamics during autophagy [22]. As soon as Atg8 is translated, the carboxy-terminal Arg residue of Atg8 is cleaved off by a cysteine protease, Atg4, exposing a critical Glycine residue at the C terminus [23]. Mediated by the E1 protein Atg7 and the E2 protein Atg3, Atg8 is then covalently conjugated to phosphatidylethanolamine (PE) through an amide bond between the C-terminal glycine and the amino group of phosphatidylethanolamine [22]. Lipidation of Atg8 converts the soluble Atg8 (mammalian LC3-I) into a membrane bound, autophagosome associated form (mammalian LC3-II), which is widely used as the biomarker of autophagy. In this manner, Atg8 proteins recruit lipid molecules to expand autophagosome membrane. Targeting Atg8 to isolate membrane needs the Atg12-Atg5-Atg16 complex [24]. Besides the Atg12-Atg5 conjugation can act as a ubiquitin-protein ligase (E3)-like enzyme for Atg8-PE conjugation reaction, promoting the lipidation of Atg8 [25]. Once the autophagosome expansion is complete, Atg8 detaches from PE under the assistance of Atg4 and then released to cytosol [23].

Autolysosome maturation further requires docking and fusion of autoghagosome with endolysosomal compartments and is accompanied by the breakdown of autophagosomal contents. Successful docking and fusion involve microtubules and endolysosomal molecules, such as small GTPase Rab7, lysosomal-associated proteins 1 and 2 (LAMP1 and LAMP2), AAA ATPase SKD1, SNARE protein Vti1b, and ESCRT complex [26, 27]. Some autophagy mediators such as UVRAG, functioning in the early steps, can regulate autophagosome maturation [28]. After docking and fusion, the autophagosomal cargoes are digested by acidic hydrolases, and then nutrient and energy are recycled. As mentioned above, mTOR signaling is inhibited during initiation of autophagy, but inactivated by prolonged starvation. Interestingly, numerous recent reports have indicated that increased mTOR activity may trigger the disassembly of autolysosome, and result in generating proto-lysosome tubules and vesicles that mainly contain lysosome membrane components extruded from autolysosomes [29]. Proto-lysosomes ultimately mature into functional lysosomes after restoring lumen proteins of lysosome in the cell [30]. Thus, this process named as autophagic lysosome reformation (ALR) cycle, is one of the key roles of autophagy through degradation of cellular content.

3. Survival or Death Signaling Pathways in Autophagy-Related Cancer Networks

3.1. Janus-Like Role of Autophagy in Cancer

Autophagy is the evolutionarily conserved degradative process to lysosome/vacuole resulted in irreversible elimination or turnover of intracellular constituents. In view of autophagic morphological change, double- or multiple-membrane-delimited autophagosome non-specific sequestrates of cytoplasm, and then fuses with lysosome/vacuolar for degradation. Besides housekeeping, autophagy also participates in different intracellular biological processes, for example, cellular differentiation, growth control, cell defense, tissue remodeling, acclimatization, and etc. Under nutrient deprivation or starvation situation, transferring of injured cells or cumulative damaged inclusions to self-degradative pathways serves as autophagic pro-survival mechanisms. But cell may suicide by means of pro-death mechanisms of autophagy, which is correlated with PCD closely, autophagic cell death is also known as type II PCD, which grapples with prolonging survival all the way [31].

From the standpoint of physiological mechanism, autophagy can act as a pathway of temporary survival for tumor cells. On one hand, autophagy is crucial for cell survival under extreme conditions through degradation of intracellular macromolecules, which may provide the energy required for minimal cell functioning when nutrients are deprived [32]. Also, autophagy-mediated elimination of altered cytosolic constituents, such as aggregated proteins or damaged organelles, preserves cells from further damages, indicating that autophagy plays a protective role in early stages of cancer. On the other hand, autophagy plays a death-promoting role as type II programmed cell death (type II PCD), suggesting a *bona fide* tumor suppressor mechanism in cancer [33]. That is to say, autophagy plays different roles in cancer as either guardian or executioner; therefore, these end-effects depend on different stages of cancer, or different extracellular environments. As mentioned above, these autophagic pro-survival or pro-death genes or proteins would be integrated into the cancer networks (Figure 2).

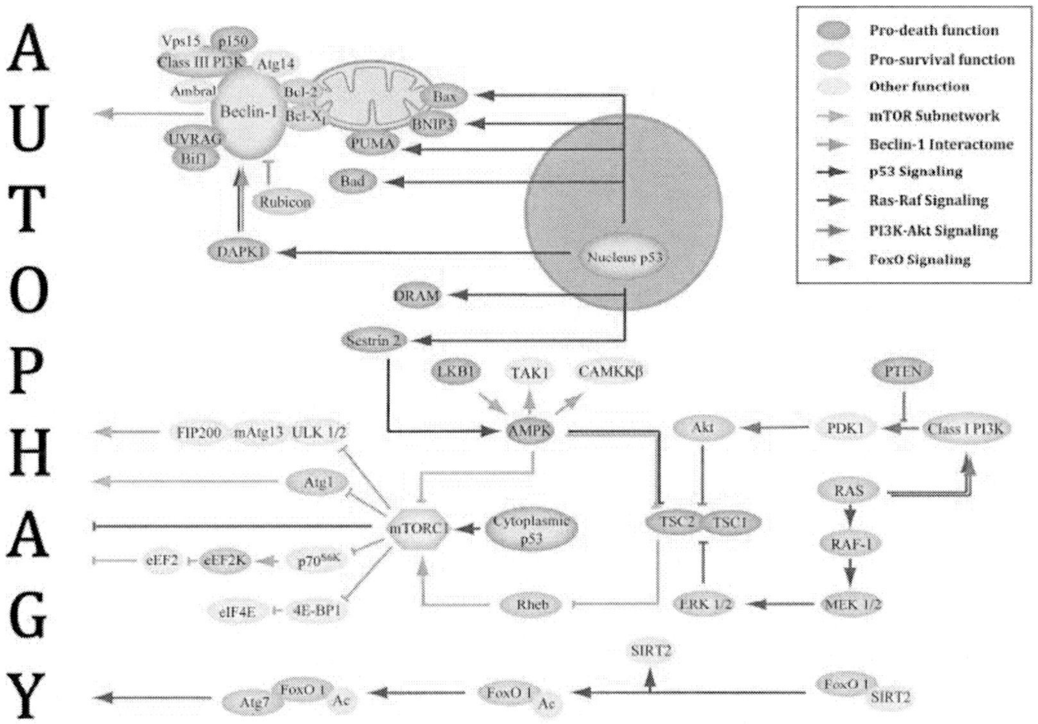

Figure 2. Core signaling networks involved in autophagy regulation.

The proteins with pro-death function are presented in red, and the proteins with pro-survival function are presented in green. Besides, the proteins, whose functions are not known, are shown in yellow. The core signaling networks include mTOR subnetwork, p53 signaling, Beclin-1 interactome, Ras-Raf signaling, PI3K-Akt signaling and FoxO signaling.

3.2. mTOR Subnetwork

The target of rapamycin TOR (mTOR in mammals), an evolutionarily conserved serine/ threonine kinase, can regulate autophagy as well [33]. Two types of mTOR exist in mammals, but we only focus on mTORC1 which contains a raptor due to its sensitivity to rapamycin [34]. mTOR1 plays a pivotal role in autophagy repression; it integrates different signal inputs including growth factor signaling, cytokines, nutrient and metabolic stresses. Three major mTORC1-inducing pathways have been figured out as follows: PI3K-Akt pathway and extracellular signal-regulated kinase (ERK)-90-kDa ribosomal S6 kinase (RSK)-death-associated protein kinase (GAPK) kinase cascades can activate mTORC1 inhibitory pathway via activation of AMP-activated protein kinase (AMPK) and glycogen synthase kinase 3 (Gsk 3) [35]. The signaling pathways promoting mTORC1 activity is often induced by oncoproteins and /or loss of tumor suppressors. Therefore, mTORC1-inhibited autophagy is often observed in cancer cells.

TSC2/TSC1 tumor suppressor complex existing in various cancer types is a key-point at the upstream of mTORC1 which is converged by many signaling pathways regulating autophagy [36]. It can suppress mTORC1 by inactivating mTORC1-interacting protein, Rheb. Akt, as well as ERK mitogen activated protein kinase (MAPK), and its downstream effectors can similarly destabilize TSC2, disrupt its interaction with TSC1 through phosphorylation and allow Rheb activation of mTORC1 [36, 37]. However, phosphorylation of TSC2 by AMPK can increase its GTPase activity, stabilize the TSC2/TSC1 complex, and inactivate Rheb, resulting in inactivation of mTORC1 to trigger autophagy. AMPK is also regarded as an intracellular energy status sensor to detect and respond to the changes in AMP/ATP ratios. Serine-threonine kinase, liver kinaseB1(LKB1) which is a tumor suppressor mutated in Peuz-Jeghers syndrome is activated by high intracellular AMP level to phosphorylate AMPK, thus offering nutrient to starved cells by inducing autophagy [38]. However, the initiation of autophagy by AMPK does not only include the starvation-induced activation of LKB1. When encountering the cytosolic Ca^{2+} concentration and cytokines via Ca^{2+}/calmodulin-dependent kinase kinase-b (CaMKKb) or transforming growth factor-beta-activated kinase1 (TAK1) pathways, AMPK can mediate autophagy while genotoxic stresses can trigger AMPK-mediated autophagy in a p53-dependent manner [39, 40].

A number of autophagy-related genes and proteins involved in cell physiology and cancer pathology comprise downstream executors of mTOR. Under nutrient rich condition of yeast, mTORC1 hyper phosphorylates Atg13, reducing its association with Atg1 thus attenuating Atg1 activity which is essential for autophagy initiation [10]. In mammals, uncoordinated 51-like kinase1 (UK1) and UK2 are two homologs of Atg1 and scaffold protein FIP200 have been identified [10, 41]. Under nutrient starvation condition, mTORC1 disrupts the binding of Atg13 with ULK and destabilizes ULK inhibiting the ULK-dependent phosphorylation of FIP200 and autophagy induction by phosphorylation of ULK and Atg13. Moreover, activated FIP200 localizes ULK to PAS where it regulates vesicle nucleation by recruitment of other Atg proteins to that location [42]. Summarizing, mTORC1 establishes a link with the core autophagy machinery in mammals thus regulating autophagy more directly with the help of ULK complex.

mTORC1 can also mediate protein translation and cell growth through 4E-BP1 and $p70^{S6K}$ phosphorylation to regulate autophagy. Phosphorylation of 4E-binding protein 1 (4E-BP1) leads to its detachment from eukaryotic translation initiation factor 4E (eIF4E) and upregulates cap-dependent translation [43]. Further detailed information can illustrate that mTORC1 certainly plays a central part in autophagy-regulated network. It is obvious that a huge number of oncogenes and tumor suppressors can be involved in mTORC1-centered signaling network which indicates a close connection between those oncogenes and tumor suppressors with cancer.

3.3. Ras-Raf-MAPK Signaling

The mitogen-activated protein kinase (MAPK) pathways are belong to the primordial signalling systems. In MAPK pathways, a G-protein works upstream of a core module, which consists of three kinases: a MAPK kinase kinase (MAPKKK) phosphorylates and activates a MAPK kinase (MAPKK), which, in turn activates MAPK [44]. The Ras-Raf signaling features Ras as the G-protein, Raf as MAPKKK, MEK as MAPKK and ERK as MAPK. The Ras, Raf, MEK (mitogen-activated protein kinase/ERK kinase), ERK (extracellular-signal-regulated kinase) pathway is at the heart of signaling networks that regulate proliferation, differentiation and cell survival [45].

Both Ras and Raf are proto-oncogenes targeting cell proliferation transformation, differentiation, apoptosis and etc [46]. Diversified hormones, growth factors, differentiation factors, and tumor-promoting substances can activate Ras proteins through inducing the exchange of GDP with GTP. Functioning as an adapter, the activated Ras binds to Raf kinases, which causes their translocation to the cell membrane. Raf protein is a serine/threonine-specific kinase that integrates the upstream input signals, and participates in regulation of different biological processes [47]. There are three kinds of Raf proteins in mammals: Raf-1, A-Raf and B-Raf, only the ubiquitously expressed Raf-1 has been well studied. As the predominant MEK activator in most cell types, Raf kinase can activate both MEK-1 and MEK-2 by phosphorylation of two serine residues in the activation loop. And both MEK isoforms can activate the downstream ERK kinases. MEK belongs to the rare breed of dual-specificity kinases which can phosphorylate both threonine and tyrosine residues [48]. The MEK isoforms activate ERK-1 and ERK-2 via selectively phosphorylating the Tyr and Thr regulatory sites to improve accuracy of signaling transduction and avoid wrong activation of ERK. The pathways of ERK involves in the core signaling subnetwork of regulating cell growth, development and division, as well as, the Ras-Raf-MEK/Erk pathways always function as promoting cell survival in cancer.

3.4. PI3K-Akt Signaling

In human cancer, PI3K-Akt pathway is considered as the key regulator of a large number of cellular processes [49]. Belonging to a large family of PI3K-related kinases or PIKK, phosphoinositide 3-kinase (PI3K) acts as an effector of PtdIns (4, 5) P2 signaling cascade. The PI3K family consists of eight different isoforms which can be divided into three classes based on the sequence homology and substrate preference, namely class I PI3K, class II PI3K, and class III PI3K. The class III PI3K VPS34p produces the majority of the cellular PtdIns-3-P and involves in protein trafficking through the lysosome [50]. As a lipid phosphatase, PTEN (phosphatase and tensin homolog deleted on chromosome 10) antagonizes signaling from the PI3K pathway by dephosphorylation of the D3 position of PtdIns-3, 4-P2 and PtdIns-3,4,5-P3, the lipid products of the PI3K lipid kinase activity. Besides, PTEN can also dephosphorylate tyrosine-, serine-, and threonine-phosphorylated peptides, the activity of which dependent on regulation of cell adhesion and spreading.

The key downstream effector of PI3K is the serine-threonine kinase Akt [51]. After PI3K activation, Akt phosphorylates and regulates the activity of a number of targets including kinases, transcription factors and other regulatory molecules. Activation of Akt is a multi-step process involving both membrane binding and phosphorylation [52]. Then, the activation needs a conformational change and phosphorylation on two residues where one of phosphorylation sites lies in the Akt kinase domain activation loop and is phosphorylated by another PH-domain containing protein, PDK1. In cancer cells, PDK1 activates Akt and subsequently downregulates the expression of TSC1/TSC2 complex, thereby playing a pro-survival role [53].

3.5. p53 Signaling

Transcriptional factor p53 is widely regarded as tumor suppressor protein. When cells are exposed to a variety of stress-factors producing DNA damage, oxidative injury and etc, p53 can become activated. However, it has been shown recently that p53-gene is spontaneously mutated in more than 50% of human cancers. In mammalian cells, p53 is represented in two pools, namely cytoplasmic p53 and nucleus p53. Following activation, cytoplasm p53 translocates to the nucleus and regulates target gene expression, causing DNA repair, cell cycle arrest and apoptosis. Although recent reports indicate that multiple p53 target genes stimulate autophagy, the role of p53 in autophagy seems to be paradoxical depending on its different subcellular localization [35].

In nucleus, p53 abets autophagy mainly by interacting with its targets, damage-regulated autophagy modulator (DRAM) and sestrin 1/2. DRAM, a p53 target gene encoding a lysosomal protein, which induces macroautophagy, is found to be an effector of p53 mediated cell death pointing at the direct link between p53 and autophagy [54]. In response to DNA-damaging agents, DRAM can trigger autophagy under the control of p53. It has been recently reported that DRAM can be down-regulated by a self-hypermethylation mechanism toward epithelial cancer cells. DRAM, as a tumor suppressor, triggers autophagy under the control of p53 in response to DNA damage agents. Also, other targets of p53, Sestrin1/2, whose expressions are usually induced upon DNA damage and oxidative stresses, are negative regulators of mTORC1, and execute their function through activation of AMPK and TSC2. Consistently, disruption of Sestrin1/2 in mice attenuated its ability to inhibit mTORC1 [55]. In addition, disruption of Sestrin2 in mice attenuated its ability to inhibit mTOR signaling upon genotoxic challenge [56]. Therefore, the exploration of molecular mechanisms involving sestrin1/2, demonstrates a connection between p53 and autophagy that can be regarded as p53-mediated tumor expression via mTORC1 signaling pathways.

On the other hand, cytoplasmic p53 can inhibit autophagy via transcription-independent mechanism [55]. When some mammalian cells are subjected to knockout, knockdown or pharmacological inhibition of p53, depletion or inhibitions of p53 can encourage autophagy. Moreover, under the condition of nutrient depletion and hypoxia, autophagy may prefer to survival of the p53-deficient cancer cells, indicating the involvement of autophagy in cancer-associated dysregulation of p53 and its cytoprotective role [35].

3.6. Beclin-1 Interactome

Beclin-1, the mammalian homolog of yeast Atg6 and a Bcl-2 interacting coiled-coil protein, enhances autophagy by combining with PI3K III at the initiating stage of autophagy. It has been reported that Beclin-1 is mono-allelically deleted in 75% ovarian, 50% breast and 40% prostate cancers [57, 58]. There are numerous observations that autophagy-promoting activity of Beclin-1 is highly linked with its tumor suppression function [57, 59]. An evolutionarily conserved domain (ECD) of Beclin 1 can interact with PI3III/Vps34 thus helping Beclin-1 to enhance autophagy while the lack of ECD can result in inability of Beclin-1 to improve autophagy as well as the loss of its tumor suppressor function. These findings suggest that autophagy promotion and tumor inhibition function of Beclin-1 are interconnected. Additionally, the mediators that are implicated in tumorigenesis can demonstrate the tumor suppressor function of Beclin-1. By interacting with Beclin-1 via their coiled-coil domain, UVRAG, a major Beclin-1 positive mediator, can markedly enhance PI3KC3 lipid kinase activity to facilitate autophagy [28]. Moreover, UARAG is also mutated in various human colon cancer cells and tissues [60]. After promoting autophagy by mediating Beclin-1-PI3III complex, UVRAG can inhibit tumorigenesis of human colon cancer cells Bif-1 [28], another Beclin-1 activator, which can interact with Beclin-1 through UVRAG to regulate autophagy and suppress tumorigenesis [35]. The loss of Bif-1 can result in the extending development of spontaneous tumor in mice while downregulation of Bif-1 is found in many kinds of cancer [14, 61]. Although details of the Bif-1 mechanisms in suppressing tumor have not been fully understood, we may expect that the tumor inhibitory activity of Bif-1 is associated with its roles in regulating autophagy. There are also some negative regulators of Beclin-1 such as Bcl-2 proteins that we are going to discuss below. Summarizing, Beclin 1 can be a crucial part in autophagy and tumorigenesis by forming a Beclin-1-PI3KIII/Vps34 complex, mediated by its positive regulators, UVRAG and Bif-1, as well as its negative regulator Bcl-2.

3.7. FoxO1 Signaling

The mammalian FoxO transcription factors, i.e., FoxO1, FoxO3a, FoxO4 and FoxO6, are homologues of the *Caenorhabditis elegans* transcription factor abnormal Dauer Formation-16 (DAF-16). Besides binding to the same DNA target sequences, most FoxO proteins also exert influence on crossover patterns of gene expression. In mammals, FoxO members can serve as potential tumor suppressors [62], due to the ability to induce cell cycle arrest, DNA repair and apoptosis.

The cytosolic FoxO1, a fork-head O family protein, is a mediator of autophagy. Under stress conditions of stress (e.g., serum starvation, oxidative stress), FoxO1 dissociates from a complex with sirtuin-2 (SIRT2) that is followed by its acetylation. The acetylation status of FoxO1 is balanced by histone acetylases and HDAC. While SIRT1 is located in the nucleus, SIRT2 is primarily located in the cytosol, and is the main deacetylase of cytosolic FoxO1.

Then, the acetylated FoxO1 binds to the E1-like Atg7; Atg7 plays an important role in ubiquitin-like conjugation (Ublc) systems for autophagosome expansion [63]. The Foxo1-Atg7 complex influences the autophagic process, which culminates in cell to death. This FoxO1-modulated cell death is associated with tumor suppressor activity.

4. The Key Proteins in Autophagic Networks

4.1. p53

Encoded by the TP53 gene, p53 is a tumor suppressor protein which regulates the cell cycle, and thus, functions preventing cancer. The p53 plays a vital role in conserving stability by preventing genome mutations [64, 65].

In the great majority of human cancer cells, p53 is frequently functionally inactive, mutant or defective. Several proteins can interact with p53, which leads to lose the normal biological function [66]. DNA tumor viruses (e.g., HPV16, 18, SV40 and adenovirus encoding tumor proteins) can cause malignant pathological changes, these tumor proteins, such as SV40T antigen, adenovirus ELa, adenovirus ELb and HPVE6, can bind to Rb and p53. After binding to p53, the proteins like HPVE6 degrade p53 by intracellular enzymes. Besides, the gene products, such as MDM2, inactivate p53 through binding to p53.

The p53 gene mutation can be divided into 3 forms: the zero mutation, the negative mutation, and the positive mutation. The p53 zero mutant has no new regulatory functions. The p53 negative mutant loses the function of negative regulation, and inactivation of the wild type, which do not have direct relationship with tumor. The p53 positive mutant gains the transferring function, and may take the place of oncogenes in carcinogenesis [67]. In the inactivation mechanism, the wild type p53 binds to specific site as tetramer, and then trans-activates the downstream expression of growth suppression genes.

4.2. Beclin-1

Beclin-1, the mammalian homolog of Atg6, is defined as a myosin-like Bcl-2 interacting coiled-coil protein. It is localized on Golgi apparatus and Golgi membrane. The protein is required for several biological processes such as autophagy, cytokinesis and cell cycle regulation; it may also participate in antiviral host defense [59]. The autophagic function of Beclin-1 is evolutionarily conserved from yeast to human, and involves the binding of Bcl-2 to other autophagy regulators, such as Vps34 (PI3K), p150, UVRAG, Bif1, Atg14L and Rubicon, to form protein complexes and initiate the formation of double-membrane autophagosome [68].

Beclin-1 is encoded by the haploinsuffcient tumor suppressor gene BECN1. Allelic BECN1 deletion is frequently observed in different kinds of human cancers and leads to tumor development, whereas ectopic BECN1 expression restores autophagy, inhibits tumorigenesis.

After dissociation of BECN1 from Bcl-2, autophagy is activated, in response to nutrient limitation, which depends on Bcl-2 phosphorylation. BECN1 and Bcl-2 complex formation is primarily dependent on the BH3 domain of BECN1 and the BH3-receptor domain of Bcl-2. By Bcl-2 knockdown, BH3-mimetics or JNK activation, pharmacologic interference with this interaction results in autophagy induction [69].

4.3. Bcl-2 Family

Bcl-2 family, a representative of Beclin 1 negative regulators, which consists of B cell CLL/lymphoma-2 (Bcl-2) as well as its relatives, is originally defined as a controller of outer mitochondrial membrane integrity and apoptosis. Those included proteins are functionally classified as either antiapoptotic or proapoptotic proteins [35]. Antiapoptotic Bcl-2 members (e.g., Bcl-2 and Bcl-XL) contain four Bcl-2 homology domains (BH), while proapoptotic Bcl-2 proteins can be effector proteins (e.g., Bax and BAK) which contain three BH domains or the BH3-only proteins (e.g., BAD and Noxa). Bcl-2, the oncogene which negatively regulates Beclin 1-PI3KIII/Vps34, plays an important role in suppressing autophagy. Beclin-1 provides a BH3 domain which sustains Bcl-2 binding affinity to Beclin-1, therefore when interacting with Beclin-1, Bcl-2 can block Belin1 interaction with PI3KIII/Vps34, decrease PI3KIII activity and downregulate autophagy [70, 71]. Accordingly, mutations in BH3 domain of Beclin-1or BH3 receptor domains of Bcl-2 would abolish the Bcl-2 mediated inhibition autophagy [71]. Although the precise mechanism by which Bcl-2 blocks Belcin1 and PI2KIII/Vps 34 interaction is unclear, the binding of Bcl-2 with Beclin-1 seems to be constitutive and its detachment from Beclin-1 is of great importance in autophagy induction. Two models are proposed to explain the disassociation of Beclin-1 with Bcl-2. The second model proposes a mechanism involving replacement of Beclin-1 in the Belcin1/Bcl-2 complex by BH-only proteins [35]. The typical example involves BAD, which when facing starvation, can competitively displace Beclin1 from Bcl-2 [69].

4.4. mTOR

Encoded by the FRAP1 gene, mTOR is a serine/threonine protein kinase targeting regulation of cell growth, cell proliferation, cell motility, cell survival, protein synthesis, and transcription [35]. As a member of phosphatidylinositol 3-kinase-related kinase protein family, mTOR is postulated to serve as a multichannel processor that integrates different signal inputs. Growth factors, cytokines, nutrient and metabolic stresses are all the signals participating in mTOR pathway. mTOR inactivation/depletion, rapamycin treatment and starvation can elicit varieties of responses, covering gene transcription, protein synthesis, autophagy induction, cell proliferation, cell growth, cell cycle arrest at the G1/S boundary, then entry into G0 [73]. There are two forms of mTOR: mTORC1 and mTORC2. Due to the sensitivity to rapamycin, mTORC1 plays a more important role than mTORC2 does [74]. mTORC1 is characterized by the classic features of mTOR by functioning as a nutrient/energy/redox sensor and controlling protein synthesis, its activity is stimulated by insulin, growth factors, serum, phosphatidic acid, amino acids (particularly leucine), and oxidative stress [75]. mTORC2 is insensitive to rapamycin, functioning as an important regulator of the cytoskeleton through its stimulation of F-actin stress fibers, paxillin, RhoA, Rac1, Cdc42, and protein kinase C α (PKCα). mTORC2 also appears to possess the activity of PDK2. mTORC2 appears to be regulated by insulin, growth factors, serum, and nutrient levels. Originally, mTORC2 was identified as a rapamycin-insensitive entity, while acute exposure to rapamycin did not affect mTORC2 activity or Akt phosphorylation [76].

4.5. DAPK

Death-associated protein kinase (DAPK), a stress-regulated protein kinase, can mediate a wide range of processes including signal-induced cell death and autophagy [77]. DAPK is a large 160 kDa protein composed of several functional domains including a kinase domain. The autophagy signaling activity can be either pro-survival or lead to or join in cell death. In some cell types, the over expression of DAPK can lead to the appearance of autophagic vesicles [78]. Although little is known about how DAPK exerts its effects on autophagy, recent studies identified MAP1B, a major component of the neuronal cytoskeleton, as a DAPK interacting protein that exerts effects as a positive cofactor in DAPK-mediated autophagic vesicle formation and membrane blebbing [79]. During the induction of membrane blebbing, the co-transfection of both genes stimulated the disruption of microtubules, which indicates MAP1B-DAPK-induced blebbing can involve changes in both the dynamics of microtubules as well as of contractile cortical actin [79]. However, we need further studies to demonstrate whether MAP1B is a key factor that switches DAPK activity towards autophagy induced by certain stresses such as INF-γ [77]. A recent report demonstrated that a constitutively activated form of DAPK triggers autophagy in a Beclin-1-dependent manner [80]. Phosphorylating velin-1 on the Thr 119 located at crucial position within its BH3 domain, DAPK can promote the dissociation of belin-1 from Bcl-X_L and the induction of autophagy. Therefore, DAPK can act as one of the core proteins in the autophagy and the new substrate provides a new phosphorylation-based mechanism illustrating that DAPK can activate autophagy by reducing the interaction of Beclin-1 with its inhibitor Bcl-X_L. DAPK may also be able to regulate autophagy via syntaxin-1A because the former one can phosphorylate the latter which is the key component of the soluble N-ethylmaleimide-sensitive factor (NSF) attachment protein receptors complex essential for synaptic vesicles docking and fusion [81]. It is acknowledged that autophagy contribute to cell death in process termed type II PCD and DAPK can be defined as an important component in ER stress-induced cell death. Irreparable damage to ER can cause ER-specific cell death mechanisms [82]. In addition, the DAPK-dependent signaling can regulate starvation-associated autophagy via mTORC1 mechanism [83].

5. Autophagy in Human Diseases

Autophagy, an intracellular degradation pathway of impaired organelles and long-life proteins, has drawn increasing attention for the past several decades. Although first observed in the late 1950s, the intricate relevance between autophagy and human physiology still has not been clearly unfolded. As a matter of fact, those connections still remained to be researched until quite recently [84]. With the identification of autophagy-related genes (ATGs) in yeast as well as higher eukaryotes, an increasing number of evidence has been emerged, indicating that autophagy involves in various aspects of human health and plays a crucial role in a wide range of human diseases [85]. Several serious human diseases, including lysosomal storage diseases, neurodegenerative diseases and cancer, could be ascribed to autophagic defect. Meanwhile, autophagy has been reported implicating in the innate and adaptive immune response to various types of pathogens infection and the mutations in ATG genes are deemed to repeal autophagy and subsequent cause diseases. Also, deregulation of autophagy can seriously affect human health [86].

5.1. Autophagy in Cancer: An Executioner or Guardian?

A seeming paradox is that autophagy functions primarily in cancer cell survival, and yet also in cell death. To alleviate this puzzle, two hypotheses have been put forward. One hypothesis suggests that the role of autophagy varies depending on different stages of tumor development. Autophagy limits tumor formation in the early stage, while favors tumor cell survival and invasion as soon as cancer has formed [87]. The other hypothesis indicates that autophagy regulates carcinogenesis in a cell- and tissue- specific manner. At molecular level, autophagy plays the pro-survival or pro-death role by regulating some tumor suppressor genes including Beclin 1, ultraviolet irradiation resistance-associated gene (UVRAG), Bax-interacting factor-1 (Bif-1) and p53, or other oncogenes including B-cell lymphoma 2 (Bcl-2), Ras, Class I PI3K (PI3KI) and protein kinase B (Akt) in cancer [88]. As mentioned above, a number of core signaling pathways of survival/death would be integrated into the autophagy-related cancer networks that may ultimately seal the fate of cancer cell.

5.1.1. The Two-Faced Role of Autophagy in Cancer

On one hand, the observation of reduced autophagy and the absent of autophagic genes in different kinds of cancers showed that autophagy, as a pro-death mechanism, may inhibit cancer. Beclin-1, the mammalian homologue of the yeast Atg6, which is a crucial factor in the early stage of autophagosome formation and has a Bcl-2 homolgy-3(BH3) domain, was rediscovered in a yeast two-hybrid screen of an adult mouse brain interacting to Bcl-2 [89]. Monoallelic deletion of Beclin-1 has been identified in 40% (9/22) of breast cancer and the Beclin-1 gene maps to a tumor-susceptibility locus on human chromosome 17Q21 that is monallelically deleted in up to 75% of ovarian cancer, 50% of breast cancer and 40% of prostate cancers [90]. However, over expression of Beclin-1 in a breast cancer cell line increase autophagy and decrease the growth of tumorigenicity of these cells [59]. It indicates that Beclin-1 and some other autophagy genes may function as tumor suppressors. Bcl-2 and Bcl-X_L, which are known as anti-apoptotic proteins, may inhibit autophagy by blocking the binding of Vps34 to Beclin-1 [70], demonstrating these proteins promote tumorigenesis not only by blocking apoptosis, but also by inhibiting autophagy. Furthermore, it is showed the UV irradiation resistance-associated gene protein (UVRAG) also can bind to Beclin-1 and PI3K to promote autophagosome formation and inhibit human colon cancer cell proliferation and tumorigenicity [13]. Some other known suppressors like PTEN also enhance autophagy by decreasing the concentration of class I PI3K products [91] (phosphatase and tensin homolog deletion chromosome) and down-regulation of mTOR which increase protein synthesis, tumor cell proliferation and inhibit autophagy through mTORC1 [92].

Some kinds of metabolic stresses in autophagy-deficient tumor cells lead to genome damage [93, 94]. This can be provoked by damage to proteins, organelles, and/or DNA or by low level of ATP available for basic cellular functions. Genetic instability and an enhanced mutation rate will promote tumor-cell evolution and resistance to therapy. While DNA damage facilitates both cancer and ageing, autophagy plays a general role in protecting cellular and genome fitness to prevent cancer and extend lifespan.

On the other hand, the investigation of high level of autophagy in cancer indicates autophagy, as a pro-survival mechanism to tumor cells, providing nutrients during starvation [2] by "self-eating" and helping tumor cells survive from hypoxia, oxidative damage or other stress. Experiments with melanoma biopsies show high level of autophagy in response to chemotherapy and it seems more aggressive when it growing to xenografts [95]. In addition, further studies have revealed that autophagy may help cells adapt survival mechanism, protecting cells from cytotoxicity and radiation [96].Thus, defective autophagy, which is found in many types of tumors, is more susceptible than others to necrotic cell death in response to hypoxia and nutrient deprivation. Thus, inhibition of autophagy may increase susceptibility to radiation or chemotherapy. Stimulation of autophagy in tumor with constitutive activation of PI3K/AKT pathway can render them more susceptible to apoptosis particularly in the absence of p53 and p27 [97]. Genetic inactivation of autophagy, either indirectly by constitutive activation of the phosphatidylinositol 3-kinase (PI3K) pathway or directly by allelic loss of Beclin 1 or deficiency in Atg5, or by RNA interference (RNAi), decreases survival under metabolic deprivation even when apoptosis is inactivated [93, 98]. So the roles of autophagy in cell damage control and its response to stress are equally important [3].

5.1.2. microRNAs Act as Regulators of Cancers through Autophagy

MicroRNAs, a class of conserved short single stranded RNA with the ability of not coding proteins but regulating the expression of genes, have been recognized to play a critical role in cell survival and death. MicroRNAs (miRNAs) are small endogenous non-coding RNAs which regulate the gene expression at the post-transcriptional level. Recently, miRNAs have been recognized to play a critical role in cell survival and death.

And, many researches are being done to relate miRNAs to apoptosis and autophagy, involved in cancer. Beclin-1, the positive regulator of autophagy, was firstly found relevant with miRNAs. And, miR-30a inhibits the expression of Beclin-1 to modulate autophagy [99].

In the further researches, miR-130, 98, 124, 204 and 142 were found having functions on autophagy-lysosomal via regulating the expression of genes SREBP1, USF, AP-1and NFE2 at posttranscriptional level [100]. MiR-196 is also found overexpressed in Crohn's disease deregulating IRGM-dependent xenophagy, down-regulating IRGM. And IRGM is supposed to promote autophagy by stimulating mitochondrial fission [101]. Recently, an experiment showed that formation of LC3-II, a marker of autophagy progress, can be inhibited by miR-204 in cardiomyocytes during hypoxia-reoxygenation [102]. This suggests that miR-204 may regulate autophagy in the same way as to miR-34a. Thus, the generic effects of microRNAs in autophagy can involve regulation of expression of key autophagic proteins. Generally, miRNAs effect on autophagy mainly through regulating the expression of important autophagic regulator proteins.

5.1.3. The Connections between Apoptosis and Autophagy in Cancer Progression

Autophagy, a kind of programmed cell death (PCD), is crucial in normal physiology and pathology. Positive or negative links between apoptosis and autophagy in some specific cases provide the possibility of 'molecular switch' between them. p53, an important common regulator of apoptosis and autophagy, positively regulates apoptosis under irremediable damages, while plays a dual role in autophagy depending on different subcellular localizations. When faced with genomic instability or oncogenic activation, pro-autophagic effect of p53 can be activated. While in the p53 defective mutations, p53 is down-regulated inhibiting autophagy in the condition of starvation or ER stress. PI3K/Akt/mTOR is another vital cellular pathway involved in both apoptosis and autophagy. Recent reports have shown that Akt inhibits apoptosis by phosphorylation of Bad, a member from Bcl-2 proteins family, allowing for cell survival. Proteins from the Bcl-2 family, associated with Beclin-1 function as inducers or inhibitors on autophagy. Under specific stimuli, BH3-only proteins like Bad can competitively antagonize the interactions between Beclin-1 and Bcl-2/Bcl-X_L. On the mitochondrial membrane, BH3-only proteins trigger apoptosis by inhibiting the mitochondrion permeabilizing of pro-apoptotic multi-domain proteins such as Bax and Bak. In ER, BH3-only proteins promote autophagosome formation by releasing Beclin-1 from Bcl-2/Bcl-X_L [84]. Additionally, a variety of biochemical cascades are indicated as core regulators function on the cross-talks between apoptosis and autophagy. For example, on one hand, ER stress can result in signaling pathways through PERK, IRE1 and ATF6 and thus triggering pro-apoptotic signals via the activation of downstream molecules like CHOP, JNK or members of the Bcl-2 proteins family; on the other hand, ER stress can also promote autophagosome formation with LC3I/II conversion through either PERK-eIF2a or IRE1-JNK pathway [103].

5.1.4. Autophagy in Cancer Therapy

Chemotherapeutic agents have been reported to induce autophagic cell death by targeting mTORC1, which can stimulate autophagosome formation when inhibited. So far, there have been many agents that can inhibit mTORC1 such as rapamycin and its derivative temsirolimus, everolimus [104], perhexiline, niclosamide, amiodarone, and rottlerin [92].

Due to the dual roles in tumorigenesis and cancer treatment, autophagy contributes to targeting-therapy in cancer. On one hand, autophagy defects may lead to the DNA damage accumulation of normal cells, genomic instability, and accelerated tumorigenicity. On the other hand, high level of autophagy promotes cell survival, metastasis and resistance to cancer therapy. The role of autophagy depends on the particular type of cancer, the progression of the disease and the treatment methods being employed. When trying to regulate autophagy by drugs for cancer therapy, we should take full consideration of all these effect aroused by autophagy [85].

Depriving tumor cells of their essential proliferation mechanism through the inhibition of autophagy would enhance the efficacy of anticancer drug. Some tumor cells lose its apoptosis death pathway and rely on autophagy for survival under metabolic stress, so autophagy inhibition may have a remarkable effect on these apoptosis defect tumor cells. Moreover, the high metabolic demands of proliferation of tumor make the tumor cells have more dependency of autophagy than other normal cells. In that case, when we inhibit autophagy, the undesired treatment side-effect will be acceptable. For example, chloroquine, a lysosomotropic agent which inhibits autophagy by raising PH intralysosomal and interfering with autophagosome degradation within lysosomes was proved that it enhances the anticancer activity of Alkylating agent cyclophosphamide in amycinduced lymphoma model [105].

Interestingly, autophagy protects not only the tumor cell itself, but also its resistance mechanism. A serious clinical problem is the frequent re-emergence of tumors after treatment, often after prolonged dormancy [106]. The tumor cells escape from the death of metabolic stress and other adverse situation may have their special escaping mechanism to cancer treatment, so eliminate these cells is really necessary in achieving durable treatment responses. Autophagy may be the key mechanism which contributes to a long-term tumor cell survival and eventual re-growth and relapse. In that case, autophagy may allow residual or metastasizing tumor cells to tolerate metabolic stress even metabolic deprivation. This capacity for recovery afforded by autophagy vaguely resembles the process of sporulation in microorganisms, and its suppression may be essential to provide a new perspective in cancer eradication.

6. Anti-Tumor Drugs Aiming at Key Targets in Autophagic Pathways

While many details of autophagy pathway remain unclear, novel emerging strategies are focused on implementation of autophagic targets for cancer therapies. Thus, a number of autophagy-inducing agents are proposed for potential utilization in cancer therapeutics (see in Table 2 and Table 3).

6.1. Autophagy Inhibitors

Autophagy supports genomic integrity and stability under circumstances of metabolic stress, drug administration or radiation treatment. However, in pathophysiological conditions autophagy can be employed by the tumor cells to supply nutrients and energy thus promoting tumor survival. Thus, some drugs have been developed to blockade autophagy processing so as to decrease tumor development. A certain level of autophagy is demanded for tumor growth; therefore treatment with autophagy inhibitor alone could achieve anti-tumor effect, and exert a synergistic effect while combined with other inhibitors. To compensate a high energy demand and low oxidative phosphorylation level, tumor cells can activate autophagy supporting oxidative phosphorylation. So combination of autophagy inhibitors and angiogenesis inhibitors or glucose uptake inhibitors might serve as a potential cancer therapy. Furthermore, autophagy is also an effective mechanism to eliminate damaged and potentially dangerous organelles; thus joint administration with organelle-harmful drugs is another possible method to induce cell death. According to the process in which autophagy could be divided into several stages, the inhibition of autophagy could also be classified into several phases, including the initiation period, docking and fusion of the autophagosome with lysosomes, and catalytic degradation of the cytoplasmic material inside the autolysosomes.

6.1.1. Inhibition of the Autophagosome Formation

The majority of drugs under this category belong to the group of PI3K inhibitors, e.g., 3-MA, wortmannin; they can interfere or block the autolysosome formation [150]. Since PI3K is essential for the initiation of autophagy, these PI3K inhibitors can specifically disrupt the formation of autolysosome by suppressing the formation of autophagosome. So far, many other drugs are designed to target components of PI3K-Akt-mTOR pathway [151].

6.1.2. Inhibition of Fusion of the Autophagosome with Lysosomes

Although there are no available reports confirming the anti-tumor effects derived directly from autophagy inhibition by lysosomotrophic agents, it has been observed that cancer cells did acquire increased level of autophagy and were more sensitive to, and also died in response to lysosomotrophic agents. Several agents that inhibit the fusion process can interrupt autophagy, such as Bafilomycin A1, Vinblastine, Nuokaodazuo. Bafilomycin A1, a sort of macrolide antibiotics derived from Streptomyces griseus, is a specific inhibitor of H^+-ATP enzyme and bears antibacterial, antifungal, anti-tumor effects. When synaptic vesicles are secreted through exocytosis, Bafilomycin A1 could avoid the vesicle re-acidification. It is reported that treating tumor cells that have autophagy processing with Bafilomycin A1 could inhibit the degradation of proteins and the number of autophagosome is increased while autolysosome decreased, and the activity of acid phosphatase also drops remarkably, inferring that Bafilomycin A1 blockade the fusion of autophagosome with lysosome.

6.1.3. Inhibition of Autolysosome Degradation

After autophagosome fuses with lysosome, the autophagosome content is subjected to degradation by lysosomal hydrolases and degraded material could be recycled in the cell. Specific inhibition of this process can also inhibit autophagy and protease inhibitors are such candidates that effectively terminate autophagy via inhibiting autolysosome degradation. Both E64D and pepstatin A are protease inhibitors and when used with 1:1 ration, autophagy was curbed. It was reported that combined treatment with these two drugs in colon cancer cell lines, the processing of autolysosome degradation was significantly repressed while autophagosome formation was not affected.

The anti-malarial drugs, hydroxychloroquine and chloroquine, can block lysosomal acidification and inhibit the degradation of autophagosomes, thereby suppressing autophagy in Myc-driven lymphoma and increasing the anti-tumor effect of cyclophosphamide [105, 152]. In imatinib-resistant BCR-ABL-positive CML cell lines chloroquine enhanced cell death by inhibiting autophagy and strengthening the activity of vorinostat, the HDAC inhibitor [5, 149]. In combination with antimalarial drug quinacrine, chloroquine remarkably sensitized gastrointestinal stromal tumor cells to administration with imatinib, both *in vitro* and *in vivo* [153] and enforced the efficiency of quinacrine.

As the autophagic degradation system and proteasome are partly complementary processes, and proteasome absence induces autophagy and autophagy suppression causes the accumulation of polyubiquitinated proteins, it would be attractive in cancer treatment by inhibiting autophagy and proteasome synergetically [143, 149]. Of interest, treatment with chloroquine prevents tumor dormancy and subsequent regrowth in ovarian cancer xenografts, suggesting that inhibiting autophagy might exert therapeutic applications in treating with the residual cancer cell population after primary chemotherapy [154].

Table 1. A list of yeast Atg proteins

Name	Autophagy step	Function	Complex or interaction	Homologues in higher eukaryotes
Atg1	Induction	Serine/threonine protein kinase regulates magnitude of autophagy.	ULK1/2–Atg13–FIP200	Yes ULK1/2
Atg2	Nucleation	Interacts with Atg18, mediates the retrieval of Atg9 from the PAS back to peripheral sites.	Atg2–Atg18	Yes
Atg3	Elongation	E2-like conjugating enzyme, functions in the conjugation of Atg8.	Atg3–Atg7–LC3	Yes
Atg4	Elongation	Cysteine protease, removes the C-terminal arginine residue from newly synthesized Atg8 or PE from Atg8-PE conjugate.	LC3	Yes
Atg5	Elongation	Conjugation target of Atg12, part of the Atg12-Atg5-Atg16 complex.	Atg12–Atg5–Atg16L	Yes
Atg6/Vps30	Nucleation	Common component of class III PI3K complex I and II, functions in autophagy and the VPS pathway.	Atg14L–Beclin1–PI3KCIII–p150; UVRAG–Beclin1–PI3KCIII–p150	Yes Beclin1
Atg7	Elongation	E1-like activating enzyme, functions in the conjugation of Atg8 and Atg12.	Atg3–Atg7–LC3	Yes
Atg8	Elongation	Ubiquitin-like protein, conjugated to PE, controls the expansion of the phagophore.	Atg3–Atg7–LC3; Atg12–Atg5–Atg16L; Atg4	Yes LC3, GABARAP
Atg9	Nucleation	Transmembrane protein, shuttles between the PAS and other peripheral sites.	ULK1, PI3KCIII, Atg2, Atg18	Yes
Atg10	Elongation	E2-like conjugating enzyme, functions in the conjugation of Atg12.	Atg12–Atg5–Atg16L	Yes
Atg11	Elongation	Scaffold protein, required for selective autophagy, participates in the assembly of PAS proteins.	Atg19	
Atg12	Elongation	Ubiquitin-like protein, conjugated to Atg5, part of the Atg12-Atg5-Atg16 complex.	Atg12–Atg5–Atg16L	Yes
Atg13	Induction	Subunit of Atg1 complex, phosphorylation status controls the activity of Atg1 and the magnitude of autophagy.	ULK1/2–Atg13–FIP200	Yes
Atg14	Nucleation	Autophagy-specific component of class III PI3K complex I, targets the complex to the PAS.	Atg14L–Beclin1–PI3KCIII–p150	Yes
Atg15		Vacuolar lipase, functions in the breakdown of autophagic bodies.		
Atg16	Elongation	Component of the Atg12-Atg5-Atg16 complex, mediates the oligomerization of the complex.	Atg12–Atg5–Atg16L	Yes
Atg17	Induction	Subunit of Atg1 complex, regulates the magnitude of autophagy, participates in the assembly of PAS proteins.	ULK1/2–Atg13–FIP200	
Atg18	Nucleation	WIPI protein, binds PI3P, interacts with Atg2 to mediate the retrieval of Atg9 from the PAS back to peripheral sites.	Atg2–Atg18	Yes WIPI^{+}
Atg19	Nucleation	Cargo receptor in the Cvt pathway.	Atg11	

Name	Autophagy step	Function	Complex or interaction	Homologues in higher eukaryotes
Atg20	Induction	Sorting nexin, binds PI3P, functions in the Cvt pathway.	ULK1/2–Atg13–FIP200	
Atg21		WIPI protein, binds PI3P.		Yes WIPI[+]
Atg22		Amino acid permease on the vacuolar membrane.		
Atg23		Peripheral membrane protein, transits with Atg9.		
Atg24/Snx4		Sorting nexin, binds PI3P, functions in the Cvt pathway.		
Atg25		Functions during macrophexophagy in *Hansenula polymorpha*.		
Atg26		Functions during pexophagy in *Pichia pastoris*.		
Atg27		Integral membrane protein, transits with Atg9.		
Atg28		Functions during pexophagy in *Pichia pastoris*.		
Atg29		Part of the Atg17-Atg29-Atg31 complex, functions in non-selective autophagy.		
Atg30		Receptor of peroxisomes in *Pichia pastoris*.		
Atg31/Cis1		Part of the Atg17-Atg29-Atg31 complex, functions in non-selective autophagy.		
Atg32		Receptor of mitochondria, mitochondrial outer-membrane protein.		
Atg33		Functions in mitophagy under certain conditions.		
Atg34		Paralogue of Atg19, receptor of Ams1.		

[#] Proteins labeled with bold font are collectively referred to as the "core autophagy machinery" proteins, reflecting the fact that these proteins are essential for autophagosomes formation and are conserved during evolution. In most cases, homologues of yeast autophagy genes found in other species are named after their yeast counterparts. Only those named differently are listed in the homologue column.

[+] In general, there are multiple WIPI family members in a given species; not all WIPI proteins function in autophagy.

Table 2. FDA approved drugs modulating the key targets in autophagy pathway

Drug names	Active ingredients	Targets/pathways	Cancer/cell types	Company	FDA Application number(NDA or ANDA or BLA)	Availability	References
IRESSA (Gefitinib, ZD1839)	gefitinib	EGFR (Ras/Raf/MEK, PI3K/Akt)	advanced lung cancer	AstraZeneca	(NDA) 021399	FDA Approved in May 5, 2003	[107]
TARCEVA (Erlotinib)	erlotinib hydrochloride	EGFR, ABL, FLT (Ras/Raf/MEK, PI3K/Akt)	non-small cell lung cancer (NSCLC), advanced lung cancer	OSI PHARMS	(NDA) 021743	FDA Approved in November 18, 2004	[108]
NEXAVAR (Sorafenib)	sorafenib tosylate	VEGFR2, PDGFR, FLT3, c-Kit, Erk, Raf (Ras/Raf/MEK, PI3K/Akt)	non-small cell lung cancer (NSCLC), renal carcinoma, hepatoma, melanoma	BAYER HLTHCARE	(NDA) 021923	FDA Approved in December 20, 2005	[109]

Table 2. (Continued)

Drug names	Active ingredients	Targets/pathways	Cancer/cell types	Company	FDA Application number(NDA or ANDA or BLA)	Availability	References
Sutent	sunitinib malate	FLT3, c-Kit, PDGFR, VEGFR2 (Ras/Raf/MEK, PI3K/Akt, PKCs)	Renal cell carcinoma	Pfizer	(NDA) 021938	FDA Approved in January 26, 2006	[110]
SPRYCEL (Dasatinib, BMS-354825)	dasatinib	Src, BCR-ABL, TCR (Ras/Raf/MEK, PI3K /Akt, JAK/STAT)	Chronic myelogenous leukemia (CML), Acute lymphocytic leukemia (ALL)	Bristol-Myers Squibb	(NDA) 021986	FDA Approved in June 8, 2006	[111]
TYKERB (Lapatinib, GW572016)	lapatinib	EGFR,HER2,p95HER2 (Ras/Raf/MEK, PI3K/Akt)	metastatic breast cancer (MBC)	GlaxoSmith Kline		FDA Approved in March 13, 2007	[112]
TASIGNA (Nilotinib)	nilotinib hydrochloride monohydrate	ABL (Ras/Raf/MEK, PI3K/Akt, JAK/STAT)	Chronic myelogenous leukemia (CML), Acute lymphocytic leukemia (ALL)	Novartis	(NDA) 022068	FDA Approved in October 29, 2007	[113]
AFINITOR	everolimus	AKT/mTOR	Advanced renal carcinoma	NOVARTIS	(NDA) 022334	FDA Approved in March30, 2009	
ZORTRESS	everolimus	AKT/mTOR	Advanced renal carcinoma	Novartis	(NDA) 021560	FDA Approved in April20, 2010	
TORISEL	temsirolimus	AKT/mTOR	Renal cell carcinoma (RCC)	WYETH PHARMS INC	(NDA) 022088	FDA Approved in May 30, 2007	[114]
GLEEVEC	imatinib mesylate	PDGFR, c-Kit, BCR-ABL (Ras/Raf/MEK, PI3K/Akt)	Chronic myelogenous leukemia (CML), gastrointestinal stromal tumors (GISTs)	Novartis	(NDA) 021588	FDA Approved in April 18, 2003	[115]
VANDETANIB	vandetanib	VEGFR, EGFR, RET (Ras/Raf/MEK, PI3K/Akt)	Advanced medullary thyroid carcinoma	IPR PHARMS INC	(NDA) 022405	FDA Approved in April 6, 2011	

Drug names	Active ingredients	Targets/pathways	Cancer/cell types	Company	FDA Application number(NDA or ANDA or BLA)	Availability	References
VOTRIENT (Pazopanib)	pazopanib hydrochloride	c-Kit, PDGFR, VEGFR2 (Ras/Raf/MEK, PI3K/Akt)	Advanced renal carcinoma	GlaxoSmith Kline	(NDA) 022465	FDA Approved in October 19,2009	
TRISENOX	arsenic trioxide	BNIP3	Leukemia	CEPHALON	(NDA) 021248	FDA Approved in December 25,2000	[116]
PACERONE	amiodarone hydrochloride	mTORC1		UPSHER SMITH	(ANDA) 075135	FDA Approved in April 30, 1998	[117]
Metformin Hydrochloride		AMPK	Non-insulin-dependent diabetes mellitus			FDA Approved in 2004	[118]
Niclosamide		mTORC1	Most tapeworm infections	Bayer Healthcare			[119]
RAPAMUNE	sirolimus	mTOR	kidney transplant recipients	WYETH PHARMS INC	(NDA) 021110	FDA Approved in August 25, 2000	[120]
VELCADE	bortezomib	proteasome	Relapsed multiple myeloma, mantle cell lymphoma	MILLENNIU M PHARMS	(NDA) 021602	FDA Approved in May 13, 2003	[121]
SPRYCEL	dasatinib	Ephrins, SRC, GFR	Chronic myelogenous leukemia (CML)	Bristol Myer Squibb	(NDA) 021986	FDA Approved in June 28, 2006	[122]
Nexavar (sorafenib)	sorafenib tosylate	Raf kinase, PDGF, VEGF receptor 2 & 3 kinases	Renal cell carcinoma, advanced hepatocellular carcinoma	BAYER HLTHCARE	(NDA) 021923	FDA Approved in December 20, 2005	[123]
Zolinza (vorinostat)		HDAC	Cutaneous T cell lymphoma (CTCL)	Merck			[124]
RITUXAN	rituximab	CD20, EGFR	Non-Hodgkin's lymphoma (NHL)	GENENTECH	(BLA) 103705)		[125]
				IDEC PHARMACE UTICALS CORP	(BLA) 103737)		
VECTIBIX	panitumumab	EGFR	Colorectal cancer	AMGEN	(BLA) 125147	FDA Approved in September 27, 2006	[126]

Table 2. (Continued)

Drug names	Active ingredients	Targets/pathways	Cancer/cell types	Company	FDA Application number(NDA or ANDA or BLA)	Availability	References
NOLVADEX	tamoxifen citrate	Estrogen receptor	Breast cancer	ASTRAZENECA	(NDA) 017970	FDA Approved in December 30, 1977	[127]
FARESTON	toremifene citrate	Hormone receptors	Breast cancer	GTX INC	(NDA) 020497	FDA Approved in May 29, 1997	[128]
HYDROXYCHLOROQUINE SULFATE (Generic Drug)	hydroxychloroquine sulfate	Inhibition of lysosomal degradation	Malaria, systemic lupus erythematosus, rheumatoid arthritis	WATSON LABS	(ANDA) 040133	FDA Approved in November 30, 1995	[105]

Table 3. The agents ongoing clinical trials and preclinical studies

Drugs	Targets/pathways	Cancer/cell types	Company	Availability	References
Vandetanib (ZD6474)	VEGFR, EGFR, RET (Ras/Raf/MEK, PI3K/Akt)	thyroid cancer, non-small cell lung cancer (NSCLC)	AstraZeneca	Phase III	[129]
Pazopanib (GW-786034)	c-Kit,DGFR,VEGFR2 (Ras /Raf / MEK, PI3K/Akt)	renal carcinoma cell (RCC)	GlaxoSmithKline	Phase III	[130]
CP-690550	JAK(JAK/STAT)	Rheumatoid Arthritis	Pfizer	Phase III	[131]
Midostaurin (PKC412)	PKC, FLT3, KIT, VEGFR, PDGFR(Ras/Raf/MEK, PI3K/Akt, PKCs, JAK/STAT)	leukemia	Novartis	Phase III	[132]
Vatalanib (PTK-787)	PDGFR, c-Kit (Ras/Raf/MEK, PI3K/Akt)	Colon cell, Primary tumor, solid tumor	Novartis and Bayer	Phase III	[133]
Ruboxistaurin (LY333531)	PKC (Ras/Raf/MEK, PI3K/Akt, PKCs)	Diabetic Retinopathy	Eli Lilly	Phase III	[134]
Drugs	Targets/pathways	Cancer/cell types	Company	Availability	References
AMG-706	EGFR1/FLT1,VEGFR2/Flk-1,VEGFR3/FLT4,PDGF	malignancy	Amgen	Phase III	[135]
Axitinib (AG013736)	VEGFR-1,2,3,PDGFR (Ras/Raf PI3K/Akt)	nephroblastoma	Pfizer	Phase III	[136]
XL184	VEGFR2 MET (PI3K/Akt)	non-small cell lung cancer (NSCLC), thyroid cancer	Exelixis	Phase III	[137]

Drugs	Targets/pathways	Cancer/cell types	Company	Availability	References
Ridaforolimus (AP23573, deforolimus, MK-8669)	Akt/mTOR	advanced solid tumors, Leukemia,Non-Small Cell Lung Cancer	Arad and merck	Phase II	[138]
Advexin (contusugene ladenovec)	P53	platinum- and paclitaxel- resistant epithelial ovarian cancer, Breast Cancer	Introgen Therapeutics	Phase III completed	[139]
Zarnestra	Ras-Raf	Non–Small Cell Lung Cancer, Leukemia	Johnson&Johnson(J&J)	Phase III completed	[140]
SAHA (Vorinostat)	HDAC	Lymphoma	Merck	Phase I	[5]
Perifosine	Akt	Renal Cell Carcinoma	Aeterna Zentaris	Phase III	[141]
Triciribine	Akt	Leukemia		Phase I / II	[141]
NV-128	Akt /mTOR	Paclitaxel- and Carboplatin-resistant EOC cells		Preclinical research	[142]
NPI-0052	Proteasome	Non-small Cell Lung Cancer, Pancreatic Cancer, Melanoma, Lymphoma		Phase I	[143]
Epoxomicin	Proteasome, NF-κB	human B cell lymphoma cell lines		Preclinical research	[144]
Sarasar (lonafarnib)	Ras-Raf	Prostate Cancer, Breast Cancer, Ovarian Cancer, Lung Cancer	ScheringPlough	Phase III Terminated	[145]
ABT-888 (Veliparib)	PARP1	Colorectal Cancer, Prostate Cancer		Phase II	[146]
EB-1089	Beclin-1	Liver Neoplasms, MCF7 cells		Preclinical research	[147]
Obatoclax mesylate (GX15-070MS)	Bcl-2	Chronic Lymphocytic Leukemia		Phase II Completed	[123]
2-deoxyglucose	P62	Prostate Cancer, Non Small Cell Lung Cancer		Phase I / II	[148]
3-methyladenine		human mammary carcinoma cells (MCF-7), glioma		Preclinical research	[149]

6.1.4. Other Potential Autophagy-Inducing Molecules

In addition to the above-mentioned reported drugs, there are also other autophagy inhibitors with anti-tumor properties. They are mainly small-molecule compounds, such as nitric oxide [155], 3-methyladenine [156], Ly294002 [156], aurintricarboxylic acid [157], Nvp-bez235, and some bioactive macromolecules, such as Cst-1.

6.2. Drugs Activating Autophagy in Tumor Therapy

In many types of tumors, the deficiency of autophagy is observed, indicating that tumor progression might involve with defective autophagy [9]. Experimental data indicate that metabolic stress in tumors stimulates autophagy to meet enormous energy demands along cancer progression [158, 159]. However, if unfavorable metabolic conditions surpass a crucial point needed for cell survival, autophagic cell death might occur, which can be a target of anticancer therapy [98]. Accumulating researches indicate that multiple anti-tumor drugs are related to the regulation of autophagy, and some of them could directly or indirectly activate autophagy and thus induce cell death, or release some death-inducing factors from lysosomal protease within autolysosome, leading to apoptosis or necrosis. For some other drugs, the tumor cells manifest themselves as up-regulated activity of autophagy or even autophagic cell death. To some extent, it is a novel anti-tumor therapy to enhance autophagy activity of tumor cells, especially on occasion when tumor cells gain resistance to apoptosis-inducing factors.

Hitherto, various chemotherapeutic agents (Alkylating agents, actinomycind, and Arsennic trioxide), hormone therapies (tamoxifen and Vitamin D analogues [160], natural compounds (resveratrol), cytokines (IFN), genetic therapies (p53 and p27 KIP 1 gene), radiation and photodynamic therapies have been demonstrated to be able to induce tumor cells to go through cell death by using autophagy mechanism *in vitro*, but more experimental evidence is necessary to identify the anti-tumor role of autophagy *in vivo*.

The vitamin D analog EB1089 induces Beclin-1-dependent autophagy and sensitizes MCF7 breast cells to radiation [147, 160], and augments the influx of Ca^{2+} which putatively stimulates CaMKK, and eventually activates TSC2 and inhibits mTOR [39]. Arsenic trioxide, an apoptosis inducer in treating acute promyelocytic leukemia, also triggered autophagy through activation of Beclin-1 in leukemia cells and glioma cells *in vitro* as well [161]. Resveratrol, a polyphenol existed in grapes, peanuts and other plants, that bears robust antioxidant and anti-tumorigenic properties could also switch on autophagy, but contribute to cell death and prosurvival function in cultured breast cancer [162] and glioma cells [163], respectively. Additionally, autophagy induced by resveratrol in breast cancer cells was independent on Beclin-1, and the contradictory role of autophagy concerning molecular pathways remains to be elucidated.

Induction of autophagy exhibits an attractive prospect in cancer treatment; however, autophagy-inducing agents might impede tumor cells from death that are triggered by nutrient-deficiency or other therapies. For instance, rapamycin exhibits an inhibitory effect in radiation-caused apoptosis. Besides, combination of diverse autophagy-inducing therapies might increase the sensitivity of tumor cells toward autophagic cell death. Therefore, when combined therapy involve with autophagy-related drugs, appropriate design of the therapy is very crucial to obtain a better effect.

6.2.1. Inhibiting Negative Regulators of Autophagy

Induction of autophagy can also be achieved through repressing the negative regulators of autophagy, for instance, mTOR [104]. mTOR is a primary regulator of cell growth that has been implicated in tumorigenesis. Rapamycin inhibits the mTORC1complexs through binding the 12 kDa immunophilin FK506-binding protein (FKBP12). Its derivatives, CCI-779 and RAD001, have been applied in clinical trials for treating tumor [150]. It has been demonstrated that tumor-suppression following rapamycin administration is involved in the induction of autophagic cell death [164]. For example, treatment of RAD001 combined with the caspase-3 inhibitor Devd radio sensitizes non-small cell lung cancer cells in mice models, leading to increased cytotoxicity through autophagy-induction and tumor growth delay. Thus, inhibition of mTOR by rapamycin or its derivatives provides a powerful therapeutic tool for the treatment of various malignancies.

Another possible mechanism of inducing autophagy is allosteric inhibitors of mTORC1 (rapalogs), such as sirolimus. Sirolimus is an anti-rejection drug used in organ transplantation and bear anticancer activity presumably by activating autophagy in malignant glioma and breast cancer cell lines [150, 165]. The water-soluble esters of sirolimus, temsirolimus and everolimus, have been adopted in conquering renal-cell carcinoma. Treatment of mantle-cell lymphoma cell lines with temsirolimus is shown to downregulate p21 and to enhance autophagy [166]. Temsirolimus and everolimus can also upregulate autophagy and suppress acute lymphoblastic leukemia, prostate cancer and non-small-cell lung cancer (NSCLC) cells and xenografts [167]. Additionally, everolimus sensitize tumor cells to radiation when combined with apoptosis inhibitors [104, 164]. Current extensive research in this area is focused on the specific function of autophagy in treating cancer with mTOR inhibitors in multiple clinical trials.

6.2.2. Inducing Autophagy with Apoptosis Inhibitors

Autophagy might serve as a primary cell death program when apoptosis mechanism is deficient. Although autophagy plays a pro-survival role in normal physiological conditions, it can also drive tumor cells to death if triggered with certain agents. Therefore, induction of autophagic cell death opens a new perspective for solving the problem of apoptosis resistance in tumor tissues.

6.2.2.1. Bak/Bax Inhibitors

During the past several years, Bak/Bax-mediated apoptosis induced by radiation was widely investigated but it can only explain less than 20% cell death. Surprisingly, recent research examined that cells without Bak/Bax could elevate radiation sensitization via inducing autophagy, which could be repressed by 3-MA or Beclin-1 specific inhibitor Sirna. Taken together, Bak/Bax inhibitors can be employed as anti-tumor drugs mediating autophagic cell death.

6.2.2.2. Caspase Inhibitors

Endoplasmic reticulum stress could be activated by radiation, that blocks apoptotic pathway, and thus leading to autophagic cell death. Exploration of this mechanism can provide another novel strategy for cancer therapy.

6.2.3. Enhancing Abnormal Genes' Expression to Increase Autophagic Ability

In the process of tumor formation, a variety of genes that regulate autophagy are abnormal, which is the major cause of reduced autophagy activity. These abnormalities can be corrected with anti-tumor drugs. A mammalian autophagy gene, Beclin-1 controls cell growth and inhibits tumor formation. In 40%-75% diffused breast and ovarian cancer, there are deletions of Beclin-1 single-allele. In lots of cancers, somatic gene PTEN is often found to be mutated or deleted. Over-expression of PTEN gene favors autophagy, while PTEN mutation or constitutive activation mutant type can reduce the incidence of autophagy. Thus, some drugs that can regulate PTEN gene may be utilized to achieve the purpose of conquering cancer.

6.2.4. Angiogenesis Inhibitors as Endothelial Cell-Autophagy Inducers

Neovascularization of the tumor stroma could be achieved through angiogenesis of solid tumor, which plays an important role in tumor growth and metastasis. Andostatin, a fragment of collagen, inhibits proliferation of endothelial cell, thereby prevents tumor angiogenesis. That is accompanied by, formation of enormous number of autophagic vesicles. In solid tumors, angiogenesis contributes to tumor metastasis and growth while induction of autophagic cell death represses tumor development via inhibiting angiogenesis. Human blood plasminogen Kringle 5 (K5) is effective angiogenesis inhibitors that can also up-regulate apoptosis in endothelial cells. In endothelial cells, K5 and endostatin can induce autophagy in a selective manner, i.e., independently of such factors as nutritional deficiencies, hypoxia, or vascular endothelial growth factor (VEGF). Therefore, K5 and endostatin analogues can be considered as an autophagy targets in tumor treatment.

6.2.5. Inhibitors of Kinases and Enzymes

Farnesyltransferase Inhibitors

Farnesyltransferase inhibitors (FTIs) were originally developed to oppose the mitogenic property of oncogene RAS. It has been shown recently that a combination of FTIs with onafarnib can induce autophagic death he U20S human osteosarcoma cell line and human pancreatic cancer cell line [145]. We believe that FTIs can be involved in autophagy via interaction with autophagy regulators, including Rheb and the PI3K/Akt/mTOR pathway [168].

Tyrosine Kinase Inhibitors

Preclinical models showed that autophagy could be stimulated by some tyrosine kinase inhibitors. In various cancer cell lines, imatinib could competitively inhibit BCR-ABL, KIT, and PDGFR and induce autophagy, but the specific role of autophagy in treating tumor is unclear [169]. Similarly, dasatinib, the inhibitor of BCR-ABL, KIT, and the proto-oncogene tyrosine-protein kinase SRC, induced autophagic cell death, thereby intensified the effect of chemotherapy in glioma cell line. Moreover, inhibition of autophagy with 3-methyladenine could mitigate this effect [170]. Sorafenib, a multi-tyrosine-kinase inhibitor combined with HDAC inhibitor vorinostat, triggered autophagy in hepatocellular carcinoma and pancreatic cancer cell lines, but this autophagy played a cytoprotective role in this context [123]. Combined with lapatinib, GX15-070, a Bcl-2 inhibitor, it facilitated the anti-tumor effect via inducing autophagy in breast cancer cell lines [171]. Besides, treatment with GX15-070, solely or combined with sorafenib and vorinostat, could stimulate tumor cell death via autophagic pathways [123].

Histone Deacetylase (HDAC) Inhibitors

HDAC inhibitors are confirmed to spark off caspase-mediated apoptosis and caspase-independent autophagic cell death on the occasion that apoptosis is blocked. HDAC inhibitors, vorinostat and sodium butyrate assist cell death via autophagy in HeLa cells [172] and similarly, LAQ824 and panobinostat bring forward autophagic cell death in lymphoma cell lines and mouse xenografts when apoptosis was blocked.

6.2.6. Natural Compounds to Induce Tumor Autophagy

Numerous natural compounds listed in the traditional Chinese medicine, e.g., oridonin, protein *Polygonatum Cyrtonema* Lectin (PCL) and Concanavalin A (ConA), can promote autophagic death of cancer cells. Oridonin was reported to be able to induce both apoptosis and autophagy depending on nature of cancer cells and therefore it can be a promising candidate for cancer therapy [173]. PCL has been reported to induce autophagy via a mitochondria-mediated-ROS-p38–p53 pathway in human melanoma A375 cell line [174] and Ras-Raf-MAPK and PI3K-Akt signaling pathways in murine fibrosarcoma L929 cells [175, 176]. The full list of natural compounds inducing autophagic tumor cell death is shown in Table 4.

Overall, recent growing interest to biology of natural products and nutritional and alternative medicine opens new horizons for development integrative strategies for the target cancer treatment that could explore capacity of natural products in regulating programmed death mechanisms, in particular autophagy-dependent mechanisms. However, considering complicity of autophagy pathway and its controversial effects in cell death and survival it needs extensive research and development of prior autophagy-based therapy meets clinical requirements.

Table 4. Natural compound-mediated autophagy in cancer

Active molecules	Targets	Cell types	References
ConA	BINP3	Hepatoma cells	[177, 178]
Polygonatum cyrtonema lectin (PCL)	ROS–p38–p53, Ras-Raf, PI3K/Akt	Murine fibrosarcoma L929 cell, human melanoma A375 cells	[174-176]
G. lucidum triterpene extract (GLT)	Beclin-1, p38 MAPK	Human colon cancer cells HT-29, colon cancer	[179]
Licochalcone-a	Bcl-2, mTOR	Incap prostate cancer cells	[180]
Areca nut extract	mTOR	Carcinoma cells	[181]
Tocotrienols		Rat pancreatic stellate cells	[182]
Sesamin	EphA1 and EphB2	Human colon cell lines HT29 and LS180	[183]
Pancratistatin (PST)		Metastatic prostate cancer cells,	[184]
Plumbagin	Akt/mTOR	Breast cancer cells	[185]
Avicin D	AMPK, mTOR	Human osteosarcoma cells	[186]
Dihydrocapsaicin (DHC)		HCT116, MCF-7, WI38 cell lines	[187]

Table 4. (Continued)

Active molecules	Targets	Cell types	References
Timosaponin A-III	mitochondria-dependent pathway	HeLa cells	[188]
Δ^9-tetrahydrocannabinol (THC)	Akt/mTOR	Human glioma cells	[189]
Pentagalloylglucose	mTOR	Human PC-3 and mouse TRAMP-C2 prostate cancer cells	[190]
6-shogaol	Akt/mTOR	Human non-small cell lung cancer A549 cells	[191]
Alpha-eleostearic acid	Akt/mTOR,ERK1/2	Human cervix carcinoma cells	[192]
α-Mangostin	AMPK	Glioblastoma Cells	[193]
Alisol B	CaMKK-AMPK- mTOR	MCF-7 cells	[194]
Apogossypolone	LC3-II, Beclin-1	Human prostate cancer cells	[195]
Lipopolysaccharide	BIRC2, p53	Human umbilical vein endothelial cells	[196]
Bromelain	P38, JNK	Mammary carcinoma cells	[197]
Pterostilbene	mTOR/p70^{S6K}	Human bladder cancer cells	[198]
Caffeine	PI3K/Akt/mTOR	HeLa cells	[199]

Conclusion

Emerging molecular pharmacology of autophagy is faced to challenging problems resulted from (1) complicity of autophagy driving mechanisms, and (2) diverse roles of autophagy in determination of death and survival of cancer and normal cells. The understanding of these problems is based on a broad analysis of etiology and pathogenesis of a variety of human diseases that include cancer, neurodegeneration, ageing, cardiac disease, muscular and immune disorders, as well as infections and inflammation where a crucial role of key autophagic mediators including, i.e., ATGs, Bcl-2, PI3K, mTOR and p53, have been determined.

Many details of patho-physiological role of autophagy in these diseases were revealed using powerful agents such as rapamycin, erlotinib, sorafenib, nilotinib, temsirolimus. In spite of many disadvantages these and other compounds (such as Tamoxifen, Erlotinib, Sorafenib, Nilotinib, Temsirolimus, Imatinib, Pazopanib, Cisplatin, Trisenox, Sprycel, Rituximab, Vectibix, Fareston, Polyamine spermidine), currently they are the only effective tools in discerning the vital and deleterious effects of autophagy. In this light, we expect that integration of the previous autophagy R&D data would provide a fundamental basis for launching a next round of small molecule drug discovery.

This new approach should integrate (i) structural information of the potential targets obtained from X-ray crystallography, nuclear magnetic resonance (NMR) and structural bioinformatics-docking techniques, and (ii) sophisticated mathematics modeling based on Naïve Bayesian framework and support vector machine (SVM) for the disruption of protein-protein interactions.

The small molecules with predicted inhibiting or activating effects in these protein–protein interactions (PPIs) can be further screened as potential candidate drugs. The best expectations from autophagy target therapy will be based on the discovery of appropriate small molecules that can ultimately correct the physiological processes impaired by compromised key metabolic mechanisms (e.g., autophagy), rather than correct their regulatory genes or incorrectly expressed or improperly folded proteins.

Thus, the human autophagy PPI network can be considered as an emerging effective tool for discovery of the autophagy-related functional targets associated with different human diseases.

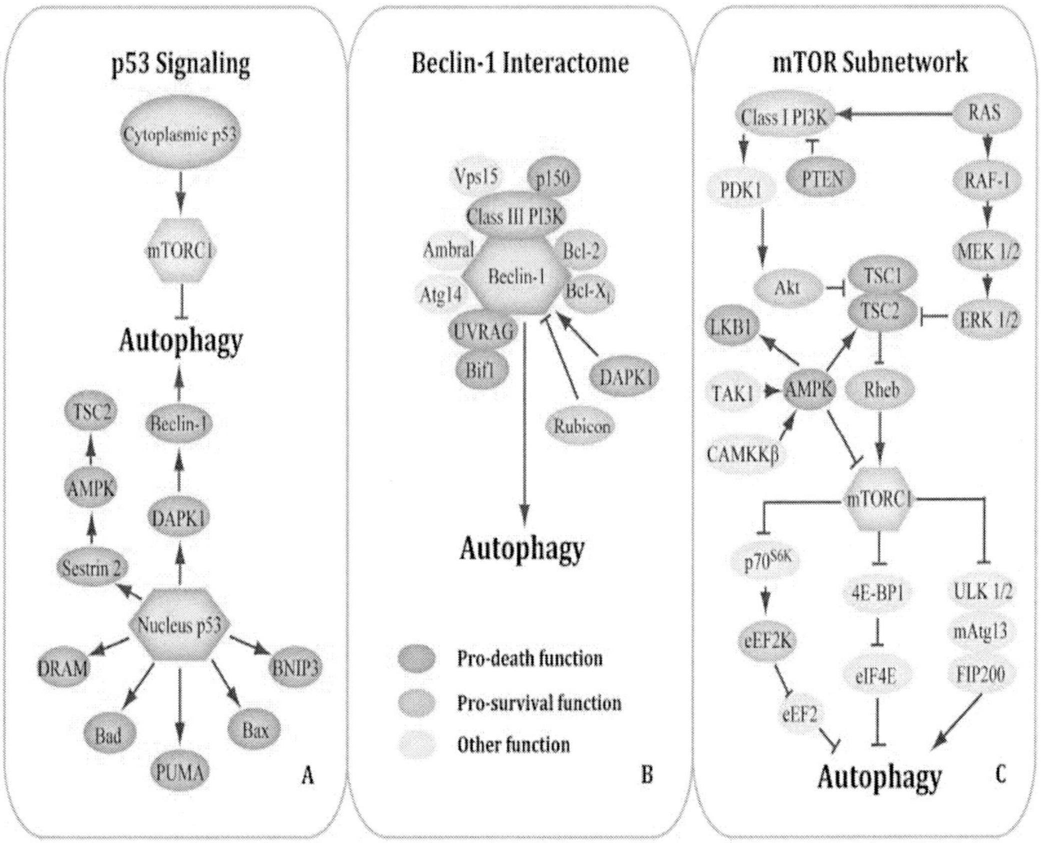

Figure 3. Major regulators of autophagy pathways.

A. p53, the best-characterized tumor suppressor protein, plays a dual role in autophagy. On one hand, nuclear p53 induces autophagy through transcriptional effects by activating DAPK1 and then Beclin-1; On the other hand, cytoplasmic p53 may act as a master repressor of autophagy via the activation of mTORC1 which suppresses autophagy response. B. Beclin-1 (Atg6) complex involves the binding of Bcl-2 to other autophagy regulators, including Vps34 (PI3K), p150, UVRAG, Bif1, Atg14L and Rubicon; thereby forming protein complexes and initiating the formation of double-membrane autophagosome. C. mTOR (mammalian target of rapamycin) has made great progress since it began as a kinase of unknown function. As part of the mTORC1 and mTORC2 complexes, mTOR has the crucial roles in autophagic pathways in human cancer.

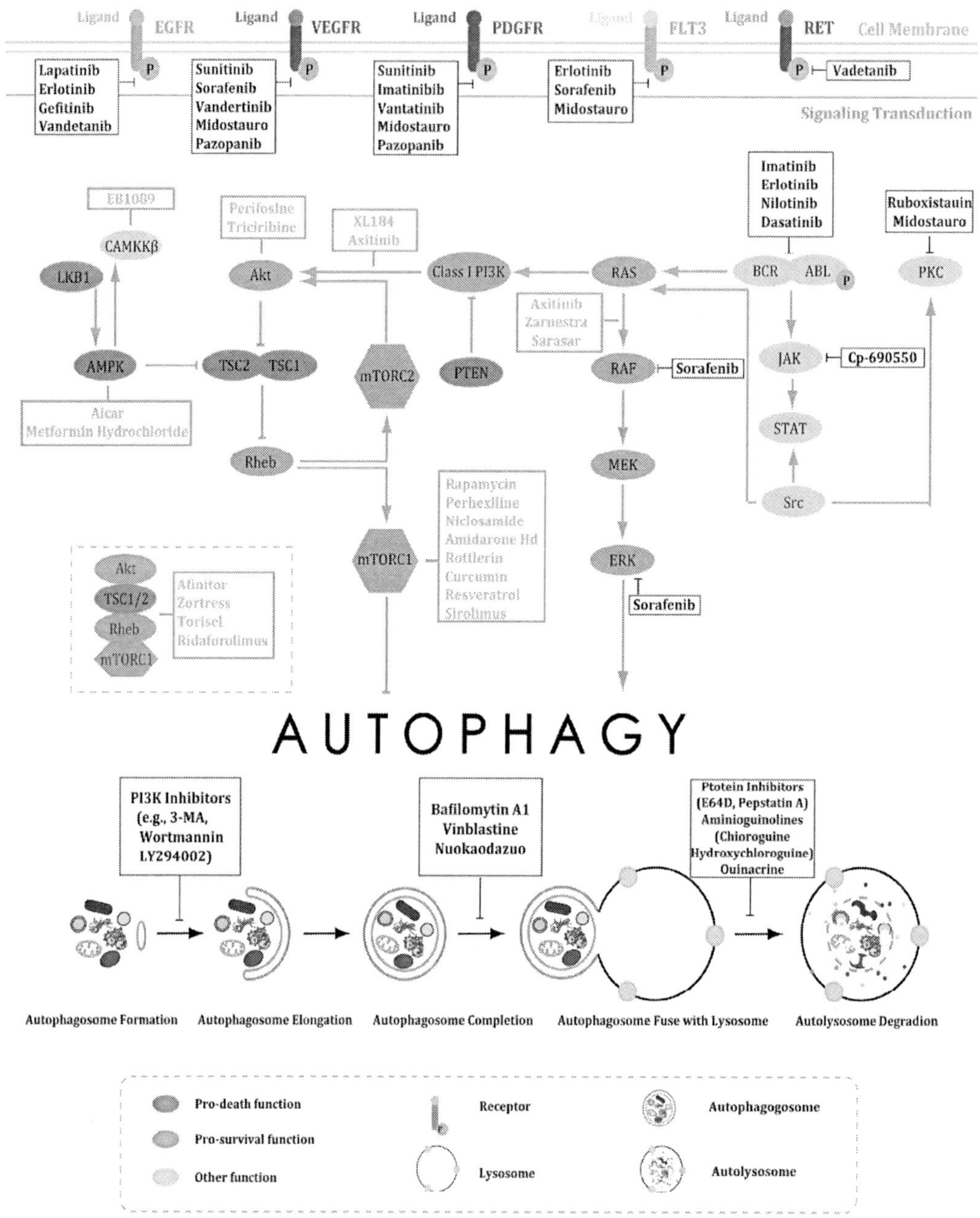

Figure 4. Anti-cancer drugs targeting autophagy.

(A) Certain drugs can target certain cascades in autophagic network. The drugs indicated in black are multiple-target drugs that function through inhibiting signaling transduction, and the drugs shown in grey target either signal pathway or key proteins in the network. (B) Several drugs inhibit autophagic process targeting autophagosome formation, autophagosome fusion with the lysosome, and the degradation of autophagic body.

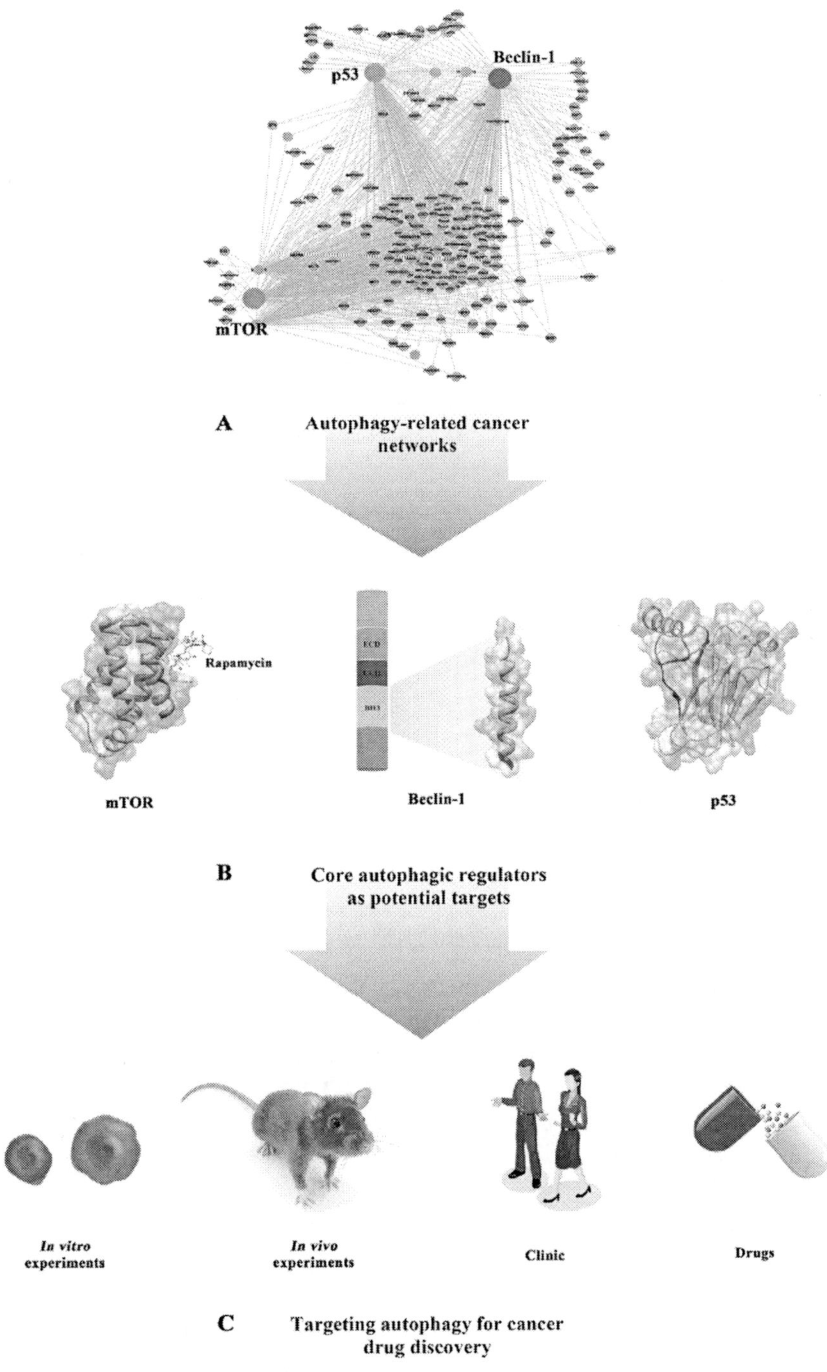

Figure 5. The schematic flowchart depicts that targeting autpophagy-related cancer networks for discovery of potential anti-cancer drugs.

Often autophagy operates as a tumor suppressor by activating pro-autophagic genes and blocking anti-autophagic genes. A few of key genes such as mTOR, p53 and Beclin-1 play the crucial roles in regulating the complicated autophagic network in cancer.

Acknowledgments

We are grateful to Dr. Yan Cheng (Pennsylvania State University), He-jiao Bian (Boston University), Ming-wei Min (University of Cambridge), Qian Liu (National University of Singapore) and Jun-jie Liu (Tsinghua University) for their critical reviews on this manuscript. We also thank Jia-ying Yu, En-qin Wang, Shu-ya Wang, Zi-yi Zhao, Xin Wen, Zi-qiang Lin, Hao Peng, Feng Gao, Yi Wang, Nan Wang, Cheng-cheng Zhou, Miao Ming, Shun Yao, Qi-jia Yu, Zi-yue Li, Chun-yang Li, Xue-mei He, Xiao-yue Zhang, Ke-qin Niu, Li-li Xu, Xiao-min Bao, Hang Zhang, Jinsong Xu and Jing Li (Sichuan University) for their good suggestions on this work. Additionally, this work was supported in part by grants from the National Natural Science Foundation of China (No.30970643), Young teacher's fund of Sichuan University (No. 2010SCU11066) and the Science Foundation for Post Doctorate Research of China (No. 20110491725).

References

[1] Mizushima N, Levine B, Cuervo AM, Klionsky DJ. Autophagy fights disease through cellular self-digestion. *Nature.* 2008;451(7182):1069-75.

[2] Levine B, Kroemer G. Autophagy in the pathogenesis of disease. *Cell.* 2008;132(1):27-42.

[3] Mathew R, Karantza-Wadsworth V, White E. Role of autophagy in cancer. *Nat Rev Cancer.* 2007;7(12):961-7.

[4] Abedin MJ, Wang D, McDonnell MA, Lehmann U, Kelekar A. Autophagy delays apoptotic death in breast cancer cells following DNA damage. *Cell Death Differ.* 2007;14(3):500-10.

[5] Carew JS, Nawrocki ST, Kahue CN, Zhang H, Yang C, Chung L, et al. Targeting autophagy augments the anticancer activity of the histone deacetylase inhibitor SAHA to overcome Bcr-Abl-mediated drug resistance. *Blood.* 2007;110(1):313-22.

[6] Tsukada M, Ohsumi Y. Isolation and characterization of autophagy-defective mutants of *Saccharomyces cerevisiae*. *FEBS Lett.* 1993;333(1-2):169-74.

[7] Fader CM, Colombo MI. Autophagy and multivesicular bodies: two closely related partners. *Cell Death Differ.* 2009;16(1):70-8.

[8] Yoshimori T, Noda T. Toward unraveling membrane biogenesis in mammalian autophagy. *Curr Opin Cell Biol.* 2008;20(4):401-7.

[9] Wullschleger S, Loewith R, Hall MN. TOR signaling in growth and metabolism. *Cell.* 2006;124(3):471-84.

[10] Kamada Y, Funakoshi T, Shintani T, Nagano K, Ohsumi M, Ohsumi Y. Tor-mediated induction of autophagy via an Apg1 protein kinase complex. *J Cell Biol.* 2000;150(6):1507-13.

[11] Kihara A, Noda T, Ishihara N, Ohsumi Y. Two distinct Vps34 phosphatidylinositol 3-kinase complexes function in autophagy and carboxypeptidase Y sorting in *Saccharomyces cerevisiae*. *J Cell Biol.* 2001;152(3):519-30.

[12] Stack JH, DeWald DB, Takegawa K, Emr SD. Vesicle-mediated protein transport: regulatory interactions between the Vps15 protein kinase and the Vps34 PtdIns 3-kinase essential for protein sorting to the vacuole in yeast. *J Cell Biol.* 1995;129(2):321-34.
[13] Liang C, Feng P, Ku B, Dotan I, Canaani D, Oh BH, et al. Autophagic and tumour suppressor activity of a novel Beclin1-binding protein UVRAG. *Nat Cell Biol.* 2006;8(7):688-99.
[14] Takahashi Y, Coppola D, Matsushita N, Cualing HD, Sun M, Sato Y, et al. Bif-1 interacts with Beclin 1 through UVRAG and regulates autophagy and tumorigenesis. *Nat Cell Biol.* 2007;9(10):1142-51.
[15] Obara K, Sekito T, Ohsumi Y. Assortment of phosphatidylinositol 3-kinase complexes--Atg14p directs association of complex I to the pre-autophagosomal structure in *Saccharomyces cerevisiae*. *Mol Biol Cell.* 2006;17(4):1527-39.
[16] Itakura E, Kishi C, Inoue K, Mizushima N. Beclin 1 forms two distinct phosphatidylinositol 3-kinase complexes with mammalian Atg14 and UVRAG. *Mol Biol Cell.* 2008;19(12):5360-72.
[17] Wishart MJ, Taylor GS, Dixon JE. Phoxy lipids: revealing PX domains as phosphoinositide binding modules. *Cell.* 2001;105(7):817-20.
[18] Gillooly DJ, Simonsen A, Stenmark H. Cellular functions of phosphatidylinositol 3-phosphate and FYVE domain proteins. *Biochem J.* 2001;355(Pt 2):249-58.
[19] Mizushima N, Noda T, Yoshimori T, Tanaka Y, Ishii T, George MD, et al. A protein conjugation system essential for autophagy. *Nature.* 1998;395(6700):395-8.
[20] Tanida I, Mizushima N, Kiyooka M, Ohsumi M, Ueno T, Ohsumi Y, et al. Apg7p/Cvt2p: A novel protein-activating enzyme essential for autophagy. *Mol Biol Cell.* 1999;10(5):1367-79.
[21] Mizushima N, Kuma A, Kobayashi Y, Yamamoto A, Matsubae M, Takao T, et al. Mouse Apg16L, a novel WD-repeat protein, targets to the autophagic isolation membrane with the Apg12-Apg5 conjugate. *J Cell Sci.* 2003;116(Pt 9):1679-88.
[22] Ichimura Y, Kirisako T, Takao T, Satomi Y, Shimonishi Y, Ishihara N, et al. A ubiquitin-like system mediates protein lipidation. *Nature.* 2000;408(6811):488-92.
[23] Kirisako T, Ichimura Y, Okada H, Kabeya Y, Mizushima N, Yoshimori T, et al. The reversible modification regulates the membrane-binding state of Apg8/Aut7 essential for autophagy and the cytoplasm to vacuole targeting pathway. *J Cell Biol.* 2000;151(2):263-76.
[24] Mizushima N, Yamamoto A, Hatano M, Kobayashi Y, Kabeya Y, Suzuki K, et al. Dissection of autophagosome formation using Apg5-deficient mouse embryonic stem cells. *J Cell Biol.* 2001;152(4):657-68.
[25] Hanada T, Noda NN, Satomi Y, Ichimura Y, Fujioka Y, Takao T, et al. The Atg12-Atg5 conjugate has a novel E3-like activity for protein lipidation in autophagy. *J Biol Chem.* 2007;282(52):37298-302.
[26] Fass E, Shvets E, Degani I, Hirschberg K, Elazar Z. Microtubules support production of starvation-induced autophagosomes but not their targeting and fusion with lysosomes. *J Biol Chem.* 2006;281(47):36303-16.
[27] Corcelle EA, Puustinen P, Jaattela M. Apoptosis and autophagy: Targeting autophagy signalling in cancer cells -'trick or treats'? *FEBS J.* 2009;276(21):6084-96.

[28] Liang C, Lee JS, Inn KS, Gack MU, Li Q, Roberts EA, et al. Beclin1-binding UVRAG targets the class C Vps complex to coordinate autophagosome maturation and endocytic trafficking. *Nat Cell Biol.* 2008;10(7):776-87.

[29] Kang R, Zeh HJ, Lotze MT, Tang D. The Beclin 1 network regulates autophagy and apoptosis. *Cell Death Differ.* 2011;18(4):571-80.

[30] Tong J, Yan X, Yu L. The late stage of autophagy: cellular events and molecular regulation. *Protein Cell.* 2010;1(10):907-15.

[31] Dalby KN, Tekedereli I, Lopez-Berestein G, Ozpolat B. Targeting the prodeath and prosurvival functions of autophagy as novel therapeutic strategies in cancer. *Autophagy.* 2010;6(3):322-9.

[32] Kroemer G, Levine B. Autophagic cell death: the story of a misnomer. *Nat Rev Mol Cell Biol.* 2008;9(12):1004-10.

[33] Fingar DC, Blenis J. Target of rapamycin (TOR): an integrator of nutrient and growth factor signals and coordinator of cell growth and cell cycle progression. *Oncogene.* 2004;23(18):3151-71.

[34] Loewith R, Jacinto E, Wullschleger S, Lorberg A, Crespo JL, Bonenfant D, et al. Two TOR complexes, only one of which is rapamycin sensitive, have distinct roles in cell growth control. *Mol Cell.* 2002;10(3):457-68.

[35] Liu B, Cheng Y, Liu Q, Bao JK, Yang JM. Autophagic pathways as new targets for cancer drug development. *Acta Pharmacol Sin.* 2010;31(9):1154-64.

[36] Inoki K, Li Y, Zhu T, Wu J, Guan KL. TSC2 is phosphorylated and inhibited by Akt and suppresses mTOR signalling. *Nat Cell Biol.* 2002;4(9):648-57.

[37] Roux PP, Ballif BA, Anjum R, Gygi SP, Blenis J. Tumor-promoting phorbol esters and activated Ras inactivate the tuberous sclerosis tumor suppressor complex via p90 ribosomal S6 kinase. *Proc Natl Acad Sci U S A.* 2004;101(37):13489-94.

[38] Corradetti MN, Inoki K, Bardeesy N, DePinho RA, Guan KL. Regulation of the TSC pathway by LKB1: evidence of a molecular link between tuberous sclerosis complex and Peutz-Jeghers syndrome. *Genes Dev.* 2004;18(13):1533-8.

[39] Hoyer-Hansen M, Bastholm L, Szyniarowski P, Campanella M, Szabadkai G, Farkas T, et al. Control of macroautophagy by calcium, calmodulin-dependent kinase kinase-beta, and Bcl-2. *Mol Cell.* 2007;25(2):193-205.

[40] Budanov AV, Karin M. p53 target genes sestrin1 and sestrin2 connect genotoxic stress and mTOR signaling. *Cell.* 2008;134(3):451-60.

[41] Hara T, Takamura A, Kishi C, Iemura S, Natsume T, Guan JL, et al. FIP200, a ULK-interacting protein, is required for autophagosome formation in mammalian cells. *J Cell Biol.* 2008;181(3):497-510.

[42] Suzuki K, Kubota Y, Sekito T, Ohsumi Y. Hierarchy of Atg proteins in pre-autophagosomal structure organization. *Genes Cells.* 2007;12(2):209-18.

[43] Gingras AC, Raught B, Sonenberg N. Regulation of translation initiation by FRAP/mTOR. *Genes Dev.* 2001;15(7):807-26.

[44] Roux PP, Blenis J. ERK and p38 MAPK-activated protein kinases: a family of protein kinases with diverse biological functions. *Microbiol Mol Biol Rev.* 2004;68(2):320-44.

[45] Kolch W. Coordinating ERK/MAPK signalling through scaffolds and inhibitors. *Nat Rev Mol Cell Biol.* 2005;6(11):827-37.

[46] Howe CL, Valletta JS, Rusnak AS, Mobley WC. NGF signaling from clathrin-coated vesicles: evidence that signaling endosomes serve as a platform for the Ras-MAPK pathway. *Neuron.* 2001;32(5):801-14.

[47] Buday L, Downward J. Many faces of Ras activation. *Biochim Biophys Acta.* 2008;1786(2):178-87.

[48] Roberts PJ, Der CJ. Targeting the Raf-MEK-ERK mitogen-activated protein kinase cascade for the treatment of cancer. *Oncogene.* 2007;26(22):3291-310.

[49] Engelman JA. Targeting PI3K signalling in cancer: opportunities, challenges and limitations. *Nat Rev Cancer.* 2009;9(8):550-62.

[50] Burman C, Ktistakis NT. Regulation of autophagy by phosphatidylinositol 3-phosphate. *FEBS Lett.* 2010;584(7):1302-12.

[51] Steelman LS, Pohnert SC, Shelton JG, Franklin RA, Bertrand FE, McCubrey JA. JAK/STAT, Raf/MEK/ERK, PI3K/Akt and BCR-ABL in cell cycle progression and leukemogenesis. *Leukemia.* 2004;18(2):189-218.

[52] Rommel C, Clarke BA, Zimmermann S, Nunez L, Rossman R, Reid K, et al. Differentiation stage-specific inhibition of the Raf-MEK-ERK pathway by Akt. *Science.* 1999;286(5445):1738-41.

[53] Kwon DS, Kwon CH, Kim JH, Woo JS, Jung JS, Kim YK. Signal transduction of MEK/ERK and PI3K/Akt activation by hypoxia/reoxygenation in renal epithelial cells. *Eur J Cell Biol.* 2006;85(11):1189-99.

[54] Crighton D, Wilkinson S, O'Prey J, Syed N, Smith P, Harrison PR, et al. DRAM, a p53-induced modulator of autophagy, is critical for apoptosis. *Cell.* 2006;126(1):121-34.

[55] Tasdemir E, Maiuri MC, Galluzzi L, Vitale I, Djavaheri-Mergny M, D'Amelio M, et al. Regulation of autophagy by cytoplasmic p53. *Nat Cell Biol.* 2008;10(6):676-87.

[56] Maiuri MC, Malik SA, Morselli E, Kepp O, Criollo A, Mouchel PL, et al. Stimulation of autophagy by the p53 target gene Sestrin2. *Cell Cycle.* 2009;8(10):1571-6.

[57] Qu X, Yu J, Bhagat G, Furuya N, Hibshoosh H, Troxel A, et al. Promotion of tumorigenesis by heterozygous disruption of the beclin 1 autophagy gene. *J Clin Invest.* 2003;112(12):1809-20.

[58] Yue Z, Jin S, Yang C, Levine AJ, Heintz N. Beclin 1, an autophagy gene essential for early embryonic development, is a haploinsufficient tumor suppressor. *Proc Natl Acad Sci U S A.* 2003;100(25):15077-82.

[59] Liang XH, Jackson S, Seaman M, Brown K, Kempkes B, Hibshoosh H, et al. Induction of autophagy and inhibition of tumorigenesis by beclin 1. *Nature.* 1999;402(6762):672-6.

[60] Ionov Y, Nowak N, Perucho M, Markowitz S, Cowell JK. Manipulation of nonsense mediated decay identifies gene mutations in colon cancer Cells with microsatellite instability. *Oncogene.* 2004;23(3):639-45.

[61] Coppola D, Oliveri C, Sayegh Z, Boulware D, Takahashi Y, Pow-Sang J, et al. Bax-interacting factor-1 expression in prostate cancer. *Clin Genitourin Cancer.* 2008;6(2):117-21.

[62] Hoekstra AV, Ward EC, Hardt JL, Lurain JR, Singh DK, Buttin BM, et al. Chemosensitization of endometrial cancer cells through AKT inhibition involves FOXO1. *Gynecol Oncol.* 2008;108(3):609-18.

[63] Nakae J, Biggs WH, 3rd, Kitamura T, Cavenee WK, Wright CV, Arden KC, et al. Regulation of insulin action and pancreatic beta-cell function by mutated alleles of the gene encoding forkhead transcription factor Foxo1. *Nat Genet.* 2002;32(2):245-53.

[64] Tasdemir E, Chiara Maiuri M, Morselli E, Criollo A, D'Amelio M, Djavaheri-Mergny M, et al. A dual role of p53 in the control of autophagy. Autophagy. 2008;4(6):810-4.

[65] Ryan KM. p53 and autophagy in cancer: guardian of the genome meets guardian of the proteome. *Eur J Cancer.* 2011;47(1):44-50.

[66] Vogelstein B, Lane D, Levine AJ. Surfing the p53 network. *Nature.* 2000;408 (6810):307-10.

[67] Levine B, Abrams J. p53: The Janus of autophagy? *Nat Cell Biol.* 2008;10(6):637-9.

[68] Kondo Y, Kondo S. Autophagy and cancer therapy. *Autophagy.* 2006;2(2):85-90.

[69] Maiuri MC, Criollo A, Tasdemir E, Vicencio JM, Tajeddine N, Hickman JA, et al. BH3-only proteins and BH3 mimetics induce autophagy by competitively disrupting the interaction between Beclin 1 and Bcl-2/Bcl-X(L). *Autophagy.* 2007;3(4):374-6.

[70] Pattingre S, Tassa A, Qu X, Garuti R, Liang XH, Mizushima N, et al. Bcl-2 antiapoptotic proteins inhibit Beclin 1-dependent autophagy. *Cell.* 2005;122(6):927-39.

[71] Maiuri MC, Le Toumelin G, Criollo A, Rain JC, Gautier F, Juin P, et al. Functional and physical interaction between Bcl-X(L) and a BH3-like domain in Beclin-1. *EMBO J.* 2007;26(10):2527-39.

[72] Zalckvar E, Berissi H, Eisenstein M, Kimchi A. Phosphorylation of Beclin 1 by DAP-kinase promotes autophagy by weakening its interactions with Bcl-2 and Bcl-XL. *Autophagy.* 2009;5(5):720-2.

[73] Crespo JL, Powers T, Fowler B, Hall MN. The TOR-controlled transcription activators GLN3, RTG1, and RTG3 are regulated in response to intracellular levels of glutamine. *Proc Natl Acad Sci U S A.* 2002;99(10):6784-9.

[74] Jacinto E, Hall MN. Tor signalling in bugs, brain and brawn. *Nat Rev Mol Cell Biol.* 2003;4(2):117-26.

[75] Kamada Y, Yoshino K, Kondo C, Kawamata T, Oshiro N, Yonezawa K, et al. Tor directly controls the Atg1 kinase complex to regulate autophagy. *Mol Cell Biol.* 2010;30(4):1049-58.

[76] Levine B, Klionsky DJ. Development by self-digestion: molecular mechanisms and biological functions of autophagy. *Dev Cell.* 2004;6(4):463-77.

[77] Lin Y, Hupp TR, Stevens C. Death-associated protein kinase (DAPK) and signal transduction: additional roles beyond cell death. *FEBS J.* 2010;277(1):48-57.

[78] Inbal B, Bialik S, Sabanay I, Shani G, Kimchi A. DAP kinase and DRP-1 mediate membrane blebbing and the formation of autophagic vesicles during programmed cell death. *J Cell Biol.* 2002;157(3):455-68.

[79] Harrison B, Kraus M, Burch L, Stevens C, Craig A, Gordon-Weeks P, et al. DAPK-1 binding to a linear peptide motif in MAP1B stimulates autophagy and membrane blebbing. *J Biol Chem.* 2008;283(15):9999-10014.

[80] Zalckvar E, Berissi H, Mizrachy L, Idelchuk Y, Koren I, Eisenstein M, et al. DAP-kinase-mediated phosphorylation on the BH3 domain of beclin 1 promotes dissociation of beclin 1 from Bcl-XL and induction of autophagy. *EMBO Rep.* 2009;10(3):285-92.

[81] Tian JH, Das S, Sheng ZH. Ca^{2+}-dependent phosphorylation of syntaxin-1A by the death-associated protein (DAP) kinase regulates its interaction with Munc18. *J Biol Chem.* 2003;278(28):26265-74.

[82] Wu J, Kaufman RJ. From acute ER stress to physiological roles of the Unfolded Protein Response. *Cell Death Differ.* 2006;13(3):374-84.
[83] Scott RC, Schuldiner O, Neufeld TP. Role and regulation of starvation-induced autophagy in the Drosophila fat body. *Dev Cell.* 2004;7(2):167-78.
[84] Klionsky DJ. The molecular machinery of autophagy and its role in physiology and disease. *Semin Cell Dev Biol.* 2010;21(7):663.
[85] Huang J, Klionsky DJ. Autophagy and human disease. *Cell Cycle.* 2007;6(15):1837-49.
[86] Todde V, Veenhuis M, van der Klei IJ. Autophagy: principles and significance in health and disease. *Biochim Biophys Acta.* 2009;1792(1):3-13.
[87] Kundu M, Thompson CB. Autophagy: basic principles and relevance to disease. *Annu Rev Pathol.* 2008;3:427-55.
[88] Eskelinen EL, Saftig P. Autophagy: a lysosomal degradation pathway with a central role in health and disease. *Biochim Biophys Acta.* 2009;1793(4):664-73.
[89] Liang XH, Kleeman LK, Jiang HH, Gordon G, Goldman JE, Berry G, et al. Protection against fatal Sindbis virus encephalitis by beclin, a novel Bcl-2-interacting protein. *J Virol.* 1998;72(11):8586-96.
[90] Aita VM, Liang XH, Murty VV, Pincus DL, Yu W, Cayanis E, et al. Cloning and genomic organization of beclin 1, a candidate tumor suppressor gene on chromosome 17q21. *Genomics.* 1999;59(1):59-65.
[91] Arico S, Petiot A, Bauvy C, Dubbelhuis PF, Meijer AJ, Codogno P, et al. The tumor suppressor PTEN positively regulates macroautophagy by inhibiting the phosphatidylinositol 3-kinase/protein kinase B pathway. *J Biol Chem.* 2001;276(38):35243-6.
[92] Turcotte S, Giaccia AJ. Targeting cancer cells through autophagy for anticancer therapy. *Curr Opin Cell Biol.* 2010;22(2):246-51.
[93] Karantza-Wadsworth V, Patel S, Kravchuk O, Chen G, Mathew R, Jin S, et al. Autophagy mitigates metabolic stress and genome damage in mammary tumorigenesis. *Genes Dev.* 2007;21(13):1621-35.
[94] Jin S, White E. Role of autophagy in cancer: management of metabolic stress. *Autophagy.* 2007;3(1):28-31.
[95] Ma XH, Piao S, Wang D, McAfee QW, Nathanson KL, Lum JJ, et al. Measurements of tumor cell autophagy predict invasiveness, resistance to chemotherapy, and survival in melanoma. *Clin Cancer Res.* 2011;17(10):3478-89.
[96] Eskelinen EL. The dual role of autophagy in cancer. *Curr Opin Pharmacol.* 2011;11(4):294-300.
[97] Liang J, Shao SH, Xu ZX, Hennessy B, Ding Z, Larrea M, et al. The energy sensing LKB1-AMPK pathway regulates p27(kip1) phosphorylation mediating the decision to enter autophagy or apoptosis. *Nat Cell Biol.* 2007;9(2):218-24.
[98] Lum JJ, Bauer DE, Kong M, Harris MH, Li C, Lindsten T, et al. Growth factor regulation of autophagy and cell survival in the absence of apoptosis. *Cell.* 2005;120(2):237-48.
[99] Zhu H, Wu H, Liu X, Li B, Chen Y, Ren X, et al. Regulation of autophagy by a beclin 1-targeted microRNA, miR-30a, in cancer cells. *Autophagy.* 2009;5(6):816-23.
[100] Jegga AG, Schneider L, Ouyang X, Zhang J. Systems biology of the autophagy-lysosomal pathway. *Autophagy.* 2011;7(5):477-89.

[101] Brest P, Lapaquette P, Souidi M, Lebrigand K, Cesaro A, Vouret-Craviari V, et al. A synonymous variant in IRGM alters a binding site for miR-196 and causes deregulation of IRGM-dependent xenophagy in Crohn's disease. *Nat Genet.* 2011;43(3):242-5.

[102] Jian X, Xiao-yan Z, Bin H, Yu-feng Z, Bo K, Zhi-nong W, et al. MiR-204 regulate cardiomyocyte autophagy induced by hypoxia-reoxygenation through LC3-II. *Int J Cardiol.* 2011;148(1):110-2.

[103] Liu JJ, Lin M, Yu JY, Liu B, Bao JK. Targeting apoptotic and autophagic pathways for cancer therapeutics. *Cancer Lett.* 2011;300(2):105-14.

[104] Cao C, Subhawong T, Albert JM, Kim KW, Geng L, Sekhar KR, et al. Inhibition of mammalian target of rapamycin or apoptotic pathway induces autophagy and radiosensitizes PTEN null prostate cancer cells. *Cancer Res.* 2006;66(20):10040-7.

[105] Amaravadi RK, Yu D, Lum JJ, Bui T, Christophorou MA, Evan GI, et al. Autophagy inhibition enhances therapy-induced apoptosis in a Myc-induced model of lymphoma. *J Clin Invest.* 2007;117(2):326-36.

[106] Aguirre-Ghiso JA. The problem of cancer dormancy: understanding the basic mechanisms and identifying therapeutic opportunities. *Cell Cycle.* 2006;5(16):1740-3.

[107] She QB, Solit D, Basso A, Moasser MM. Resistance to gefitinib in PTEN-null HER-overexpressing tumor cells can be overcome through restoration of PTEN function or pharmacologic modulation of constitutive phosphatidylinositol 3'-kinase/Akt pathway signaling. *Clin Cancer Res.* 2003;9(12):4340-6.

[108] Eimer S, Belaud-Rotureau MA, Airiau K, Jeanneteau M, Laharanne E, Veron N, et al. Autophagy inhibition cooperates with erlotinib to induce glioblastoma cell death. *Cancer Biol Ther.* 2011;11(12):1017-27.

[109] Huether A, Hopfner M, Baradari V, Schuppan D, Scherubl H. Sorafenib alone or as combination therapy for growth control of cholangiocarcinoma. *Biochem Pharmacol.* 2007;73(9):1308-17.

[110] Faivre S, Demetri G, Sargent W, Raymond E. Molecular basis for sunitinib efficacy and future clinical development. *Nat Rev Drug Discov.* 2007;6(9):734-45.

[111] Quintas-Cardama A, Kantarjian H, Jones D, Nicaise C, O'Brien S, Giles F, et al. Dasatinib (BMS-354825) is active in Philadelphia chromosome-positive chronic myelogenous leukemia after imatinib and nilotinib (AMN107) therapy failure. *Blood.* 2007;109(2):497-9.

[112] Xia W, Mullin RJ, Keith BR, Liu LH, Ma H, Rusnak DW, et al. Anti-tumor activity of GW572016: a dual tyrosine kinase inhibitor blocks EGF activation of EGFR/erbB2 and downstream Erk1/2 and AKT pathways. *Oncogene.* 2002;21(41):6255-63.

[113] Kantarjian H, Giles F, Wunderle L, Bhalla K, O'Brien S, Wassmann B, et al. Nilotinib in imatinib-resistant CML and Philadelphia chromosome-positive ALL. *N Engl J Med.* 2006;354(24):2542-51.

[114] Boni J, Leister C, Burns J, Cincotta M, Hug B, Moore L. Pharmacokinetic profile of temsirolimus with concomitant administration of cytochrome p450-inducing medications. *J Clin Pharmacol.* 2007;47(11):1430-9.

[115] Deininger MW, Druker BJ. Specific targeted therapy of chronic myelogenous leukemia with imatinib. *Pharmacol Rev.* 2003;55(3):401-23.

[116] Lu J, Chew EH, Holmgren A. Targeting thioredoxin reductase is a basis for cancer therapy by arsenic trioxide. *Proc Natl Acad Sci U S A.* 2007;104(30):12288-93.

[117] Singh BN, Vaughan Williams EM. The effect of amiodarone, a new anti-anginal drug, on cardiac muscle. *Br J Pharmacol.* 1970;39(4):657-67.
[118] Marchesini G, Brizi M, Bianchi G, Tomassetti S, Zoli M, Melchionda N. Metformin in non-alcoholic steatohepatitis. *Lancet.* 2001;358(9285):893-4.
[119] Balgi AD, Fonseca BD, Donohue E, Tsang TC, Lajoie P, Proud CG, et al. Screen for chemical modulators of autophagy reveals novel therapeutic inhibitors of mTORC1 signaling. *PLoS ONE. 2009*;4(9):e7124.
[120] Chan S. Targeting the mammalian target of rapamycin (mTOR): a new approach to treating cancer. *Br J Cancer.* 2004;91(8):1420-4.
[121] Adams J, Kauffman M. Development of the proteasome inhibitor Velcade (Bortezomib). *Cancer Invest.* 2004;22(2):304-11.
[122] Talpaz M, Shah NP, Kantarjian H, Donato N, Nicoll J, Paquette R, et al. Dasatinib in imatinib-resistant Philadelphia chromosome-positive leukemias. *N Engl J Med.* 2006;354(24):2531-41.
[123] Martin AP, Park MA, Mitchell C, Walker T, Rahmani M, Thorburn A, et al. BCL-2 family inhibitors enhance histone deacetylase inhibitor and sorafenib lethality via autophagy and overcome blockade of the extrinsic pathway to facilitate killing. *Mol Pharmacol.* 2009;76(2):327-41.
[124] Munshi A, Tanaka T, Hobbs ML, Tucker SL, Richon VM, Meyn RE. Vorinostat, a histone deacetylase inhibitor, enhances the response of human tumor cells to ionizing radiation through prolongation of gamma-H2AX foci. *Mol Cancer Ther.* 2006;5(8):1967-74.
[125] Edwards JC, Szczepanski L, Szechinski J, Filipowicz-Sosnowska A, Emery P, Close DR, et al. Efficacy of B-cell-targeted therapy with rituximab in patients with rheumatoid arthritis. *N Engl J Med.* 2004;350(25):2572-81.
[126] Wu M, Rivkin A, Pham T. Panitumumab: human monoclonal antibody against epidermal growth factor receptors for the treatment of metastatic colorectal cancer. *Clin Ther.* 2008;30(1):14-30.
[127] Howell A, Cuzick J, Baum M, Buzdar A, Dowsett M, Forbes JF, et al. Results of the ATAC (Arimidex, Tamoxifen, Alone or in Combination) trial after completion of 5 years' adjuvant treatment for breast cancer. *Lancet.* 2005;365(9453):60-2.
[128] Taneja SS, Smith MR, Dalton JT, Raghow S, Barnette G, Steiner M, et al. Toremifene--a promising therapy for the prevention of prostate cancer and complications of androgen deprivation therapy. *Expert Opin Investig Drugs*. 2006;15(3):293-305.
[129] Arao T, Fukumoto H, Takeda M, Tamura T, Saijo N, Nishio K. Small in-frame deletion in the epidermal growth factor receptor as a target for ZD6474. *Cancer Res.* 2004;64(24):9101-4.
[130] Podar K, Tonon G, Sattler M, Tai YT, Legouill S, Yasui H, et al. The small-molecule VEGF receptor inhibitor pazopanib (GW786034B) targets both tumor and endothelial cells in multiple myeloma. *Proc Natl Acad Sci U S A*. 2006;103(51):19478-83.
[131] Kudlacz E, Perry B, Sawyer P, Conklyn M, McCurdy S, Brissette W, et al. The novel JAK-3 inhibitor CP-690550 is a potent immunosuppressive agent in various murine models. *Am J Transplant.* 2004;4(1):51-7.
[132] Millward MJ, House C, Bowtell D, Webster L, Olver IN, Gore M, et al. The multikinase inhibitor midostaurin (PKC412A) lacks activity in metastatic melanoma: a phase IIA clinical and biologic study. *Br J Cancer.* 2006;95(7):829-34.

[133] Drevs J, Muller-Driver R, Wittig C, Fuxius S, Esser N, Hugenschmidt H, et al. PTK787/ZK 222584, a specific vascular endothelial growth factor-receptor tyrosine kinase inhibitor, affects the anatomy of the tumor vascular bed and the functional vascular properties as detected by dynamic enhanced magnetic resonance imaging. *Cancer Res.* 2002;62(14):4015-22.

[134] Bartlett S, Beddard GS, Jackson RM, Kayser V, Kilner C, Leach A, et al. Comparison of the ATP binding sites of protein kinases using conformationally diverse bisindolylmaleimides. *J Am Chem Soc.* 2005;127(33):11699-708.

[135] Polverino A, Coxon A, Starnes C, Diaz Z, DeMelfi T, Wang L, et al. AMG 706, an oral, multikinase inhibitor that selectively targets vascular endothelial growth factor, platelet-derived growth factor, and kit receptors, potently inhibits angiogenesis and induces regression in tumor xenografts. *Cancer Res.* 2006;66(17):8715-21.

[136] Wilmes LJ, Pallavicini MG, Fleming LM, Gibbs J, Wang D, Li KL, et al. AG-013736, a novel inhibitor of VEGF receptor tyrosine kinases, inhibits breast cancer growth and decreases vascular permeability as detected by dynamic contrast-enhanced magnetic resonance imaging. *Magn Reson Imaging.* 2007;25(3):319-27.

[137] Durante C, Russo D, Verrienti A, Filetti S. XL184 (cabozantinib) for medullary thyroid carcinoma. *Expert Opin Investig Drugs.* 2011;20(3):407-13.

[138] Rivera VM, Squillace RM, Miller D, Berk L, Wardwell SD, Ning Y, et al. Ridaforolimus (AP23573; MK-8669), a potent mTOR inhibitor, has broad antitumor activity and can be optimally administered using intermittent dosing regimens. *Mol Cancer Ther.* 2011;10(6):1059-71.

[139] Wolf JK, Bodurka DC, Gano JB, Deavers M, Ramondetta L, Ramirez PT, et al. A phase I study of Adp53 (INGN 201; ADVEXIN) for patients with platinum- and paclitaxel-resistant epithelial ovarian cancer. *Gynecol Oncol.* 2004;94(2):442-8.

[140] Bissett D, O'Byrne KJ, von Pawel J, Gatzemeier U, Price A, Nicolson M, et al. Phase III study of matrix metalloproteinase inhibitor prinomastat in non-small-cell lung cancer. *J Clin Oncol.* 2005;23(4):842-9.

[141] Kim D, Cheng GZ, Lindsley CW, Yang H, Cheng JQ. Targeting the phosphatidylinositol-3 kinase/Akt pathway for the treatment of cancer. *Curr Opin Investig Drugs.* 2005;6(12):1250-8.

[142] Alvero AB, Montagna MK, Chen R, Kim KH, Kyungjin K, Visintin I, et al. NV-128, a novel isoflavone derivative, induces caspase-independent cell death through the Akt/mammalian target of rapamycin pathway. *Cancer.* 2009;115(14):3204-16.

[143] Zhu K, Dunner K, Jr., McConkey DJ. Proteasome inhibitors activate autophagy as a cytoprotective response in human prostate cancer cells. *Oncogene.* 2010;29(3):451-62.

[144] Mathew R, Karp CM, Beaudoin B, Vuong N, Chen G, Chen HY, et al. Autophagy suppresses tumorigenesis through elimination of p62. *Cell.* 2009;137(6):1062-75.

[145] Pan J, Chen B, Su CH, Zhao R, Xu ZX, Sun L, et al. Autophagy induced by farnesyltransferase inhibitors in cancer cells. *Cancer Biol Ther.* 2008;7(10):1679-84.

[146] Albert JM, Cao C, Kim KW, Willey CD, Geng L, Xiao D, et al. Inhibition of poly(ADP-ribose) polymerase enhances cell death and improves tumor growth delay in irradiated lung cancer models. *Clin Cancer Res.* 2007;13(10):3033-42.

[147] Hoyer-Hansen M, Bastholm L, Mathiasen IS, Elling F, Jaattela M. Vitamin D analog EB1089 triggers dramatic lysosomal changes and Beclin 1-mediated autophagic cell death. *Cell Death Differ.* 2005;12(10):1297-309.

[148] Stein M, Lin H, Jeyamohan C, Dvorzhinski D, Gounder M, Bray K, et al. Targeting tumor metabolism with 2-deoxyglucose in patients with castrate-resistant prostate cancer and advanced malignancies. *Prostate*. 2010;70(13):1388-94.

[149] Carew JS, Medina EC, Esquivel JA, 2nd, Mahalingam D, Swords R, Kelly K, et al. Autophagy inhibition enhances vorinostat-induced apoptosis via ubiquitinated protein accumulation. *J Cell Mol Med.* 2010;14(10):2448-59.

[150] Takeuchi H, Kondo Y, Fujiwara K, Kanzawa T, Aoki H, Mills GB, et al. Synergistic augmentation of rapamycin-induced autophagy in malignant glioma cells by phosphatidylinositol 3-kinase/protein kinase B inhibitors. *Cancer Res.* 2005;65(8):3336-46.

[151] Fujiwara K, Iwado E, Mills GB, Sawaya R, Kondo S, Kondo Y. Akt inhibitor shows anticancer and radiosensitizing effects in malignant glioma cells by inducing autophagy. *Int J Oncol.* 2007;31(4):753-60.

[152] Maclean KH, Dorsey FC, Cleveland JL, Kastan MB. Targeting lysosomal degradation induces p53-dependent cell death and prevents cancer in mouse models of lymphomagenesis. *J Clin Invest.* 2008;118(1):79-88.

[153] Gupta A, Roy S, Lazar AJ, Wang WL, McAuliffe JC, Reynoso D, et al. Autophagy inhibition and antimalarials promote cell death in gastrointestinal stromal tumor (GIST). *Proc Natl Acad Sci U S A.* 2010;107(32):14333-8.

[154] Lu Z, Luo RZ, Lu Y, Zhang X, Yu Q, Khare S, et al. The tumor suppressor gene ARHI regulates autophagy and tumor dormancy in human ovarian cancer cells. *J Clin Invest.* 2008;118(12):3917-29.

[155] Esumi H, Izuishi K, Kato K, Hashimoto K, Kurashima Y, Kishimoto A, et al. Hypoxia and nitric oxide treatment confer tolerance to glucose starvation in a 5'-AMP-activated protein kinase-dependent manner. *J Biol Chem.* 2002;277(36):32791-8.

[156] Kim R. Recent advances in understanding the cell death pathways activated by anticancer therapy. *Cancer.* 2005;103(8):1551-60.

[157] Pattingre S, Bauvy C, Codogno P. Amino acids interfere with the ERK1/2-dependent control of macroautophagy by controlling the activation of Raf-1 in human colon cancer HT-29 cells. *J Biol Chem.* 2003;278(19):16667-74.

[158] Degenhardt K, Mathew R, Beaudoin B, Bray K, Anderson D, Chen G, et al. Autophagy promotes tumor cell survival and restricts necrosis, inflammation, and tumorigenesis. *Cancer Cell.* 2006;10(1):51-64.

[159] Jin S, DiPaola RS, Mathew R, White E. Metabolic catastrophe as a means to cancer cell death. *J Cell Sci.* 2007;120(Pt 3):379-83.

[160] Demasters G, Di X, Newsham I, Shiu R, Gewirtz DA. Potentiation of radiation sensitivity in breast tumor cells by the vitamin D3 analogue, EB 1089, through promotion of autophagy and interference with proliferative recovery. *Mol Cancer Ther.* 2006;5(11):2786-97.

[161] Kanzawa T, Kondo Y, Ito H, Kondo S, Germano I. Induction of autophagic cell death in malignant glioma cells by arsenic trioxide. *Cancer Res.* 2003;63(9):2103-8.

[162] Scarlatti F, Maffei R, Beau I, Codogno P, Ghidoni R. Role of non-canonical Beclin 1-independent autophagy in cell death induced by resveratrol in human breast cancer cells. *Cell Death Differ.* 2008;15(8):1318-29.

[163] Li J, Qin Z, Liang Z. The prosurvival role of autophagy in Resveratrol-induced cytotoxicity in human U251 glioma cells. *BMC Cancer.* 2009;9:215.

[164] Kim KW, Hwang M, Moretti L, Jaboin JJ, Cha YI, Lu B. Autophagy upregulation by inhibitors of caspase-3 and mTOR enhances radiotherapy in a mouse model of lung cancer. *Autophagy.* 2008;4(5):659-68.

[165] Paglin S, Lee NY, Nakar C, Fitzgerald M, Plotkin J, Deuel B, et al. Rapamycin-sensitive pathway regulates mitochondrial membrane potential, autophagy, and survival in irradiated MCF-7 cells. *Cancer Res.* 2005;65(23):11061-70.

[166] Yazbeck VY, Buglio D, Georgakis GV, Li Y, Iwado E, Romaguera JE, et al. Temsirolimus downregulates p21 without altering cyclin D1 expression and induces autophagy and synergizes with vorinostat in mantle cell lymphoma. *Exp Hematol.* 2008;36(4):443-50.

[167] Crazzolara R, Bradstock KF, Bendall LJ. RAD001 (Everolimus) induces autophagy in acute lymphoblastic leukemia. *Autophagy.* 2009;5(5):727-8.

[168] Pan J, Yeung SC. Recent advances in understanding the antineoplastic mechanisms of farnesyltransferase inhibitors. *Cancer Res.* 2005;65(20):9109-12.

[169] Ertmer A, Huber V, Gilch S, Yoshimori T, Erfle V, Duyster J, et al. The anticancer drug imatinib induces cellular autophagy. *Leukemia.* 2007;21(5):936-42.

[170] Milano V, Piao Y, LaFortune T, de Groot J. Dasatinib-induced autophagy is enhanced in combination with temozolomide in glioma. *Mol Cancer Ther.* 2009;8(2):394-406.

[171] Martin AP, Mitchell C, Rahmani M, Nephew KP, Grant S, Dent P. Inhibition of MCL-1 enhances lapatinib toxicity and overcomes lapatinib resistance via BAK-dependent autophagy. *Cancer Biol Ther.* 2009;8(21):2084-96.

[172] Shao Y, Gao Z, Marks PA, Jiang X. Apoptotic and autophagic cell death induced by histone deacetylase inhibitors. *Proc Natl Acad Sci U S A.* 2004;101(52):18030-5.

[173] Cheng Y, Qiu F, Ikejima T. Molecular mechanisms of oridonin-induced apoptosis and autophagy in murine fibrosarcoma L929 cells. Autophagy. 2009;5(3):430-1.

[174] Liu B, Cheng Y, Zhang B, Bian HJ, Bao JK. *Polygonatum cyrtonema* lectin induces apoptosis and autophagy in human melanoma A375 cells through a mitochondria-mediated ROS-p38-p53 pathway. *Cancer Lett.* 2009;275(1):54-60.

[175] Liu B, Cheng Y, Bian HJ, Bao JK. Molecular mechanisms of *Polygonatum cyrtonema* lectin-induced apoptosis and autophagy in cancer cells. Autophagy. 2009;5(2):253-5.

[176] Liu B, Wu JM, Li J, Liu JJ, Li WW, Li CY, et al. *Polygonatum cyrtonema* lectin induces murine fibrosarcoma L929 cell apoptosis and autophagy via blocking Ras-Raf and PI3K-Akt signaling pathways. *Biochimie.* 2010;92(12):1934-8.

[177] Lei HY, Chang CP. Induction of autophagy by concanavalin A and its application in anti-tumor therapy. *Autophagy.* 2007;3(4):402-4.

[178] Liu B, Min MW, Bao JK. Induction of apoptosis by Concanavalin A and its molecular mechanisms in cancer cells. *Autophagy.* 2009;5(3):432-3.

[179] Thyagarajan A, Jedinak A, Nguyen H, Terry C, Baldridge LA, Jiang J, et al. Triterpenes from *Ganoderma Lucidum* induce autophagy in colon cancer through the inhibition of p38 mitogen-activated kinase (p38 MAPK). *Nutr Cancer.* 2010;62(5):630-40.

[180] Yo YT, Shieh GS, Hsu KF, Wu CL, Shiau AL. Licorice and licochalcone-A induce autophagy in LNCaP prostate cancer cells by suppression of Bcl-2 expression and the mTOR pathway. *J Agric Food Chem.* 2009;57(18):8266-73.

[181] Liu SY, Lin MH, Hsu YR, Shih YY, Chiang WF, Lee CH, et al. Arecoline and the 30-100 kDa fraction of areca nut extract differentially regulate mTOR and respectively induce apoptosis and autophagy: a pilot study. *J Biomed Sci.* 2008;15(6):823-31.

[182] Rickmann M, Vaquero EC, Malagelada JR, Molero X. Tocotrienols induce apoptosis and autophagy in rat pancreatic stellate cells through the mitochondrial death pathway. *Gastroenterology.* 2007;132(7):2518-32.

[183] Tanabe H, Kuribayashi K, Tsuji N, Tanaka M, Kobayashi D, Watanabe N. Sesamin induces autophagy in colon cancer cells by reducing tyrosine phosphorylation of EphA1 and EphB2. *Int J Oncol.* 2011;39(1):33-40.

[184] Griffin C, McNulty J, Pandey S. Pancratistatin induces apoptosis and autophagy in metastatic prostate cancer cells. *Int J Oncol.* 2011;38(6):1549-56.

[185] Kuo PL, Hsu YL, Cho CY. Plumbagin induces G2-M arrest and autophagy by inhibiting the AKT/mammalian target of rapamycin pathway in breast cancer cells. *Mol Cancer Ther.* 2006;5(12):3209-21.

[186] Xu ZX, Liang J, Haridas V, Gaikwad A, Connolly FP, Mills GB, et al. A plant triterpenoid, avicin D, induces autophagy by activation of AMP-activated protein kinase. *Cell Death Differ.* 2007;14(11):1948-57.

[187] Oh SH, Kim YS, Lim SC, Hou YF, Chang IY, You HJ. Dihydrocapsaicin (DHC), a saturated structural analog of capsaicin, induces autophagy in human cancer cells in a catalase-regulated manner. *Autophagy.* 2008;4(8):1009-19.

[188] Sy LK, Yan SC, Lok CN, Man RY, Che CM. Timosaponin A-III induces autophagy preceding mitochondria-mediated apoptosis in HeLa cancer cells. *Cancer Res.* 2008;68(24):10229-37.

[189] Salazar M, Carracedo A, Salanueva IJ, Hernandez-Tiedra S, Lorente M, Egia A, et al. Cannabinoid action induces autophagy-mediated cell death through stimulation of ER stress in human glioma cells. *J Clin Invest.* 2009;119(5):1359-72.

[190] Hu H, Chai Y, Wang L, Zhang J, Lee HJ, Kim SH, et al. Pentagalloylglucose induces autophagy and caspase-independent programmed deaths in human PC-3 and mouse TRAMP-C2 prostate cancer cells. *Mol Cancer Ther.* 2009;8(10):2833-43.

[191] Hung JY, Hsu YL, Li CT, Ko YC, Ni WC, Huang MS, et al. 6-Shogaol, an active constituent of dietary ginger, induces autophagy by inhibiting the AKT/mTOR pathway in human non-small cell lung cancer A549 cells. *J Agric Food Chem.* 2009;57 (20):9809-16.

[192] Eom JM, Seo MJ, Baek JY, Chu H, Han SH, Min TS, et al. Alpha-eleostearic acid induces autophagy-dependent cell death through targeting AKT/mTOR and ERK1/2 signal together with the generation of reactive oxygen species. *Biochem Biophys Res Commun. 2010;391(1):903-8.*

[193] Chao AC, Hsu YL, Liu CK, Kuo PL. alpha-Mangostin, a dietary xanthone, induces autophagic cell death by activating the AMP-activated protein kinase pathway in glioblastoma cells. *J Agric Food Chem.* 2011;59(5):2086-96.

[194] Law BY, Wang M, Ma DL, Al-Mousa F, Michelangeli F, Cheng SH, et al. Alisol B, a novel inhibitor of the sarcoplasmic/endoplasmic reticulum Ca(2+) ATPase pump, induces autophagy, endoplasmic reticulum stress, and apoptosis. *Mol Cancer Ther.* 2010;9(3):718-30.

[195] Zhang XQ, Huang XF, Hu XB, Zhan YH, An QX, Yang SM, et al. Apogossypolone, a novel inhibitor of antiapoptotic Bcl-2 family proteins, induces autophagy of PC-3 and LNCaP prostate cancer cells in vitro. *Asian J Androl.* 2010;12(5):697-708.

[196] Meng N, Wu L, Gao J, Zhao J, Su L, Su H, et al. Lipopolysaccharide induces autophagy through BIRC2 in human umbilical vein endothelial cells. *J Cell Physiol.* 2010;225(1):174-9.
[197] Bhui K, Tyagi S, Prakash B, Shukla Y. Pineapple bromelain induces autophagy, facilitating apoptotic response in mammary carcinoma cells. *Biofactors.* 2010;36(6):474-82.
[198] Chen RJ, Ho CT, Wang YJ. Pterostilbene induces autophagy and apoptosis in sensitive and chemoresistant human bladder cancer cells. *Mol Nutr Food Res.* 2010;54(12):1819-32.
[199] Saiki S, Sasazawa Y, Imamichi Y, Kawajiri S, Fujimaki T, Tanida I, et al. Caffeine induces apoptosis by enhancement of autophagy via PI3K/Akt/mTOR/p70S6K inhibition. *Autophagy.* 2011;7(2):176-87.

Part III.
Autophagy in Aging, Metabolic and Degenerative Disorders, and Responses to Toxic Agents and Injury

In: Autophagy: Principles, Regulation and Roles in Disease ISBN: 978-1-61942-266-7
Editor: Nikolai Gorbunov

Chapter VIII

Autophagy in the Eye: A Double-Edged Sword

***Goran Petrovski*[1,2*], *Réka Albert*[1,2], *Kai Kaarniranta*[3,4], *Morten C. Moe*[5], *László Fésüs*[1], *András Berta*[2] and *Dipak K. Das*[6]**

[1] University of Debrecen, Department of Biochemistry and Molecular Biology, The Apoptosis and Genomics Research Group of the Hungarian Academy of Sciences, Debrecen, Hungary
[2] University of Debrecen, Medical and Health Science Center, Department of Ophthalmology, Debrecen, Hungary
[3] Department of Ophthalmology, Institute of Clinical Medicine, University of Eastern Finland, Kuopio, Finland
[4] Department of Ophthalmology, Kuopio University Hospital, Kuopio, Finland
[5] Department of Ophthalmology, Oslo University Hospital, Center for Eye Research, University of Oslo, Oslo, Norway
[6] Cardiovascular Research Center, University of Connecticut School of Medicine, Farmington, CT, US

Abstract

Autophagy, a cytoplasmic recycling process active from early development throughout lifetime and aging has turned from largely a morphological phenomenon to a well described molecular mechanism in health and disease. The eye as an organ in which an interplay of three germ layers is responsible for achieving complex visual response, is the place where battles for cellular life or death takes place continuously – an organ in which the highest dead cell turnover occurs in the human body. The role of autophagy in the different segments of the eye remains not so well understood or described, making it difficult to classify autophagy as a protective or disease causing process. This chapter

* Corresponding author: Goran Petrovski MD, PhD; Department of Ophthalmology, Medical and Health Science Center; University of Debrece; Nagyerdei krt. 98, 4012 Debrecen, Hungary; Tel/Fax: + 36 52 255 626, Email: gokip@indi.dote.hu.

summarizes the current knowledge about the role of autophagy in eye health and disease as well as the present drug targets applying autophagy as a protective mechanism against ophthalmic neurodegeneration and aging.

1. Introduction

The recycling machinery of the cells relies upon two major degradative pathways: the ubiquitin-proteasome system (UPS) on one hand, responsible for the degradation of short-lived or soluble proteins and the autophago-lysosomal pathway on the other, responsible for the degradation of long-lived or aggregated proteins and cytoplasmic organelles (Figure 1). Autophagy under normal physiologic conditions plays a recycling role converging upon the lysosomal pathway as its final step. Three types of autophagy have been described so far based upon the way the recycled cargo gets carried into the lysosome: 1) macroautophagy, utilizing a double membraned autophagosome for sequestering and transporting material – both soluble proteins and cytoplasmic organelles (from here on referred as autophagy), 2) chaperone-mediated autophagy (CMA) using chaperones selectively recognizing and binding the cargo for direct transport into the lysosomes via LAMP2A (lysosomal-associated membrane protein 2A) and 3) microautophagy, a less described phenomemenon involving lysosomal invagination of the degraded material without necessarily forming autophagic vacuoles - not discussed further in this chapter [61, 69, 79]. Besides the known interplay among the different types of autophagy [42, 71], more recently, a connection between autophagy and the UPS has been shown as a proof of the complexity and redundancy cells rely upon during survival or death [105, 134]. It is the dual, pro-survival or pro-death, role autophagy plays in cells, the threshold for which still remains unclear, what makes it a double-edge-sworded entity in health and disease. The present chapter summarizes the role autophagy has in the eye as an organ as well as the therapeutic applications of it for prevention of ophthalmic neurodegeneration and aging. Before going into detail of describing autophagy in the different eye segments, a short overview of the role of autophagy during development, cellular life and aging, as well as the anatomy and germ layer-development of the eye have to be presented for better understanding.

2. Autophagy during Development, Life and Aging

As a process of self-eating, autophagy begins with formation of isolation membrane which entraps cytosolic components destined for degradation into a double-membraned structure or autophagosome (Figure 1). The nucleation step involves a complex of proteins which become further conjugated with other autophagy-specific proteins into a multimeric complex of the phagophore. Lipidation of a microtubule-associated protein 1 light chain 3 (LC3), proteolytic activation and conjugation with phosphatidyl-ethanolamine follow with an insertion into the autophagosomal membrane. In the maturation step, the autophagic vacuole fuses with the lysosome via a large multivesicular structure or amphisome. After the fusion with the lysosome, the low pH and the lysosomal degradation enzymes or hydrolases digest the cargo, the final product of which gets transported into the cytoplasm for recycling (Figure 1) [24].

The importance of autophagy during development has been described widely from yeast to *Caenorhabditis elegans* and *Drosophila melanogaster*, as well as lower and higher vertebrates in general [14]. Autophagy as a process of cell structuring and renovation gets activated as early as four hours after fertilization, working closely with the UPS degradation pathway during the maternal oocyte-to-embryo transition [22]. Oocyte-specific Autophagy-related gene (Atg) 5-deficient female mice can ovulate normally upon external hormonal stimulation and be fertilized *in vivo*, however, the deficient oocytes fail to develop beyond the four- and eight-cell stages when fertilized by Atg5-null sperm [117]. In *Drosophila*, inhibition of the class-I phosphatidyl-inositol 3-kinase (PI3K) pathway, a known inhibitor pathway for autophagy, causes elimination of organs, such as the fat body at the end of larval stage [1, 2].

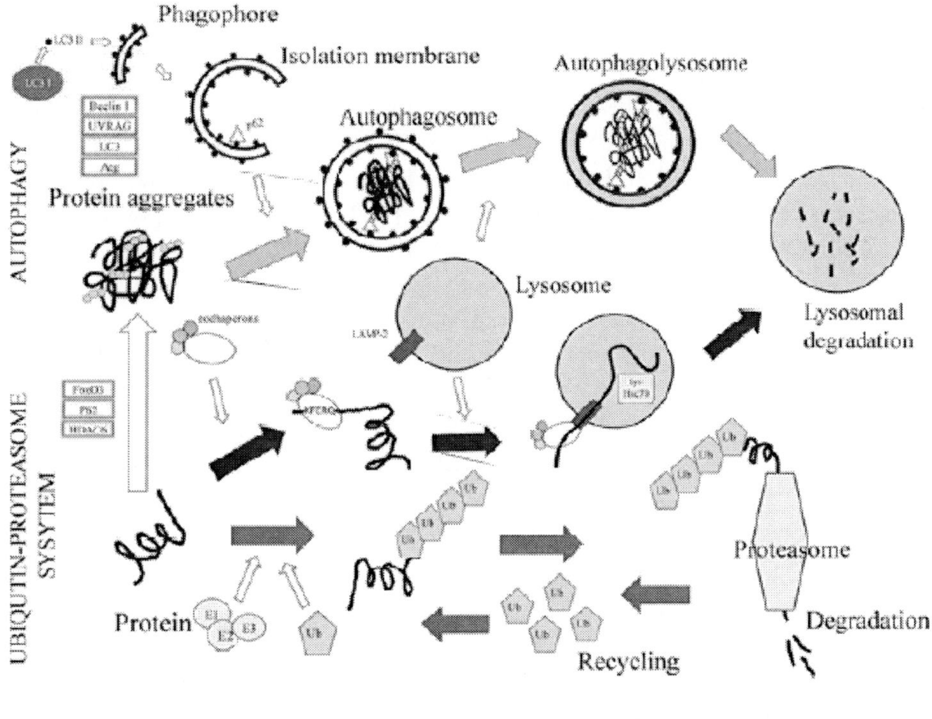

Figure 1. Major degradative pathways in the cell: autophago-lysosomal pathway and ubiquitin-proteasomal system (Atg: autophagy-related gene; Beclin 1: autophagy-related gene becn1/myosin-like BCL-2 interacting protein; E1/E2/E3: cascade of ubiquitin transferases; FOXO3a: mammalian homolog of forkhead box O transcription factor protein/DAF-16 transcription factor; HDAC6: histone deacetylase 6; Hsc 7 (heat shock cognate protein 70); KFERQ: Lys-Phe-Glu-Arg-Gln/target sequence for lysosomal degradation; UVRAG: LC3: mammalian homolog of Atg8/microtubule-associated proteins 1A/1B light chains 3A/LC3A and 3B/LC3B (MAP1LC3A/B); Uq: ubiquitin; UVRAG: UV irradiation resistance-associated gene)).

The importance of autophagy in the further development and life has been highlighted by a number of authors in the past, summarized concisely by Cecconi et al. [14]. Interestingly, the complex maintenance of the nervous system has been shown to be exceptionally sensitive to dysfunctions of autophagy, manifested by neural tube defects, axonal growth and migration impairment. Autophagy contributes to the development of the nervous system by providing energy for cell corpse removal after physiological cell death, a process associated with retinal neurogenesis [7]. Certainly, the eye being a continuation of the central nervous system (CNS), is very much prone to any dysregulation in the autophagic pathway.

The loss of Beclin 1 (autophagy-related gene *becn1*; myosin-like BCL-2 interacting protein) produces lethality at embryonic day 7.5 (E7.5) due to failed visceral endoderm formation [63]. Failure of the Activating molecule in Beclin 1-regulated autophagy (Ambra1), a regulator of autophagy expressed only in vertebrates, to form a multiprotein complex with Beclin and Vps34.6 results in death at E14.5, manifested by defects in the neural tube closure, increased proliferation and cell death during the nervous system development [68].

Knockout of a fundamental autophagy gene, Atg5, in mice causes increased cell death in the retina at E18.5 [91]. In addition, defects in autophagy can affect the chick retinal neuroepithelium at E4, affecting the retinal ganglion cells (RGCs) while undergoing differentiation [88].

Throughout life and aging, autophagy has been described as induced during ischemia/reperfusion (I/R) or pressure overload in the heart [21, 73, 86, 114], as well as during early stages of neurodegenerative diseases such as Alzheimer's, Parkinson's and Huntington's disease [3, 5, 28, 55, 81, 101-103]. In addition, autophagy has been found to have a role in cancer [58, 59, 62], atherosclerosis [70], diabetes [35], muscular and [80] liver [131] disorders, pancreatitis and Crohn's disease [100, 127]. The role of autophagy in the delivery of microbial antigens to the innate and adaptive immune system has gained crucial importance and become a matter of indepth study more recently [24, 60]. Finally, the decrease in autophagic activity during aging and its possible boosting by calorie restriction may give the answer on how to achieve better health and longevity [17, 85, 107].

3. Development of the Eye

At the end of the 4th week of fetal life, a pair of optic vesicles appear on both sides of the forebrain as an outpocketing. Upon contact with the surface ectoderm, the outpocketings induce lens formation and upon invagination, the pigmented and neural layers of the retina form (Figure 2). The hyaloid artery supplies the lens and the eye with nutrition during development, later this artery becomes rudimentary in the inner vitreous or gel-like filling of the eye bulb and turns into central artery of the retina at the periphery. The vitreous itself has a similar mesenchymal origin as the surrounding eye primordium. The nerve fibers become entrapped into the choroidal groove of the forming invagination and represent a continuation of the CNS. The cornea, having the highest refractive power in the whole eye gets formed by a fusion of the surface ectodermal layer - inner stromal layer continuous with the fibrous sclera, and epithelial layer bordering the anterior chamber of the eye. All the spaces in the eye are formed by a process of vacuolization and that is how the anterior and posterior chambers of the eye develop [106].

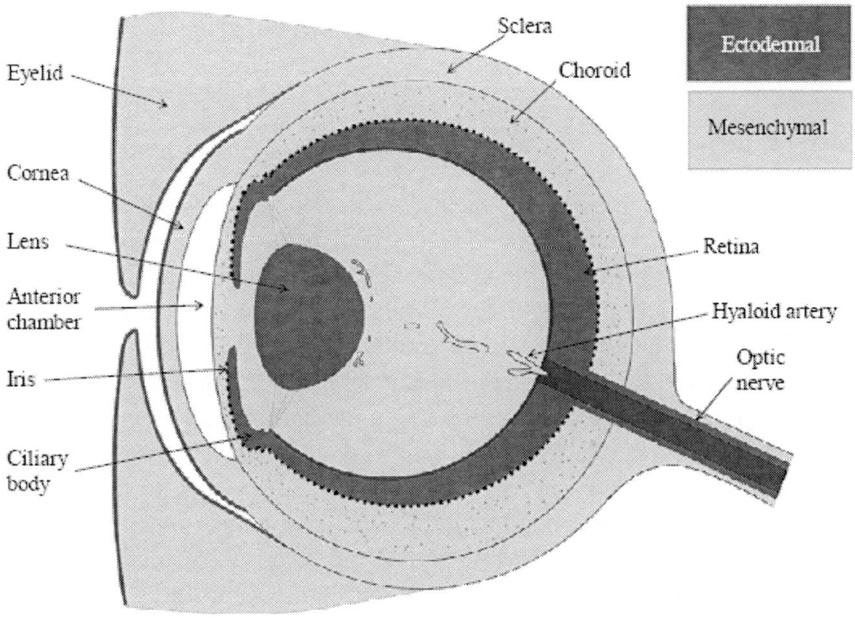

Figure 2. Development of the eye and origin of the different cell or tissue layers.

4. Autophagy in the Ocular Surface and the Orbit

Tears and Lacrimal Gland

There are constant environmental insults and exposures of the ocular surface to dust, smoke or airborne pathogens, the tear film being the only physical and protective barrier of the surface nonkeratinized epithelium [115]. Inadequate production of tears results in chronic ocular surface irritation leading to dry eye syndrome. Insufficient lacrimal gland secretion is still poorly understood, but chronic inflammation of the lacrimal gland (due to either autoimmune diseases (Sjögren syndrome, sarcoidosis, diabetes), after organ transplantation (chronic graft versus host disease) or due to aging) has been considered a major contributor [136].

Tissue remodeling after injury or trauma often resembles very much the cellular events present during embryogenesis. Type I programmed cell death or apoptosis, has been described as being crucial for the embryonic tissue development and maintenance of cellular and organ homeostasis [18]. Its occurrence under both physiologic and pathophysiologic conditions such as inflammation has been well documented [18], and especially its role in tissue repair and atrophy. Not mutually exclusive with type I programmed cell death are type II or autophagic cell death and type III or necrosis, very often occurring simultaneously in tissues or conjointly in cells [23, 27].

Previous animal studies (BALB/c or C57BL/6 mice) have shown that IL-1 (IL-1α or IL-1β) injection in the lacrimal gland can induce severe, but transient inflammatory response in it, causing destruction of the acinar cells and reduced aqueous tear production [136, 137]. The reversibility of the inflammatory response in this case was speculated to be due to the autophagic component of the cell death present in the lacrimal gland.

Conjunctiva and Cornea

Some of the earliest reports of presence of cell death and autophago-lysosomal degradation in the basal and intermediate layers of normal guinea pig conjunctiva epithelium date back to the late 70's [54]. Since then, other reports of cellular damage of the conjunctiva and corneal surface epithelium by chronic exposure to preservatives in topical drugs have been described [11, 82]. The most common preservative used in anti-glaucoma drugs being benzalkonium chloride (BAC) – a potent inducer of reactive oxygen species (ROS) at high concentrations, is capable of causing morphologic disruption of the corneal epithelium and apoptosis in Chang's conjunctival cells [19, 20]. The ocular surface being the first barrier of the eye exposed directly to ultraviolet (UV) irradiation is also prone to light-induced cellular toxicity or photoconjunctivitis, photokeratitis and pterygium. All these reversible or irreversible ocular surface disease states potentially involve UV-induced oxidative stress and ROS signaling cascades [116]. When the repair mechanisms fail, programmed cell death is activated [50]. Both BAC and UV irradiation can induce *cytochrome c* release from mitochondria, caspase activation, and nuclear chromatin condensation, suggesting caspase-dependent apoptosis. BAC has been found to specifically activate a caspase-independent pathway by inducing the mitochondrial release of apoptosis-inducing factor. In addition, BAC-induced autophagy has been shown to be protective in Chang conjunctival cells, since knocking-down of the Beclin-1 gene could increase the BAC toxicity [10].

Besides the conjunctiva, the cornea, in need of keeping itself clear or transparent, at the same time serving a protective, yet immunopriviledged role, is under constant invasion of non-infectious and infectious agents from the environment or its vicinity. A direct or primary herpes simplex virus type 1 (HSV-1) infection of the cornea or alternatively, a secondary infection from the oral mucosa via the superior cervical ganglion to the cornea nerve fibers, may be a sight-threatening event, prone to chronicity and need for regular check-ups and treatment. Experiments with HSV-1 innoculating corneal infections on 129 Ev/Sv and C57BL/6 mice showed induction of autophagy, an important component of the host innate and adaptive immunity to viruses, linking the HSV-1 with the essential autophagy protein Beclin 1 (Atg6) [56]. Our own experiments show similar autophago-lysosomal degradation in the Staatens Seruminstitute Rabbit Cornea cell line (SIRC) as a model of HSV-1 induced keratitis (manuscript under preparation). In brief, while the SIRC cells undergo apoptosis, autophagy is induced by the HSV-1 infection and an active autophagic flux could be verified by chloroquine treatment.

A non-infectious disease state which shows signs of induction of autophagy in the cornea compared to the control is keratoconus (our own unpublished preliminary data). Probably, the hypoxic state which arises in keratoconic patients due to contact lens wear is responsible for the induction of autophagy, primarily to serve a protective role on the cornea epithelium and stroma. Indeed, a non-lethal hypoxia preconditioning has been shown to be protective against apoptosis induced by UV, Fas or epithelial scrape in bovine corneal stromal cells [128]. This protective effect of hypoxia preconditioning has been claimed to be due to induction of hypoxia inducible factor-1α (HIF-1α), main transcriptional factor activated by hypoxia [90].

Extraocular Muscles (EOMs)

The constant activity and metabolic stress present in the EOMs makes them prone to age-related mitochondrial defects, linked to apoptosis and autophagy. Direct evidence for it was found in a recent study of muscles from male Fisher 344-Brown Norway rats: while the apoptosis as measured by the TUNEL positivity and presence of caspase-3, -8, -9, and -12 activity showed no change with aging, the autophagy as measured by the LC3 protein abundance showed a decrease in the rat EOMs [74].

5. Autophagy in the Trabecular Meshwork, Iris and Ciliary Body

Trabecular meshwork (TM), is the part of the eye responsible for maintaining proper level of intraocular pressure (IOP), with its anatomical location at the iridocorneal junction, is involved in the development or pathogenesis of primary open angle glaucoma (POAG) - the most prevalent type of glaucoma [57]. Glaucomas in general cause progressive damage of the optic nerve and therefore irreversible loss of vision and blindness [92]. Although a specific cause of the different forms of glaucoma has not yet been found, aging of the TM is suspected to be a major risk factor in the development of the disease. Prolonged exposure of porcine TM cells to a hyperoxic environment has been shown to increase the ROS production and oxidized material as revealed by the cytoplasmic accumulation of autophagic vacuoles and lipofuscin (LS)-loaded lysosomes [64]. Chronic exposure of the TM cells to oxidative stress was shown in this study to cause accumulation of nondegradable material within the lysosomal compartment, leading to diminished lysosomal activity and increased senescence-associated β galactosidase (SA-β-gal) expression, therefore causing progressive decline in the TM cell function [64].

Iris, is the part of the eye responsible for controlling the diameter and the size of the pupil and the amount of light reaching it. The iris pigmented epithelial cells contain different amount of melanosomes that are responsible for the pigmented nature of the iris. In dogs, inherited tapetal degeneration or abnormally light pigmentation has been shown to be due to abnormal melanosome content and presence of organelles resembling secondary lysosomes or residual bodies [9]. It has been proposed that a defect in the synthesis of matrix component of the melanosomes, which can result in absent or abnormal deposition of melanin, is responsible for initiating autophagy or self-eating of these organelles [9].

Fully differentiated iris epithelial cells of adult newts could be dedifferentiated and converted into proliferating lens cells when put into culture via activation of the autophagic machinery affecting the melanosomes. [129, 130].

Our own sections from adult human cadaver eyes show increased basal LC3 expression in the iris pigmented epithelial cell layer, very likely due to the aging process (Figure 3a). Mouse iris sections, on the contrary, showed no basal LC3 expression (Figure 3c, middle)

Ciliary body (CB), is the thin vascular middle layer of the eye, located posteriorly and in continuum to the iris, between the outside sclera and the inner retina. Aqueous humour is produced by the non-pigmented (pars plicata) epithelium of the ciliary body and secreted into the posterior eye chamber. The humour flows through a narrow cleft between the front of the lens and the back of the iris, to escape through the pupil into the anterior chamber, and then to drain out of the eye via the TM into the Schlemm's canal and the episcleral and orbital veins consecutively. So far, no clear description of the role autophagy has in the CB has been made. Our sections from adult human cadaver eyes show increased basal LC3 expression in the CB, but unlike the iris, it is in the non-pigmented aqueous humour producing epithelium, raising a speculation of the involvement of autophagy in maintaining the IOP balance during aging (Figure 3b). Similarly, the mouse CB shows basal LC3 expression in the non-pigmented epithelium (Figure 3c, middle).

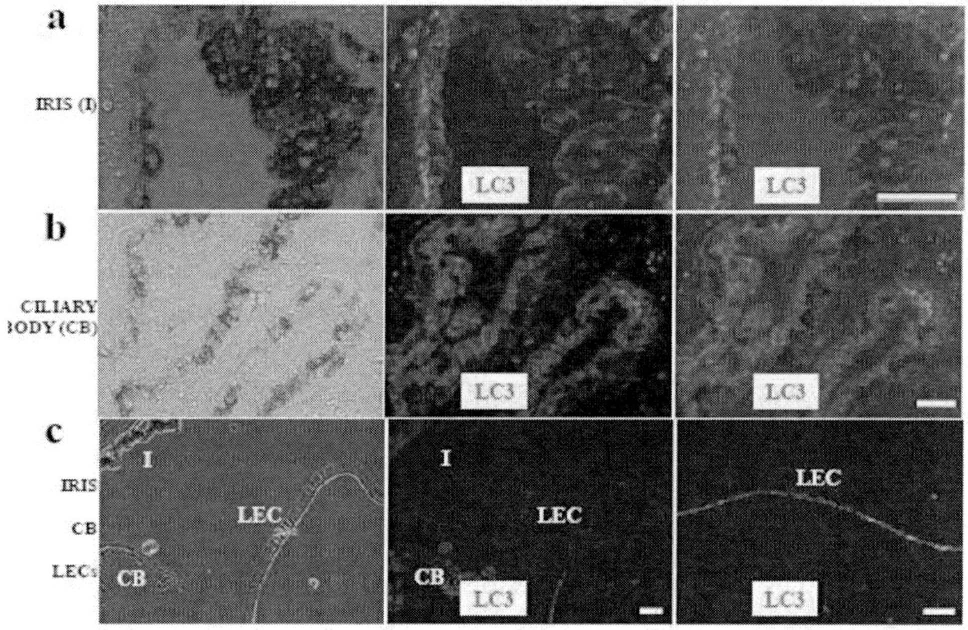

Figure 3. Eye sections from adult human cadaver a) iris and b) ciliary body, and c) mouse from left to right: (REMOVE: phase contrast)) (WRITE IN: light) microscopy, immunostaining with anti-LC3 antibody (red color, 1:100 dilution, Novus Biologicals NB100-2331) and merged image of the two, respectively; in c) the right image is higher magnification of the lens epithelial cells (LECs) stained for LC3 as shown in the middle image. The nuclei were counterstained with Hoechst.

6. Autophagy in the Lens

After the cornea, the lens has the second highest refractive power in the eye, being transparent throughout younger life, but loosing its clarity and refractive power over the aging. Cataract is the loss of transparency of the lens, the most common type being progressive senile cataract. Specific for the lens is the unusual outside-in growth of the lens epithelial cells (LECs) from the lens capsule towards the center or the nucleus, as well as the long, onion-like layered, nucleus-free cells. It is actually this process of cytoplasmic and nuclear content clearance in lens cells that has been described to involve autophagy.

Selective elimination of mitochondria by occlusion into membranous mitochondrial clusters, decomposition and exocytosis has been suggested to be involved in the maturation of lens fiber cells [67]. Similarly, degradation and quality control of proteins found within the nucleus and its compartments, as well as nucleic acids, has been described during differentiation of the lens [78]. Interestingly, another study observed that autophagy occurs in embryonic lens cells, but organelle degradation in the lens cells was not affected in autophagy-deficient Atg5(-/-) mice [72]. Our own mouse eye sections show basal LC3 expression in the lens epithelial cells, which may be attributed to the aging process itself (Figure 3c).

Following cataract surgery, an attempt to solve the formation of secondary cataract or posterior capsule opacification (PCO) has been made via induction of autophagy in posterior capsule lens cellular remnants. Furthermore, trypan blue (0.1%) has been found to cause LEC death through apoptosis and autophagy, therefore reducing the incidence of PCO after cataract surgery [87].

7. Autophagy in the Optic Nerve, Retina and Choroid

Basal levels of autophagy in all cells can be rapidly upregulated in response to starvation and cellular stress. The optic nerve being an ocular extension of the CNS, similar to other neurodegenerative diseases, can involve activation of autophagy. Whether autophagy has a destructive or protective role in optic nerve degenerations still remains unclear.

Optic Nerve and Retinal Ganglion Cells (RGCs)

Axonal degeneration is a pathological hallmark of many traumatic and neurodegenerative disorders. Crush lesion experiments of the optic nerve in rats could show calcium-dependent induction of autophagy during active axonal degradation, possibly serving as a future therapeutic target for delaying such degradative processes [46, 47]. During optic nerve transection under serum deprivation, autophagy could be activated in RGC-5 cell line-, once again serving as a protective mechanism [44]. Axotomy of neonatal RGCs could trigger the intrinsic apoptotic pathway (activating caspase-3/9), balancing or counteracting the pro-survival autophagic response [111].

Normal prion protein in *Drosophila* have been shown to enhance the neurotoxicity of pathogenic polyglutamine proteins and alter the susceptibility to oxidative and autophagy signaling modulators [83]. The neurotoxicity of pathogenic Machado-Joseph Disease (MJD) glutamine (Q) 78 and 127Q proteins could be enhanced by the co-expression of Mo-PrP^3(F)^4 in the fly eyes, while the eyes of control flies and flies expressing Mo-PrP^3(F)^4 alone or together with MJD-Q27 or 20Q proteins showed no defects. Similarly, mutation in the fly small heat shock protein HSP67Bc (Dm-HSP67Bc) involved in the pathogenesis of polyglutamine disorders, could cause induction of autophagy via the eukaryotic translation initiation factor 2 alpha (eIF2α) pathway. Down-regulation of endogenous Dm-HSP67Bc significantly worsens the spinocerebellar ataxia 3-mediated eye degeneration in flies [13]. In ATG5-/- cells, defective in autophagy, the anti-aggregation activity of HSPB7 could be substantially reduced, hence it could prevent toxicity of polyQ proteins at an early stage of aggregate formation via a non-canonical activation of the autophagic machinery [120].

Photoreceptors

The photoreceptor layer is a continuation of the neural layer in the eye within the neuroretina. Autophagy in this cell layer has been described first by Remé *et al.* [96, 99]. For demonstration, we show presence of LC3 positivity in the photoreceptor layer of aged human cadaver eye sections (Figure 4). Autophagy in the rod inner segments was shown by Remé *et al.* to follow a diurnal variation, the peak period of autophagic activity occurring after the peak period of disk-shedding during daytime, suggesting an important role of autophagy in the visual cell degradative mechanisms [96]. The same group further described that autophagic vacuoles per cone were significantly higher in frog and vertebrate retinas incubated in white light than in retinas incubated in red light, while in rods they were equal under both conditions, indicating an intracellular mechanism of degradation in visual cells dependent on the metabolic conditions and the level of illumination [95]. The monoamineoxidase type A inhibitor drug clorgyline could phase delay the circadian rest-activity cycle, but not the rhythms of the autophagic degradation nor of disk-shedding in rats and hamsters [93, 94, 97]. Similarly, the temporal pattern of cone outer-segment disc shedding was examined in the retina of California ground squirrel, *Spermophilus beecheyi*: dense granules were often found in the photoreceptors, retinal neurons, Müller cells, microglia and vascular cells, primarily during the dark period and early light period, probably being of autophagic origin [66]. Human photoreceptor inner segments show LS granules which represent an accumulation of residual bodies of autophagy [37]. The adjustment to new lighting environment is performed by changes in rhodopsin levels in rod outer segments, the autophagic degradation of which may partially subserve the adaptation to abruptly increased illuminance by removing surplus visual pigment [98].

Photoreceptor degeneration in human photoreceptor dystrophies and in animal models is executed by caspase-mediated apoptosis and caspase-independent cell death mechanisms such as the activation of cystein-proteases, lysosomal proteases, autophagy and complement-mediated lysis. The rd/rd mouse provides evidence that irrespective of how photoreceptor degeneration is triggered (gene defects or environmental stress (light-damage)), a number of pro-apoptotic mechanisms are triggered leading to photoreceptor degeneration, and other caspase-independent mechanisms may actively participate in the photoreceptor demise as well [52, 65]. Autophagy can initiate apoptosis during oxidative stress of photoreceptors in the light-damage mouse model and 661W photoreceptor cells challenged with H_2O_2; the up-stream regulators of autophagy may therefore serve as potential therapeutic targets in photoreceptor degeneration [51]. The activated rhodopsin in some *Drosophila* mutants which leads to retinal degeneration, is degraded by autophagy in collaboration with endosomal pathways in normal photoreceptor cells to prevent retinal degeneration [77]. Knockdown or mutation of the autophagy-essential components, autophagy-related protein 7 and 8 (atg-7/atg-8), or genes essential for PE (phosphatidylethanolamine) biogenesis and autophagosome formation, including Phosphatidylserine decarboxylase (Psd) and CDP-ethanolamine:diacylglycerol ethanolaminephosphotransferase (Ept) indicate a cross talk between the autophagic and endosomal/lysosomal pathways. This study clearly pinpoints the autophagy as suppressor of light-dependent retinal degeneration, the rhodopsin being a key substrate for autophagic degradation.

Hyperactivation of the target of rapamycin (TOR) or suppressing autophagy leads to photoreceptor cell death in an age- and light-dependent manner in *Drosophila melanogaster* providing direct genetic evidence that autophagy alleviates cell death in several common types of neurodegenerative disease [124].

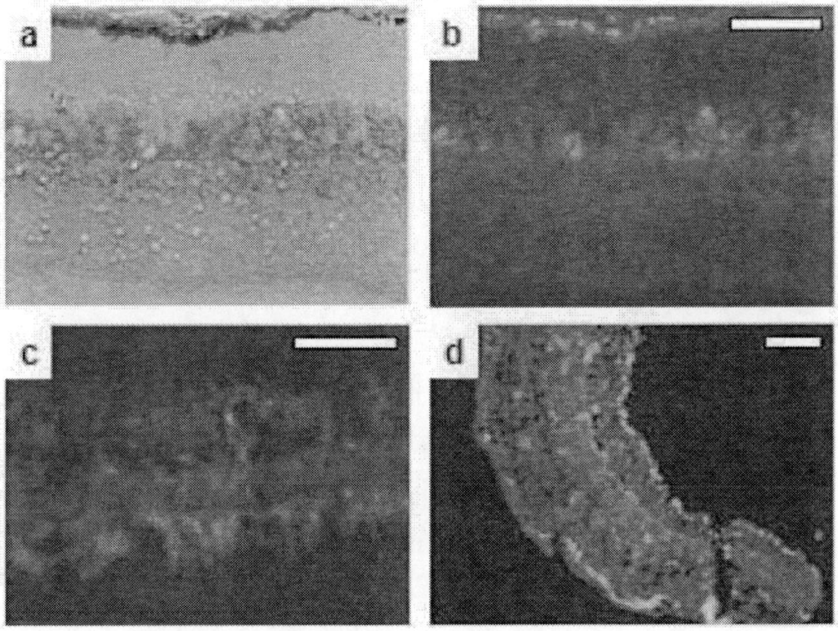

Figure 4. Adult human cadaver eye sections of the retina: a) (REMOVE: phase contrast) (WRITE IN: light) microscopy combined with immunostaining for LC3 (red) to detect autophagy, b) immunostaining for RPE65 (green) to detect the RPE cell layer combined with LC3 (red), c) immunostaining for Nestin (green) to detect MÃ¼ller glial cells combined with LC3 (red); d) Wildtype mouse eye sections immunostaining for rhodopsin to detect photoreceptors (green) combined with LC3 (red).

Stainings: anti-LC3 antibody (red color, 1:100 dilution, Novus Biologicals NB100-2331), anti-Nestin antibody (green color, 1:200 dilution, Sigma-Aldrich N5413), anti-Rhodopsin antibody (green color, 1:1500 dilution, Sigma-Aldrich R5403); anti-RPE65 antibody (green color, 1:800 dilution, Millipore MAB5428). The nuclei were counterstained with Hoechst 33342.

Retina

Developmentally, the retinal tissue shows ultrastructural features of both apoptosis (TUNEL and caspases-3/9-positivity) and autophagy, the later preceding activation of the first [33]. Studies in retinal tissues undergoing transient ischemia, experimentally implicating ocular pathologies such as glaucoma, anterior ischemic optic neuropathy and retinal vessels occlusion show variations in the autophagic response also. In adult rats, increasing the IOP causes retinal ischemia characterized by reduction in the phosphatidylethanolamine-modified form of LC3 (LC3II) and significant decrease in Beclin-1 [104]. Interestingly, this event can be prevented by intravitreal treatment with a non-competitive N-methyl-D-spartate antagonist and calpain inhibition of the cleavage of Beclin-1. Pharmacological inhibition or Beclin-1 silencing in retinal ganglion cell model line RGC-5 can increase the cell death, suggesting a pro-survival role of autophagic during rat retina I/R injury. Further proof of the protective role of ischemic preconditioning after severe ischemic injury was demonstrated in rats undergoing acute ligation of retinal arteries followed by reperfusion: the rat autophagocytosis protein (Apg3p) was upregulated by the ischemic preconditioning but downregulated by prolonged ischemia, the autophagy contributing to the neuroprotective effect of the preconditioning [126]. Retinas from comatose patients sustained with a respirator for couple of days before death, show selective characteristic alterations ascribed to oligemia (ischemia) typical for autophagy, cell swelling, and coagulation necrosis of the inner nuclear layer of the retina – the boundary zone between the choroidal and retinal circulations [25].

Retinal Pigment Epithelium (RPE)

The development of the RPE and photoreceptor layer is divided into first establishing the structure of the two prospective layers - RPE and neuronal retina – followed by invagination of the optic cup [112]. The neuronal retina is built into two layers separated by a thin lumen remnant filled with a so called interphotoreceptor matrix [29-31]. Proper development of the retina appears to be dependent on the genes involved in the melanogenesis pathway in RPE cells [38]. The polarized RPE cells constitute a polygonal monolayer between the retina rod and cones and the fenestrated capillaries of the choroid. The RPE has multiple functions: absorption of light energy, transport of metabolites and nutrients between photoreceptors and choriocapillaris, expression of growth factors for photoreceptors, regulation of homeostasis of the ionic environment, phagocytosis of the shed tips of photoreceptor outer segments (POS), regulation of visual cycle, and creation of the blood–retinal barrier [110]. Selective loss of the RPE affects the neuronal retina and photoreceptors the most [4].

First morphological reports of presence of autophagic vacuoles and small lipid droplets in the RPE come from normal adult golden Syrian hamsters [12]. Later on, human RPE cultures in which postconfluency was followed for more than 2 years were used to examine the molecular and biologic properties of a model for developing RPE LS with autophagocytic and heterophagic component, similar to the *in situ* situation [8]. LS accumulation in the RPE is associated with age and miscellaneous retinal diseases such as age-related macular degeneration (AMD) and Stargardt disease, the toxic compounds and oxidative protein modifications of which may interfere with normal RPE function [108]. LF formation occurs inside the lysosomal compartment partly as a result of iron-catalyzed peroxidation and polymerization [53]. Lipid peroxidation-derived protein modifications can also result in increased lipofuscinogenesis and reduced autophagy activity in RPE cells *in vitro* [49]. The LS fluorophor A2-E can severely impair lysosomal function of RPE cells, causing a potent inhibition of the ATP-driven proton pump in the lysosomal membrane with subsequent congestion of the photoreceptor outer segments phagocytosis and autophagic sequestration of cytoplasmic material [6].

For monitoring autophagy in cells, the golden-standard still remains detection of double-membraned autophagic vacuoles by electron microscopy (Figure 5). More recently, monitoring of the autophagic flux in RPE cells has been established for following the dynamic involvement of the autophagic machinery *in vivo* and *in vitro* in AMD disease models [36, 121].

RPE Phagocytosis

In the retina, various forms of cell death and phagocytosis occur on a daily basis [32]. RPE cell apoptosis, necrosis, paraptosis and autophagy have all been described extensively [34, 36, 84, 105, 118]. Failure of dead cells to clear may lead to drusen formation and age-related macular degeneration (AMD) [26]. Phagocytosis of the shed photoreceptor outer segments by the RPE is impaired in βA3/A1-crystallin, a major structural protein found in RPE cells and the ocular lens as well, mutated mice. The mutated βA3/A1-crystallin in the RPE of these mice fails to translocate to the lysosomes due to misfolding of the mutant protein and improper trafficking. In addition, a defect in the autophagy process is present in these mice, probably due to impaired lysosomal function and sharing of common degradation mechanisms between phagocytosis and autophagy [135]. Although autophagy is essential for phosphatidyl-serine presentation by apoptotic cells during the phase of cell death associated to neurogenesis, this is not a universal requirement for all phases of cell death occurring during retinal development [75, 76].

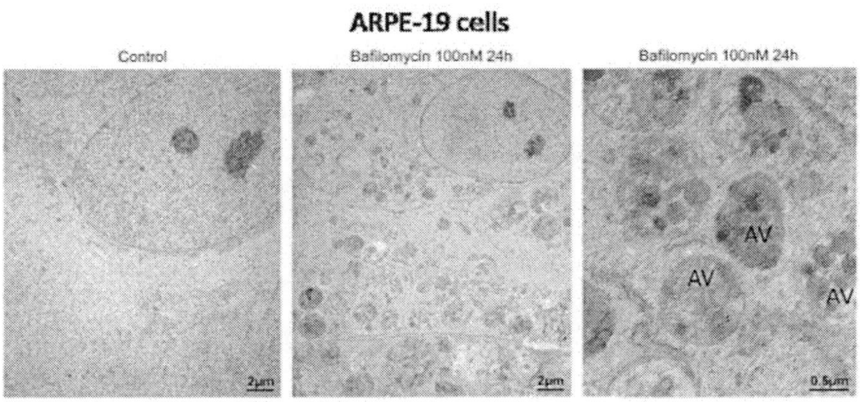

Figure 5. Transmission electron microscopy of the formation of autophagic vacuoles (AV) upon treatment of ARPE-19 cells with bafilomycin over 24 hours (Bafilomycin is considered an inhibitor of the vacuolar type H($^+$)-ATPase (V-ATPase) in cells, inhibiting the acidification of organelles such as lysosomes and endosomes, therefore inhibiting the autophagolysosomal fusion).

Choroid

Although no direct evidence has been shown about the presence of autophagy in the choroidal layer, our aged cadaver sections show *in situ* LC3 positivity in this tissue (Figure 4). However, presence of autophagy during vascular development of the retina has been described previously [43]. Failure of the hyaloid vessel (Figure 1), a transient intraocular circulatory system, to undergo complete involution causes pathologically persistent hyperplastic primary vitreous (PHPV) and severe ocular pathologies. Autophagy contributes to the regression of the hyaloid vessels - vascular endothelial cells under hypoxic conditions, express higher LC3-II/LC3-I ratio along with caspase-3 activation or autophagy-mediated cell death, suggesting a possible therapeutic approach to PHPV [43].

8. Ocular Diseases with Activated Autophagy

Autosomal Dominant Optic Atrophy (ADOA)

Mutations in the OPA1 gene can cause loss of RGCs due to increased autophagy or mitophagy, progressive myelin degeneration in the optic nerve and decreased visual acuity [125].

Similarly, in a ***dominant type of inherited optic atrophy/neuropathy (Charcot-Marie-Tooth type 2A)***, a primary deficiency of mitochondrial fusion and fission through the autophagic degradative pathway or mitophagy occurs as part of a protective mechanism [15].

Glaucoma

E50K mutation in the optineurin (ON) gene has been identified in glaucoma patients, suggesting involvement of the UPS and autophagy pathways in the ON processing in RGC5 and neuronal PC12 cells [109]. This study showed that upregulation or mutation of ON compromises the function of UPS, therefore activating autophagy as demonstrated *in vivo* in RGCs of E50K transgenic mice also.

Retinitis Pigmentosa (RP)

Retinal degeneration such as RP leads to rod apoptotic cell death initiated by mutations in genes encoding proteins with crucial photoreceptors functions; cone death is due to nutritional deficiencies, starvation, and autophagy driven by the insulin/mTOR pathway providing alternatives for future therapeutic cone rescue [48]. Changes in the insulin/mammalian TOR pathway coincide with the activation of autophagy during the period of cone death. Mice treated systemically with insulin show prolonged cone survival, suggest a non-autonomous cone death in retinitis pigmentosa could, in part, due to starvation of cones [89]. Recently, rapamycin (sirolimus), an analogue and antagonist of the mTOR signaling pathway was approved for the treatment of several cancers [113] and is currently in clinical trials for other ocular uses.

AMD is the leading cause of loss of central vision in developed countries, characterized by progressive degeneration of the macular region of the retina. AMD essentially involves chronic oxidative stress and increased accumulation of cytotoxic protein aggregates and LS in senescent, postmitotic RPE cells, accompanied by a decreased lysosomal enzyme activity [41] and impaired autophagic clearance machinery. Drusen formation, a type of extracellular deposits that accumulate on the Bruch's membrane below the RPE, is the hallmark of early or dry type of AMD. Presence of chronic inflammation in the retina is the other hallmark of AMD. [36, 40].

Nuclear factor erythroid 2-related factor 2 (NRF2) knockout mice deficient in the transcriptional factor that plays a key role in retinal antioxidant and detoxification responses develop age-dependent degenerative pathology in the RPE: drusen-like deposits, accumulation of LS, spontaneous choroidal neovascularization (CNV) and sub-RPE deposition of inflammatory proteins, along with autophagy-related vacuoles and multivesicular bodies within the RPE and the Bruch's membrane [133]. The UPS pathway and the auto-phago-lysosomal pathway work closely together with heat shock proteins (HSPs) in an attempt to to refold stress-induced misfold, the p62/sequestosome 1 (p62) being here a key player linking the proteasomal and the lysosomal clearance systems (Figure 1) [41, 119].

An increased mitochondrial DNA (mtDNA) damage and decreased DNA repair enzyme capabilities has also been shown in the rodent RPE/choroid with age, wile the autophagy and exosome markers were upregulated in old mice [122, 123]. Postnatal ablation of RPE mitochondrial oxidative phosphorylation in mice triggers gradual epithelium dedifferentiation, decreased response to light and eventual degeneration of the photoreceptors via stimulation of the AKT/mammalian TOR pathway. Treatment with rapamycin could blunt the dedifferentiation process and preserve the photoreceptor function after oxidative stress providing a rationale for developing therapeutic strategies for retinal degenerative diseases [132].

Lysosomal Storage Diseases (LSDs)

Mucolipidosis type IV is a genetic lysosomal storage disease due to mutations in the gene MCOLN1, which codes for the TRP family ion channel, mucolipin, associated with degenerative processes mainly in the brain and the eye. LSD diseases are typically associated with significant mitochondrial fragmentation and decreased mitochondrial Ca2+ buffering efficiency. Inhibition of autophagy with 3-methyladenine suggests inefficient autophagolysosomal recycling of mitochondria during mucolipidosis and increased sensitivity to apoptosis via the caspase-8-dependent pathway [39, 45].

Niemann-Pick type C (NPC) disease is an inherited LSD caused by mutations in Npc1 or Npc2, which mediate cooperatively the egress of cholesterol from lysosomes, manifested by a progressive neurodegeneration. Npc1 deficiency in the mouse retina causes striking retinal degeneration with signs of age-related maculopathies (impaired visual function, accumulation of LS in the RPE, degeneration of photoreceptor outer segments and a disruption of the synaptic layers). An increase in the autophagy markers in the RGC is described in this mouse model for studying age-related retinal degeneration [16].

Conclusion

Although a single comprehensive study of the functional involvement of autophagy in physiological processes of the eye does not exist as of yet, obvious proofs of its presence during eye development, lifetime and aging do exist in literature. The eye being a visual sense organ in which all three germ layers interact to achieve a healthy visual response, diseases affecting these layers in other organs with documented involvement of autophagy, will probably be relevant in the eye as well. To this date, most of the studies in the eye show autophagy as being protective, but a general consensus of the role of autophagy in this organ will probably show a balance between protection and damage. It is very likely that at cellular level there is a threshold beyond which less autophagy means more damage and more autophagy means better protection and vice versa. Further studies of the involvement of autophagy in the different segments of the eye and in ophthalmic diseases in general are needed to fill in the missing gaps and pave the roads towards finding better therapeutic targets utilizing autophagic components for maintaining healthy vision.

References

[1] Baehrecke EH. Autophagic programmed cell death in Drosophila. *Cell death and differentiation*, 2003 10(9), 940-5.

[2] Baehrecke EH. Autophagy: dual roles in life and death? *Nature reviews Molecular cell biology*, 2005 6(6), 505-10.

[3] Bandhyopadhyay U, Cuervo AM. Chaperone-mediated autophagy in aging and neurodegeneration: lessons from α-synuclein. *Experimental gerontology*, 2007 42(1-2), 120-8.

[4] Bell ET. Experimental studies on the development of the eye and the nasal cavities in frog embryos. *Anatomischer Anzeiger*, 1906 29, 185-94.

[5] Berger Z, Ravikumar B, Menzies FM, Oroz LG, Underwood BR, Pangalos MN, et al. Rapamycin alleviates toxicity of different aggregate-prone proteins. *Human molecular genetics*, 2006 15(3), 433-42.

[6] Bergmann M, Schütt F, Holz FG, Kopitz J. Inhibition of the ATP-driven proton pump in RPE lysosomes by the major lipofuscin fluorophore A2-E may contribute to the pathogenesis of age-related macular degeneration. *The FASEB journal: official publication of the Federation of American Societies for Experimental Biology*, 2004 18(3), 562-4.

[7] Boya P, Mellén MA, de la Rosa EJ. How autophagy is related to programmed cell death during the development of the nervous system. *Biochemical Society transactions*, 2008 36(5), 813-7.

[8] Burke JM, Skumatz CM. Autofluorescent inclusions in long-term postconfluent cultures of retinal pigment epithelium. *Investigative ophthalmology & visual science*, 1998 39(8), 1478-86.

[9] Burns MS, Tyler NK, Bellhorn RW. Melanosome abnormalities of ocular pigmented epithelial cells in beagle dogs with hereditary tapetal degeneration. *Current eye research*, 1988 7(2), 115-23.

[10] Buron N, Micheau O, Cathelin S, Lafontaine PO, Creuzot-Garcher C, Solary E. Differential mechanisms of conjunctival cell death induction by ultraviolet irradiation and benzalkonium chloride. *Investigative ophthalmology & visual science*, 2006 47(10), 4221-30.

[11] Burstein NL. Preservative cytotoxic threshold for benzalkonium chloride and chlorhexidine digluconate in cat and rabbit corneas. *Investigative ophthalmology & visual science*, 1980 19(3), 308-13.

[12] Buyukmihci N, Goehring-Harmon F. Histology and fine structure of the hamster retinal pigment epithelium. *Acta anatomica*, 1982 112(1), 36-46.

[13] Carra S, Boncoraglio A, Kanon B, Brunsting JF, Minoia M, Rana A, et al. Identification of the Drosophila Ortholog of HSPB8. *The Journal of biological chemistry*, 2010 285(48), 37811-22.

[14] Cecconi F, Levine B. The role of autophagy in mammalian development: cell makeover rather than cell death. *Developmental cell*, 2008 15(3), 344-57.

[15] Chen H, Chan DC. Mitochondrial dynamics--fusion, fission, movement, and mitophagy--in neurodegenerative diseases. *Human molecular genetics*, 2009 18(R2), R169-R76.

[16] Claudepierre T, Paques M, Simonutti M, Buard I, Sahel J, Maue RA, et al. Lack of Niemann-Pick type C1 induces age-related degeneration in the mouse retina. *Molecular and cellular neurosciences*, 2010 43(1), 164-76.

[17] Cuervo AM, Bergamini E, Brunk UT, Dröge W, Ffrench M, Terman A. Autophagy and aging: the importance of maintaining "clean" cells. *Autophagy*, 2005 1(3), 131-40.

[18] Danial NN, Korsmeyer SJ. Cell death: critical control points. *Cell*, 2004 116(2), 205-19.

[19] De Saint Jean M, Debbasch C, Brignole F, Rat P, Warnet JM, Baudouin C. Toxicity of preserved and unpreserved antiglaucoma topical drugs in an in vitro model of conjunctival cells. *Current eye research*, 2000 20(2), 85-94.

[20] Debbasch C, Brignole Fo, Pisella P-J, Warnet J-M, Rat P, Baudouin C. Quaternary ammoniums and other preservatives' contribution in oxidative stress and apoptosis on chang conjunctival cells. *Investigative ophthalmology & visual science*, 2001 42(3), 642-52.

[21] Decker RS, Wildenthal K. Lysosomal alterations in hypoxic and reoxygenated hearts. I. Ultrastructural and cytochemical changes. *The American journal of pathology*, 1980 98(2), 425-44.

[22] DeRenzo C, Seydoux G. A clean start: degradation of maternal proteins at the oocyte-to-embryo transition. *Trends in cell biology*, 2004 14(8), 420-6.

[23] Edinger AL, Thompson CB. Death by design: apoptosis, necrosis and autophagy. *Current opinion in cell biology*, 2004 16(6), 663-9.

[24] Fésüs L, Demény MÁ, Petrovski G. Autophagy shapes inflammation. *Antioxidants & redox signaling*, 2011 14(11), 2233-43.

[25] Foos RY, Rhodes RH. Respirator retina. *Archives of ophthalmology*, 1984 102(2), 296-303.

[26] Forrester JV. Macrophages eyed in macular degeneration. *Nature medicine*, 2003 9(11), 1350-1.

[27] Galluzzi L, Maiuri MC, Vitale I, Zischka H, Castedo M, Zitvogel L, et al. Cell death modalities: classification and pathophysiological implications. *Cell death and differentiation*, 2007 14(7), 1237-43.

[28] Gil JM, Rego AC. Mechanisms of neurodegeneration in Huntington's disease *The European journal of neuroscience*, 2008 27(11), 2803-20.

[29] Gonzalez-Fernandez F. Interphotoreceptor retinoid-binding protein--an old gene for new eyes. *Vision research*, 2003 43(28), 3021-36.

[30] Gonzalez-Fernandez F, Healy JI. Early expression of the gene for interphotoreceptor retinol-binding protein during photoreceptor differentiation suggests a critical role for the interphotoreceptor matrix in retinal development. *The Journal of cell biology*, 1990 111(6), 2775-84.

[31] Gordon WC, Bazan NG. Visualization of [3H]docosahexaenoic acid trafficking through photoreceptors and retinal pigment epithelium by electron microscopic autoradiography. *Investigative ophthalmology & visual science*, 1993 34(8), 2402-11.

[32] Green WR, Enger C. Age-related macular degeneration histopathologic studies. The 1992 Lorenz E. Zimmerman lecture. *Ophthalmology*, 1993 100(10), 1519-35.

[33] Guimarães CA, Benchimol M, Amarante-Mendes GP, Linden R. Alternative programs of cell death in developing retinal tissue. *The Journal of biological chemistry*, 2003 278(43), 41938-46.

[34] Hafezi F, Marti A, Munz K, Remé CE. Light-induced apoptosis: Differential timing in the retina and pigment epithelium. *Experimental eye research*, 1997 64(6), 963-70.

[35] Hartley T, Brumell J, Volchuk A. Emerging roles for the ubiquitin-proteasome system and autophagy in pancreatic β-cells. *American journal of physiology Endocrinology and metabolism*, 2009 296(1), E1-E10.

[36] Hyttinen JMT, Petrovski G, Salminen A, Kaarniranta K. AMPK-mTOR-axis as therapeutic target for age-related macular degeneration (AMD). *Rejuvenation Reseearch* 2011, in press.

[37] Iwasaki M, Inomata H. Lipofuscin granules in human photoreceptor cells. *Investigative ophthalmology & visual science*, 1988 29(5), 671-9.

[38] Jeffery G. The retinal pigment epithelium as a developmental regulator of the neural retina. *Eye*, 1998 12(3B), 499-503.

[39] Jennings JJ, Zhu J-h, Rbaibi Y, Luo X, Chu CT, Kiselyov K. Mitochondrial aberrations in mucolipidosis Type IV. *The Journal of biological chemistry*, 2006 281(51), 39041-50.

[40] Kaarniranta K, Hyttinen J, Ryhanen T, Viiri J, Paimela T, Toropainen E, et al. Mechanisms of protein aggregation in the retinal pigment epithelial cells. *Frontiers in bioscience (Elite edition)*, 2010 2, 1374-84.

[41] Kaarniranta K, Salminen A, Eskelinen E-L, Kopitz J. Heat shock proteins as gatekeepers of proteolytic pathways--Implications for age-related macular degeneration (AMD). *Ageing research reviews*, 2009 8(2), 128-39.

[42] Kaushik S, Massey AC, Mizushima N, Cuervo AM. Constitutive activation of chaperone-mediated autophagy in cells with impaired macroautophagy. *Molecular biology of the cell*, 2008 19(5), 2179-92.

[43] Kim JH, Yu YS, Mun JY, Kim KW. Autophagy-induced regression of hyaloid vessels in early ocular development. *Autophagy*, 2010 6(7), 922-8.

[44] Kim SH, Munemasa Y, Kwong JMK, Ahn JH, Mareninov S, Gordon LK, et al. Activation of autophagy in retinal ganglion cells. *Journal of neuroscience research*, 2008 86(13), 2943-51.

[45] Kiselyov K, Jennings JJ, Rbaibi Y, Chu CT. Autophagy, mitochondria and cell death in lysosomal storage diseases. *Autophagy*, 2007 3(3), 259-62.

[46] Knöferle J, Koch JC, Ostendorf T, Michel U, Planchamp Vr, Vutova P, et al. Mechanisms of acute axonal degeneration in the optic nerve in vivo. *Proceedings of the National Academy of Sciences of the United States of America*, 2010 107(13), 6064-9.

[47] Koch JC, Knöferle J, Tönges L, Ostendorf T, Bähr M, Lingor P. Acute axonal degeneration in vivo is attenuated by inhibition of autophagy in a calcium-dependent manner. *Autophagy*, 2010 6(5), 658-9.

[48] Koenekoop RK. Why do cone photoreceptors die in rod-specific forms of retinal degenerations? *Ophthalmic genetics*, 2009 30(3), 152-4.

[49] Krohne TU, Stratmann NK, Kopitz J, Holz FG. Effects of lipid peroxidation products on lipofuscinogenesis and autophagy in human retinal pigment epithelial cells. *Experimental eye research*, 2010 90(3), 465-71.

[50] Kulms D, Schwarz T. Molecular mechanisms involved in UV-induced apoptotic cell death. *Skin pharmacology and applied skin physiology*, 2002 15(5), 342-7.

[51] Kunchithapautham K, Rohrer Br. Apoptosis and autophagy in photoreceptors exposed to oxidative stress. *Autophagy*, 2007 3(5), 433-41.

[52] Kunchithapautham K, Rohrer Br. Autophagy is one of the multiple mechanisms active in photoreceptor degeneration. *Autophagy*, 2007 3(1), 65-6.

[53] Kurz T, Karlsson M, Brunk UT, Nilsson SE. ARPE-19 retinal pigment epitheliaL CELLS are highly resistant to oxidative stress and exercise strict control over their lysosomal redox-active iron. *Autophagy*, 2009 5(4), 494-501.

[54] Latkovic S, Nilsson SE. The ultrastructure of the normal conjunctival epithelium of the guinea pig. I. The basal and intermediate layers of the perilimbal zone. *Acta Ophthalmologica*, 1979 57(1), 106-22.

[55] Lee JA, Gao FB. Regulation of Aβ pathology by beclin 1: a protective role for autophagy? *The Journal of clinical investigation*, 2008 118(6), 2015-8.

[56] Leib DA, Alexander DE, Cox D, Yin J, Ferguson TA. Interaction of ICP34.5 with beclin 1 modulates herpes simplex virus type 1 pathogenesis through control of CD4+ T-cell responses. *Journal of virology*, 2009 83(23), 12164-71.

[57] Leske MC, Wu SY, Hennis A, Honkanen R, Nemesure B. Risk factors for incident open-angle glaucoma. The Barbados Eye Studies. *Ophthalmology*, 2008 115(1), 85-93.

[58] Levine B. Unraveling the role of autophagy in cancer. *Autophagy*, 2006 2(2), 65-6.

[59] Levine B. Cell biology: autophagy and cancer. *Nature*, 2007 446(7137), 745-7.

[60] Levine B, Deretic V. Unveiling the roles of autophagy in innate and adaptive immunity. *Nature reviews Immunology*, 2007 7(10), 767-77.

[61] Levine B, Klionsky DJ. Development by self-digestion: molecular mechanisms and biological functions of autophagy. *Developmental cell*, 2004 6(4), 463-77.

[62] Levine B, Kroemer G. Autophagy in the pathogenesis of disease. *Cell*, 2008 132(1), 27-42.

[63] Liang XH, Jackson S, Seaman M, Brown K, Kempkes B, Hibshoosh H, et al. Induction of autophagy and inhibition of tumorigenesis by beclin 1. *Nature*, 1999 402(6762), 672-6.

[64] Liton PB, Lin Y, Luna C, Li G, Gonzalez P, Epstein DL. Cultured porcine trabecular meshwork cells display altered lysosomal function when subjected to chronic oxidative stress. *Investigative ophthalmology & visual science*, 2008 49(9), 3961-9.

[65] Lohr HR, Kuntchithapautham K, Sharma AK, Rohrer B. Multiple, parallel cellular suicide mechanisms participate in photoreceptor cell death. *Experimental eye research*, 2006 83(2), 380-9.

[66] Long KO, Fisher SK, Fariss RN, Anderson DH. Disc shedding and autophagy in the cone-dominant ground squirrel retina. *Experimental eye research*, 1986 43(2), 193-205.

[67] Lyamzaev KG, Nepryakhina OK, Saprunova VB, Bakeeva LE, Pletjushkina OY, Chernyak BV, et al. Novel mechanism of elimination of malfunctioning mitochondria (mitoptosis): Formation of mitoptotic bodies and extrusion of mitochondrial material

from the cell. *Biochimica et biophysica acta (BBA) - Bioenergetics*, 2008 1777(7-8), 817-25.

[68] Maria Fimia G, Stoykova A, Romagnoli A, Giunta L, Di Bartolomeo S, Nardacci R, et al. Ambra1 regulates autophagy and development of the nervous system. *Nature*, 2007 447(7148), 1121-5.

[69] Martinet W, Agostinis P, Vanhoecke B, Dewaele M, De Meyer GRY. Autophagy in disease: a double-edged sword with therapeutic potential. *Clinical science*, 2009 116(9), 697-712.

[70] Martinet W, De Meyer GR. Autophagy in atherosclerosis. *Current atherosclerosis reports*, 2008 10(3), 216-23.

[71] Massey AC, Kaushik S, Sovak G, Kiffin R, Cuervo AM. Consequences of the selective blockage of chaperone-mediated autophagy. *Proceedings of the National Academy of Sciences of the United States of America*, 2006 103(15), 5805-10.

[72] Matsui M, Yamamoto A, Kuma A, Ohsumi Y, Mizushima N. Organelle degradation during the lens and erythroid differentiation is independent of autophagy. *Biochemical and biophysical research communications*, 2006 339(2), 485-9.

[73] Matsui Y, Takagi H, Qu X, Abdellatif M, Sakoda H, Asano T, et al. Distinct roles of autophagy in the heart during ischemia and reperfusion. *Circulation research*, 2007 100(6), 914-22.

[74] McMullen CA, Ferry AL, Gamboa JL, Andrade FH, Dupont-Versteegden EE. Age-related changes of cell death pathways in rat extraocular muscle. *Experimental gerontology*, 2009 44(6-7), 420-5.

[75] Mellén MA, de la Rosa EJ, Boya P. The autophagic machinery is necessary for removal of cell corpses from the developing retinal neuroepithelium. *Cell death and differentiation*, 2008 15(8), 1279-90.

[76] Mellén MA, de la Rosa EJ, Boya P. Autophagy is not universally required for phosphatidyl-serine exposure and apoptotic cell engulfment during neural development. *Autophagy*, 2009 5(7), 964-72.

[77] Midorikawa R, Yamamoto-Hino M, Awano W, Hinohara Y, Suzuki E, Ueda R, et al. Autophagy-dependent rhodopsin degradation prevents retinal degeneration in Drosophila. *The Journal of neuroscience: the official journal of the Society for Neuroscience*, 2010 30(32), 10703-19.

[78] Mijaljica D, Prescott M, Devenish R. Nibbling within the nucleus: turnover of nuclear contents. *Cellular and molecular life sciences : CMLS*, 2007 64(5), 581-8.

[79] Mizushima N, Levine B, Cuervo AM, Klionsky DJ. Autophagy fights disease through cellular self-digestion. *Nature*, 2008 451(7182), 1069-75.

[80] Nishino I. Autophagic vacuolar myopathies. *Current neurology and neuroscience reports*, 2003 3(1), 64-9.

[81] Nixon RA. Autophagy, amyloidogenesis and Alzheimer disease. *Journal of cell science*, 2007 120(23), 4081-91.

[82] Noecker RJ, Herrygers LA, Anwaruddin R. Corneal and conjunctival changes caused by commonly used glaucoma medications. *Cornea*, 2004 23(5), 490-6.

[83] Park Y, Kim W, Kim AY, Choi HJ, Choi JK, Park N, et al. Normal prion protein in Drosophila enhances the toxicity of pathogenic polyglutamine proteins and alters susceptibility to oxidative and autophagy signaling modulators. *Biochemical and Biophysical Research Communications*, 2011 404(2), 638-45.

[84] Petrovski G, Berényi E, Moe MC, Vajas A, Fésüs L, Berta A, et al. Clearance of dying ARPE-19 cells by professional and nonprofessional phagocytes in vitro– implications for age-related macular degeneration (AMD). *Acta Ophthalmologica*, 2010 89(1), e30–4.

[85] Petrovski G, Das DK. Does autophagy take a front seat in lifespan extension? *Journal of cellular and molecular medicine*, 2010 14, 2543–51.

[86] Petrovski G, Das S, Juhasz B, Kertesz A, Tosaki A, Das DK. Cardioprotection by endoplasmic reticulum stress-induced autophagy. *Antioxidants & redox signaling*, 2011 14(11), 2191-200.

[87] Portes ALF, Almeida AC, Allodi S, Monteiro MLR, Miguel NCO. Trypan blue staining for capsulorhexis: ultrastructural effect on lens epithelial cells and capsules. *Journal of cataract and refractive surgery*, 2010 36(4), 582-7.

[88] Prada C, Puga J, Pérez-Méndez L, López R, Ramírez G. Spatial and temporal patterns of neurogenesis in the chick retina. *The European journal of neuroscience*, 1991 3(6), 559-69.

[89] Punzo C, Kornacker K, Cepko CL. Stimulation of the insulin/mTOR pathway delays cone death in a mouse model of retinitis pigmentosa. *Nature neuroscience*, 2009 12(1), 44-52.

[90] Pursiheimo JP, Rantanen K, Heikkinen PT, Johansen T, Jaakkola PM. Hypoxia-activated autophagy accelerates degradation of SQSTM1/p62. *Oncogene*, 2008 28(3), 334-44.

[91] Qu X, Zou Z, Sun Q, Luby-Phelps K, Cheng P, Hogan RN, et al. Autophagy gene-dependent clearance of apoptotic cells during embryonic development. *Cell*, 2007 128(5), 931-46.

[92] Quigley HA. Number of people with glaucoma worldwide. *The British journal of ophthalmology*, 1996 80(5), 389-93.

[93] Remé C, Wirz-Justice A, Aeberhard B, Rhyner A. Chronic clorgyline dampens rat retinal rhythms. *Brain research*, 1984 298(1), 99-106.

[94] Remé C, Wirz-Justice A, Rhyner A, Hofmann S. Circadian rhythm in the light response of rat retinal disk-shedding and autophagy. *Brain research*, 1986 369(1-2), 356-60.

[95] Remé CE, Knop M. Autophagy in frog visual cells in vitro. *Investigative ophthalmology & visual science*, 1980 19(5), 439-56.

[96] Remé CE, Sulser M. Diurnal variation of autophagy in rod visual cells in the rat. *Albrecht von Graefe's archive for clinical and experimental ophthalmology*, 1977 203(3-4), 261-70.

[97] Remé CE, Wirz-Justice A, Da Prada M. Retinal rhythms in photoreceptors and dopamine synthetic rate: and the effect of a monoamineoxidase inhibitor. *Transactions of the ophthalmological societies of the United Kingdom*, 1983 103(4), 405-10.

[98] Remé CE, Wolfrum U, Imsand C, Hafezi F, Williams TP. Photoreceptor autophagy: effects of light history on number and opsin content of degradative vacuoles. *Investigative ophthalmology & visual science*, 1999 40(10), 2398-404.

[99] Remé CE, Young RW. The effects of hibernation on cone visual cells in the ground squirrel. *Investigative ophthalmology & visual science*, 1977 16(9), 815-40.

[100] Ropolo A, Grasso D, Pardo R, Sacchetti ML, Archange C, Re AL, et al. The pancreatitis-induced vacuole membrane protein 1 triggers autophagy in mammalian cells. *The Journal of biological chemistry*, 2007 282(51), 37124-33.

[101] Ross CA, Poirier MA. Protein aggregation and neurodegenerative disease. *Nature medicine*, 2004 10(SUPPL.), S10-S7.

[102] Ross CA, Poirier MA. What is the role of protein aggregation in neurodegeneration? *Nature reviews Molecular cell biology*, 2005 6(11), 891-8.

[103] Rubinsztein DC. The roles of intracellular protein-degradation pathways in neurodegeneration. *Nature*, 2006 443(7113), 780-6.

[104] Russo R, Berliocchi L, Adornetto A, Varano GP, Cavaliere F, Nucci C, et al. Calpain-mediated cleavage of Beclin-1 and autophagy deregulation following retinal ischemic injury in vivo. *Cell death & disease*, 2011 2, e144.

[105] Ryhänen T, Hyttinen JMT, Kopitz J, Rilla K, Kuusisto E, Mannermaa E, et al. Crosstalk between Hsp70 molecular chaperone, lysosomes and proteasomes in autophagy-mediated proteolysis in human retinal pigment epithelial cells. *Journal of cellular and molecular medicine*, 2009 13(9 B), 3616-31.

[106] Sadler TW, editor. Langman's Medical Embryology 11th ed: Lippincott Williams & Wilkins; 2009.

[107] Salminen A, Kaarniranta K. Regulation of the aging process by autophagy. *Trends in Molecular Medicine*, 2009 15(5), 217-24.

[108] Schutt F, Bergmann M, Holz FG, Kopitz J. Proteins Modified by Malondialdehyde, 4-Hydroxynonenal, or Advanced Glycation End Products in Lipofuscin of Human Retinal Pigment Epithelium. *Investigative ophthalmology & visual science*, 2003 44(8), 3663-8.

[109] Shen X, Ying H, Qiu Y, Park J-S, Shyam R, Chi Z-L, et al. Processing of optineurin in neuronal cells. *The Journal of biological chemistry*, 2011 286(5), 3618-29.

[110] Steinberg RH. Interactions between the retinal pigment epithelium and the neural retina. *Documenta Ophthalmologica*, 1985 60(4), 327-46.

[111] Sternberg C, Benchimol M, Linden R. Caspase dependence of the death of neonatal retinal ganglion cells induced by axon damage and induction of autophagy as a survival mechanism. *Brazilian journal of medical and biological research*, 2010 43(10), 950-6.

[112] Strauss O. The retinal pigment epithelium in visual function. *Physiological reviews*, 2005 85(3), 845-81.

[113] Sudarsanam S, Johnson DE. Functional consequences of mTOR inhibition. *Current opinion in drug discovery & development*, 2010 13(1), 31-40.

[114] Sybers HD, Ingwall J, DeLuca M. Autophagy in cardiac myocytes. *Recent advances in studies on cardiac structure and metabolism*, 1976 12, 453-63.

[115] Tiffany JM, Bron AJ. Role of tears in maintaining corneal integrity. *Transactions of the ophthalmological societies of the United Kingdom*, 1978 98(3), 335-8.

[116] Tsai YY, Cheng YW, Lee H, Tsai FJ, Tseng SH, Lin CL, et al. Oxidative DNA damage in pterygium. *Molecular vision*, 2005 11, 71-5.

[117] Tsukamoto S, Kuma A, Murakami M, Kishi C, Yamamoto A, Mizushima N. Autophagy is essential for preimplantation development of mouse embryos. *Science*, 2008 321(5885), 117-20.

[118] Valamanesh F, Torriglia A, Savoldelli M, Gandolphe C, Jeanny JC, BenEzra D, et al. Glucocorticoids induce retinal toxicity through mechanisms mainly associated with paraptosis. *Molecular vision*, 2007 13, 1746-57.

[119] Viiri J, Hyttinen JM, Ryhanen T, Rilla K, Paimela T, Kuusisto E, et al. p62/sequestosome 1 as a regulator of proteasome inhibitor-induced autophagy in human retinal pigment epithelial cells. *Molecular vision*, 2010 16, 1399-414.

[120] Vos MJ, Zijlstra MP, Kanon B, van Waarde-Verhagen MAWH, Brunt ERP, Oosterveld-Hut HMJ, et al. HSPB7 is the most potent polyQ aggregation suppressor within the HSPB family of molecular chaperones. *Human molecular genetics*, 2010 19(23), 4677-93.

[121] Wang AL, Boulton ME, Dunn WA, Rao HV, Cai J, Lukas TJ, et al. Using LC3 to monitor autophagy flux in the retinal pigment epithelium. *Autophagy*, 2009 5(8), 1190-3.

[122] Wang AL, Lukas TJ, Yuan M, Du N, Tso MO, Neufeld AH. Autophagy and exosomes in the aged retinal pigment epithelium: possible relevance to drusen formation and age-related macular degeneration. *PloS one*, 2009 4(1), e4160.

[123] Wang AL, Lukas TJ, Yuan M, Du N, Tso MO, Neufeld AH. Autophagy, exosomes and drusen formation in age-related macular degeneration. *Autophagy*, 2009 5(4), 563-4.

[124] Wang T, Lao U, Edgar BA. TOR-mediated autophagy regulates cell death in Drosophila neurodegenerative disease. *The Journal of cell biology*, 2009 186(5), 703-11.

[125] White KE, Davies VJ, Hogan VE, Piechota MJ, Nichols PP, Turnbull DM, et al. OPA1 Deficiency Associated with Increased Autophagy in Retinal Ganglion Cells in a Murine Model of Dominant Optic Atrophy. *Investigative ophthalmology & visual science*, 2009 50(6), 2567-71.

[126] Wu BX, Darden AG, Laser M, Li Y, Crosson CE, Hazard ESr, et al. The rat Apg3p/Aut1p homolog is upregulated by ischemic preconditioning in the retina. *Molecular vision*, 2006 12, 1292-302.

[127] Xavier RJ, Huett A, Rioux JD. Autophagy as an important process in gut homeostasis and Crohn's disease pathogenesis. *Gut*, 2008 57(6), 717-20.

[128] Xing D, Sun X, Li J, Cui M, Tan-Allen K, Bonanno JA. Hypoxia preconditioning protects corneal stromal cells against induced apoptosis. *Experimental eye research*, 2006 82(5), 780-7.

[129] Yamada T, Dumont JN, Moret R, Brun JP. Autophagy in dedifferentiating newt iris epithelial cells in vitro. *Differentiation; research in biological diversity*, 1978 11(3), 133-47.

[130] Yamada T, Moret R, Brun JP. Telolysomes in cultured iris epithelial cells and in the TVI cell-line. *Differentiation; research in biological diversity*, 1979 12(3), 191-3.

[131] Yin X-M, Ding W-X, Gao W. Autophagy in the liver. *Hepatology*, 2008 47, 1773–85.

[132] Zhao C, Yasumura D, Li X, Matthes M, Lloyd M, Nielsen G, et al. mTOR-mediated dedifferentiation of the retinal pigment epithelium initiates photoreceptor degeneration in mice. *The Journal of clinical investigation*, 2011 121(1), 369-83.

[133] Zhao Z, Chen Y, Wang J, Sternberg P, Freeman ML, Grossniklaus HE, et al. Age-related retinopathy in NRF2-deficient mice. *PLoS one*, 2011 6(4), e19456.

[134] Zheng Q, Li J, Wang X. Interplay between the ubiquitin-proteasome system and autophagy in proteinopathies. *International journal of physiology, pathophysiology and pharmacology*, 2009 1(2), 127-42.

[135] Zigler JS, Zhang C, Grebe R, Sehrawat G, Hackler L, Adhya S, et al. Mutation in the βA3/A1-crystallin gene impairs phagosome degradation in the retinal pigmented epithelium of the rat. *Journal of cell science*, 2011 124(4), 523-31.

[136] Zoukhri D. Effect of inflammation on lacrimal gland function. *Experimental eye research*, 2006 82(5), 885-98.

[137] Zoukhri D, Macari E, Kublin CL. A single injection of interleukin-1 induces reversible aqueous-tear deficiency, lacrimal gland inflammation, and acinar and ductal cell proliferation. *Experimental eye research*, 2007 84(5), 894-904.

In: Autophagy: Principles, Regulation and Roles in Disease ISBN: 978-1-61942-266-7
Editor: Nikolai Gorbunov

Chapter IX

Regulation of Autophagy by Insulin/IGF-1 Signaling Pathways

Rodrigo Troncoso[1,*], César Vásquez-Trincado[1,*], Andrea del Campo[1], Damián Gatica[1], Pablo E. Morales[1], Felipe Paredes[1], Natalia Torrealba[1], Camila Lopez-Crisosto[1], Zully Pedrozo[1], Lorena García[1], Guillermo Díaz-Araya[1], Pavan K. Battiprolu[3], Andriy Nemchenko[3], Anwarul Ferdous[3], Thomas G. Gillette[3], Joseph A. Hill[3] and Sergio Lavandero[1,2,3,#]

[1] Centro de Estudios Moleculares de la Celula, Facultad de Ciencias Quimicas y Farmaceuticas/Facultad Medicina, Chile

[2] Instituto de Ciencias Biomedicas, Facultad de Medicina, Universidad de Chile, Santiago, Chile

[3] Department of Internal Medicine (Cardiology), University of Texas Southwestern Medical Center, Dallas, Texas, US

Abstract

Insulin and insulin-like growth factor-1 (IGF-1) share a high degree of structural homology and biological actions. Both peptides mediate several cellular processes involved mostly in metabolism and growth. Insulin is mainly implicated in the maintenance of energy balance whereas IGF-1 is essential for normal growth, development and differentiation. Autophagy is one of the key catabolic process regulated by insulin/IGF-1. In recent years, autophagic activity has been linked with multiple pathological conditions in patients. However, the underlying signaling mechanisms mediating the regulation of autophagy by insulin/IGF-1 and its consequences in the settings of cancer, atrophy, aging and diabetes are less understood. This review is focused

* Both authors contributed equally to this work.

Corresponding author: Sergio Lavandero. Facultad de Ciencias Quimicas y Farmaceuticas, Universidad de Chile, Olivos 1007, Santiago 838-0492, Chile. Telephone: +56 (2) 978-2919; Fax: +56 (2) 978-2912; E-mail: slavander@uchile.cl.

on the interface between insulin/IGF-1 pathways and autophagy with emphasis on these pathologies.

Keywords: Growth factors, Autophagy, Metabolism, Cancer, Longevity, Diabetes

1. Introduction

Diabetes, cardiovascular diseases, cancer and chronic lung diseases are responsible for the majority of deaths caused by non-communicable diseases (NCDs)[1]. Among the risk factors responsible for the majority of these deaths are smoking, physical inactivity, harmful use of alcohol and unhealthy diet[1].

Diverse studies have shown that dysregulation of autophagy contributes to the pathogenesis of diabetes, cardiovascular diseases and cancer[2-4]. Autophagy is a catabolic process involved in protein degradation, organelle turnover and a nonselective breakdown of cytoplasmic components during nutrient starvation or stress. Autophagy begins with the formation of the autophagosome, a double-membrane structure of reticular origin that engulfs cytoplasmic contents and then fuses with lysosomes for cargo delivery and degradation. Materials degraded within the newly formed autolysosomes are recruited to anabolic reactions in order to maintain energy levels and to provide macromolecules for the synthesis of higher structures (nucleic acids, proteins or organelles), thereby sustaining cell metabolism, homeostasis and survival[5]. Despite this key role, autophagy also contributes to cell death when it is executed excessively or inefficiently, as occurs during tissue and organ development, or pathologic states. Indeed, autophagic cell death is claimed as a morphological term derived from electron microscopic observations that denote a cell death phenotype characterized by abundant autophagic vacuoles in the cytoplasm[6].

In the presence of abundant nutrient supply, some growth factors stimulate anabolic processes. In this context autophagy is moderated. In contrast, autophagy is triggered by nutritional stress or by defects in growth factor-dependent signaling to replenish intracellular stores of key components of energy metabolism. Logically, two key growth factors, insulin and insulin-like growth factor-1 (IGF-1) serve as important regulators of autophagy. This review is focused on the interface between insulin/IGF-1 pathway and autophagy with emphasis on cancer, muscle pathologies and diabetes.

2. Insulin and IGF-1

Insulin and IGF-1 are closely related peptides that mediate several cellular processes involved mostly in metabolism and growth. Insulin is mainly implicated in the maintenance of energy balance, regulating carbohydrate, lipid and protein metabolism, while IGF-1 is essential for normal growth, development and differentiation events in specific tissues[7] (Figure 1). Both peptides initiate signaling pathways by triggering their respective tyrosine kinase receptors which share 84% of homology[8]. Upon binding of ligand to their respective receptor, they trigger autophosphorylation of the receptor's tyrosine residue, resulting in the activation of two downstream signaling pathways: phosphatidylinositol 3-kinase (PI3K)-Akt and Ras-Raf-extracellular signal-regulated kinase (ERK) cascades[9,10] (Figure 1).

The PI3K-Akt pathway starts with the autophosphorylation of tyrosine residues in the β-subunits of insulin receptor (IR) or IGF-1 receptor (IGF-1R). This leads to the recruitment of insulin receptor substrate-1/2 (IRS 1/2) to the receptor phosphotyrosines for subsequent phosphorylation of IRS1/2 by the intrinsic tyrosine kinase activity of the IR/IGF-1R. Next, the PI3K binds to IRS1/2 and catalyzes addition of a third phosphate moiety to the phosphatidylinositol 4,5-biphosphate (PIP2), to produce a second messenger of lipid origin, phosphatidylinositol 3,4,5-triphosphate (PIP3). This last event is necessary for the activation of the serine/threonine kinase Akt, a critical node in this signaling pathway. When PIP3 is formed, Akt is translocated to the plasma membrane and is phosphorylated by 3-phosphoinositide-dependent protein kinase-1 (PDK-1). This phosphorylation leads to incomplete Akt activation. An additional phosphorylation by mammalian target of rapamycin complex 2 (mTORC2)[11] is necessary to fully activate Akt. Recently, it has also been reported that additional ubiquitination of Akt by the E3-ligase TRAF6 is required to achieve its complete activation[12]. Finally, activation of Akt after insulin stimulation also triggers phosphorylation of many other substrates such as glycogen synthase kinase-3β (GSK-3β), mammalian target of rapamycin (mTOR), forkhead box-containing protein O subfamily 1/3 (FoxO1/3) and 160-kDa substrate of Akt (AS160), resulting in increased synthesis of glycogen and proteins, and stimulation of glucose uptake[13] (Figure 1).

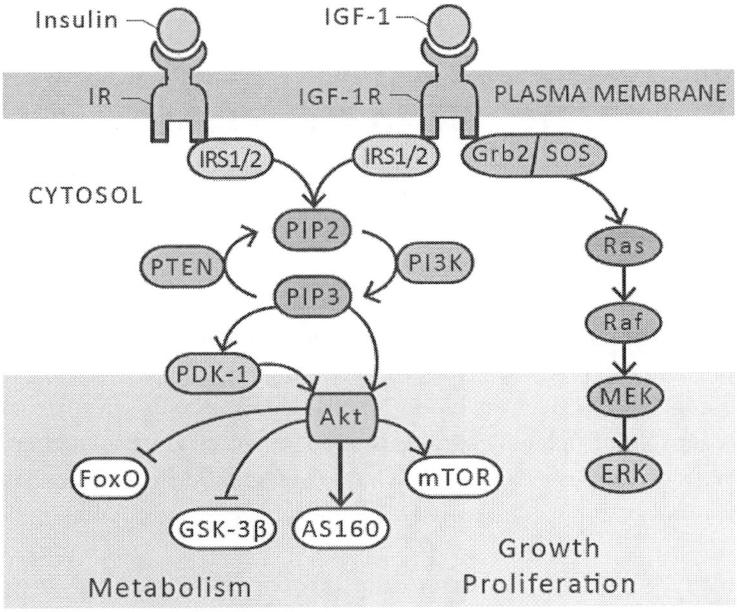

Figure 1. *Differences and similarities between the signaling pathways for IGF-1 and insulin.* IGF-1 and insulin turn on the same signaling pathways only with differences in the emphasis they exert in each part of it. While insulin is markedly more metabolic, IGF-1 is mainly responsible for growth and cell proliferation. The metabolic effects of insulin are mostly controlled by the serine threonine kinase Akt, which is the central node of the phosphorylation cascade that includes AS160, mTOR and GSK-3β among others. Proliferation and growth effects are, in turn, essentially controlled by IGF-1 through the activation of the Akt/mTOR and ERK1/2.

In contrast to insulin, stimulation with IGF-1 has different effects, although they activate the same protein kinases. This is likely due to the differences in the intensity of the signal and the specific Akt isoforms involved in each process. Akt1 is required for normal cardiac growth while Akt2 regulates cardiac metabolism and cardiomyocyte survival[14]. Therefore, IGF-1 exerts its action through the Akt1 isoform and controls cell growth, whereas insulin preferentially activates Akt2 regulating metabolism, thus establishing the fundamental difference between these related peptides[14]. Additionally, preferential stimulation of the Ras-Raf-MAPK pathway by IGF-1 and not insulin may account for distinct net effects of the two hormones.

The Ras-Raf-MAPK pathway mainly regulates cell growth, survival and differentiation. Its initial step requires binding of growth factor receptor-bound protein-2 (Grb2) and son-of-sevenless (SOS) to the phosphotyrosines on IRS proteins: Grb2-associated-binding protein 1 (Gab1) and Src homology 2 domain-containing protein (Shc). This induces Ras and Raf activity, followed by activation of several kinases. At this point phosphorylation of mitogen-activated protein kinase (MAPK)/ERK kinase 1/2 (MEK1/2) leads to phosphorylation of its substrate, ERK 1/2, which then activates several of its own targets[15]. IGF-1 strongly induces cell growth and proliferation through the Ras-Raf-ERK pathway in cooperation with the Akt/mTOR pathway that mediates protein synthesis[16] (Figure 1).

Thus, although insulin and IGF-1 are related peptides that share high homology in their respective receptors and activate similar signaling pathways, they exert distinct effects, i.e. insulin triggering to the metabolic actions and IGF-1 regulating development. These processes will be explained in detail below.

3. Insulin in Metabolism

As mentioned before, insulin plays a critical role in the regulation of glucose, lipids and protein metabolism, participating in uptake, synthesis and degradation of these compounds to maintain cell energy homeostasis. Glucose, for example, must be maintained at a constant plasma range from 4 to 7 mM in healthy individuals even during a fasting state. This regulation of glucose levels is governed by enforcing strict control over glucose production in the liver and its uptake and metabolism in peripheral organs[17].

To exert its regulatory effect on blood glucose concentration, insulin acts either in the liver where it is capable of inducing synthesis and storage of glucose, and enhances glucose uptake in skeletal and adipose tissue[18]. Thus, dysregulation of insulin levels or cellular responses to insulin, as in insulin resistance, may increase circulating glucose and lipid concentrations[19].

Insulin directly regulates the expression of enzymes involved in glycolysis and gluconeogenesis[20]. For example, insulin suppresses the transcription of the gene coding for phosphoenolpyruvate carboxylase (PEPC), the rate-limiting enzyme of gluconeogenesis[20]. Insulin represses fructose-1,6-bisphosphatase and glucose-6-phosphatase gene expression, but induces the expression of the glycolytic (glucokinase and pyruvate kinase) and lipogenic (fatty acid synthase and acetyl-CoA carboxylase) enzymes[20].

Energy metabolism is regulated by several feedback mechanisms and specific sensors measure circulating levels of lipids, amino acids and glucose and send this information to hormone-secreting cells. Glucagon and catecholamines oppose insulin's metabolic actions, stimulating the production and/or secretion of fatty acids or glucose when these are needed. It remains poorly understood whether or not glucagon antagonizes any of the biological effects of IGF-1.

4. Insulin/IGF-1 and Growth

IGF-1 is a powerful regulator of protein synthesis and cell growth, while its effects on nutrient metabolism are less significant. Growth can be defined as an increase in the mass of a given tissue due to an increment in the number of cells and/or cell size. This phenomenon begins in the early stages of development, where IGF-1 plays a vital role. For example, mice lacking *Igf-1* are born with a 40% decrease in body weight in comparison to wild type mice[21]. Moreover, mice with double knockout of *Igf-1* and *Igf2* are born 70% smaller in body size compared to their *WT* littermates[22]. This growth-promoting effect of IGF-1 lasts through post-natal life, as observed by a reduction of almost 70% of body weight in $IGF\text{-}1^{-/-}$ mice at day 60 and 65% of normal body length at day 130[23]. Furthermore, it has been described that growth hormone (GH)-resistance, as in Laron Syndrome, impairs IGF-1 secretion resulting in newborns with smaller body size, retarded skeletal muscle maturation and delayed organ growth in childhood[7]. Although GH does not play an important role in fetus development[24], the observations described above support the idea that the GH/IGF-1 axis is pivotal for normal growth in mammals.

In addition to maintaining normal growth, IGF-1 also promotes overgrowth (or hypertrophy) of high-demand tissues. For instance, both IGF-1 and IGF-1R are up-regulated in a skeletal muscle cell line undergoing differentiation, whereas local overexpression of IGF-1 in skeletal and cardiac muscle results in increased protein content and cell size[25]. Notably, studies where mice overexpressing heart-specific IGF-1R were subjected to exercise show that IGF-1-dependent cardiac hypertrophy mimics physiological, not pathological, cardiac growth[26]. It has been demonstrated that although IGF-1 promotes activation of several signaling pathways, muscle hypertrophy is achieved mostly by the stimulation of the PI3K/Akt/mTOR pathway[25].

Like IGF-1, insulin also plays a role in physiological growth. For example, hearts from 6-12 week old mice with cardiomyocyte-specific knockout of the insulin receptor (CIRKO) are smaller than hearts from WT mice[27], while rats subjected to chronic hyperinsulinemia (7 weeks) had larger hearts, kidneys and adipose tissue than non-treated rats[28]. These growth-promoting effects of insulin are likely mediated by its characteristic signaling pathway. In agreement with this are two lines of evidence: mice down-regulated for PDK-1 are nearly 30% smaller at birth than wild-type mice[29], and the presence of dominant negative Akt1 totally abolishes these insulin-mediated effects on heart[30]. Also, insulin is able to restore the protein content and weight of diabetic rats, likely by stimulating ribosome synthesis in heart and skeletal muscle[31].

As noted, IGF-1 is a considerably more potent growth inducer than insulin, which mainly controls metabolism. This highlights the functional differences between these two closely related hormones.

5. Insulin/IGF-1 and Autophagy

The insulin signaling pathway blunts autophagy via activation of PI3K/Akt which then induces mTOR (a known negative regulator of autophagy) and inhibits the autophagy-activating transcription factor FoxO3[32] (Figure 2). This regulation of autophagy by insulin is of great importance since it affects energy metabolism, cancer development, muscle atrophy, and aging (see below).

Unlike the effects of insulin on PI3K/Akt, diverging evidence exists regarding the effects of IGF-1 on autophagy, and the mechanistic pathways involved in these effects. An inhibitory role for IGF-1 on autophagy through the Akt/mTOR pathway has been observed in different cell types, including human osteocarcinoma cells[33], vascular cells from patients with atherosclerotic lesions stimulated with TNF-α[34], and mammary epithelial cell involution[35]. Conversely, IGF-1 has been demonstrated to promote autophagy in the H9c2 cell line[36] and in Purkinje neurons[37]. This distinction may be due to a requirement for autophagy in growth and remodeling of some cell types.

On another hand, it has been also shown that insulin and IGF-1 induces an inositol 1,4,5-trisphosphate (InsP3)-dependent Ca^{2+} release due to a positive regulation of the inositol 1,4,5-triphosphate receptor (InsP3R) that increases its Ca^{2+} releasing activity[38,39]. Besides, there is an essential role for constitutive InsP3R-dependent Ca^{2+} release in the control of mitochondrial bioenergetics and autophagy[40-42], that could be modulated by the Insulin/IGF-1 pathway.

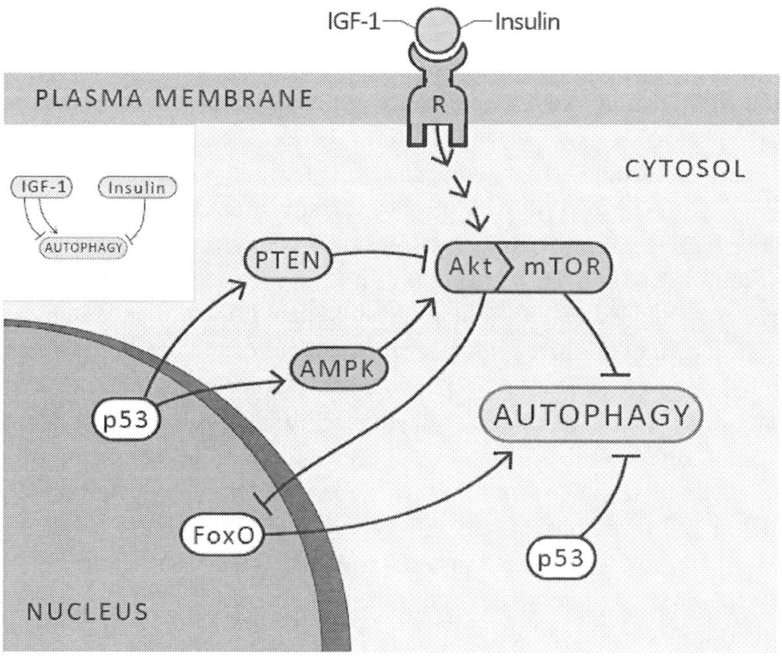

Figure 2. *Regulation of autophagy by insulin and IGF-1 signaling pathways.* Insulin and IGF-1 control autophagy at the level of Akt, mTOR and FoxO. The transcription factor p53 also regulates autophagy by controlling the expression of phosphatase PTEN and AMPK.

6. Insulin/IGF-1 and Cancer

In addition to acting as potent mitogen factors and accelerating cell division and proliferation, insulin and IGF-1 can promote cell survival by stimulating anti-apoptotic signaling. Several studies have suggested these peptides as risk factors for cancer development. Epidemiological studies have shown a relationship between high concentrations of circulating IGF-1 in breast, lung, colorectal and prostate cancer[43].

The transcription factor p53 protects cells against cancer through the induction of cell cycle arrest and activation of DNA repair signals. In the setting of excessive or irreversible DNA damage, p53 induces apoptosis[44]. At least 50 % of cancers carry mutations in p53[45]. In addition to its own tumor-suppressing activity, p53 activates the transcription of another well-known tumor suppressor protein, PTEN, which interestingly, is also a potent suppressor of the PI3K/Akt pathway (Figure 2). Compellingly, p53 mutations lead to a reduced expression of PTEN resulting in Akt activation by IGF-1, leading to pro-survival signaling and proliferation of cancer tissue[45]. Furthermore, p53 is involved in the transcription of the IGF-BP3 gene, a major regulator of IGF-1 bioactivity. IGF-1 BP3 binds IGF-1 in the plasma membrane and thereby blocks IGF-1 from binding its own receptor. IGF-BP3, independently of IGF-1, can stimulate cell death by apoptosis[43]. p53 mutations reduce IGF-BP3 transcription, thus relieving repression from IGF-1 – IGF-1R interaction, and triggering pro-survival signaling[46]. Supporting this, wild-type p53 was shown to inhibit IGF-1R gene expression, decreasing the effect of IGF-1 and insulin signaling. Lack of p53 activity due to mutations, however, leads to accelerated cell growth via inhibition of apoptosis and autophagic (type II) cell death[47].

The effect of p53 on autophagy is either activating or inhibitory, depending on the subcellular localization (Figure 2). Nuclear p53 induces autophagy by promoting the transcription of pro-autophagic genes including AMP-activated protein kinase (AMPK), tuberous sclerosis protein 2 (TSC2) and damage-regulated autophagy modulator (DRAM), while cytoplasmic p53 inhibits autophagy[48]. A number of studies have shown that standard pro-autophagic stimuli such as nutrient deprivation and rapamycin induce proteasomal degradation of p53. Additionally, p53 inactivation by pharmacological inhibitors, siRNA or gene deletion strategies induces autophagy[49]. IGF-1 overexpression in cancer could lead to increased PI3K/Akt signaling leading to the inhibition of autophagy and cell survival. Since FoxO is a key transcriptional activator of pro-autophagic proteins, its inactivation by Akt acts as a suppressor of autophagy. This dual role of autophagy in cancer development is already a focus of the ongoing research.

Recently, the role of autophagy in cancer development and therapy has become an important area of research. Some studies show that cancer cells induce autophagy as a means of obtaining nutrients for the growth and proliferation of tumor cells. Autophagy is particularly instrumental in fulfilling metabolic needs of the hypoxic regions of some tumors that otherwise cannot obtain nutrients from blood vessels. These findings open the possibility of manipulating autophagy for therapeutic gain. Other investigations show that apoptosis is inhibited in various types of cancers, which explains why autophagic cell death could also be an attractive target for cancer therapy[50].

7. Insulin/IGF-1 in Muscle Atrophy

Muscle atrophy is a hallmark feature of various physiological and pathological states. For example, prolonged bed rest and weightlessness cause physiological muscle atrophy while pathological muscle atrophy is manifested in various diseases, including cancer and arthritis. A tight balance between the synthesis and degradation of muscle proteins may be impaired during disease or muscle disuse increasing the degradation rate and provoking atrophy[51].

There are two major pathways that regulate protein degradation in skeletal muscle. One acts through the induction of the two muscle-specific ubiquitin ligases, Atrogin1 and MuRF1, while the second is mediated by autophagy by a still unknown mechanism[52]. These two components target different proteins but also act coordinately to augment protein degradation in different atrophy conditions[53]. Understanding the underlying mechanisms enabling such a dual role of autophagy in atrophy will be essential to understand the role of autophagy in skeletal muscle. It is still unclear whether autophagy is beneficial or detrimental in myocyte responses to atrophy. The genetic suppression of autophagy using knockout *in vivo* models sets off myopathy and weakness as a result of massive degradation of muscle proteins, suggesting a requirement for basal autophagy in maintaining muscle mass[54]. Conversely, it has been described that several autophagy-related genes that are under the control of FoxO3 also serve as atrogenes. Furthermore, knockdown of the microtubule-associated protein 1 light chain 3 (LC3) that participates directly in the formation of autophagosomes partially preserves muscle mass under atrophy conditions[55]. These data suggest that strict regulation of autophagy activity is required for growth and maintenance of muscle mass.

It is clear now that IGF-1 is a major regulator of the delicate balance between synthesis and degradation of muscle proteins. This growth factor stimulates synthesis of proteins, blocks their breakdown[56], and stimulates the replication of satellite cells and myotube formation after the injury[57]. In addition, IGF-1 transgenic mice are resistant to muscle atrophy induced by angiotensin II treatment or in a mouse model of cardiac cachexia[58,59]. Also, local IGF-1 injection is sufficient to block disuse atrophy[60]. In these models of muscle loss, IGF-1 robustly suppressed the induction of atrogin-1 and MuRF1 that are normally blocked by the Akt-mediated inhibition of the FoxO family[60]. Specifically, phosphorylation of FoxO factors by Akt promotes their nuclear export, thus blunting FoxO-mediated transactivation. As expected, the reduced activity of the Akt pathway observed in models of muscle atrophy results in shuttling FoxO back into the nucleus. Active FoxO then initiates the expression of atrogin-1, MuRF1, and pro-autophagic genes[61]. However, the role of the insulin/IGF-1 pathway in regulation of autophagy in different muscle atrophy models is currently unknown. It remains to be determined whether this catabolic process is maladaptive or beneficial in the setting of muscle atrophy.

8. Insulin/IGF-1 in Aging

The last stages of life in all species are generally characterized by a progressive decline in the ability to resist environmental stressors. This change is caused by the loss of normal homeostasis and an increase in systemic diseases[62]. Here, insulin/IGF-1 can be viewed as a natural line of defense as it has been linked with longevity in many organisms[63]. In fact,

IGF-1 signaling has been associated with an extension of lifespan through its regulation of autophagy[64]. This control of longevity by the IGF-1 pathway uses mechanisms that are conserved from yeast to mammals[65]. For instance, mutations in the insulin/IGF-1 signaling pathway reduce growth and body size, and paradoxically, prolong longevity in worms, flies and mice,[66,67]. Moreover, studies in *C. elegans* demonstrate that inhibition of daf-2 (an insulin receptor homologue in worms) extends longevity[68]. Supporting this, another study shows that heterozygous knockout mice for IGF-1R live 26% longer than wild type mice[69].

Likewise, IGF-1 deficiency retards the aging processes and prolongs lifespan in animal models[70]. However, a basal level of this growth factor appears to be necessary for optimal survival[71]. The IGF-1 signaling pathway involves the activation of mTOR whose role in aging has been extensively studied. For example, reduced mTOR signaling by exposure to rapamycin or through genetic modulation of *dTsc1*, *dTsc2*, *dTOR* and *dS6K* has been previously associated with extended lifespan in flies, *C. elegans* and mice[72,73 74]. In addition, the over-expression of FoxO in adipose tissue was also shown to increase the lifespan in the *Drosophila* and *C. elegans*[75].

The positive effect of autophagy on lifespan has been demonstrated in various systems. For instance, pharmacological stimulation of autophagy decreases the accumulation of cytotoxic protein aggregates. Furthermore, accumulation of α-synuclein and β-amyloid peptides associated with Parkinson and Alzheimer disease can be successfully reversed by autophagy inducers[2]. Inhibition of autophagy on the other hand, leads to the accumulation of inclusion bodies that are correlated with neurodegeneration, one of the most typical pathology associated with old age[76].

Autophagy also removes damaged mitochondria, that can become a source of harmful reactive oxygen species (ROS), known to be a major cause of aging[2], and participates in the elimination of infectious pathogens via the process called xenophagy. Autophagy aids in the elimination of leukocytes after an immune response, obviating unnecessary inflammatory reactions. In addition, autophagy, along with apoptosis, takes part in the removal of old or dying cells. A decrease in autophagy leads to the accumulation of damaged cells which may trigger secondary necrosis, resulting in a harmful inflammatory response[2].

9. Insulin Resistance, Diabetes and Autophagy

The early stages of diabetes, obesity or insulin resistance are associated with increased ER stress and mitochondrial uncoupling and dysfunction. Recently, the importance of autophagy, in preserving the function of the insulin-producing β-cell, was linked to its role in organelle homeostasis and protein degradation[77]. Autophagy in β-cells is induced during the introduction of high fat diet, while obesity is accompanied by decreased autophagy in hepatocytes[78,79]. Since insulin and IGF-1 regulate autophagy through both FoxO and mTOR [61], it is tempting to speculate that impairment of autophagy may also promote insulin resistance. An alternative possibility, that insulin resistance may disrupt autophagic flux, cannot be discarded either. Nevertheless, both phenomena are closely related and require further studies.

Conclusion

Insulin and IGF-1 are well known first messengers controlling a host of intracellular signaling cascades and biological functions. Despite the structural similarity between insulin and IGF-1, their cellular effects differ. Insulin preferentially functions in maintaining energy balance, regulating carbohydrate, lipid and protein metabolism, whereas the main roles of IGF-1 are regulation of growth, development and differentiation. Furthermore, there is a consensus that insulin negatively regulates autophagy through the Akt/mTOR pathway, while IGF-1 may induce or inhibit autophagy depending on the context. Future studies will shed light on the molecular mechanism underlying these differences and how insulin and IGF-1 regulate autophagy in normal cells and in cancer, atrophy, aging and diabetes.

Acknowledgments

This work was supported by CONICYT, Chile (FONDECYT grant 1080436 and FONDAP grant 15010006, S.L.) and by grants from the NIH (HL-075173, J.A.H.; HL-080144, J.A.H.; HL-090842, J.A.H.), AHA (0640084N, J.A.H.), ADA (7-08-MN-21-ADA, J.A.H.), and the AHA-Jon Holden DeHaan Foundation (0970518N, J.A.H.). We thank CONICYT, Chile for the PhD fellowships to A.d.C., and F.P., for MSc fellowship to P.M., C.L.C., and for the postdoctoral fellowships 3110114 to R.T. and 3110039 to Z.P.

References

[1] Alwan A, et al. Global status report on noncommunicable diseases 2010. World Health Organization 2010.

[2] Levine B, Kroemer G. Autophagy in the pathogenesis of disease. *Cell* 2008;132:27-42.

[3] Battiprolu PK, Gillette TG, Wang ZV, Lavandero S, Hill JA. Diabetic Cardiomyopathy: Mechanisms and Therapeutic Targets. *Drug Discov Today Dis Mech* 2010;7:e135-e43.

[4] Nemchenko A, Chiong M, Turer A, Lavandero S, Hill JA. Autophagy as a therapeutic target in cardiovascular disease. *J Mol Cell Cardiol* 2011;51:584-93.

[5] Mizushima N. Autophagy: process and function. *Genes Dev* 2007;21:2861-73.

[6] Galluzzi L, Maiuri MC, Vitale I, et al. Cell death modalities: classification and pathophysiological implications. *Cell Death Differ* 2007;14:1237-43.

[7] Laron Z. Insulin-like growth factor 1 (IGF-1): a growth hormone. *Mol Pathol* 2001;54:311-6.

[8] Abbott AM, Bueno R, Pedrini MT, Murray JM, Smith RJ. Insulin-like growth factor I receptor gene structure. *J Biol Chem* 1992;267:10759-63.

[9] Becker AB, Roth RA. Insulin receptor structure and function in normal and pathological conditions. *Annu Rev Med* 1990;41:99-115.

[10] Foncea R, Andersson M, Ketterman A, et al. Insulin-like growth factor-I rapidly activates multiple signal transduction pathways in cultured rat cardiac myocytes. *J Biol Chem* 1997;272:19115-24.

[11] Sarbassov DD, Guertin DA, Ali SM, Sabatini DM. Phosphorylation and regulation of Akt/PKB by the rictor-mTOR complex. *Science* 2005;307:1098-101.

[12] Yang WL, Wang J, Chan CH, et al. The E3 ligase TRAF6 regulates Akt ubiquitination and activation. *Science* 2009;325:1134-8.

[13] Siddle K. Signalling by insulin and IGF receptors: supporting acts and new players. *J Mol Endocrinol* 2011;47:R1-10.

[14] Muslin AJ. Akt2: a critical regulator of cardiomyocyte survival and metabolism. *Pediatr Cardiol* 2011;32:317-22.

[15] Pouyssegur J, Volmat V, Lenormand P. Fidelity and spatio-temporal control in MAP kinase (ERKs) signalling. *Biochem Pharmacol* 2002;64:755-63.

[16] Fuentes EN, Bjornsson BT, Valdes JA, et al. The IGF-I/PI3K/Akt and the IGF-I/MAPK/ERK pathways in vivo in skeletal muscle are regulated by nutrition and contribute to somatic growth in the fine flounder. *Am J Physiol* 2011. 2011;R1532-42.

[17] Saltiel AR, Kahn CR. Insulin signalling and the regulation of glucose and lipid metabolism. *Nature* 2001;414:799-806.

[18] Bergman RN, Steil GM, Bradley DC, Watanabe RM. Modeling of insulin action in vivo. *Annu Rev Physiol* 1992;54:861-83.

[19] Klip A, Ramlal T, Bilan PJ, Cartee GD, Gulve EA, Holloszy JO. Recruitment of GLUT-4 glucose transporters by insulin in diabetic rat skeletal muscle. *Biochem Biophys Res Commun* 1990;172:728-36.

[20] Sutherland C, O'Brien RM, Granner DK. New connections in the regulation of PEPCK gene expression by insulin. *Philos Trans R Soc Lond B Biol Sci* 1996;351:191-9.

[21] Wang J, Zhou J, Powell-Braxton L, Bondy C. Effects of Igf1 gene deletion on postnatal growth patterns. *Endocrinology* 1999;140:3391-4.

[22] Liu JP, Baker J, Perkins AS, Robertson EJ, Efstratiadis A. Mice carrying null mutations of the genes encoding insulin-like growth factor I (Igf-1) and type 1 IGF receptor (Igf1r). *Cell* 1993;75:59-72.

[23] Lupu F, Terwilliger JD, Lee K, Segre GV, Efstratiadis A. Roles of growth hormone and insulin-like growth factor 1 in mouse postnatal growth. *Dev Biol* 2001;229:141-62.

[24] Randhawa R, Cohen P. The role of the insulin-like growth factor system in prenatal growth. *Mol Genet Metab* 2005;86:84-90.

[25] Glass DJ. Skeletal muscle hypertrophy and atrophy signaling pathways. *Int J Biochem Cell Biol* 2005;37:1974-84.

[26] McMullen JR, Shioi T, Huang WY, et al. The insulin-like growth factor 1 receptor induces physiological heart growth via the phosphoinositide 3-kinase(p110alpha) pathway. *J Biol Chem* 2004;279:4782-93.

[27] Belke DD, Betuing S, Tuttle MJ, et al. Insulin signaling coordinately regulates cardiac size, metabolism, and contractile protein isoform expression. *J Biol Chem* 2002;109:629-39.

[28] Holmang A, Yoshida N, Jennische E, Waldenstrom A, Bjorntorp P. The effects of hyperinsulinaemia on myocardial mass, blood pressure regulation and central haemodynamics in rats. *Eur J Clin Invest* 1996;26:973-8.

[29] Lawlor MA, Mora A, Ashby PR, et al. Essential role of PDK1 in regulating cell size and development in mice. *EMBO J* 2002;21:3728-38.

[30] Shiojima I, Yefremashvili M, Luo Z, et al. Akt signaling mediates postnatal heart growth in response to insulin and nutritional status. *J Biol Chem* 2002;277:37670-7.

[31] Ashford AJ, Pain VM. Insulin stimulation of growth in diabetic rats. Synthesis and degradation of ribosomes and total tissue protein in skeletal muscle and heart. *J Biol Chem* 1986;261:4066-70.
[32] Sandri M. Signaling in muscle atrophy and hypertrophy. *Physiology* (Bethesda) 2008;23:160-70.
[33] Gu Y, Wang C, Cohen A. Effect of IGF-1 on the balance between autophagy of dysfunctional mitochondria and apoptosis. *FEBS Lett* 2004;577:357-60.
[34] Jia G, Cheng G, Gangahar DM, Agrawal DK. Insulin-like growth factor-1 and TNF-alpha regulate autophagy through c-jun N-terminal kinase and Akt pathways in human atherosclerotic vascular smooth cells. *Immunol Cell Biol* 2006;84:448-54.
[35] Sobolewska A, Gajewska M, Zarzynska J, Gajkowska B, Motyl T. IGF-I, EGF, and sex steroids regulate autophagy in bovine mammary epithelial cells via the mTOR pathway. *Eur J Cell Biol* 2009;88:117-30.
[36] Aki T, Yamaguchi K, Fujimiya T, Mizukami Y. Phosphoinositide 3-kinase accelerates autophagic cell death during glucose deprivation in the rat cardiomyocyte-derived cell line H9c2. *Oncogene* 2003;22:8529-35.
[37] Bains M, Florez-McClure ML, Heidenreich KA. IGF-I prevents the accumulation of autophagic vesicles and cell death in Purkinje neurons by increasing the rate of autophagosome-to-lysosome fusion and degradation. *J Biol Chem* 2009;284:20398-407.
[38] Contreras-Ferrat AE, Toro B, Bravo R, et al. An inositol 1,4,5-triphosphate (IP3)-IP3 receptor pathway is required for insulin-stimulated glucose transporter 4 translocation and glucose uptake in cardiomyocytes. *Endocrinology* 2010;151:4665-77.
[39] Ibarra C, Estrada M, Carrasco L, et al. Insulin-like growth factor-1 induces an inositol 1,4,5-trisphosphate-dependent increase in nuclear and cytosolic calcium in cultured rat cardiac myocytes. *J Biol Chem* 2004;279:7554-65.
[40] Criollo A, Vicencio JM, Tasdemir E, Maiuri MC, Lavandero S, Kroemer G. The inositol trisphosphate receptor in the control of autophagy. *Autophagy* 2007;3:350-3.
[41] Vicencio JM, Ortiz C, Criollo A, et al. The inositol 1,4,5-trisphosphate receptor regulates autophagy through its interaction with Beclin 1. *Cell Death Differ* 2009;16:1006-17.
[42] Cardenas C, Miller RA, Smith I, et al. Essential regulation of cell bioenergetics by constitutive InsP3 receptor Ca2+ transfer to mitochondria. *Cell* 2010;142:270-83.
[43] Furstenberger G, Senn HJ. Insulin-like growth factors and cancer. *Lancet Oncol* 2002;3:298-302.
[44] Feng Z. p53 regulation of the IGF-1/AKT/mTOR pathways and the endosomal compartment. *Cold Spring Harb Perspect Biol* 2010;2:a001057.
[45] LeRoith D, Roberts CT, Jr. The insulin-like growth factor system and cancer. *Cancer Lett* 2003;195:127-37.
[46] Pollak M. Insulin and insulin-like growth factor signalling in neoplasia. *Nat Rev Cancer* 2008;8:915-28.
[47] Kurmasheva RT, Houghton PJ. IGF-I mediated survival pathways in normal and malignant cells. *Biochim Biophys Acta* 2006;1766:1-22.
[48] Maiuri MC, Galluzzi L, Morselli E, Kepp O, Malik SA, Kroemer G. Autophagy regulation by p53. *Curr Opin Cell Biol* 2010;22:181-5.
[49] Tasdemir E, Maiuri MC, Galluzzi L, et al. Regulation of autophagy by cytoplasmic p53. *Nat Cell Biol* 2008;10:676-87.

[50] Yang ZJ, Chee CE, Huang S, Sinicrope F. Autophagy modulation for cancer therapy. *Cancer Biol Ther* 2011;11:169-76.
[51] Jackman RW, Kandarian SC. The molecular basis of skeletal muscle atrophy. *Am J Physiol* 2004;287:C834-43.
[52] Bodine SC, Latres E, Baumhueter S, et al. Identification of ubiquitin ligases required for skeletal muscle atrophy. *Science* 2001;294:1704-8.
[53] Sandri M. Autophagy in health and disease. 3. Involvement of autophagy in muscle atrophy. *Am J Physiol* 2010;298:C1291-7.
[54] Masiero E, Agatea L, Mammucari C, et al. Autophagy is required to maintain muscle mass. *Cell Metab* 2009;10:507-15.
[55] Dobrowolny G, Aucello M, Rizzuto E, et al. Skeletal muscle is a primary target of SOD1G93A-mediated toxicity. *Cell Metab* 2008;8:425-36.
[56] Sacheck JM, Ohtsuka A, McLary SC, Goldberg AL. IGF-I stimulates muscle growth by suppressing protein breakdown and expression of atrophy-related ubiquitin ligases, atrogin-1 and MuRF1. *Am J Physiol* 2004;287:E591-601.
[57] Barton-Davis ER, Shoturma DI, Sweeney HL. Contribution of satellite cells to IGF-I induced hypertrophy of skeletal muscle. *Acta Physiol Scand* 1999;167:301-5.
[58] Schulze PC, Fang J, Kassik KA, et al. Transgenic overexpression of locally acting insulin-like growth factor-1 inhibits ubiquitin-mediated muscle atrophy in chronic left-ventricular dysfunction. *Circ Res* 2005;97:418-26.
[59] Song YH, Li Y, Du J, Mitch WE, Rosenthal N, Delafontaine P. Muscle-specific expression of IGF-1 blocks angiotensin II-induced skeletal muscle wasting. *J Biol Chem* 2005;115:451-8.
[60] Stitt TN, Drujan D, Clarke BA, et al. The IGF-1/PI3K/Akt pathway prevents expression of muscle atrophy-induced ubiquitin ligases by inhibiting FOXO transcription factors. *Mol Cell* 2004;14:395-403.
[61] Mammucari C, Milan G, Romanello V, et al. FoxO3 controls autophagy in skeletal muscle in vivo. *Cell Metab* 2007;6:458-71.
[62] Sohal RS. Role of oxidative stress and protein oxidation in the aging process. *Free Radic Biol Med* 2002;33:37-44.
[63] Tatar M, Bartke A, Antebi A. The endocrine regulation of aging by insulin-like signals. *Science* 2003;299:1346-51.
[64] Cavallini G, Donati A, Gori Z, Bergamini E. Towards an understanding of the anti-aging mechanism of caloric restriction. *Curr Aging Sci* 2008;1:4-9.
[65] Bartke A. Mutations prolong life in flies; implications for aging in mammals. *Trends Endocrinol Metab* 2001;12:233-4.
[66] Kimura KD, Tissenbaum HA, Liu Y, Ruvkun G. daf-2, an insulin receptor-like gene that regulates longevity and diapause in Caenorhabditis elegans. *Science* 1997;277:942-6.
[67] Tatar M, Kopelman A, Epstein D, Tu MP, Yin CM, Garofalo RS. A mutant Drosophila insulin receptor homolog that extends life-span and impairs neuroendocrine function. *Science* 2001;292:107-10.
[68] Kenyon C, Chang J, Gensch E, Rudner A, Tabtiang R. A C. elegans mutant that lives twice as long as wild type. *Nature* 1993;366:461-4.
[69] Holzenberger M, Dupont J, Ducos B, et al. IGF-1 receptor regulates lifespan and resistance to oxidative stress in mice. *Nature* 2003;421:182-7.

[70] Brown-Borg HM, Borg KE, Meliska CJ, Bartke A. Dwarf mice and the ageing process. *Nature* 1996;384:33.

[71] Shimokawa I, Higami Y, Utsuyama M, et al. Life span extension by reduction in growth hormone-insulin-like growth factor-1 axis in a transgenic rat model. *Am J Pathol* 2002;160:2259-65.

[72] Fontana L, Partridge L, Longo VD. Extending healthy life span--from yeast to humans. *Science* 2010;328:321-6.

[73] Kapahi P, Chen D, Rogers AN, et al. With TOR, less is more: a key role for the conserved nutrient-sensing TOR pathway in aging. *Cell Metab* 2011;11:453-65.

[74] Bjedov I, Toivonen JM, Kerr F, et al. Mechanisms of life span extension by rapamycin in the fruit fly Drosophila melanogaster. *Cell Metab* 2010;11:35-46.

[75] Broughton S, Partridge L. Insulin/IGF-like signalling, the central nervous system and aging. *Biochem J* 2009;418:1-12.

[76] Madeo F, Eisenberg T, Kroemer G. Autophagy for the avoidance of neurodegeneration. *Genes Develop* 2009;23:2253-9.

[77] Scheuner D, Kaufman RJ. The unfolded protein response: a pathway that links insulin demand with beta-cell failure and diabetes. *Endocr Rev* 2008;29:317-33.

[78] Liu HY, Cao SY, Hong T, Han J, Liu Z, Cao W. Insulin is a stronger inducer of insulin resistance than hyperglycemia in mice with type 1 diabetes mellitus (T1DM). *J Biol Chem* 2009;284:27090-100.

[79] Ebato C, Uchida T, Arakawa M, et al. Autophagy is important in islet homeostasis and compensatory increase of beta cell mass in response to high-fat diet. *Cell Metab* 2008;8:325-32.

In: Autophagy: Principles, Regulation and Roles in Disease ISBN: 978-1-61942-266-7
Editor: Nikolai Gorbunov

Chapter X

Autophagy: Regulation and Role in Degenerative Diseases

***Maylin Almonte-Becerril*[1*] *and Juan B. Kouri*[2*]**
[1,2] Laboratorio de Infectómica y Patogénesis Molecular del Centro de investigación y de estudios avanzados del IPN (Cinvestav-IPN). Av. Instituto Politécnico Nacional 2508 Col. San Pedro Zacatenco C.P. 07360 México, D.F. Apartado postal 14-740, 07000 México, D.F.

Abstract

Autophagy is a ubiquitous process that persists at low basal levels in all eukaryotic cells to support homeostasis. According current paradigm, autophagy is an evolutionary conserved and strictly regulated lysosomal pathway to degrade unnecessary or compromised organelles, cytoplasmic and endocytosed materials and even invaded pathogens. Overall autophagy is considered to be adaptive and protective cellular mechanisms. Interestingly, as many other crucial pathways, autophagy plays dual role under normal and pathological conditions. In addition to the well-known role of autophagy in cell survival, autophagy-mediated type II programmed cell death has long been proposed. The autophagic cell death was originally reported in tissues subjected active development and remodeling; that effect seems to be analogous to apoptosis in similar metamorphoses. However, the proposed function of autophagy in cell death is not restricted to the developmental programmed cell death; it also extends to cell death that occurs under various pathological conditions, e.g., immune disorders, infections, and degenerative disorders such as osteoarthritis. It seems that under these circumstances, uncontrolled activation of autophagy capable of massive removal of subcellular structures such as mitochondria can be detrimental to the cell and tissue functions.

Osteoarthritis (OA) is a slowly progressive degenerative disease characterized by gradual loss of articular cartilage, as a consequence of the degradation of extracellular matrix (ECM) components. Presumably, the pathogenesis of this degeneration can be driven by different cellular/molecular mechanisms and the effects of these mechanisms

[*] maylin_ab@hotmail.com.
[*] bkouri@cinvestav.mx.

can be gradually altered upon progress of the disease. Thus, it has been proposed recently that autophagy at a low-level activity can protective against ECM degradation during the early stages of OA. In contrast, in late stages of OA, when a high expression of both autophagic and apoptotic markers can co-exist in the cells constituting both the superficial and middle zones of the cartilage, activation of autophagy is associated with the cell death process termed chondroptosis. Based on these observations, it is possible to conclude, that autophagy plays an important role in both cell survival and cell death mechanism, depending on dominance of the cell signaling pathways. One would expect that a pattern of cellular signals that supports cellular homeostasis under normal conditions would over-activate autophagy under a pathological when cellular signals are over-exaggerated due to failure of homeostasis. This activation of autolysosomal mechanism can exacerbate a slowly-developing degenerative process and lead to cell death.

The objective of this chapter is to discuss a potential role of autophagy in degenerative type of diseases such as osteoarthritis.

1. Introduction

Cell homeostasis depends on the balance between the production and destruction of macromolecules and organelles. There are two major systems in eukaryotic cells that degrade cellular components: the ubiquitin proteasome system (UPS) and the lysosome. Both mechanisms are able to degrade proteins, but only through the lysosomal system, can also degrade macromolecules, and even entire organelles by autophagy [1, 2].

In 1963, soon after the discovery of lysosomes, de Duve introduced the term autophagy (from the Greek words auto=self and phagein=to eat), to describe the occurrence in rat liver cells under starvation conditions, in which distinct intracellular, membranous vesicles that contained degraded cytoplasmic material were identified [3]. Since then, autophagy denotes an evolutionary mechanism conserved in eukaryotic cells, which starts with the formation of an autophagosome, enclosed within a double membrane that engulfs part of the cytoplasm as a quality control mechanism to eliminate protein aggregates, damaged organelles, and other nonactive structures. One of its primary roles in unicellular organisms is to regulate intracellular homeostasis and to adjust organelle numbers in response to stress such as changes in nutrient availability. In higher eukaryotes, autophagy plays also an important role in stress-response, development, cell differentiation, immunity and tumor suppression [1-6].

Under vegetative growth conditions, autophagy occurs at a basal level, in all cells to perform homeostatic functions. However, this process is rapidly upregulated when cells need to generate intracellular nutrients and energy, undergo architectural remodeling, or rid themselves of damaging cytoplasmic components by nutrient deprivation in mammalian cells. Accordingly, there must be a mechanism for sensing the extracellular milieu and transducing an appropriate signal regulatory elements that allow autophagy to be induced [6, 7].

2. Mechanisms and Molecular Components

Although autophagy was initially identified in mammals, its molecular components were elucidated in yeast [1, 3, 6]. The autophagy-related genes (ATG) encode proteins needed for the induction of autophagy, and the generation, maturation, and recycling of autophagosomes. These proteins are composed of 4 functional groups, including a protein serine/threonine kinase complex that responds to upstream signals such as TOR kinase (Atg1, Atg13, Atg17), a lipid kinase signaling complex that mediates vesicle nucleation (Atg6, Atg14, ps34, and Vps15), 2 novel ubiquitin-like conjugation pathways that mediate vesicle expansion (the Atg8 and Atg12 systems), and a recycling pathway that mediates the disassembly of Atg proteins from matured autophagosomes (Atg2, Atg9, Atg18) [7]. More than 32 ATG have been identified to participate in macroautophagy in *S. cerevisiae*, and mammalian orthologs exist for almost all of them [8, 9]. However, the yeast have only one form of each Atg, whereas in mammals, some Atgs have several isoforms and paralogs [1].

3. Types of Autophagy

More than a dozen of autophagic processes have been described in recent years based on the nature of the cargo degraded by each of them. A number of organelles are specifically degraded by macroautophagy including the ER (ER-phagy), peroxisomes (pexophagy) and mitochondria (mitophagy), in which, selectivity depends in part on a subset of cytosolic proteins that function as cargo-recognition molecules. Other specific macroautophagy-associated processes include the degradation of pathogens (xenophagy), lipids (macrolipidophagy) and protein aggregates (aggrephagy), wherein the ability to degrade protein aggregates is critical for the maintenance of a stable proteome, and alterations in this process underlie the basis of common protein conformational disorders, including age-related neurodegeneration diseases such as Alzheimer's disease or Parkinson's disease. Nevertheless, all of these processes can be catalogued under one of the following three main types of autophagy according to the mechanism used for delivery of cargo into lysosomes: macroautophagy, microautophagy and chaperone-mediated autophagy (CMA) [1, 9].

Macroautophagy, the most extensively described type of autophagy, involves the sequestration of cytosolic regions by the elongation of a phagophore (or pre-autophagosomal structure) to form a double-membrane vacuole called the autophagosome or early autophagic vacuoles (AV-I) [9]. The outer membrane of the autophagosome fuses with the lysosome (in mammalian cells) or vacuole (in yeast and plants); upon the fusion, AV-I form single membrane structures called autolysosomes or late autophagic vacuoles (AV-II) in which luminal content is degraded and resulting elements are returned into the cytosol for metabolic reactions [7, 10].

A similar process of sequestration characterizes microautophagy, but in this case the lysosomal membrane is the one that surrounds the regions of cytosol destined for degradation and internalizes into the lysosomal lumen in the form of single-membrane vesicles. Although microautophagy occurs in yeast at the vacuole (equivalent to the lysosome), in mammalian cells this process has been recently shown to take place at the surface of the late endosomes (endosomal microautophagy) and to utilize the machinery required for the biogenesis of multivesicular bodies, where is dependent on GTP hydrolysis and on calcium [9]. On the other hand, even though selective forms of microautophagy require different sets of Atg proteins [3].

Chaperone-mediated autophagy (CMA) is a secondary response to starvation and, unlike the other two processes, involves direct translocation of the targeted proteins across the lysosomal membrane. Currently, there is no known equivalent of this process in yeast [6]. Furthermore, during CMA, cytosolic proteins containing a specific pentapeptide domain (consensus sequence=KFERQ) are unfolded and directly translocated across the lysosomal membrane by the cytosolic constitutive chaperone hsc70 [11]. In addition, LAMP-2A, serves as a receptor in the translocation of unfolded polypeptides across the lysosomal membrane[1].

4. Induction of Autophagy

4.1. Induction or Cargo Packaging

With macroautophagy, pexophagy, and the CMA pathway, cells use largely the same machinery to accomplish a similar goal, delivery of cytoplasmic components into the interior of the vacuole. Growth conditions dictate which components must be targeted for delivery [4]. The origin of the phagophore has been a controversial topic until recently, when it has become evident that these structures form by recruitment of lipids and proteins from different pre-formed organelles, containing high levels of PI3K known as the omegasome (from its ω-like shape), which is dynamically connected to the endoplasmic reticulum (ER). Nevertheless, it was reported that mitochondria, plasma membrane and Golgi could participate as a donor of the phagophore constituents [1, 9].

Two evolutionarily conserved nutrient sensors play roles in autophagy regulation: (a) the target of rapamycin (TOR) kinase is the major inhibitory signal that shuts off autophagy during nutrient abundance, and (b) the eukaryotic initiation factor 2α (eIF2 α) kinase Gcn2 and its downstream target Gcn4, a transcriptional transactivator of autophagy genes, which turn on this process during nutrient depletion. The class I PI3K/Akt signaling molecules link receptor tyrosine kinases to TOR activation and thereby repress autophagy in response to insulin-like and other growth factor signals[7].

Tor is a serine/threonine kinase that, in response to amino acids and growth factors, coordinates different aspects of cell growth. Starvation inhibits Tor activity, provoking various cellular responses, including cell arrest in the early G1 phase, inhibition of protein synthesis, nutrient transporter turnover, transcriptional changes and autophagy. Tor signaling negatively regulates the association between Atg1 and Atg13 proteins. Under nutrient-rich conditions, active Tor causes hyperphosphorylation of Atg13, preventing or moderating its association with Atg1. It is not known whether Tor directly phosphorylates Atg13. In contrast, Tor inactivation by starvation or rapamycin treatment promotes the rapid dephosphorylation of Atg13,which binds to Atg1; this association promotes autophosphorylation and activation of Atg1, leading to the induction of macroautophagy [4].

The autophagic process in mammalian cells is initiated by the ULK1 (mammalian ortholog of the yeast Atg1) complex. This complex is formed by ULK1 Ser/Thr protein kinase, Atg13 and FIP200 (FIP200 is the mammalian homolog of the yeast Atg17). The phosphorylation of Atg13 and FIP200 by ULK1 is an important step in triggering autophagy. Moreover, ULK1 may have other substrates that are involved in the early stage of autophagy[1]. The nucleation and assembly of the initial phagophore membrane are dependent on the Beclin 1:class III phosphatidylinositol 3-kinase (PI3K) complex. Therefore, the activation of the ULK1/Atg1/FIP200 complex, induce the macroautophagy initiation. However, the interaction of Beclin 1 with antiapoptotic proteins of the Bcl-2 family blocks the induction of autophagy by inhibiting the formation of PI3K by hVps34 (Atg 12). Thus, Bcl-2 is not only antiapoptotic but also an antiautophagic protein.[1, 9].

4.2. Vesicle Formation and Completion

Vesicle formation starts with the enwrapping of the selected cargo, organelle, or cytosol by a membrane. This process ends with the fusion of the extremities of the surrounding membrane, leading to the completion of a double-membrane vesicle, an autophagosome or a cytoplasm-to-vacuole targeting (Cvt) pathway. These processes operate under different nutritional conditions, and depending on size and on nutritional conditions[4].

The expansion and closure of the autophagosome in mammalian cells are dependent on two ubiquitin-like conjugation systems: one pathway involves the covalent conjugation of Atg12 to Atg5, with the help of the E1-like enzyme (Atg7) and the E2-like enzyme (Atg10). The second pathway involves the conjugation of phosphatidylethanolamine (PE) to LC3 (the mammalian homologues of Atg8) by the sequential action of the protease Atg4, the E1-like enzyme and the E2-like enzyme. Lipid conjugation (from the COOH terminus of LC3 to the polar head to PE) leads to the conversion of the soluble form of LC3 (named LC3-I) to the autophagic vesicle- associated form (LC3-II). LC3-II is used as a marker of autophagy because its lipidation and specific recruitment to autophagosomes provides a shift from diffuse to punctate staining of the protein and increases its electrophoretic mobility on gels compared with LC3-I. Autophagosomes undergo maturation by fusion with lysosomes to create autolysosomes. In the autolysosomes, the inner membrane as well as the luminal content of the autophagic vacuoles is degraded by lysosomal enzymes that act optimally within this acidic compartment. After closure of the autophagosome, the LC3 located on the external surface is released from PE by Atg4, whereas the LC3-PE on the inner surface is transported to the lysosomal compartment to be degraded along with other cargos [1, 12, 13]. In all of these scenarios, cells use an identical fusion machinery, which consists of SNARE proteins, a rab-GTPase, and the class C vps protein complex, also known as the HOPS (homotypic fusion and vacuole protein sorting) complex [4].

4.3. Vesicle Breakdown

Once the autophagosome has fuse with the lysosomal membrane, the interior single-membrane vesicles are release into the lumen from the lysosomes. These vesicles, now termed autophagic, are subsequently consumed by hydrolases. This breakdown requires normal acidification of the vacuole, presumably to maintain the optimal pH of the degradative enzymes but also because it is essential for the autocatalytic cleavage and subsequent activation of proteinase A (Pep4) and proteinase B (the product of the PRB1 gene, allelic to CVT1). These proteases are at the apex of a proteolytic cascade that leads to the activation of most vacuolar hydrolases [4].

5. Health and Disease

5.1. Autophagy in Immunity and Inflammation

Autophagy contributes to innate immunity by protecting the cytosol from colonization by intracellular microbial pathogens. Many bacteria, such as group *Streptococcus, Ricketssia coronii, Staphylococcus aureus, or Mycobacterium tuberculosis,* and also parasites, such as *Toxoplasma gondii*, are targeted to be autophagositize and therefore removed. Although autophagy acts as a cellular antimicrobial defense system in most models of infection studied so far, autophagy is also exploited by certain pathogens, such as poliovirus, the replication of which is increased when autophagy is stimulated. Furthermore, it has been reported that macrophages infected with human immunodeficiency virus (HIV-1) accumulate autophagosomes, due the HIV-1 blocks fusion of autophagosomes with lysosomes thereby avoiding degradation, possibly by the interaction of the viral protein Nef with Beclin 1[1].

Apart from being involved in the elimination of intracellular microorganisms, autophagy can also contribute to adaptive immunity. Lysosomes generate the antigenic peptides that are exposed on the cell surface in association with major histocompatibility complex class II (MHC II) for presentation to CD4-positive (CD4+) T cells. Autophagy can also influence adaptive immunity by promoting the development, survival, and proliferation of B and T cells [1].

5.2. Cancer

Accumulating evidence reveals that alterations in autophagy occur in various human diseases. Beclin 1 is considered a haplo-insufficient tumor suppressor. Binding of Beclin 1 to the anti-apoptotic protein Bcl-2 decreases Beclin 1 activity and thereby inhibits autophagy. These observations suggest that autophagy is important for tumor suppression. However, autophagy may also promote tumor cell survival during metabolic stress in the tumor microenvironment, under conditions of hypoxia and low nutrients. Accordingly, genetic or pharmacological inhibition of autophagy was shown to enhance the cytotoxicity of cancer chemotherapy agents and to promote tumour regression. Thus, autophagy can act both positively or negatively with regard to cancer cell survival; autophagy probably functions to prevent cancer initially, but once a tumour develops, the cancer cells utilize autophagy for their own cytoprotection[3].

5.3. Neurodegeneration

Morphologic evidence of autophagy has been reported in neurodegenerative diseases including Parkinson, Huntington, and Alzheimer diseases, and transmissible spongiform encephalopathies. A contrasting view is that autophagy may be a protective mechanism to degrade mutant or toxic proteins. According to this model, defects in autophagy related pathways contribute to the accumulation of neurotoxic proteins and the consequent neuronal cell death [7]. Furthermore, two parallel mouse studies using neuronal-specific knockouts of

Atg5 or Atg7 demonstrate that basal autophagy controls the constitutive turnover of soluble, cytosolic proteins to prevent the accumulation of abnormal neuroproteins that may cause symptoms of neurodegeneration. Increasing evidence suggests that p62 (marker for autophagic flux) and NBR1 (neighbor of BRCA1 gene 1), an apparent autophagy receptor that contains an LC3- interacting region (LIR) and an ubiquitin-associated (UBA) domain, serve as cargo receptors to selectively deliver polyubiquitylated, misfolded, aggregated proteins and potentially deleterious organelles for clearance through autophagy in both mammals and Drosophila[3].

5.4. Ageing and Longevity

A common feature of all ageing cells is a progressive accumulation of damaged proteins and organelles (such as defective mitochondria) and decreased autophagic activity could be could be importantly involve in these circumstances. Early studies showed that autophagy function declines with age *in vivo* in rodents and *in vitro* in isolated hepatocytes. Current efforts to avoid the decline of autophagy function with age include the practice of using an anti-lipolytic drug that mimics the beneficial effect of caloric restriction on autophagy [3].

6. Autophagy: Cell Survival and Cell Death Mechanisms

In most circumstances, autophagy (macroautophagy) promotes cell survival by adapting cells to stress conditions. In this context it appears paradoxical that autophagy has also been considered as a non-apoptotic programmed cell death that is referred to as 'autophagic' or 'type-II' cell death. This is based on the finding that some cases of cell death are preceded or accompanied by massive autophagic vacuolization [6, 10].

Programmed cell death type II has some characteristics in common with apoptosis (programmed cell death type I); both mechanisms have some characteristics in common such as: ATP consumption and the absence of inflammation during the death process. The differences between apoptosis and autophagy are that during apoptosis caspase cascade is activated, the internucleosomal fragmentation of DNA is present and the organelles are maintained apparently intact until late stages. On the other hand, during the autophagy, the presence of double membrane vesicles or vacuoles (autophagy and/or digestive vacuoles), which contain cytoplasmic residues and the high number of lysosomes in the cytoplasm are the central features in this process [7, 14-17].

In addition, autophagy represents a prominent mechanism of cellular adaptation to stress, such as increased temperature, high population density and nutrient deprivation. Under carbon and nitrogen starvation, for instance, the activity of the mTOR is rapidly shut down, and the ensuing up-regulation of the autophagic pathway supplies cells with metabolic substrates to meet their bioenergetic demands. In addition, autophagy is activated in response to invasion by intracellular pathogens and plays a versatile role in innate and adaptive immunity, by promoting the degradation of invading microbes (xenophagy), by delivering viral nucleic acids to TLRs (Toll-like receptors), by contributing to the phagocytic pathway, and by feeding antigens to MHC class II compartments [11].

6.1. Autophagy in Senescence

In senescent and postmitotic cells, autophagy serves as a mechanism of adaptation to stress. In fact, autophagosomes have been shown to accumulate in senescent fibroblasts to facilitate the renewal of cytosolic compounds and organelles. It should be noted that age-associated neurodegenerative diseases share some characteristics with the pathology induced by knockout of autophagy-related (atg) genes in the brain, e.g. the accumulation of ubiquitinated proteins and inclusion bodies in the cytoplasm, an increase in apoptosis of neuronal cells and progressive neuronal cell loss[10].

6.2. Autophagy When Apoptosis Is Inhibited

In general terms, it appears that similar stimuli can induce either apoptosis or autophagy; whereas a mixed phenotype of both can sometimes be detected in response to these common stimuli, in many other instances, autophagy and apoptosis develop in a mutually exclusive manner, perhaps as a result of variable initiation for both processes, or as a result of a cellular "decision" between the two responses that may be linked to a mutual inhibition of the two phenomena. In general, "short-living" cells such as limphocytes, granulocytes and epithelial cells are subjected to apoptosis for repopulation while long-living cells such as neurons, stromal cells, myocytes etc, employ autophagy for renovation and remodeling. Recently, several pathways that link the apoptotic and autophagic machineries or that polarize the cellular response between autophagy and apoptosis have been deciphered at the molecular level [1, 12].

Some types of disorders in the apoptotic machinery, such as caspase inhibition, have been reported to induce autophagic cell death. Notably, when cells are exposed to stress signals that cause decreased intracellular metabolite concentrations (resulting from the loss of growth-factor signaling), the autophagy that arises from the inhibition of apoptosis actually protects cells from the death. In this respect, it was elegantly shown that immortalized interleukin-3 (IL3)-dependent cell lines generated from the bone marrow of Bax–/– Bak–/– mice failed to undergo apoptosis upon IL3 withdrawal (which is the default pathway of wild-type cells) and entered a month-long autophagic process that caused a severe reduction in cell size with the removal of most of the cytoplasm. Furthermore, autophagy in this cellular setting is the catabolic process that mobilizes nutrients by macromolecular degradation, thus replenishing the vanishing energy reserves of the starving cell and, hence, preventing a bioenergetic catastrophe that would otherwise culminate in cell death [12].

6.3. Apoptosis due to Inhibited Autophagy

When autophagy is blocked at an early stage by depletion of beclin-1, ATG5, ATG10, ATG12 or VPS34, no autophagic vacuoles are found in the cells, which undergo a typical type I cell death (apoptosis), probably as a result of a bioenergetic shortage that triggers apoptosis. In contrast, when the fusion of autophagosomes and lysosomes is blocked by the addition of lysosomal inhibitors (such as chloroquine or the vacuolar ATPase inhibitor bafilomycin A) or by depletion of the lysosomal protein LAMP2, autophagic vacuoles accumulate and the cells display a mixed type I–type II morphology before the cell succumbs to death. This result shows that the inhibition of autophagy, either at early or late stages of the process, may lead to apoptosis as a result of the failure to adapt to starvation. Moreover, it illustrates that autophagic vacuolization is not always the result of enhanced autophagy but may also result from the failed elimination of autophagic debris. It should be noted that inhibition of autophagy can also induce the type III cell death (necrosis) and not only apoptosis [12].

6.4. Autophagy as a Cell Survival Mechanism

The pro-survival function of autophagy is an evolutionarily ancient process, conserved from yeast to mammals, and best characterized in nutrient deficiency. During nutrient deficiency, degradation of membrane lipids and proteins by the autolysosome generates free fatty acids and amino acids that can be reused to fuel mitochondrial ATP energy production and maintain protein synthesis. Thus, the mechanism by which autophagy genes promote survival during nutrient deprivation may involve suppression of the canonical apoptotic cell death pathway. Autophagy is also involved in removing damaged mitochondria and other organelles, in degrading intracellular pathogens, and in degrading protein aggregates too large to be removed by the ubiquitin-proteosomal system. It is not known whether autophagy alters the production of death-promoting signals, prevents the movement of death-promoting signals into uninfected tissues, or protects uninfected tissues against death induced by these signals, or whether a similar function of autophagy plays a role in the spatial restriction of development and stress-induced programmed cell death in other eukaryotic organisms [7].

7. Autophagy and Osteoarthritis

Autophagy is now recognized as an important process in physiological and pathological situations. Studies in animal models, as well as observations and genetic analyses in patients, clearly implicate autophagy in numerous diseases, including cancer, neurodegenerative and muscle disease, hepatic disorders, cardiac failure, infectious and inflammatory diseases, diabetes, obesity and chronic degenerative diseases as Osteoarthritis [1].

Osteoarthritis (OA) is the most common degenerative chronic disease; it has a multifactorial origin and is characterized by the progressive loss of articular cartilage, synovial inflammation, pain, and subchondral bone alterations. Chondrocytes are the only cell type present in the articular cartilage; they are responsible for maintaining the dynamic equilibrium between production of the extracellular matrix (ECM) and its enzymatic degradation [15, 18, 19].

During the OA pathogenesis, the ECM loss and chondrocytes death are major features of the cartilage degeneration. Although there are observations showing incidence of apoptotic [20-23] or non apoptotic [24-27] mechanisms in OA cartilage . Thus, it was recently communicated the presence of both apoptotic and autophagic pathways during the OA pathogenesis in an animal model. This phenomenon was first reported by Roach and co-workers and named "chondroptosis"[28]. Chondroptosis include some commin characteristics of apoptosis, e.g., TUNEL positive staining, the presence of shrunken cells and condensed chromatin nuclei. However, it was also reported the presence of some changes in the normal development of the endoplasmic reticulum (ER) and Golgi complex. In addition, it was identified an increased expression of LC3II, which indicate autophagy activation [15, 28, 29].

Previously reports, demonstrates that ULK1, Beclin1 and LC3 are expressed in murine articular cartilage, suggesting activation of autophagy and the presence of LC3-II directly indicates autophagosome formation [30]. In addition, the presence of LC3II since early stages of OA indicates its role in the synthetic activity of chondrocytes, due to the fact that chondrocytes to express LC3 marker, where active caspase 3 (AC3) is negative [15].

OA start as a mechanic injury to induce the degradation of extracellular matrix (ECM) components. Afterward, chondrocytes increase its metabolic activity since early stages of OA induction, showed a development of ER and Golgi complex, probably related to the fruitless attempt to repair the injuries produced by OA pathogenesis (Figure 1). However, as the degenerative process progress, the restored capacity is surpass by the increased catabolic activity, which induce the matrix components degradation and also death of chondrocytes. Afterward, in late stages of OA, both autophagic (LC3II) and apoptotic (Caspase 3 active) markers were observed within the same cells from the SZ and MZ, which indicate that autophagy plays an important role, in the cell death process, due autophagy has been characterized as a highly regulated cellular mechanism with both beneficial and pathogenic effects. Furthermore, all these mechanisms guide to Chondroptosis activation, since the survival respond mechanism are inefficient to avoid the OA degenerative process (Figure 2) [15, 30]. In contrast, in the deep zone (DZ) of the cartilage, only apoptosis was observed, maybe related with the abnormal calcification of the cartilage present in late stage of OA [15, 28, 31, 32].

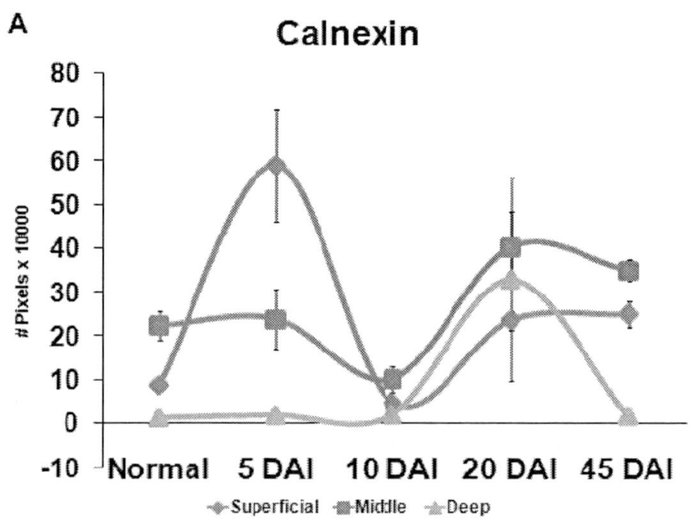

Figure 1. (Continued).

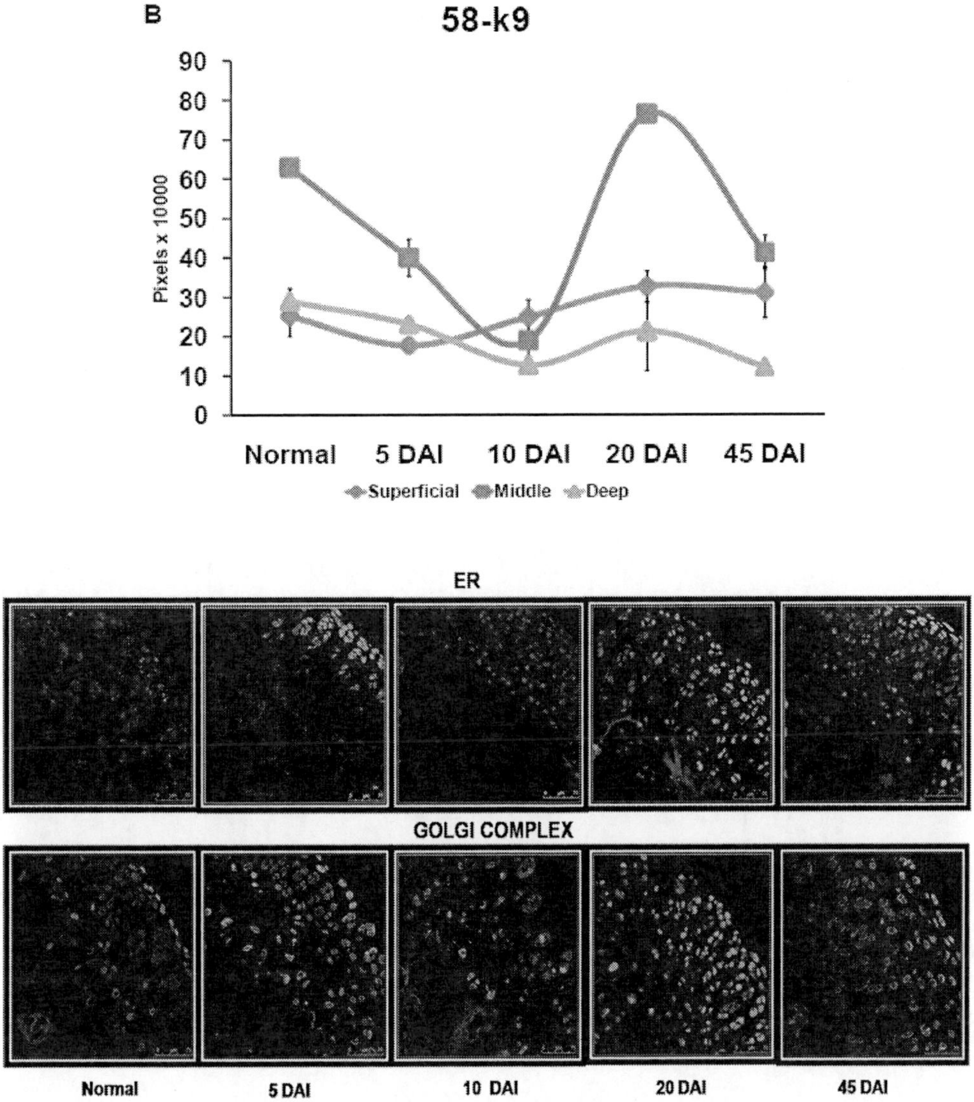

Figure 1. Calnexin and 58k-9 expression in OA cartilage. (A and B) Calnexin and 58k-9 were represented by number of pixels/ area, from superficial (SZ), middle (MZ) and deep (DZ) zones of the cartilage at 5, 10, 20 and 45 days after OA induction (DAI). (C and D) Representative IHC assays to detect calnexin (C, FITC) and 58k-9 protein (D, FITC) in OA cartilage; nuclei were stained with DAPI. All the OA samples were compared with a normal cartilage as an internal control. (A and C) In the SZ (♦), calnexin expression was increased since 5 DAI ($P \leq 0.01$), showed without differences at 10 DAI compared with the normal samples and maintained it highest expression at 20 DAI ($P \leq 0.001$) and 45 DAI ($P \leq 0.01$). In the MZ (■), the calnexin expression was increased at 20 and 45 DAI ($P \leq 0.05$) and in the DZ (▲) it did not show significant differences compared to normal cartilage. (B and D) The expression of 58k-9 in SZ (♦) and DZ (▲) did not show significant differences in comparison with normal samples. However, in MZ (■), it expression was increased since 5 DAI ($P \leq 0.05$), showed it highest expression at 20 DAI ($P \leq 0.001$), and maintained at 10 DAI without differences compared with normal samples. The results show means ± S.E.M. of 3 independent experiments using Tukey–Kramer multiple comparison test. Significant differences were expressed by an asterisk (*). Scale bar= 75μm. (figure taken from reference [15]).

Finally, it was reported that all these diversity of cell death within the same tissue might be related to the specific zones of the cartilage according to the role of each zone in the physiopathology of the cartilage during OA pathogenesis [15].

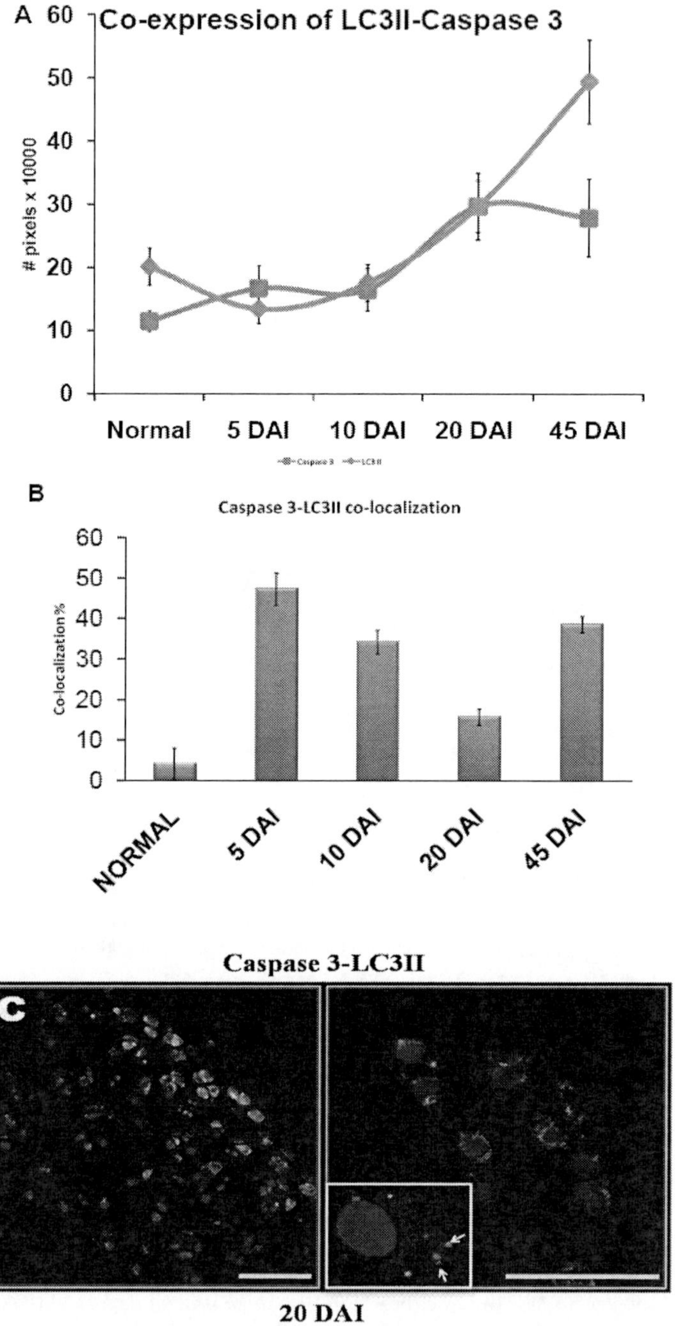

Figure 2. (Continued).

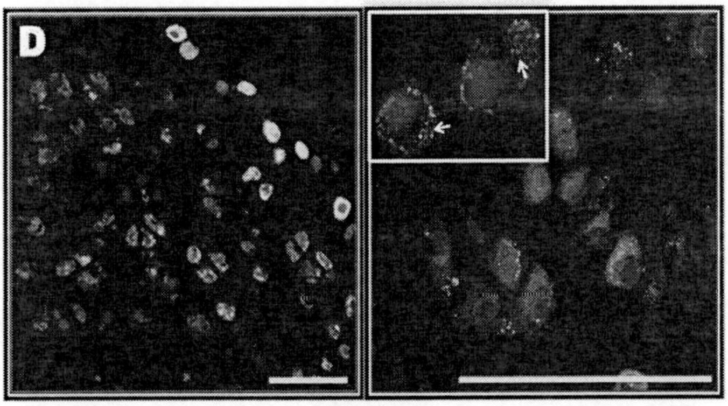

45 DAI

Figure 2. Identification of cell death in OA cartilage. (A) Shows the total number of pixels which represent the expression of LC3II (♦) and CCA (■) in cartilage at different DAI, in which was observed that both markers had a tendency to increase their expression. In the case of LC3II (TRITC) it highest expression was observed at 45 DAI ($P \leq 0.01$), while CCA (FITC) expression was increased since 20 DAI ($P \leq 0.01$). (B) Represents the percentage of co-localization between LC3II and CCA markers, expressed in OA cartilage at different DAI and compared with a normal samples, in which the co-localization percentage was increased since 5 DAI ($P \leq 0.001$) and it was maintained during all the degenerative process, showed its highest expression at 45 DAI ($P \leq 0.001$). (C and D) Show representative IHC assays from CA-LC3II co-localizations at 20 and 45 DAI; in which nuclei were stained with DAPI, and co-localizations was identified by yellow marks. Left charts from C and D figures show the 3 zones of the cartilage, while the right charts are representative amplifications from MZ. White arrows indicate autophagic vacuoles. The results show means ± S.E.M. of 3 independent experiments using at Tukey–Kramer multiple comparison test. Scale bar= 50μm. Inset= magnification from SZ cells (Figure taken from the reference [15]).

References

[1] Mehrpour, M., et al., Autophagy in health and disease. 1. Regulation and significance of autophagy: an overview. *Am. J. Physiol. Cell Physiol.* 298(4): p. C776-85.

[2] Codogno, P. and A.J. Meijer, Autophagy and signaling: their role in cell survival and cell death. *Cell Death Differ.* 2005. 12 Suppl 2: p. 1509-18.

[3] Yang, Z. and D.J. Klionsky, Eaten alive: a history of macroautophagy. *Nat. Cell Biol.* 12(9): p. 814-22.

[4] Reggiori, F. and D.J. Klionsky, Autophagy in the eukaryotic cell. *Eukaryot. Cell.* 2002. 1(1): p. 11-21.

[5] Edinger, A.L. and C.B. Thompson, Death by design: apoptosis, necrosis and autophagy. *Curr. Opin. Cell Biol.* 2004. 16(6): p. 663-9.

[6] Klionsky, D.J., The molecular machinery of autophagy: unanswered questions. *J. Cell Sci.* 2005. 118(Pt 1): p. 7-18.

[7] Levine, B. and J. Yuan, Autophagy in cell death: an innocent convict? *J. Clin. Invest.* 2005. 115(10): p. 2679-88.

[8] Amar, N., et al., Two newly identified sites in the ubiquitin-like protein Atg8 are essential for autophagy.*EMBO Rep.* 2006. 7(6): p. 635-42.

[9] Hubbard, V.M., et al., Selective autophagy in the maintenance of cellular homeostasis in aging organisms. *Biogerontology*.
[10] Vicencio, J.M., et al., Senescence, apoptosis or autophagy? When a damaged cell must decide its path--a mini-review. *Gerontology*. 2008. 54(2): p. 92-9.
[11] Galluzzi, L., et al., Life, death and burial: multifaceted impact of autophagy. *Biochem. Soc. Trans.* 2008. 36(Pt 5): p. 786-90.
[12] Maiuri, M.C., et al., Self-eating and self-killing: crosstalk between autophagy and apoptosis. *Nat. Rev. Mol. Cell Biol.* 2007. 8(9): p. 741-52.
[13] Tsujimoto, Y. and S. Shimizu, Another way to die: autophagic programmed cell death. *Cell Death Differ.* 2005. 12 Suppl 2: p. 1528-34.
[14] Gomez-Camarillo, M.A., et al., Chondrocyte proliferation in a new culture system. *Cell Prolif.* 2009. 42(2): p. 207-18.
[15] Almonte-Becerril, M., et al., Cell death of chondrocytes is a combination between apoptosis and autophagy during the pathogenesis of Osteoarthritis within an experimental model. *Apoptosis*. 15(5): p. 631-8.
[16] Kroemer, G. and S.J. Martin, Caspase-independent cell death. *Nat. Med.* 2005. 11(7): p. 725-30.
[17] Lockshin, R.A. and Z. Zakeri, Apoptosis, autophagy, and more. *Int. J. Biochem. Cell Biol.* 2004. 36(12): p. 2405-19.
[18] Kuhn, K., et al., Cell death in cartilage. *Osteoarthritis Cartilage.* 2004. 12(1): p. 1-16.
[19] Kouri, J.B., et al., Apoptotic chondrocytes from Osteoarthrotic Human Articular Cartilage and Abnormal Calcification of Subcondral Bone. *Journal of Rheumatology*. 2000. 27(4): p. 1005-1019.
[20] Blanco, F.J., et al., Osteoarthritis chondrocytes die by apoptosis. A possible pathway for osteoarthritis pathology. *Arthritis Rheum.* 1998. 41(2): p. 284-9.
[21] Chen, C.T., et al., Chondrocyte necrosis and apoptosis in impact damaged articular cartilage. *J. Orthop. Res.* 2001. 19(4): p. 703-11.
[22] D'Lima, D., et al., Caspase inhibitors reduce severity of cartilage lesions in experimental osteoarthritis. *Arthritis Rheum.* 2006. 54(6): p. 1814-21.
[23] Thomas, C.M., et al., Chondrocyte death by apoptosis is associated with cartilage matrix degradation. *Osteoarthritis Cartilage.* 2007. 15(1): p. 27-34.
[24] Kapitonova, M.Y. and O. Mansor, Ultrastructural changes of the articular cartilage in some arthropathies with special reference to chondrocyte cell death. *Malays. J. Pathol.* 2003. 25(1): p. 15-27.
[25] Aigner, T., et al., Apoptotic cell death is not a widespread phenomenon in normal aging and osteoarthritis human articular knee cartilage: a study of proliferation, programmed cell death (apoptosis), and viability of chondrocytes in normal and osteoarthritic human knee cartilage. *Arthritis Rheum.* 2001. 44(6): p. 1304-12.
[26] Mistry, D., et al., Chondrocyte death during murine osteoarthritis. *Osteoarthritis Cartilage.* 2004. 12(2): p. 131-41.
[27] Wei, L., et al., CD95-induced osteoarthritic chondrocyte apoptosis and necrosis: dependency on p38 mitogen-activated protein kinase. *Arthritis Res. Ther.* 2006. 8(2): p. R37.
[28] Roach, H.I., T. Aigner, and J.B. Kouri, Chondroptosis: a variant of apoptotic cell death in chondrocytes? *Apoptosis.* 2004. 9(3): p. 265-77.

[29] Kouri-Flores, J.B., K.A. Abbud-Lozoya, and L. Roja-Morales, Kinetics of the ultrastructural changes in apoptotic chondrocytes from an osteoarthrosis rat model: a window of comparison to the cellular mechanism of apoptosis in human chondrocytes. *Ultrastruct. Pathol.* 2002. 26(1): p. 33-40.

[30] Carames, B., et al., Autophagy is a protective mechanism in normal cartilage, and its aging-related loss is linked with cell death and osteoarthritis. *Arthritis Rheum*. 62(3): p. 791-801.

[31] Kouri, J.B. and C. Lavalle, Do chondrocytes undergo "activation" and "transdifferentiation" during the pathogenesis of osteoarthritis? A review of the ultrastructural and immunohistochemical evidence. *Histol. Histopathol.* 2006. 21(7): p. 793-802.

[32] Kouri, J.B., et al., Modifications of Golgi complex in chondrocytes from osteoarthrotic (OA) rat cartilage. *J. Histochem. Cytochem.* 2002. 50(10): p. 1333-40.

In: Autophagy: Principles, Regulation and Roles in Disease ISBN: 978-1-61942-266-7
Editor: Nikolai Gorbunov

Chapter XI

Altered Autophagy in Amyotrophic Lateral Sclerosis: Mechanisms and Therapeutic Opportunities?

***Jean-Paul Decuypere*[1], *Kim A. Staats*[2], *Kirsten Welkenhuyzen*[1], *Ludwig Missiaen*[1], *Humbert De Smedt*[1], *Ludo Van Den Bosch*[2], *Jan B. Parys*[1] *and Geert Bultynck*[1,*]**

[1] Lab. Molecular and Cellular Signaling,
Department Molecular Cell Biology, K.U. Leuven,
Campus Gasthuisberg O/N-1 bus 802,
Herestraat 49, BE-3000 Leuven, Belgium
[2] Lab. Neurobiology, Vesalius Research Center,
K.U. Leuven and VIB, Campus Gasthuisberg O/N-2 bus 1022,
Herestraat 49, BE-3000 Leuven, Belgium

Abstract

Amyotrophic lateral sclerosis (ALS) is caused by selective degeneration of upper and lower motor neurons, resulting in muscle weakness, paralysis and death of the patient on average 2 to 5 years after diagnosis. In the majority of ALS cases (~90 %) no clear hereditary component is present (sporadic ALS), while the remaining ~10 % of the cases have a familial form (fALS). Most studies have used transgenic mice expressing mutant superoxide dismutase 1 (SOD1) as a model system for the *in vivo* study of ALS. Clearly, ALS is a multi-factorial disease and one of the best characterized hallmarks is the accumulation of toxic mutant protein aggregates and damaged organelles. These are key features of many neurodegenerative diseases, including ALS. Several of those neurodegenerative diseases are characterized by alterations in autophagic activity, leading

* Correspondence to Geert Bultynck; Laboratory of Molecular and Cellular Signaling; Department Molecular Cell Biology, K.U. Leuven; Campus Gasthuisberg O/N-1 bus 802; Herestraat 49; BE-3000 Leuven; BELGIUM; Tel: +32-16-330215; Fax: +32-16-345991; e-mail: geert.bultynck@med.kuleuven.be.

to an impaired clearance of toxic protein aggregates and/or of damaged mitochondria. In mouse models for neurodegeneration, chemical stimulation of autophagic flux has been shown to be beneficial. Stimulation of autophagy in mice resulted in neuroprotective effects, delayed aging and increase in life span. Since there is as yet no effective treatment for ALS, targeting the altered autophagy pathway could be a promising therapeutic strategy. Initial studies in mutant-SOD1-mouse models pointed out that increasing autophagy activity may protect against motor neuron degradation by removing protein aggregates and damaged organelles. Other studies however, reported no therapeutic efficacy or even a faster progression of motor-neuron degeneration and a decrease in life span in presymptomatic mutant-SOD1 mice. Thus, the question arises whether autophagy stimulation is beneficial for the treatment of ALS. In particular, several defects in the autophagic pathway may complicate the outcome and are not fully characterized and understood for each of the different mutant genes associated with ALS. Here, we will critically address the current state-of-the-art knowledge on the role of autophagy in ALS and the modulation of autophagy activity as a potential therapeutic target in ALS.

1. Introduction to Autophagy

1.1. Autophagy: Mechanism, MolecularKey Players and Their Pathways

Macroautophagy (further referred to as autophagy) is a delivery pathway of cellular material towards lysosomes for degradation. Its most characteristic feature is the formation of double-membranous vesicles, known as autophagosomes, which engulf the soon-to-be degraded cellular material, and fuse first with endosomes to form amphisomes and finally with lysosomes to form autolysosomes [1]. The cellular cargo can consist of proteins, lipids, other macromolecules and even entire organelles. This way, fresh building blocks of the cell, like amino acids, fatty acids etc., can be generated and applied in the novel construction of cellular material. By allowing constant remodeling of the cell, autophagy plays an important role in development. In addition, the toxicity of protein aggregates or damaged organelles can be counteracted by this cellular pathway [2]. Finally, when nutrients are limiting, the upregulation of autophagy provides the cell with nutrients from its own cellular pool, which can be incorporated in stress-induced macromolecules (like molecular chaperones) [1]. Because of its importance in cellular life and death responses, autophagic deficiency leads to various diseases [3].

Molecularly, more than 30 *atg* (autophagy) genes regulate the process [1]. Downregulation of the mammalian target of rapamycin (mTOR) complex 1 (mTORC1) on the one hand, and the dissociation of Beclin 1 from Bcl-2 (B-cell lymphoma 2) to form the phosphatidylinositol 3-kinase (PI3K) complex III (PI3K III) on the other, are required for the nucleation phase. During the elongation of the autophagosomal membrane, the lipidation of Atg8/LC3 (LC3-I) towards LC3-phosphatidylethanolamine (LC3-II) occurs. With this lipid tail, LC3-II becomes inserted in both the inner and outer membrane of the forming autophagosome. This conversion can be easily assessed using Western blotting, since LC3-II (16 kDa) migrates faster on a protein gel than LC3-I (18 kDa) [4]. Moreover, GFP-LC3 constructs can be employed, which can be seen as clear punctae when recruited to autophagosomes. These LC3 assays are therefore standard and easy methods for monitoring autophagy, together with the detection of cellular vacuolization by transmission electron microscopy [5]. Caution is needed, however, since LC3-II becomes degraded in the autolysosomes; an increase of LC3-II steady-state levels therefore does not necessarily represent increased autophagy, but could be due to autophagosomal accumulation, related to defective autophagy (Fig. 1) [4-5]. Therefore, autophagic flux should be determined by monitoring LC3-II levels in the presence of lysosomal inhibitors, like bafilomycin A, or by using the tandem reporter construct RFP-GFP-LC3, in which red/green punctae represent early autophagic vesicles and red only punctae represent autolysosomes [6]. Additional evidence may be obtained by assessing the levels of the ubiquitin- and LC3-binding p62 protein, a long-lived protein selectively degraded through autophagy [7-8]. For a detailed description on measuring autophagy and autophagic flux, we wish to refer to recent reviews [5-6].

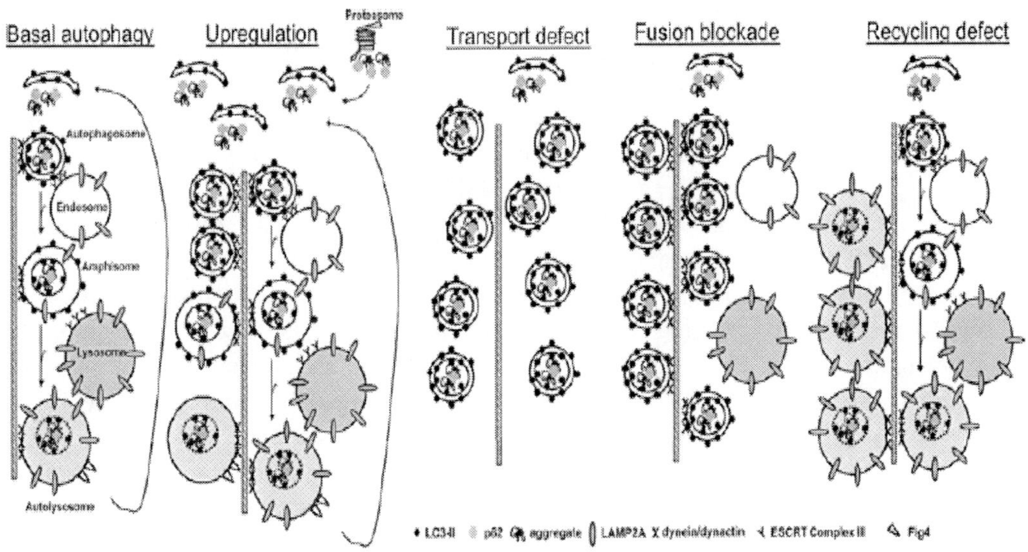

Figure 1. An overview of the different defects in the autophagy pathway that may account for the observed increase in autophagy markers. *Far left*: basal autophagy is important for clearance of protein aggregates, which are recruited to autophagosomes through a mechanism involving p62. The aggregates are included in LC3-positive autophagosomes, which are transported over the microtubuli in a dynein/dynactin-dependent manner. Autophagosomes fuse with endosomes to form amphisomes. These will then fuse with lysosomes to form autolysosomes. Fusion is mediated by the ESCRT III complex. Degradation eventually takes place in the autolysosomes, and their membranes will be recycled in a Fig4-dependent mechanism. *Middle left*: An increase of protein aggregates will lead to an upregulation of autophagy, possibly in part through inhibition of the proteasome. This will lead to an increase in all autophagic markers and features. *Middle*: Dynactin mutations result in defective autophagosome transport, leading to the accumulation of autophagosomes. *Middle right*: Mutations in the ESCRT III complex (*e.g.* in CHMP2B) will lead to defective fusion between autophagosomes and endosomes and/or lysosomes, also leading to accumulation of autophagosomes. In addition, mutations in VAPB, optineurin and ALS2 could also contribute to defective vesicle transport or fusion (see text). *Far right*: Mutations in Fig4 lead to defects in autolysosome recycling, leading to their accumulation, recognized by LAMP2A staining.

1.2. Importance of Autophagy for Neuronal Health and Survival

The importance of autophagy in development became evident thanks to knockout studies in mice. Whole-body autophagy knockout mice mostly die at embryonic stages [9-1]), and if they survive, they die within three days after birth, since these mice cannot adapt the switch from placental to breast-fed metabolism [13-16]. Therefore, tissue-specific and conditional knockout mice were generated, which developed disorders in liver, cardiac and skeletal muscle, secretory tissues, the hematopoietic system and adipose tissue [17]. This revealed autophagy as essential pahtway for proper tissue development and function.

Neurons are however much more dependent on basal autophagy than other cell types. Consequently, mice with central-nervous-system-specific knockout for autophagy genes (*e.g.* Atg5, Atg7) showed growth retardation, progressive motor deficits and neurodegeneration [18-19]. Interestingly, the most conserved feature was the accumulation of ubiquitin-positive inclusion bodies and damaged organelles. In contrast, neuronal autophagy induced by starvation is much less pronounced than in other cell types [20].

The reasons for the strong dependence of neurons on basal autophagy are probably multiple. Firstly, the fact that these cells experience no division or proliferation makes them highly susceptible to damaged organelles or misfolded protein aggregates, since they cannot be diluted over daughter cells and thus need to be removed [17]. Furthermore, the ubiquitin-proteasome system fails to degrade these protein aggregates. Thus, protein-quality control through constitutive autophagy to remove aggregate-prone proteins with aberrant structures is critically important for cellular homeostasis and neuron survival and to prevent neurodegeneration [19, 21]. Secondly, the specific shape and architecture of neurons renders them more vulnerable than others; *e.g.* accumulation of organelles and aggregates disturb axonal transport [22]. In addition, neuronal plasticity partially depends on autophagy for the remodeling of the dendrites and axons [9, 23]. Finally, a study in *Caenorhabditis elegans* has shown that neurons depend on autophagy for the clustering of neurotransmitter receptors during development, which happens when presynaptic terminals make contact with the postsynaptic cell. Before this contact, gamma-aminobutyric acid A ($GABA_A$) receptors are clustered and degraded constantly through autophagy [24]. In conclusion, the correct functioning of the autophagic process is of greater importance for neurons than for other cells. This may be one of the reasons why most neurodegenerative diseases only manifest at older ages, since autophagy has frequently been shown to be downregulated during age [25-26].

1.3. Chemical Modulation of Autophagy

1.3.1. Recent Progress

Rapamycin was one of the first identified chemical compounds that induce autophagy. The mechanism involves the inhibition of mTOR activity. More recently, rapamycin derivatives (rapalogs) have been developed. These compounds specifically destabilize mTORC1, but not mTORC2. The potency of these allosteric inhibitors depends on the mTOR stability, which differs among cell lines [27]. Therefore, ATP-competitive inhibitors, such as Torin 1, are more potent in inducing autophagy [28-29], but these compounds inhibit both mTORC1 and mTORC2. Although Torin-1-induced autophagy is independent of mTORC2 [28-29], mTORC2 inhibition may provoke adverse effects. However, a recent study circumvented these issues by combining non-effective concentrations of Torin 1 and rapamycin, which synergistically inhibited mTORC1, but not mTORC2-depedent signaling and activated autophagy [27]. Besides autophagy, mTORC1 regulates a variety of other cellular processes, including the immune responses. Thus, inhibition of mTORC1 will lead to undesirable effects, such as immunosuppression [30-31]. In addition, adverse side effects have been reported when rapamycin was applied to allograft patients [32].

PI3K modulators may be alternatives for autophagy stimulation, but PI3K I and PI3K III regulate autophagy in opposite directions. PI3K I activation stimulates mTOR through the Akt/Protein Kinase B pathway and thus negatively regulates autophagy, while Vps34 PI3K III is an essential component of the autophagic pathway. Compounds affecting the PI3K I/Akt pathway, like dexamethasone, induce autophagy [33]. PI3K III activators have not yet been discovered. PI3K III inhibitors (3-methyladenine (3-MA), wortmannin, LY294002) are known to inhibit autophagy, but these are not very specific. However, the generation of Vps34-specific inhibitors may be facilitated by the recently discovered crystal structure of the Vps34 PI3K III [34].

Other modulators of autophagy (lithium salts, carbamazepine, valproic acid) affect the phosphoinositol/Ca^{2+} signaling. The compounds reduce basal inositol 1,4,5-trisphosphate (IP_3) levels, and limits Ca^{2+} signaling from IP_3 receptors. The latter stimulates autophagy in an mTOR-independent way [35-37]. In addition, inhibitors of plasma-membrane Ca^{2+} channels (verapamil, loperamide) are established autophagy inducers [38].

Finally, some compounds, like trehalose, induce autophagy, while the mechanism remains to be elucidated [39].

1.3.2. Targeting Autophagy in Neuronal Survival and Neurodegenerative Diseases

Since neurons depend strongly on autophagy for survival and function, in particular for clearing organelles and protein aggregates, this cellular pathway is often regarded as a possible therapeutic target in proteinopathies [21, 40]. While modulating autophagy is beneficial in fly, zebrafish and mouse models [41-45], it is not yet a common strategy in patients. Moreover, the currently available autophagy-modulating compounds will not only affect autophagy, but also other cellular processes and thus likely lead to important side-effects. Thus, the development of selective autophagy modulators will be essential.

In addition, the combination of compounds working on different branches of the autophagy pathway appears more successful in inducing autophagy than their separate use, in part because these combinations may counteract mechanisms that exert negative feedback on the autophagy induction. This is the case for lithium and rapamycin. The reason underlying this phenomenon is that besides the ability of lithium ions to induce autophagy, they may also act as glycogen synthase kinase 3β (GSK-3β) inhibitors, thereby activating mTOR [46-47]. Hence, combining lithium and rapamycin might prevent this problem and causes more effective autophagy stimulation [48].

Another strategy is the application of the dual PI3K I/mTOR inhibitor PI103 [49]. Indeed, inhibition of mTOR enhances PI3K/Akt signaling and this negatively regulates autophagy. Thus, counteracting mTOR as well as PI3K I activity using PI103 seems very effective to potently induce autophagy [50].

2. Introduction to ALS

ALS is a devastating progressive neurodegenerative disease, known by a number of different names, including Lou Gehrig's disease and motor-neuron disease. The loss of motor neurons in the motor cortex, brain stem and spinal cord causes denervation of muscle fibres, resulting in increasing muscle weakness and paralysis. With an incidence of 2.7 cases per 100,000 people in Europe [51], it is diagnosed from teen years onward, but is more prevalent in the later years of life. The dire character of ALS is further illustrated by the current lack of a medical cure averaging life expectancy post diagnosis between 2 and 5 years, though 10% of patients live longer than 10 years. Patients mainly succumb to the disease by respiratory insufficiency or may opt for euthanasia where legislature permits. Although ALS is characterized by degeneration of central nervous-system tissue, mental functions remain largely unaffected resulting in a locked-in state; the patient is mentally conscious while motor function further declines [52]. Regretfully, there is but one medicine on the market today to treat the disease, riluzole, which merely slows disease progression by an estimated 12% [53].

Mutations in the ubiquitously expressed Cu/Zn-superoxide-dismutase-1 (SOD1) gene are a cause of ALS. SOD1 detoxifies potentially cell-damaging free radicals and its mutations account for 20% of the ALS patients suffering from the familial form of the disease (fALS, 10% of the total number of ALS cases) worldwide. Over 100 different missense point mutations have been identified [54], but also deletions and insertions, leading to C-terminal truncated forms [55]. The origin of the remaining 90% of ALS patients is unknown or of sporadic origin (sALS) in which gene-environment interactions are predicted. Due to the indistinguishable phenotypes of fALS and sALS patients, common disease mechanisms are hypothesized. Overexpression of mutant forms of human SOD1 causes the ALS phenotype of transgenic SOD1 mice, which has made a large contribution to ALS research [56]. Many hallmarks of the disease are shared by patients and this rodent model, including specific motor-neuron loss, aggregate formation, astrogliosis, microgliosis and lethal progressive paralysis. The disease progression in the rodent ALS models is divided into three stages: a pre-symptomatic (or asymptomatic) stage, a symptomatic stage and the end stage. Since genetic ablation of SOD1 does not produce an ALS-like phenotype in mice [57-58], the mechanism by which SOD1 causes the disease is determined as a toxic gain of function. This gain of function exerts itself by mechanisms including protein misfolding, aggregation, impaired proteasome functioning, impaired retrograde transport, excitotoxic cell death and many more (reviewed in [59]).

As ALS involves pathology of a number of different cell types, it is described as a non cell-autonomous disease [60]. Investigation of mutant SOD1 in specific cell types demonstrated that a number of cell types influence the disease, including astrocytes [61], microglia [62], Schwann cells [63] and motor neurons [64]. Although ALS is a non-cell autonomous disease, the cell type that directly leads to the loss of movement are motor neurons. Mutant SOD1 expressed solely in this cell type is sufficient to initiate the disease, although disease progression is slower than in mice ubiquitously expressing mutant SOD1 [64]. In patients, motor neurons of the motor cortex, brain stem and spinal cord undergo cell death selectively. A variety of pathogenic mechanisms underlying the selective motor-neuron degeneration in ALS have been proposed, such as oxidative stress, excitotoxicity, aggregate formation, inflammation, growth-factor deficiency and neurofilament disorganization [65].

Furthermore, since Ca^{2+} signaling plays a pivotal role in cell survival and death [66-67] and motor neurons possess unique Ca^{2+}-handling characteristics [68], dysregulation of Ca^{2+} signaling may play a prominent role in the neuronal death induced by excitatory neurotransmitters [69]. This may be related to the poor intracellular Ca^{2+}-buffering capacity and the high density of Ca^{2+}-permeable glutamate-receptor channels of the AMPA-receptor type of motor neurons. For a detailed discussion of these aspects, we wish to refer to recent reviews [68-69]. Since the role of Ca^{2+} signaling in autophagy regulation has been emerging [25, 36, 70], the importance of distorted Ca^{2+} signaling in the pathogenic mechanisms underlying motor-neuron death may also involve changes in autophagy. However, it is important to note that currently there is no direct evidence associating dysregulated Ca^{2+} signaling to altered autophagy in ALS.

Not only mutations in SOD1, but also mutations in a number of other genes cause fALS, including mutations in vesicle-associated membrane protein-associated protein B (VAPB), TAR DNA binding protein (TDP-43), fused in sarcoma/translocated in liposarcoma (FUS/TLS) and optineurin have been identified as causes of fALS [71-76]. However, the frequency of these mutations is currently unknown. Most work has been done with mutant SOD1 cells, rodent ALS models or cells/tissue from ALS patients (both familial and sporadic). However, other ALS mouse models based on recently discovered mutant genes start to emerge.

Interestingly, protein aggregation was observed in post-mortem spinal-cord tissue of both fALS and sALS patients and it is suggested to be of common pathological pattern where the above described mutant gene products (SOD1, TDP-43, FUS/TLS and optineurin) appear in motor neuron aggregates [77] and in glial cells [77-80]. Intriguingly, SOD1 aggregates are also found in patients without mutations in the SOD1 gene [77]. There are similar observations for TDP-43. Its aggregates are also detected both in patients with and without TDP-43 mutations [80]. ALS model rodents are particularly useful for investigating the effect of protein aggregation because aggregates are also detected in the spinal cords of these (a)symptomatic animals [64, 81].

Aggregates of mutant SOD1 are easily obtained and investigated by transfecting the motor neuron-like NSC-43 cell line. Addition of the proteasome inhibitor MG132 shows that mutant SOD1 can be degraded by the proteasome, as it increases the amount of aggregates in these cells [82]. This was confirmed in N2a cells by the use of another proteasome inhibitor epoxomycin [83]. Moreover, addition of bafilomycin A also increased the amount of mutant SOD1 aggregates [82, 84], as was also found when knocking down Atg5 [84], implying that both proteasome and autophagy pathways are capable of degrading mutant SOD1 *in vitro*. Similar effects were shown in N2a cells by the use of 3-MA [83]. Interestingly, this is also demonstrated for mutant TDP-43 as proteasome inhibitors and autophagy inhibitors increase the amount of TDP-43 (aggregates) *in vitro* in different cell types [82, 85-87]. Other *in vitro* work suggested that the amount of clearance by the two systems is of similar importance [83]. Moreover, evidence that SOD1 may be degraded by autophagy *in vivo* was provided by electron-microscopy images of ALS mouse spinal cords demonstrating the localization of SOD1 aggregates in autophagosomes and their co-localization with LC3 in intact cells [84] (Fig. 2).

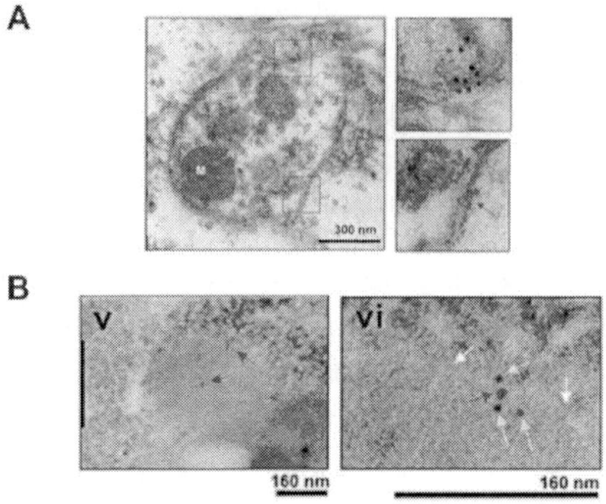

Figure 2. SOD1 accumulation in autophagosomal structures and its co-localization with the autophagy marker LC3 *in vivo*. (A) Electron-microscopy sections of motor neurons of SOD1-G86R/XBP-1Nes$^{-/-}$ mice analyzed by immunogold staining of SOD1. Autophagosomes were identified by the presence of double-membrane vesicles containing cytosol and organelles such as mitochondria (M). A higher magnification of the areas marked with red squares is shown to depict the double membrane of the vesicle and SOD1 immunogold staining (see arrows). Please note the presence of SOD1 protein inside the autophagosomal structures. Taken from [84]. Copyright permission: © 2009 by Cold Spring Harbor Laboratory Press. (B) Colocalization of SOD1 and LC3 determined by double-immunogold staining in an autophagosomal structure in a motor neuron. Panel *v* and *vi* display two different magnifications. Gold spheres were 5 nm (SOD1, yellow arrows) and 10 nm (LC3, red arrows). The white arrows indicate the double membranes of the autophagosomes. Please note the presence of both SOD1 and LC3 in the same autophagosome. Taken from [84]. Copyright permission: © 2009 by Cold Spring Harbor Laboratory Press.

Beyond the mere observation of the above described aggregates in ALS arises the question whether these aggregates are benign or harmful to cells and the organism. Aggregation may be a protective mechanism of the cell to clump mutant proteins together to limit further deleterious effects. On the other hand, the aggregates may be detrimental as they may clog the proteasome system disallowing homeostatic degradation of faulty proteins [88]. An alternative hypothesis is that the aggregates are not beneficial, nor detrimental, but are a mere side effect observed in ALS tissue without any particular function or effect on the cell or organism. Altering the levels of clearance of the aggregates, *e.g.* by inhibiting or stimulating autophagy, may shed light on the effects of the aggregates. The remainder of this chapter will address these hypotheses by discussing observations of altered autophagy in ALS (Section 3) and the state-of-the-art on the potential for manipulating autophagy as a future therapeutic strategy for ALS patients (Section 4).

3. Autophagy Alteration in the ALS Pathogenesis

New and emerging data from clinical observations and experimental models suggest that autophagy plays a crucial role in the pathogenesis of ALS. Analyses of cellular and animal ALS models [89-91] and post-mortem samples of ALS patients [84, 92] revealed a high amount of autophagic features, including high levels of LC3-II. These features are more prominent at later stages of the disease [89, 91]. Nevertheless, it has been shown that autophagic markers are upregulated at the presymptomatic stage in the spinal cord of SOD1-G93A-linked autophagy mouse models (Fig. 3) [89] and in normal-appearing anterior-horn neurons, which occasionally displayed high levels of LC3 (Fig. 4) [92]. Recently, these findings have been underpinned by a novel double transgenic mouse model for the expression of GFP-LC3 in a mutant-SOD1 mouse model [93]. *In vivo* and *ex vivo* imaging experiments confirmed the upregulation of the autophagy markers in the early symptomatic stage. From these observations, one may propose that autophagy is upregulated as a pro-survival strategy in the motor neurons activated upon stress responses, like protein aggregation, to overcome or at least delay motor-neuron death. However, this increase in autophagy does not necessarily reflect an increased autophagy initiation, since an accumulation of autophagosomes, amphisomes or autolysosomes could also arise from a blockade in the fusion events or from a defective clearance of these vesicles (Fig. 1). In this respect, a defective autophagic flux too may contribute or initiate motor-neuron degeneration. All these hypotheses have already been suggested in the literature, and the exact autophagic phenotype is possibly dependent on the various underlying mutations in the genes responsible for the fALS. Therefore, we will summarize here below the different genes known to be mutated in fALS and their relation with autophagy. It should also be pointed out that some of these mutations were also observed in cases of sALS, like the mutations in Fig4 [94], dynactin [95] and optineurin [96], although most of the sALS cases have an unknown underlying cause. Interestingly, sALS patients display similar pathologies as fALS, suggesting related pathways are affected and possible therapeutics can be translated from fALS to sALS. Alterations in autophagy could therefore be relevant for sALS as well. In this respect, recent observations revealed increased autophagic markers in post-mortem spinal cord samples of sALS patients [84, 92] (Fig. 4).

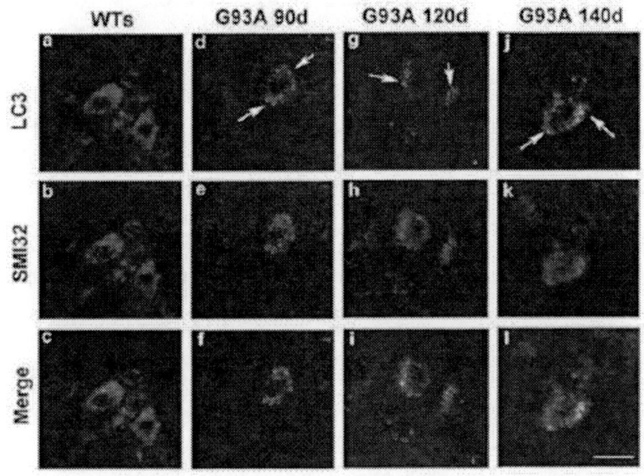

Figure 3. Altered autophagy markers in the motor neurons of SOD1-G93A mice and the presence of mutant SOD1 in autophagosomal structures. Immunohistochemical analysis of LC3 in the SMI32-labeled motor neurons in the spinal cord anterior horns of wild-type control mice (a–c) and SOD1-G93A mice (d–l) in pre-symptomatic stage (90 days), the symptomatic stage (120 days) and the end stage (140 days). The scale bar represents 30 µm. Taken from [89]. Copyright permission: ©2008 Landes Bioscience. Notice the increase in LC3 punctae, an autophagic marker, in pre-symptomatic stage in the fALS mouse model.

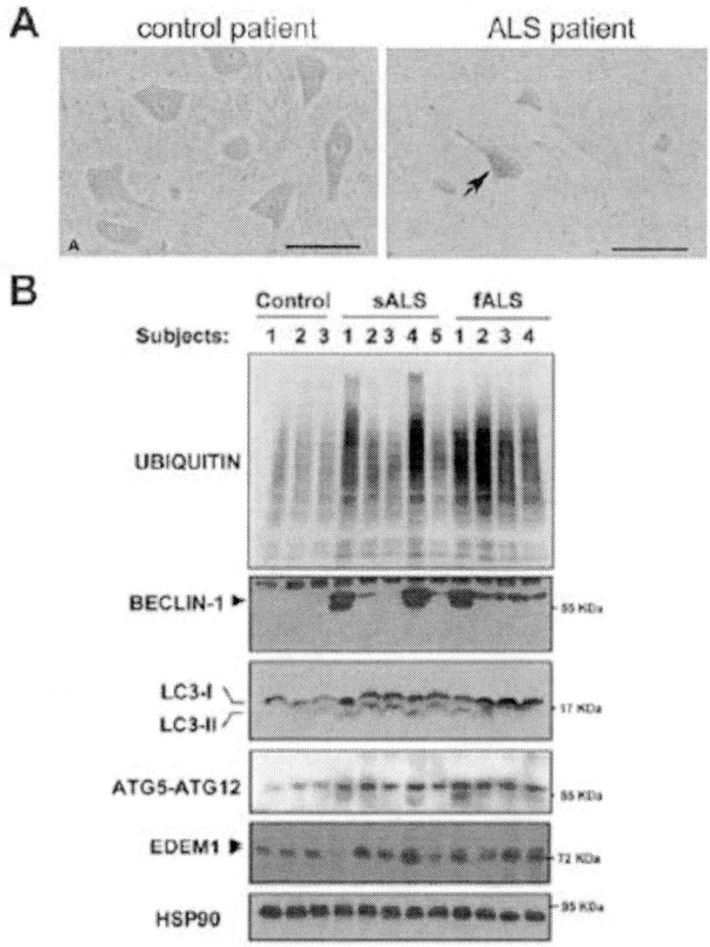

Figure 4. Elevated autophagy markers in fALS and sALS patients. (A) Immunostaining of the autophagy marker LC3 in normal-appearing motor neurons of a control and an ALS patient. Please note the presence of immunostained LC3 in the ALS patient, but not in the control patient. Taken from [9] . Copyright permission: © 2011 American Association of Neuropathologists, Inc. (B) Western-blotting analysis of a variety of autophagic markers (Beclin 1, LC3, Atg5-Atg12) and EDEM1, a key protein in ERAD in post-mortem samples of control, fALS and sALS patients. Please note the elevated levels of the autophagic markers and ERAD markers in the fALS and sALS samples. Taken from [84]. Copyright permission: © 2009 by Cold Spring Harbor Laboratory Press.

3.1. SOD1

In approximately 20% of the fALS cases, an adult-onset, autosomal dominant form of fALS has been linked to mutations in the *SOD1* gene. Since this was the first ALS-linked mutation discovered [97], mutant SOD1-based models are the most studied. A notable contribution in this research was the generation of mice expressing various SOD1 mutants, which develop similar disease features as ALS patients (see section 2).

Increased autophagy levels were observed in two different mutant SOD1 mice [84, 89-90]. Also, overexpression of mutant SOD1 in cells leads to an increase in LC3-II.

Furthermore, *in vivo* degradation of mutant SOD1 can occur through autophagy [84] (Fig. 3). In agreement with these observations, the repression of autophagy leads to enhanced cell death [42]. The increased basal levels of autophagy could be the consequence of the inhibition of proteasome activity by mutant SOD1 aggregates [98-100] (Fig. 1). The proteasome and the autophagic pathway can communicate through a mechanism involving p62, which is recruited to poly-ubiquitinated proteins (aggregates) that are unable to be degraded by the proteasome. p62 consequently is recruited to forming autophagosomes in order to degrade these proteins. Indeed, ubiquitin and p62 co-localize to mutant SOD1 aggregates [101-102]. Another explanation for the increased levels of autophagy may arise from the observation that SOD1 aggregates bind Bcl-2, thereby inhibiting its anti-apoptotic function [103]. This study however only focused on mitochondrial Bcl-2 and its function in apoptosis, but recruitment of Bcl-2 residing at the endoplasmic reticulum (ER) could also affect its anti-autophagic function through the release of the essential autophagic protein Beclin 1 [104]. Whatever the mechanism, both autophagy and the proteasome system can play a role in the clearance of the toxic SOD1 aggregates [83]. In addition, SOD1 aggregates at mitochondria damage this organelle's function [105-106], provoking their clearance *via* autophagy. It is therefore no surprise that mitochondrial structures are often observed in the cytosolic vacuoles in electron-microscopic images [107].

Other studies in different experimental models have suggested that the increase in autophagy is probably due to a defect in the autophagic pathway, leading to the accumulation of autophagosomes [91]. How mutant SOD1 might alter the autophagic machinery is less clear. Ligon *et al.* suggests an interaction between SOD1 and dynein that, together with dynactin, is important for vesicle transport across axons [108]. If SOD1 alters dynein function, it could lead to accumulation of autophagosomes in the axons (Fig. 1), and inhibition of retrograde axonal transport, which has been observed in SOD1-G93A mice [109]. Interestingly, dynactin mutations have been assigned as new possible targets for ALS mutations (see section 3.5).

Similar to the explanations for the observed increased autophagy, discrepancies are also found in studies concerning the clearance of SOD1 aggregates by activation of autophagy. Both PI3K inhibitors and rapamycin reduced the aggregates in a neuroblastoma cell line [83], while also lithium ions prolonged survival of motor neurons *in vivo* in SOD1-G93A mice [42]. Also trehalose decreased the amount of SOD1 aggregates in the motor neuronal NSC-34 cell line transfected with SOD1-G93A [110], possibly through its autophagy-activating ability. Another mechanism of autophagy induction for aggregate clearance was the overexpression of the small heat-shock protein B8 (HspB8) in SOD1-G93A NSC-34 cells [82]. Here, this study shows that HspB8 existed in a complex with BAG family molecular chaperone regulator 3 (Bag3), heat shock cognate 70kDA protein (Hsc70) and C terminus of Hsc70-interacting protein (CHIP), necessary for autophagic removal of misfolded proteins. A similar clearance of SOD1 aggregates was observed in SOD1-G86R mice deficient for transcriptional factor X-box-binding protein 1 (XBP-1). The latter is an early adaptive ER-stress-response factor activated upon ER-stress induction by inositol-requiring transmembrane kinase/endonuclease (IRE1α). XBP-1 deficiency remarkably also led to autophagy induction [84]. In contrast with the aforementioned results, the effects of lithium were not reproduced in other studies [111-112]. In addition, rapamycin has recently been reported to oppositely accelerate degeneration of motor neurons in SOD1-G93A mice, and consequently shorten lifespan, without any effect on SOD1 aggregates [91]. These findings suggest that augmenting potentially already saturated or defective autophagy pathways can also lead to detrimental effects for a cell (see Section 4).

Another report also proposed that differences in autophagic clearance depend on the cell type. Mutant SOD1 was transfected in motor neuronal NSC-34 and muscle C2C12 cells. Interestingly, SOD1 aggregates only appeared in NSC-34, while in C2C12 cells mutant SOD1 remained soluble, which was assigned to a better autophagic activation and flux [113]. Such cell-specific differences could explain why the motor neurons are more severely affected than muscle cells in the ALS pathology.

3.2. TAR DNA-Binding Protein of 43 kDa (TDP-43)

Mutations in TDP-43 account for an estimated ~5%of fALS [114-115]. Interestingly, TDP-43-positive inclusions are also present in sALS and other neurodegenerative diseases, such as Parkinson disease, Alzheimer disease and frontotemporal lobar degeneration with ubiquitin-positive inclusions (FTLD-U) [116]. In the latter, as well as for ALS, it has been shown to be a major pathological protein. TDP-43 functions in transcription and mRNA splicing. Mutations in TDP-43 result in a truncated ~25 kDa C-terminal fragment (TDP-25), missing the nuclear target domains and thus residing in the cytoplasm. It is still unclear whether a loss or gain of function is the cause of the pathological consequences.

It is very clear however, that both TDP-43 and TDP-25 are present in fALS and sALS aggregates [117], co-localizing with ubiquitin, LC3 and p62 [86, 118]. It is still unclear whether fALS SOD1 aggregates contain TDP-43 or TDP-25 [117, 119]. Notably, a study showed that TDP-25 can be toxic without cytosolic aggregation, suggesting the possibility of another mechanism, such as a loss-of-function mutation [120]. The clearance of TDP-43 aggregates has been shown to be both mediated by autophagy and the proteasome system, with autophagy having the most pronounced effect [86-87]). Inhibition of PI3K or of mTOR, as well as trehalose resulted in TDP-43 clearance [85, 87]. Also overexpression of HspB8 reduced the TDP-43 aggregates [82], similarly to its effect on SOD1 aggregates. In contrast, autophagy or proteasome inhibition led to an accumulation of TDP-25 [86-87]. In this regard, it was found that ubiquilin, a ubiquitin-binding protein, interacted with polyubiquitinated TDP-43, which is proposed to work as a co-chaperone for recruitment to autophagosomes or proteasomes [121].

3.3. Charged Multivesicular Body Protein 2B (CHMP2B)

Missense mutations were found in the gene encoding CHMP2B in patients with non-SOD1 ALS [122-123]. Similar mutations were also found in patients with frontotemporal dementia (FTD) [124]. CHMP2B's function is still unknown. But its yeast ortholog, vacuolar protein sorting 2 (VPS2), is a component of the endosomal sorting complex required for transport (ESCRT) III. The latter alleviates vesicular transport [124]. Microarray analysis from motor neurons from CHMP2B-related ALS cases revealed, among others, reduction in autophagy initiation and mTOR repression. Overexpression of the CHMP2B mutants in HEK or COS cells leads to large cytoplasmic vacuoles and impairment of autophagy [122]. In addition, ubiquitin- and p62-positive protein deposits were detected [125]. Collectively, these data imply that loss-of-function CHMP2B mutations could lead to defects in autophagosomal transport and/or fusion, resulting in accumulating autophagosomes in the cytosol (Fig. 1). This hypothesis was supported by the observation that inhibition of autophagy delayed neuronal loss in cortical neurons with CHMP2B mutations [126].

3.4. The Lipid Phosphatase Fig. 4

The protein Fig4 is the PI(3,5)P_2 5-phosphatase component of a protein complex with Vac14 and Fab1 that regulates the cellular levels of PI(3,5)P_2 [127]. Fig4 directly mediates the breakdown of PI(3,5)P_2 to PI3P, but also indirectly regulates the synthesis of PI(3,5)P_2 through a physical assembly with the machinery responsible for PI(3,5)P_2 synthesis [127-128]. It has been observed that this protein harbors mutations in 3 fALS and 6 sALS patients [94]. Work by Ferguson *et al.* provided more insight in the mechanisms linking these mutations to ALS. They observed accumulation of LC3-II, p62 and lysosomal-associated membrane protein 2 (LAMP2) in neurons and astrocytes of mice with mutant Fig4 or Vac14. These mice showed accumulation of swollen intracellular vacuoles that appeared empty and stained for LAMP2 and p62 [129]. This observation suggested that these vesicles are not just accumulated autophagosomes, caused by a blockade of fusion events. LAMP2 staining namely only occurs on late endosomes or lysosomes, indicating correct fusion. Instead, PI(3,5)P_2 is probably only present at late endosomes and lysosomes (or vacuoles in yeast) [130-131], where it might function in recycling of the autolysosomal membrane. Inhibition of this process, by Fig4 mutations for example, would then result in accumulation of empty vacuoles. In addition, this would lead to autophagic impairment, since no more membranes are available for novel autophagosome formation (Fig. 1). Interestingly, inhibition of Fig4's other partner, Fab4, also leads to LC3-II accumulation [132].

3.5. Dynactin

Missense point mutations in the dynactin gene *DCTN1* have also been observed in a few ALS [95, 133] and FTD cases [133]. Dynactin forms a complex with dynein, which is important in vesicle trafficking. In neurons, this complex promotes vesicles movement around the perykarion and axons. Its disassembly or mutations in dynein or dynactin therefore lead to inhibition of axonal transport [134-136]. Two dynactin-related mouse models have been generated, one expressing Bicaudal D2 N-terminus (BICD2-N), impairing dynein/dynactin function [136], and another expressing mutant dynactin [134]. Both of them showed impaired vesicle trafficking (Fig. 1), but only the latter led to motor-neuron degeneration. The mutant dynactin mice also revealed a higher amount of autophagosomes and increased levels of LC3-II.

3.6. Optineurin

Other mutations possibly causing fALS are located in the *OPTN* gene, encoding for optineurin [73]. Optineurin is known as a negative regulator for transcription factor NFκB

[137], which was indeed found to be overactive in cells transfected with mutant optineurin [73]. Additionally, optineurin is involved in membrane trafficking and exocytosis [138]. More interestingly, optineurin was found in cytoplasmic inclusions in one fALS case with an optineurin mutation, as well as in SOD1 fALS and sALS [73]. A study in a retinal-ganglion cell model revealed that the normal endogenous optineurin level is controlled by the proteasome system, while upregulated or mutated optineurin becomes degraded by the autophagic pathway. In this regard, mutated optineurin enhanced autophagy, while rapamycin reduced the rate of apoptosis induced by the mutation [139]. These authors also generated a transgenic mouse model with the optineurin E50K mutation. In these mice, enhanced autophagy was also demonstrated [139]. A recent study however, revealed that optineurin inclusions are not that common as first suggested, with none detected in post-mortem tissues from fALS patients with SOD1 or FUS/TLS mutations [140].

3.7. Vesicle-Associated Membrane Protein-Associated Protein B (VAPB)

VAPB is yet another protein involved in vesicle trafficking, and found to be mutated in ALS cases in 8 Brazilian families and a German pedrigree [141-144]. Although VAPB mutations led to ubiquitin-aggregate formation (141) or cytoplasmic TDP-43 accumulations [145], direct links between VAPB and autophagy have not yet been published. However, since it is involved in vesicle trafficking, like optineurin, dynactin, Fig4 and CHMP2B, VAPB is a likely candidate to be involved in autophagosome transport or fusion (Fig. 1). In addition, VAPB has been shown to be involved in ER stress, in part by modulation of ATF6. Mutant VAPB seems to inhibit ATF6 activation more strongly than the wild-type form [146]. This is of interest, since XBP-1 deficiency, one of ATF6's targets, has been shown to be beneficial for ALS through activation of autophagy, although the exact mechanism was not yet elucidated [84].

3.8. ALS2

So far 13 known mutations in the ALS2/Alsin gene lead to a recessive fALS disorder [115, 147-148]. This gene encodes for a guanine nucleotide-exchange factor for the small GTPase Rab5, which is involved in endosomal trafficking and fusion. In this regard, loss of ALS2 resulted in disturbed endo/lysosomal trafficking in SOD1-H46R mice. Subsequently, more SOD1 aggregates, poly-ubiquitinated proteins and autophagic markers were observed. Loss of ALS2 lowered the lysosome-dependent clearance of LC3 and p62 [149]. Mutant ALS2 appeared to fail in localizing to endosomes, disturbing so the formation of the amphisomes [150] (Fig. 1).

3.9. Fused in Sarcoma/Translocated in Liposarcoma (FUS/TLS)

Like TDP-43, FUS/TLS is an RNA-binding protein that was connected to ~4% fALS and ~1% sALS mutations and a subset of FTD [72, 76, 115, 151-152]. It is a nuclear protein that upon mutation appears as cytoplasmic accumulations. Although no direct links between autophagy and FUS/TLS mutations have been reported, it is a valid assumption that the autophagic clearance of these aggregates could delay motor-neuron loss. However, it should be noted that the toxicity of these FUS/TLS-aggregates is under debate [153].

3.10. Senataxin and Angiogenin

So far, 4 mutations in the senataxin gene (SETX) were found in fALS cases [154-156] and in at least one case of sALS [157]. Also angiogenin was found to be mutated in both fALS and sALS [158]. Both these proteins, like FUS/TLS and TDP-43, play important roles in RNA processing. No autophagic alterations however have so far been reported in senataxin- or angiogenin-related ALS. Cytoplasmic inclusions immunoreactive for ubiquitin, p62 and TDP-43 were however detected in oligodendrocytes and neurons from the central nervous system in an ALS case with an angiogenin mutation [159].

4. Modulating Autophagy Flux in ALS: Therapeutic Opportunities?

4.1. Basal Autophagy Markers Are Increased in ALS, but What Does It Indicate?

In basal conditions, increased levels of autophagic markers are detected in both ALS rodents (Fig. 3) [89-91, 93] and patients (Fig. 4) [84, 92].

The question remains whether these markers reflect enhanced autophagic flux initiation, defective autophagosome transport, defective fusion of autophagosomes with endo/lysosomes or deficient recycling (Fig. 1). While the answer may depend on the genetic mutation underlying the degeneration of the motor neurons, it has very important consequences. Depending on the mechanism, initiating or stimulating autophagy could have therapeutic potential for delaying or preventing motor-neuron death in ALS, but in contrast, it could also have adverse effects due to accelerated motor-neuron degeneration. When autophagic flux is intact but enhanced as a pro-survival response to remove toxic protein aggregates, the motor neurons may become addicted to the enhanced autophagy and therefore chemical stimulation of autophagic flux would be beneficial. However, when the final steps of the autophagy process are defective and arise as a pathological feature, the motor neurons may be poisoned by accumulating autophagosomes and then chemical induction of autophagy would be detrimental for the cells through the accumulation of autophagosomes or autolysosomes. The latter may be the case in ALS associated with SOD1 and Fig4 mutations. These considerations are underpinned by a study of Zhang *et al.* in which rapamycin treatment in a presymptomatic SOD1-G93A mice led to increased autophagic markers, but was neither beneficial nor did help with the clearance of SOD1 aggregates [91]. In fact, rapamycin led to elevated p62 levels in the mutant SOD1 model, suggesting defective autophagic flux. The mechanism underlying defective autophagic flux is not clear, but may involve improper lysosomal function and/or positioning. In this respect, the decreased mTOR activity may point towards dysfunctional lysosomal positioning and thus defective autophagosome-lysosome fusion. Indeed, recent evidence emerged on a proximal role for lysosomes in the autophagy process by regulating mTOR activity and thus influencing the rate of autophagosome-lysosome fusion [160-161]. Similarly, when autophagosome transport or fusion is defective, like in the case of respectively dynactin or CHMP2B mutations, or when cells are unable to recycle the membrane of autolysosomes for the formation of new autophagosomes, like in the case of Fig4 mutations, one may anticipate that chemical induction of autophagy would not be beneficial.

Hence, it is of critical importance to decipher the exact mechanism underlying the increased autophagic markers in the different forms of fALS/sALS to direct future approaches concerning the use of chemical manipulation of autophagic flux as a therapeutic tool. Below, we will critically assess the beneficial effect of decreasing or increasing autophagy with special emphasis to chemical modulation.

4.2. Inhibiting Autophagy Is Definitely Not a Good Idea

Since autophagic flux may be deficient in SOD1 ALS models, leading to the poisonous accumulation of autophagosomes, one may anticipate that reducing the autophagosomal load in the cell by decreasing autophagy could be beneficial. However, it is clear that besides the ubiquitin-dependent protease system, autophagy plays an essential role in the removal of toxic protein aggregates, including mutant SOD1 [83]. Furthermore, recent data pointed out that autophagy may in fact be the prefered pathway for the removal of disease-related protein aggregates linked to neurodegeneration, alleviating their toxicity [41, 162]. Thus, constitutive autophagy seems a *conditio sine qua non* for neuronal homeostasis, health and survival by clearing aggregate-prone proteins and thus preventing aggregate deposition [19, 21]. Hence, it is not surprising that the pharmacological blockade of autophagy is detrimental and causes mutant SOD1-mediated cell death in neurons [83]. Further evidence was provided by SOD1 and TDP-43 fALS mouse models revealing that autophagy inhibition leads to the accumulation of aggregated mutant SOD1 and TDP-43 in inclusions due to protein misfolding [82, 163]. These aggregates can perturb the ubiquitin-proteasome degradation pathway, which normally leads to the induction of autophagy [164-168]. Thus, the absence of compensatory activation of autophagy to remove these misfolded proteins due to chemical autophagy inhibition may contribute to cell death in these ALS models. These observations have been underpinned by other reports demonstrating the essential role of autophagy together with the ubiquitin-proteasome system to remove toxic TDP-43-protein aggregates [86-87]. Not only pharmacological inhibition but also *in vitro* silencing of essential autophagy genes, including *BECN1* and *ATG7*, promote the accumulation of SOD1 [83]. In this respect, in some paradigms of fALS and sALS, motor neurons may become addicted to sustained high levels of autophagy to keep up with the removal of mutant-protein aggregates to avoid their detrimental effects. Hence, one may view the elevated autophagy as a new set point allowing the cell to survive. It is expected that this comes at a cost and renders these cells more vulnerable to stress and less responsive to chemical induction of autophagy as recently observed [91].

4.3. Increasing Autophagy in ALS Models May Be a Good Idea

The strategy to genetically or chemically enhance autophagy as a therapeutic strategy for preventing or slowing down motor-neuron degeneration is far from clear, since the field has been plagued by conflicting reports on the outcome of these interventions. In addition, the different ALS paradigms may display different autophagic alterations, which can result in different efficacies of autophagy-modulating compounds.

4.3.1. Genetic Models Reveal that Autophagy Induction Is Beneficial in ALS Models

Genetic tools allow research to focus on increasing autophagy *in vitro* and *in vivo* to assess its role in ALS models. For instance, co-transfection of the autophagy inducer HspB8 and mutant SOD1 decreases the aggregates induced by SOD1 *in vitro* [82]. This also holds true for TDP-43 when co-transfecting motor neuron-like NSC-43 cells with mutant TDP-43 and HspB8 [82]. This is in accordance with a remarkable report that exploited the fact that deficiency in XBP-1 leads to upregulation of basal autophagic activity [84, 169]. According to these effects, XPB-1 or IRE1α deficiency reduced the aggregation and toxicity of mutant SOD1, largely preventing mutant SOD1-induced cell death in motor neuron-like cells. The mechanism underlying the protective role of XBP-1 deficiency in upregulating autophagy may be attributed to an impairment of ER-associated degradation (ERAD). Strikingly, mutant SOD1 mice deficient for XBP-1 expression in their neurons also lived longer than their mutant SOD1 littermates due to increased autophagy levels, thereby promoting the engulfment of SOD1 aggregates by autophagy in spinal-cord neurons [84] (Fig. 2). Hence, this study provides strong evidence that neuronal-specific induction of autophagy enhances the clearance of mutant SOD1 aggregates *in vivo*, thereby delaying the onset of the disease and extending life span. Compatible with a previous report indicating a gender-selective effect of Beclin 1-dependent autophagy in an Alzheimer disease model [170], it should be noted that the effects of XBP-1 ablation on autophagy were more prominent in females, pointing towards an influence of gender for protective autophagy [84]. Nevertheless, other genetic models that increase autophagic flux would be useful to underpin these findings.

4.3.2. Chemical Induction of Autophagy Provides Complex Outcomes in ALS

From the previous section, one may conclude that the way is open to pharmacologically induce autophagy to remove toxic protein aggregates, delay disease onset and prolong life span of ALS models and patients. While this seems a straightforward approach, experiments in ALS cell models, *in vivo* mouse models and patients demonstrated that this may turn out to be a challenging and complex task [171]. This is likely due to the lack of insight in the changes and defects present in the autophagy pathway in ALS models, which may depend on the particular model, mutation and even gender.

i) Inhibitors of mTOR

Pharmacologically increasing autophagy levels *in vitro* and *in vivo* has been applied in a variety of neurodegenerative disease models and was shown to be beneficial [45, 172-174]. In many studies, the mTORC1 inhibitor rapamycin (or sirolimus) has been shown to induce autophagy in neurons [175] to alleviate the toxicity of aggregate-forming proteins [41] and to promote neuronal survival by enhancing the clearance of mitochondria [176].

In ALS cell models, rapamycin application has been shown to i) decrease mutant-SOD1 expression [83], ii) to restore the nuclear localization of mutant TDP-43 [85], implying that mislocalized cytosolic TDP-43 can be degraded by autophagy, and iii) to decrease the amount of optineurin aggregates and of apoptotic optineurin-expressing cells [139].

Currently, only one study examined the potential of rapamycin-triggered autophagy induction for the treatment of ALS in a mutant-SOD1 fALS mouse model [91]. While rapamycin was able to increase the level of LC3-II and the number of LC3-staining autophagosomes in spinal-cord neurons of the mutant SOD1 mice, it is not clear whether the autophagic flux is still normal in these cells and whether autophagosome-lysosome fusion is still functional [171]. The observation that, in contrast to what is found in lysates of the spinal cords of the wild-type mice, p62 levels do not decline and rather increase upon rapamycin treatment in the mutant SOD1 samples, seems to point towards a defective autophagosome-lysosome fusion. Correlating with this view, the level of SOD1 aggregates in the mutant SOD1 neurons seems to increase upon rapamycin treatment. Remarkably consistent with these findings at the molecular level, the daily administration of rapamycin did not increase life span as expected (which importantly had been shown previously to increase survival of healthy mice [177]), but significantly expedited the onset of the disease and decreased life span. It cannot be excluded that the decrease in survival may be attributed to the potent immunosuppressive actions of rapamycin [178-179], since immunosuppression by lacking T-cells and/or B-cells significantly decreased survival in ALS mice [180-182]. At least in SOD1 mouse models, it therefore remains unclear whether rapamycin causes decreased life span of ALS mice by poisoning the cells with accumulating autophagosomes and/or by immunosuppression. It is interesting to note that novel drugs are being developed to provoke a more complete suppression of mTOR activity, including targeting of both mTORC1 as mTORC2 [183-185] (see Section 1.3.1).

ii) Inhibitors of IP_3 and Ca^{2+} Signaling

IP_3 receptors and Ca^{2+} signaling play an emerging role the control of autophagy [25, 36-37, 70, 186-188]. Basal IP_3-receptor activity seems to be important for mitochondrial bio-energetics, thereby suppressing basal autophagy [35]. Hence, pharmacological inhibition of IP_3 signaling by lithium [37] and of IP_3 receptors by xestospongin B [186] induced autophagy in an mTOR-independent way. This allows targeting autophagy in a manner that avoids the adverse effects of rapamycin on the immune system or at least shows potential for synergistic actions with rapamycin, thereby lowering its dose needed for application in disease treatment [48, 174, 189]. Furthermore, given the long history for some of these molecules like lithium ions in treating epilepsy, the step towards clinical applications in patients has already been taken [190].

In a mutant-SOD1 mouse model, lithium salts delayed the onset of the disease and extended the life span of the mice [42, 191]. These effects were accompanied with increased survival of the motor neurons in size and number and decreased the accumulation and aggregation of a number of proteins, including α-synuclein, ubiquitin and SOD1. Importantly, lithium salt treatment increased the number of healthy mitochondria in the spinal-cord neurons of the mutant-SOD1 mice and thus prevented mitochondrial swelling and subsequent cell death observed in untreated mutant-SOD1 mice. Further evidence was provided for an increase in autophagic markers upon chronic lithium salt administration in the mutant SOD1 mice. However, from these data, it is not clear whether autophagic flux was impaired or not. Finally, a limited clinical trial was set up in which human patients affected by ALS were treated with lithium carbonate and this significantly delayed the progression of ALS in the human patients compared to the control group treated with riluzole. These promising findings were supported by an earlier report from Sugai and co-workers who showed that valproic acid protected spinal-cord motor neurons against death from glutamate excitotoxicity and prolonged disease duration in ALS model mice [192] and by a study from Feng and co-workers who observed a marked delay in the onset of disease symptoms, a prolongation in the life span and a decrease in the neurological deficit scores in SOD1-G93A mouse models. This was obtained by combining both lithium carbonate and valproic acid, compared to the results obtained with a monotreatment with lithium carbonate or valproic acid [193].

Unfortunately, an increasing number of reports could not repeat or observe these beneficial actions of lowering IP_3 signaling in ALS models, thereby advocating against the use of lithium carbonate in clinical trials for the treatment of ALS patients. For instance, Gill and co-workers performed a rigorous survival study that included sibling matching, gender balancing, investigator blinding, and transgene copy-number verification for each experimental subject. They thereby minimized the likelihood of attaining a false positive therapeutic effect, and did not show any lithium efficacy in an SOD1-G93A mouse model for fALS [111]. Similar results were observed by Pizzasegola and co-workers who carefully analyzed two independent mouse strains for SOD1-G93A and found either no or adverse effects of lithium salt administration. In addition, no neuroprotective effect or upregulation of the autophagy marker LC3-II was observed [112]. Similar disappointing findings were reported in a pilot study in ALS patients assessing the ability of lithium carbonate to slow down disease progression. Lithium carbonate in combination with riluzole did not show any beneficial effect compared to the control group treated with riluzole alone [194]. In a trial in Italian centers, lithium carbonate seemed to be less well tolerated by ALS patients even at subtherapeutic doses and did not display any therapeutic efficacy [195]. Similar to lithium carbonate treatments, results obtained with valproic acid also seemed conflicting. While some studies showed an increase in life span of ALS mice treated with valproic acid [192-193], others did not [196-197]. Valproic acid as well has been studied as a treatment in ALS patients, but unfortunately to no avail [198].

Conclusions

While it seems that enhancement of autophagic flux is a promising approach in targeting a plethora of neurodegenerative diseases, its potential in preventing or slowing down motor-neuron degeneration in ALS still needs to be demonstrated. A large number of models demonstrated a clear upregulation of autophagic markers, which may indicate a higher demand for protein quality control in ALS, but may also point to defective autophagosomal transport, defective fusion of autophagosomes with lysosomes or defective recycling of autolysosomal membranes. Under these conditions, stimulating autophagic flux may be detrimental due to polluting the cells with accumulating autophagosomes, expediting disease onset and shortening life span. Clearly, mechanistic insights in the observed altered autophagy in ALS models are still lacking, but are critical for progress towards therapeutical applications. Therefore, future directions should include the assessment of the specific autophagic alterations corresponding to the different causes underlying ALS, in order to predict the therapeutic value of autophagy-modulating compounds.

Acknowledgments

Work performed in the author's laboratory on the topic was supported by grant GOA/09/12 and OT START1/10/044 from the Research Council of the K.U. Leuven, and by grant G073109N from the Research Foundation Flanders (FWO). JPD is recipient of a Ph.D. Fellowship from the Agency for Innovation by Science and Technology (IWT).

References

[1] Ravikumar B, Sarkar S, Davies JE, Futter M, Garcia-Arencibia M, Green-Thompson ZW, et al. Regulation of mammalian autophagy in physiology and pathophysiology. *Physiol. Rev.* 2010;90(4):1383-435.

[2] Lee JY, Yao TP. Quality control autophagy: A joint effort of ubiquitin, protein deacetylase and actin cytoskeleton. *Autophagy.* 2010;6(4):555-7.

[3] Levine B, Kroemer G. Autophagy in the pathogenesis of disease. *Cell.* 2008;132(1):27-42.

[4] Mizushima N, Yoshimori T. How to interpret LC3 immunoblotting. *Autophagy.* 2007;3(6):542-5.

[5] Klionsky DJ, Abeliovich H, Agostinis P, Agrawal DK, Aliev G, Askew DS, et al. Guidelines for the use and interpretation of assays for monitoring autophagy in higher eukaryotes. *Autophagy.* 2008;4(2):151-75.

[6] Mizushima N, Yoshimori T, Levine B. Methods in mammalian autophagy research. *Cell.* 2010;140(3):313-26.

[7] Ichimura Y, Kominami E, Tanaka K, Komatsu M. Selective turnover of p62/A170/SQSTM1 by autophagy. *Autophagy.* 2008;4(8):1063-6.

[8] Komatsu M, Waguri S, Koike M, Sou YS, Ueno T, Hara T, et al. Homeostatic levels of p62 control cytoplasmic inclusion body formation in autophagy-deficient mice. *Cell.* 2007;131(6):1149-63.

[9] Fimia GM, Stoykova A, Romagnoli A, Giunta L, Di Bartolomeo S, Nardacci R, et al. Ambra1 regulates autophagy and development of the nervous system. *Nature.* 2007;447(7148):1121-5.

[10] Gan B, Peng X, Nagy T, Alcaraz A, Gu H, Guan JL. Role of FIP200 in cardiac and liver development and its regulation of TNFalpha and TSC-mTOR signaling pathways. *J. Cell Biol.* 2006;175(1):121-33.

[11] Qu X, Yu J, Bhagat G, Furuya N, Hibshoosh H, Troxel A, et al. Promotion of tumorigenesis by heterozygous disruption of the beclin 1 autophagy gene. *J. Clin. Invest.* 2003;112(12):1809-20.

[12] Yue Z, Jin S, Yang C, Levine AJ, Heintz N. Beclin 1, an autophagy gene essential for early embryonic development, is a haploinsufficient tumor suppressor. *Proc. Natl. Acad. Sci. U. S. A.* 2003;100(25):15077-82.

[13] Komatsu M, Waguri S, Ueno T, Iwata J, Murata S, Tanida I, et al. Impairment of starvation-induced and constitutive autophagy in Atg7-deficient mice. *J. Cell Biol.* 2005;169(3):425-34.

[14] Kuma A, Hatano M, Matsui M, Yamamoto A, Nakaya H, Yoshimori T, et al. The role of autophagy during the early neonatal starvation period. *Nature.* 2004;432(7020):1032-6.

[15] Saitoh T, Fujita N, Hayashi T, Takahara K, Satoh T, Lee H, et al. Atg9a controls dsDNA-driven dynamic translocation of STING and the innate immune response. *Proc. Natl. Acad. Sci. U. S. A.* 2009;106(49):20842-6.

[16] Sou YS, Waguri S, Iwata J, Ueno T, Fujimura T, Hara T, et al. The Atg8 conjugation system is indispensable for proper development of autophagic isolation membranes in mice. *Mol. Biol. Cell.* 2008;19(11):4762-75.

[17] Marino G, Madeo F, Kroemer G. Autophagy for tissue homeostasis and neuroprotection. *Curr. Opin. Cell Biol.* 2011;23(2):198-206.

[18] Hara T, Nakamura K, Matsui M, Yamamoto A, Nakahara Y, Suzuki-Migishima R, et al. Suppression of basal autophagy in neural cells causes neurodegenerative disease in mice. *Nature.* 2006;441(7095):885-9.

[19] Komatsu M, Waguri S, Chiba T, Murata S, Iwata J, Tanida I, et al. Loss of autophagy in the central nervous system causes neurodegeneration in mice. *Nature.* 2006;441(7095):880-4.

[20] Mizushima N, Levine B, Cuervo AM, Klionsky DJ. Autophagy fights disease through cellular self-digestion. *Nature.* 2008;451(7182):1069-75.

[21] Komatsu M, Ueno T, Waguri S, Uchiyama Y, Kominami E, Tanaka K. Constitutive autophagy: vital role in clearance of unfavorable proteins in neurons. *Cell Death Differ.* 2007;14(5):887-94.

[22] Nixon RA, Yang DS, Lee JH. Neurodegenerative lysosomal disorders: a continuum from development to late age. *Autophagy.* 2008;4(5):590-9.

[23] Komatsu M, Wang QJ, Holstein GR, Friedrich VL, Jr., Iwata J, Kominami E, et al. Essential role for autophagy protein Atg7 in the maintenance of axonal homeostasis and the prevention of axonal degeneration. *Proc. Natl. Acad. Sci. U. S. A.* 2007;104(36):14489-94.

[24] Rowland AM, Richmond JE, Olsen JG, Hall DH, Bamber BA. Presynaptic terminals independently regulate synaptic clustering and autophagy of $GABA_A$ receptors in Caenorhabditis elegans. *J. Neurosci.* 2006;26(6):1711-20.

[25] Decuypere JP, Monaco G, Missiaen L, De Smedt H, Parys JB, Bultynck G. IP_3 Receptors, Mitochondria, and Ca^{2+} Signaling: Implications for Aging. *J. Aging Res.* 2011;2011:920178.

[26] Vellai T, Takacs-Vellai K. Regulation of protein turnover by longevity pathways. *Adv. Exp. Med. Biol.* 2010;694:69-80.

[27] Nyfeler B, Bergman P, Triantafellow E, Wilson CJ, Zhu Y, Radetich B, et al. Relieving autophagy and 4EBP1 from rapamycin resistance. *Mol. Cell Biol.* 2011;31(14):2867-76.

[28] Thoreen CC, Kang SA, Chang JW, Liu Q, Zhang J, Gao Y, et al. An ATP-competitive mammalian target of rapamycin inhibitor reveals rapamycin-resistant functions of mTORC1. *J. Biol. Chem.* 2009;284(12):8023-32.

[29] Thoreen CC, Sabatini DM. Rapamycin inhibits mTORC1, but not completely. *Autophagy.* 2009;5(5):725-6.

[30] Baumann G, Zenke G, Wenger R, Hiestand P, Quesniaux V, Andersen E, et al. Molecular mechanisms of immunosuppression. *J. Autoimmun.* 1992;5 Suppl A:67-72.

[31] Mota A. Sirolimus: a new option in transplantation. *Expert Opin. Pharmacother.* 2005;6(3):479-87.

[32] Cravedi P, Ruggenenti P, Remuzzi G. Sirolimus for calcineurin inhibitors in organ transplantation: contra. *Kidney Int.* 2010;78(11):1068-74.

[33] Laane E, Tamm KP, Buentke E, Ito K, Kharaziha P, Oscarsson J, et al. Cell death induced by dexamethasone in lymphoid leukemia is mediated through initiation of autophagy. *Cell Death Differ.* 2009;16(7):1018-29.

[34] Miller S, Oleksy A, Perisic O, Williams RL. Finding a fitting shoe for Cinderella: searching for an autophagy inhibitor. *Autophagy.* 2010;6(6):805-7.

[35] Cárdenas C, Miller RA, Smith I, Bui T, Molgo J, Muller M, et al. Essential regulation of cell bioenergetics by constitutive $InsP_3$ receptor Ca^{2+} transfer to mitochondria. *Cell.* 2010;142(2):270-83.

[36] Decuypere JP, Bultynck G, Parys JB. A dual role for Ca^{2+} in autophagy regulation. *Cell Calcium.* 2011;50(3):242-50.

[37] Sarkar S, Floto RA, Berger Z, Imarisio S, Cordenier A, Pasco M, et al. Lithium induces autophagy by inhibiting inositol monophosphatase. *J. Cell Biol.* 2005;170(7):1101-11.

[38] Williams A, Sarkar S, Cuddon P, Ttofi EK, Saiki S, Siddiqi FH, et al. Novel targets for Huntington's disease in an mTOR-independent autophagy pathway. *Nat. Chem. Biol.* 2008;4(5):295-305.

[39] Sarkar S, Davies JE, Huang Z, Tunnacliffe A, Rubinsztein DC. Trehalose, a novel mTOR-independent autophagy enhancer, accelerates the clearance of mutant huntingtin and alpha-synuclein. *J. Biol. Chem.* 2007;282(8):5641-52.

[40] Sarkar S, Ravikumar B, Rubinsztein DC. Autophagic clearance of aggregate-prone proteins associated with neurodegeneration. *Methods Enzymol.* 2009;453:83-110.

[41] Berger Z, Ravikumar B, Menzies FM, Oroz LG, Underwood BR, Pangalos MN, et al. Rapamycin alleviates toxicity of different aggregate-prone proteins. *Hum. Mol. Genet.* 2006;15(3):433-42.

[42] Fornai F, Longone P, Cafaro L, Kastsiuchenka O, Ferrucci M, Manca ML, et al. Lithium delays progression of amyotrophic lateral sclerosis. *Proc. Natl. Acad. Sci. U. S. A.* 2008;105(6):2052-7.

[43] Ravikumar B, Vacher C, Berger Z, Davies JE, Luo S, Oroz LG, et al. Inhibition of mTOR induces autophagy and reduces toxicity of polyglutamine expansions in fly and mouse models of Huntington disease. *Nat. Genet.* 2004;36(6):585-95.

[44] Rodriguez-Navarro JA, Rodriguez L, Casarejos MJ, Solano RM, Gomez A, Perucho J, et al. Trehalose ameliorates dopaminergic and tau pathology in parkin deleted/tau overexpressing mice through autophagy activation. *Neurobiol. Dis.* 2010;39(3):423-38.

[45] Winslow AR, Rubinsztein DC. Autophagy in neurodegeneration and development. *Biochim. Biophys. Acta.* 2008;1782(12):723-9.

[46] Jope RS. Lithium and GSK-3: one inhibitor, two inhibitory actions, multiple outcomes. *Trends Pharmacol. Sci.* 2003;24(9):441-3.

[47] Stambolic V, Ruel L, Woodgett JR. Lithium inhibits glycogen synthase kinase-3 activity and mimics wingless signalling in intact cells. *Curr. Biol.* 1996;6(12):1664-8.

[48] Sarkar S, Krishna G, Imarisio S, Saiki S, O'Kane CJ, Rubinsztein DC. A rational mechanism for combination treatment of Huntington's disease using lithium and rapamycin. *Hum. Mol. Genet.* 2008;17(2):170-8.

[49] Raynaud FI, Eccles S, Clarke PA, Hayes A, Nutley B, Alix S, et al. Pharmacologic characterization of a potent inhibitor of class I phosphatidylinositide 3-kinases. *Cancer Res.* 2007;67(12):5840-50.

[50] Degtyarev M, De Maziere A, Orr C, Lin J, Lee BB, Tien JY, et al. Akt inhibition promotes autophagy and sensitizes PTEN-null tumors to lysosomotropic agents. *J. Cell Biol.* 2008;183(1):101-16.

[51] Logroscino G, Traynor BJ, Hardiman O, Chio A, Mitchell D, Swingler RJ, et al. Incidence of amyotrophic lateral sclerosis in Europe. *J. Neurol. Neurosurg. Psychiatry.* 2009;81(4):385-90.

[52] Kotchoubey B, Lang S, Winter S, Birbaumer N. Cognitive processing in completely paralyzed patients with amyotrophic lateral sclerosis. *Eur. J. Neurol.* 2003;10(5):551-8.

[53] Bensimon G, Lacomblez L, Meininger V. A controlled trial of riluzole in amyotrophic lateral sclerosis. ALS/Riluzole Study Group. *N. Engl. J. Med.* 1994;330(9):585-91.

[54] Turner BJ, Talbot K. Transgenics, toxicity and therapeutics in rodent models of mutant SOD1-mediated familial ALS. *Prog. Neurobiol.* 2008;85(1):94-134.

[55] Watanabe Y, Yasui K, Nakano T, Doi K, Fukada Y, Kitayama M, et al. Mouse motor neuron disease caused by truncated SOD1 with or without C-terminal modification. *Brain Res. Mol. Brain Res.* 2005;135(1-2):12-20.

[56] Gurney ME, Pu H, Chiu AY, Dal Canto MC, Polchow CY, Alexander DD, et al. Motor neuron degeneration in mice that express a human Cu,Zn superoxide dismutase mutation. *Science.* 1994;264(5166):1772-5.

[57] Reaume AG, Elliott JL, Hoffman EK, Kowall NW, Ferrante RJ, Siwek DF, et al. Motor neurons in Cu/Zn superoxide dismutase-deficient mice develop normally but exhibit enhanced cell death after axonal injury. *Nat. Genet.* 1996;13(1):43-7.

[58] Shefner JM, Reaume AG, Flood DG, Scott RW, Kowall NW, Ferrante RJ, et al. Mice lacking cytosolic copper/zinc superoxide dismutase display a distinctive motor axonopathy. *Neurology.* 1999;53(6):1239-46.

[59] Bruijn LI, Miller TM, Cleveland DW. Unraveling the mechanisms involved in motor neuron degeneration in ALS. *Annu. Rev. Neurosci.* 2004;27:723-49.

[60] Boillee S, Vande Velde C, Cleveland DW. ALS: a disease of motor neurons and their nonneuronal neighbors. *Neuron.* 2006;52(1):39-59.

[61] Yamanaka K, Chun SJ, Boillee S, Fujimori-Tonou N, Yamashita H, Gutmann DH, et al. Astrocytes as determinants of disease progression in inherited amyotrophic lateral sclerosis. *Nat. Neurosci.* 2008;11(3):251-3.

[62] Boillee S, Yamanaka K, Lobsiger CS, Copeland NG, Jenkins NA, Kassiotis G, et al. Onset and progression in inherited ALS determined by motor neurons and microglia. *Science.* 2006;312(5778):1389-92.

[63] Lobsiger CS, Boillee S, McAlonis-Downes M, Khan AM, Feltri ML, Yamanaka K, et al. Schwann cells expressing dismutase active mutant SOD1 unexpectedly slow disease progression in ALS mice. *Proc. Natl. Acad. Sci. U. S. A.* 2009;106(11):4465-70.

[64] Jaarsma D, Teuling E, Haasdijk ED, De Zeeuw CI, Hoogenraad CC. Neuron-specific expression of mutant superoxide dismutase is sufficient to induce amyotrophic lateral sclerosis in transgenic mice. *J. Neurosci.* 2008;28(9):2075-88.

[65] Van Damme P, Dewil M, Robberecht W, Van Den Bosch L. Excitotoxicity and amyotrophic lateral sclerosis. *Neurodegener. Dis.* 2005;2(3-4):147-59.

[66] Mekahli D, Bultynck G, Parys JB, De Smedt H, Missiaen L. Endoplasmic-Reticulum Calcium Depletion and Disease. *Cold Spring Harb. Perspect. Biol.* 2011;3(6):pii:a004317. doi:10.1101/cshperspect. a004317.

[67] Sammels E, Parys JB, Missiaen L, De Smedt H, Bultynck G. Intracellular Ca^{2+} storage in health and disease: a dynamic equilibrium. *Cell Calcium.* 2010;47(4):297-314.

[68] Van Den Bosch L, Van Damme P, Bogaert E, Robberecht W. The role of excitotoxicity in the pathogenesis of amyotrophic lateral sclerosis. *Biochim. Biophys. Acta.* 2006;1762(11-12):1068-82.

[69] Grosskreutz J, Van Den Bosch L, Keller BU. Calcium dysregulation in amyotrophic lateral sclerosis. *Cell Calcium.* 2010;47(2):165-74.

[70] Decuypere JP, Monaco G, Bultynck G, Missiaen L, De Smedt H, Parys JB. The IP_3 receptor-mitochondria connection in apoptosis and autophagy. *Biochim. Biophys. Acta.* 2011;1813(5):1003-13.

[71] Del Bo R, Ghezzi S, Corti S, Pandolfo M, Ranieri M, Santoro D, et al. TARDBP (TDP-43) sequence analysis in patients with familial and sporadic ALS: identification of two novel mutations. *Eur. J. Neurol.* 2009;16(6):727-32.

[72] Kwiatkowski TJ, Jr., Bosco DA, Leclerc AL, Tamrazian E, Vanderburg CR, Russ C, et al. Mutations in the FUS/TLS gene on chromosome 16 cause familial amyotrophic lateral sclerosis. *Science.* 2009;323(5918):1205-8.

[73] Maruyama H, Morino H, Ito H, Izumi Y, Kato H, Watanabe Y, et al. Mutations of optineurin in amyotrophic lateral sclerosis. *Nature.* 2010;465(7295):223-6.

[74] Rutherford NJ, Zhang YJ, Baker M, Gass JM, Finch NA, Xu YF, et al. Novel mutations in TARDBP (TDP-43) in patients with familial amyotrophic lateral sclerosis. *PLoS Genet.* 2008;4(9):e1000193.

[75] Van Deerlin VM, Leverenz JB, Bekris LM, Bird TD, Yuan W, Elman LB, et al. TARDBP mutations in amyotrophic lateral sclerosis with TDP-43 neuropathology: a genetic and histopathological analysis. *Lancet Neurol.* 2008;7(5):409-16.

[76] Vance C, Rogelj B, Hortobagyi T, De Vos KJ, Nishimura AL, Sreedharan J, et al. Mutations in FUS, an RNA processing protein, cause familial amyotrophic lateral sclerosis type 6. *Science.* 2009;323(5918):1208-11.

[77] Forsberg K, Andersen PM, Marklund SL, Brannstrom T. Glial nuclear aggregates of superoxide dismutase-1 are regularly present in patients with amyotrophic lateral sclerosis. *Acta Neuropathol.* 2011;121(5):623-34.

[78] Baumer D, Hilton D, Paine SM, Turner MR, Lowe J, Talbot K, et al. Juvenile ALS with basophilic inclusions is a FUS proteinopathy with FUS mutations. *Neurology.* 2010;75(7):611-8.

[79] Huang EJ, Zhang J, Geser F, Trojanowski JQ, Strober JB, Dickson DW, et al. Extensive FUS-immunoreactive pathology in juvenile amyotrophic lateral sclerosis with basophilic inclusions. *Brain Pathol.* 2010;20(6):1069-76.
[80] Sasaki S, Takeda T, Shibata N, Kobayashi M. Alterations in subcellular localization of TDP-43 immunoreactivity in the anterior horns in sporadic amyotrophic lateral sclerosis. *Neurosci. Lett.* 2010;478(2):72-6.
[81] Bergemalm D, Forsberg K, Srivastava V, Graffmo KS, Andersen PM, Brannstrom T, et al. Superoxide dismutase-1 and other proteins in inclusions from transgenic amyotrophic lateral sclerosis model mice. *J. Neurochem.* 2010;114(2):408-18.
[82] Crippa V, Sau D, Rusmini P, Boncoraglio A, Onesto E, Bolzoni E, et al. The small heat shock protein B8 (HspB8) promotes autophagic removal of misfolded proteins involved in amyotrophic lateral sclerosis (ALS). *Hum. Mol. Genet.* 2010;19(17):3440-56.
[83] Kabuta T, Suzuki Y, Wada K. Degradation of amyotrophic lateral sclerosis-linked mutant Cu,Zn-superoxide dismutase proteins by macroautophagy and the proteasome. *J. Biol. Chem.* 2006;281(41):30524-33.
[84] Hetz C, Thielen P, Matus S, Nassif M, Court F, Kiffin R, et al. XBP-1 deficiency in the nervous system protects against amyotrophic lateral sclerosis by increasing autophagy. *Genes Dev.* 2009;23(19):2294-306.
[85] Caccamo A, Majumder S, Deng JJ, Bai Y, Thornton FB, Oddo S. Rapamycin rescues TDP-43 mislocalization and the associated low molecular mass neurofilament instability. *J. Biol. Chem.* 2009;284(40):27416-24.
[86] Urushitani M, Sato T, Bamba H, Hisa Y, Tooyama I. Synergistic effect between proteasome and autophagosome in the clearance of polyubiquitinated TDP-43. *J. Neurosci. Res.* 2010;88(4):784-97.
[87] Wang X, Fan H, Ying Z, Li B, Wang H, Wang G. Degradation of TDP-43 and its pathogenic form by autophagy and the ubiquitin-proteasome system. *Neurosci. Lett.* 2010;469(1):112-6.
[88] Cheroni C, Marino M, Tortarolo M, Veglianese P, De Biasi S, Fontana E, et al. Functional alterations of the ubiquitin-proteasome system in motor neurons of a mouse model of familial amyotrophic lateral sclerosis. *Hum. Mol. Genet.* 2009;18(1):82-96.
[89] Li L, Zhang X, Le W. Altered macroautophagy in the spinal cord of SOD1 mutant mice. *Autophagy.* 2008;4(3):290-3.
[90] Morimoto N, Nagai M, Ohta Y, Miyazaki K, Kurata T, Morimoto M, et al. Increased autophagy in transgenic mice with a G93A mutant SOD1
[91] Zhang X, Li L, Chen S, Yang D, Wang Y, Wang Z, et al. Rapamycin treatment augments motor neuron degeneration in SOD1 (G93A) mouse model of amyotrophic lateral sclerosis. *Autophagy.* 2011;7(4):412-25.
[92] Sasaki S. Autophagy in spinal cord motor neurons in sporadic amyotrophic lateral sclerosis. *J. Neuropathol. Exp. Neurol.* 2011;70(5):349-59.
[93] Tian F, Morimoto N, Liu W, Ohta Y, Deguchi K, Miyazaki K, et al. In vivo optical imaging of motor neuron autophagy in a mouse model of amyotrophic lateral sclerosis. *Autophagy.* 2011;7(9):985-92.
[94] Chow CY, Landers JE, Bergren SK, Sapp PC, Grant AE, Jones JM, et al. Deleterious variants of FIG4, a phosphoinositide phosphatase, in patients with ALS. *Am. J. Hum. Genet.* 2009;84(1):85-8.

[95] Munch C, Sedlmeier R, Meyer T, Homberg V, Sperfeld AD, Kurt A, et al. Point mutations of the p150 subunit of dynactin (DCTN1) gene in ALS. *Neurology.* 2004;63(4):724-6.

[96] Del Bo R, Tiloca C, Pensato V, Corrado L, Ratti A, Ticozzi N, et al. Novel optineurin mutations in patients with familial and sporadic amyotrophic lateral sclerosis. *J. Neurol. Neurosurg. Psychiatry.* 2011;82(11):1239-43.

[97] Rosen DR, Siddique T, Patterson D, Figlewicz DA, Sapp P, Hentati A, et al. Mutations in Cu/Zn superoxide dismutase gene are associated with familial amyotrophic lateral sclerosis. *Nature.* 1993;362(6415):59-62.

[98] Cheroni C, Peviani M, Cascio P, Debiasi S, Monti C, Bendotti C. Accumulation of human SOD1 and ubiquitinated deposits in the spinal cord of SOD1G93A mice during motor neuron disease progression correlates with a decrease of proteasome. *Neurobiol. Dis.* 2005;18(3):509-22.

[99] Kabashi E, Agar JN, Taylor DM, Minotti S, Durham HD. Focal dysfunction of the proteasome: a pathogenic factor in a mouse model of amyotrophic lateral sclerosis. *J. Neurochem.* 2004;89(6):1325-35.

[100] Urushitani M, Kurisu J, Tsukita K, Takahashi R. Proteasomal inhibition by misfolded mutant superoxide dismutase 1 induces selective motor neuron death in familial amyotrophic lateral sclerosis. *J. Neurochem.* 2002;83(5):1030-42.

[101] Bruijn LI, Becher MW, Lee MK, Anderson KL, Jenkins NA, Copeland NG, et al. ALS-linked SOD1 mutant G85R mediates damage to astrocytes and promotes rapidly progressive disease with SOD1-containing inclusions. *Neuron.* 1997;18(2):327-38.

[102] Gal J, Strom AL, Kilty R, Zhang F, Zhu H. p62 accumulates and enhances aggregate formation in model systems of familial amyotrophic lateral sclerosis. *J. Biol. Chem.* 2007;282(15):11068-77.

[103] Pasinelli P, Belford ME, Lennon N, Bacskai BJ, Hyman BT, Trotti D, et al. Amyotrophic lateral sclerosis-associated SOD1 mutant proteins bind and aggregate with Bcl-2 in spinal cord mitochondria. *Neuron.* 2004;43(1):19-30.

[104] Pattingre S, Tassa A, Qu X, Garuti R, Liang XH, Mizushima N, et al. Bcl-2 antiapoptotic proteins inhibit Beclin 1-dependent autophagy. *Cell.* 2005;122(6):927-39.

[105] Liu J, Lillo C, Jonsson PA, Vande Velde C, Ward CM, Miller TM, et al. Toxicity of familial ALS-linked SOD1 mutants from selective recruitment to spinal mitochondria. *Neuron.* 2004;43(1):5-17.

[106] Vande Velde C, Miller TM, Cashman NR, Cleveland DW. Selective association of misfolded ALS-linked mutant SOD1 with the cytoplasmic face of mitochondria. *Proc. Natl. Acad. Sci. U. S. A.* 2008;105(10):4022-7.

[107] Wong PC, Pardo CA, Borchelt DR, Lee MK, Copeland NG, Jenkins NA, et al. An adverse property of a familial ALS-linked SOD1 mutation causes motor neuron disease characterized by vacuolar degeneration of mitochondria. *Neuron.* 1995;14(6):1105-16.

[108] Ligon LA, LaMonte BH, Wallace KE, Weber N, Kalb RG, Holzbaur EL. Mutant superoxide dismutase disrupts cytoplasmic dynein in motor neurons. *Neuroreport.* 2005;16(6):533-6.

[109] Bilsland LG, Sahai E, Kelly G, Golding M, Greensmith L, Schiavo G. Deficits in axonal transport precede ALS symptoms in vivo. *Proc. Natl. Acad. Sci. U. S. A.* 2010;107(47):20523-8.

[110] Gomes C, Escrevente C, Costa J. Mutant superoxide dismutase 1 overexpression in NSC-34 cells: effect of trehalose on aggregation, TDP-43 localization and levels of co-expressed glycoproteins. *Neurosci Lett.* 2010;475(3):145-9.

[111] Gill A, Kidd J, Vieira F, Thompson K, Perrin S. No benefit from chronic lithium dosing in a sibling-matched, gender balanced, investigator-blinded trial using a standard mouse model of familial ALS. *PLoS One.* 2009;4(8):e6489.

[112] Pizzasegola C, Caron I, Daleno C, Ronchi A, Minoia C, Carri MT, et al. Treatment with lithium carbonate does not improve disease progression in two different strains of SOD1 mutant mice. *Amyotroph Lateral Scler.* 2009;10(4):221-8.

[113] Onesto E, Rusmini P, Crippa V, Ferri N, Zito A, Galbiati M, et al. Muscle cells and motoneurons differentially remove mutant SOD1 causing familial amyotrophic lateral sclerosis. *J. Neurochem.* 2011;118(2):266-80.

[114] Sreedharan J, Blair IP, Tripathi VB, Hu X, Vance C, Rogelj B, et al. TDP-43 mutations in familial and sporadic amyotrophic lateral sclerosis. *Science.* 2008;319(5870):1668-72.

[115] Ticozzi N, Tiloca C, Morelli C, Colombrita C, Poletti B, Doretti A, et al. Genetics of familial Amyotrophic lateral sclerosis. *Arch. Ital. Biol.* 2011;149(1):65-82.

[116] Warraich ST, Yang S, Nicholson GA, Blair IP. TDP-43: a DNA and RNA binding protein with roles in neurodegenerative diseases. *Int. J. Biochem. Cell Biol.* 2010;42(10):1606-9.

[117] Mackenzie IR, Bigio EH, Ince PG, Geser F, Neumann M, Cairns NJ, et al. Pathological TDP-43 distinguishes sporadic amyotrophic lateral sclerosis from amyotrophic lateral sclerosis with SOD1 mutations. *Ann. Neurol.* 2007;61(5):427-34.

[118] Brady OA, Meng P, Zheng Y, Mao Y, Hu F. Regulation of TDP-43 aggregation by phosphorylation and p62/SQSTM1. *J. Neurochem.* 2011;116(2):248-59.

[119] Robertson J, Sanelli T, Xiao S, Yang W, Horne P, Hammond R, et al. Lack of TDP-43 abnormalities in mutant SOD1 transgenic mice shows disparity with ALS. *Neurosci. Lett.* 2007;420(2):128-32.

[120] Wegorzewska I, Bell S, Cairns NJ, Miller TM, Baloh RH. TDP-43 mutant transgenic mice develop features of ALS and frontotemporal lobar degeneration. *Proc. Natl. Acad. Sci. U. S. A.* 2009;106(44):18809-14.

[121] Kim SH, Shi Y, Hanson KA, Williams LM, Sakasai R, Bowler MJ, et al. Potentiation of amyotrophic lateral sclerosis (ALS)-associated TDP-43 aggregation by the proteasome-targeting factor, ubiquilin 1. *J. Biol. Chem.* 2009;284(12):8083-92.

[122] Cox LE, Ferraiuolo L, Goodall EF, Heath PR, Higginbottom A, Mortiboys H, et al. Mutations in CHMP2B in lower motor neuron predominant amyotrophic lateral sclerosis (ALS). *PLoS One.* 2010;5(3):e9872.

[123] Parkinson N, Ince PG, Smith MO, Highley R, Skibinski G, Andersen PM, et al. ALS phenotypes with mutations in CHMP2B (charged multivesicular body protein 2B). *Neurology.* 2006;67(6):1074-7.

[124] Skibinski G, Parkinson NJ, Brown JM, Chakrabarti L, Lloyd SL, Hummerich H, et al. Mutations in the endosomal ESCRTIII-complex subunit CHMP2B in frontotemporal dementia. *Nat. Genet.* 2005;37(8):806-8.

[125] Filimonenko M, Stuffers S, Raiborg C, Yamamoto A, Malerod L, Fisher EM, et al. Functional multivesicular bodies are required for autophagic clearance of protein

aggregates associated with neurodegenerative disease. *J. Cell Biol.* 2007;179(3):485-500.

[126] Lee JA, Gao FB. Inhibition of autophagy induction delays neuronal cell loss caused by dysfunctional ESCRT-III in frontotemporal dementia. *J. Neurosci.* 2009;29(26):8506-11.

[127] Duex JE, Tang F, Weisman LS. The Vac14p-Fig4p complex acts independently of Vac7p and couples PI(3,5)P_2 synthesis and turnover. *J. Cell Biol.* 2006;172(5):693-704.

[128] Sbrissa D, Ikonomov OC, Fu Z, Ijuin T, Gruenberg J, Takenawa T, et al. Core protein machinery for mammalian phosphatidylinositol 3,5-bisphosphate synthesis and turnover that regulates the progression of endosomal transport. Novel Sac phosphatase joins the ArPIKfyve-PIKfyve complex. *J. Biol. Chem.* 2007;282(33):23878-91.

[129] Ferguson CJ, Lenk GM, Meisler MH. Defective autophagy in neurons and astrocytes from mice deficient in PI(3,5)P2. *Hum. Mol. Genet.* 2009;18(24):4868-78.

[130] Chow CY, Zhang Y, Dowling JJ, Jin N, Adamska M, Shiga K, et al. Mutation of FIG4 causes neurodegeneration in the pale tremor mouse and patients with CMT4J. *Nature.* 2007;448(7149):68-72.

[131] Jin N, Chow CY, Liu L, Zolov SN, Bronson R, Davisson M, et al. VAC14 nucleates a protein complex essential for the acute interconversion of PI3P and PI(3,5)P_2 in yeast and mouse. *EMBO J.* 2008;27(24):3221-34.

[132] Rusten TE, Vaccari T, Lindmo K, Rodahl LM, Nezis IP, Sem-Jacobsen C, et al. ESCRTs and Fab1 regulate distinct steps of autophagy. *Curr. Biol.* 2007;17(20):1817-25.

[133] Munch C, Rosenbohm A, Sperfeld AD, Uttner I, Reske S, Krause BJ, et al. Heterozygous R1101K mutation of the DCTN1 gene in a family with ALS and FTD. *Ann. Neurol.* 2005;58(5):777-80.

[134] Laird FM, Farah MH, Ackerley S, Hoke A, Maragakis N, Rothstein JD, et al. Motor neuron disease occurring in a mutant dynactin mouse model is characterized by defects in vesicular trafficking. *J. Neurosci.* 2008;28(9):1997-2005.

[135] LaMonte BH, Wallace KE, Holloway BA, Shelly SS, Ascano J, Tokito M, et al. Disruption of dynein/dynactin inhibits axonal transport in motor neurons causing late-onset progressive degeneration. *Neuron.* 2002;34:715-27.

[136] Teuling E, van Dis V, Wulf PS, Haasdijk ED, Akhmanova A, Hoogenraad CC, et al. A novel mouse model with impaired dynein/dynactin function develops amyotrophic lateral sclerosis (ALS)-like features in motor neurons and improves lifespan in SOD1-ALS mice. *Hum. Mol. Genet.* 2008;17(18):2849-62.

[137] Zhu G, Wu CJ, Zhao Y, Ashwell JD. Optineurin negatively regulates TNFalpha-induced NF-kappaB activation by competing with NEMO for ubiquitinated RIP. *Curr. Biol.* 2007;17(16):1438-43.

[138] Sahlender DA, Roberts RC, Arden SD, Spudich G, Taylor MJ, Luzio JP, et al. Optineurin links myosin VI to the Golgi complex and is involved in Golgi organization and exocytosis. *J. Cell Biol.* 2005;169(2):285-95.

[139] Shen X, Ying H, Qiu Y, Park JS, Shyam R, Chi ZL, et al. Processing of optineurin in neuronal cells. *J. Biol. Chem.* 2011;286(5):3618-29.

[140] Hortobagyi T, Troakes C, Nishimura AL, Vance C, van Swieten JC, Seelaar H, et al. Optineurin inclusions occur in a minority of TDP-43 positive ALS and FTLD-TDP

cases and are rarely observed in other neurodegenerative disorders. *Acta Neuropathol.* 2011;121(4):519-27.

[141] Chen HJ, Anagnostou G, Chai A, Withers J, Morris A, Adhikaree J, et al. Characterization of the properties of a novel mutation in VAPB in familial amyotrophic lateral sclerosis. *J. Biol. Chem.* 2010;285(51):40266-81.

[142] Landers JE, Leclerc AL, Shi L, Virkud A, Cho T, Maxwell MM, et al. New VAPB deletion variant and exclusion of VAPB mutations in familial ALS. *Neurology.* 2008;70(14):1179-85.

[143] Nishimura AL, Mitne-Neto M, Silva HC, Richieri-Costa A, Middleton S, Cascio D, et al. A mutation in the vesicle-trafficking protein VAPB causes late-onset spinal muscular atrophy and amyotrophic lateral sclerosis. *Am. J. Hum. Genet.* 2004;75(5):822-31.

[144] Funke AD, Esser M, Kruttgen A, Weis J, Mitne-Neto M, Lazar M, et al. The p.P56S mutation in the VAPB gene is not due to a single founder: the first European case. *Clin. Genet.* 2010;77(3):302-3.

[145] Tudor EL, Galtrey CM, Perkinton MS, Lau KF, De Vos KJ, Mitchell JC, et al. Amyotrophic lateral sclerosis mutant vesicle-associated membrane protein-associated protein-B transgenic mice develop TAR-DNA-binding protein-43 pathology. *Neuroscience.* 2010;167(3):774-85.

[146] Gkogkas C, Middleton S, Kremer AM, Wardrope C, Hannah M, Gillingwater TH, et al. VAPB interacts with and modulates the activity of ATF6. *Hum. Mol. Genet.* 2008;17(11):1517-26.

[147] Hentati A, Bejaoui K, Pericak-Vance MA, Hentati F, Speer MC, Hung WY, et al. Linkage of recessive familial amyotrophic lateral sclerosis to chromosome 2q33-q35. *Nat. Genet.* 1994;7(3):425-8.

[148] Hosler BA, Sapp PC, Berger R, O'Neill G, Bejaoui K, Hamida MB, et al. Refined mapping and characterization of the recessive familial amyotrophic lateral sclerosis locus (ALS2) on chromosome 2q33. *Neurogenetics.* 1998;2(1):34-42.

[149] Hadano S, Otomo A, Kunita R, Suzuki-Utsunomiya K, Akatsuka A, Koike M, et al. Loss of ALS2/Alsin exacerbates motor dysfunction in a SOD1-expressing mouse ALS model by disturbing endolysosomal trafficking. *PLoS One.* 2010;5(3):e9805.

[150] Otomo A, Kunita R, Suzuki-Utsunomiya K, Ikeda JE, Hadano S. Defective relocalization of ALS2/alsin missense mutants to Rac1-induced macropinosomes accounts for loss of their cellular function and leads to disturbed amphisome formation. *FEBS Lett.* 2011;585(5):730-6.

[151] Broustal O, Camuzat A, Guillot-Noel L, Guy N, Millecamps S, Deffond D, et al. FUS mutations in frontotemporal lobar degeneration with amyotrophic lateral sclerosis. *J. Alzheimers Dis.* 2010;22(3):765-9.

[152] Mackenzie IR, Rademakers R, Neumann M. TDP-43 and FUS in amyotrophic lateral sclerosis and frontotemporal dementia. *Lancet Neurol.* 2010;9(10):995-1007.

[153] Sun Z, Diaz Z, Fang X, Hart MP, Chesi A, Shorter J, et al. Molecular determinants and genetic modifiers of aggregation and toxicity for the ALS disease protein FUS/TLS. *PLoS Biol.* 2011;9(4):e1000614.

[154] Avemaria F, Lunetta C, Tarlarini C, Mosca L, Maestri E, Marocchi A, et al. Mutation in the senataxin gene found in a patient affected by familial ALS with juvenile onset and slow progression. *Amyotroph Lateral Scler.* 2011;12(3):228-30.

[155] Chen YZ, Bennett CL, Huynh HM, Blair IP, Puls I, Irobi J, et al. DNA/RNA helicase gene mutations in a form of juvenile amyotrophic lateral sclerosis (ALS4). *Am. J. Hum. Genet.* 2004;74(6):1128-35.

[156] Hirano M, Quinzii CM, Mitsumoto H, Hays AP, Roberts JK, Richard P, et al. Senataxin mutations and amyotrophic lateral sclerosis. *Amyotroph Lateral Scler.* 2011;12(3):223-7.

[157] Zhao ZH, Chen WZ, Wu ZY, Wang N, Zhao GX, Chen WJ, et al. A novel mutation in the senataxin gene identified in a Chinese patient with sporadic amyotrophic lateral sclerosis. *Amyotroph Lateral Scler.* 2009;10(2):118-22.

[158] Greenway MJ, Andersen PM, Russ C, Ennis S, Cashman S, Donaghy C, et al. ANG mutations segregate with familial and 'sporadic' amyotrophic lateral sclerosis. *Nat. Genet.* 2006;38(4):411-3.

[159] Seilhean D, Cazeneuve C, Thuries V, Russaouen O, Millecamps S, Salachas F, et al. Accumulation of TDP-43 and alpha-actin in an amyotrophic lateral sclerosis patient with the K17I ANG mutation. *Acta Neuropathol.* 2009;118(4):561-73.

[160] Korolchuk VI, Rubinsztein DC. Regulation of autophagy by lysosomal positioning. *Autophagy.* 2011;7(8):927-8.

[161] Korolchuk VI, Saiki S, Lichtenberg M, Siddiqi FH, Roberts EA, Imarisio S, et al. Lysosomal positioning coordinates cellular nutrient responses. *Nat. Cell Biol.* 2011;13(4):453-60.

[162] Rubinsztein DC. The roles of intracellular protein-degradation pathways in neurodegeneration. Nature. 2006;443(7113):780-6.

[163] Crippa V, Carra S, Rusmini P, Sau D, Bolzoni E, Bendotti C, et al. A role of small heat shock protein B8 (HspB8) in the autophagic removal of misfolded proteins responsible for neurodegenerative diseases. *Autophagy.* 2010;6(7):958-60.

[164] Ding WX, Ni HM, Gao W, Yoshimori T, Stolz DB, Ron D, et al. Linking of autophagy to ubiquitin-proteasome system is important for the regulation of endoplasmic reticulum stress and cell viability. *Am. J. Pathol.* 2007;171(2):513-24.

[165] Ding WX, Yin XM. Sorting, recognition and activation of the misfolded protein degradation pathways through macroautophagy and the proteasome. *Autophagy.* 2008;4(2):141-50.

[166] Kaganovich D, Kopito R, Frydman J. Misfolded proteins partition between two distinct quality control compartments. *Nature.* 2008;454(7208):1088-95.

[167] Pandey UB, Batlevi Y, Baehrecke EH, Taylor JP. HDAC6 at the intersection of autophagy, the ubiquitin-proteasome system and neurodegeneration. *Autophagy.* 2007;3(6):643-5.

[168] Pandey UB, Nie Z, Batlevi Y, McCray BA, Ritson GP, Nedelsky NB, et al. HDAC6 rescues neurodegeneration and provides an essential link between autophagy and the UPS. *Nature.* 2007;447(7146):859-63.

[169] Matus S, Nassif M, Glimcher LH, Hetz C. XBP-1 deficiency in the nervous system reveals a homeostatic switch to activate autophagy. *Autophagy.* 2009;5(8):1226-8.

[170] Pickford F, Masliah E, Britschgi M, Lucin K, Narasimhan R, Jaeger PA, et al. The autophagy-related protein beclin 1 shows reduced expression in early Alzheimer disease and regulates amyloid beta accumulation in mice. *J. Clin. Invest.* 2008;118(6):2190-9.

[171] Nassif M, Hetz C. Targeting autophagy in ALS: A complex mission. *Autophagy.* 2011;7(4):450-3.

[172] Fleming A, Noda T, Yoshimori T, Rubinsztein DC. Chemical modulators of autophagy as biological probes and potential therapeutics. *Nat. Chem. Biol.* 2011;7(1):9-17.

[173] Rubinsztein DC, Gestwicki JE, Murphy LO, Klionsky DJ. Potential therapeutic applications of autophagy. *Nat. Rev. Drug Discov.* 2007;6(4):304-12.

[174] Sarkar S, Rubinsztein DC. Small molecule enhancers of autophagy for neurodegenerative diseases. *Mol. Biosyst.* 2008;4(9):895-901.

[175] Rubinsztein DC, Nixon RA. Rapamycin induces autophagic flux in neurons. *Proc. Natl. Acad. Sci. U. S. A.* 2011;107(49):E181; author reply E2.

[176] Ravikumar B, Berger Z, Vacher C, O'Kane CJ, Rubinsztein DC. Rapamycin pre-treatment protects against apoptosis. *Hum. Mol. Genet.* 2006;15(7):1209-16.

[177] Harrison DE, Strong R, Sharp ZD, Nelson JF, Astle CM, Flurkey K, et al. Rapamycin fed late in life extends lifespan in genetically heterogeneous mice. *Nature.* 2009;460(7253):392-5.

[178] Kahan BD, Camardo JS. Rapamycin: clinical results and future opportunities. *Transplantation.* 2001;72(7):1181-93.

[179] Thomson AW, Turnquist HR, Raimondi G. Immunoregulatory functions of mTOR inhibition. *Nat. Rev. Immunol.* 2009;9(5):324-37.

[180] Beers DR, Henkel JS, Zhao W, Wang J, Appel SH. CD4+ T cells support glial neuroprotection, slow disease progression, and modify glial morphology in an animal model of inherited ALS. *Proc. Natl. Acad. Sci. U. S. A.* 2008;105(40):15558-63.

[181] Chiu IM, Chen A, Zheng Y, Kosaras B, Tsiftsoglou SA, Vartanian TK, et al. T lymphocytes potentiate endogenous neuroprotective inflammation in a mouse model of ALS. *Proc. Natl. Acad. Sci. U. S. A.* 2008;105(46):17913-8.

[182] Naor S, Keren Z, Bronshtein T, Goren E, Machluf M, Melamed D. Development of ALS-like disease in SOD-1 mice deficient of B lymphocytes. *J. Neurol.* 2009;256(8):1228-35.

[183] Shor B, Gibbons JJ, Abraham RT, Yu K. Targeting mTOR globally in cancer: thinking beyond rapamycin. *Cell Cycle.* 2009;8(23):3831-7.

[184] Yu K, Toral-Barza L, Shi C, Zhang WG, Lucas J, Shor B, et al. Biochemical, cellular, and in vivo activity of novel ATP-competitive and selective inhibitors of the mammalian target of rapamycin. *Cancer Res.* 2009;69(15):6232-40.

[185] Zask A, Verheijen JC, Curran K, Kaplan J, Richard DJ, Nowak P, et al. ATP-competitive inhibitors of the mammalian target of rapamycin: design and synthesis of highly potent and selective pyrazolopyrimidines. *J. Med. Chem.* 2009;52(16):5013-6.

[186] Criollo A, Maiuri MC, Tasdemir E, Vitale I, Fiebig AA, Andrews D, et al. Regulation of autophagy by the inositol trisphosphate receptor. *Cell Death Differ.* 2007;14(5):1029-39.

[187] Criollo A, Vicencio JM, Tasdemir E, Maiuri MC, Lavandero S, Kroemer G. The inositol trisphosphate receptor in the control of autophagy. *Autophagy.* 2007;3(4):350-3.

[188] Vicencio JM, Ortiz C, Criollo A, Jones AW, Kepp O, Galluzzi L, et al. The inositol 1,4,5-trisphosphate receptor regulates autophagy through its interaction with Beclin 1. *Cell Death Differ.* 2009;16(7):1006-17.

[189] Sarkar S, Ravikumar B, Floto RA, Rubinsztein DC. Rapamycin and mTOR-independent autophagy inducers ameliorate toxicity of polyglutamine-expanded huntingtin and related proteinopathies. *Cell Death Differ.* 2009;16(1):46-56.

[190] Siciliano G, Carlesi C, Pasquali L, Piazza S, Pietracupa S, Fornai F, et al. Clinical trials for neuroprotection in ALS. *CNS Neurol Disord Drug Targets.* 2010;9(3):305-13.

[191] Fornai F, Longone P, Ferrucci M, Lenzi P, Isidoro C, Ruggieri S, et al. Autophagy and amyotrophic lateral sclerosis: The multiple roles of lithium. *Autophagy.* 2008;4(4):527-30.

[192] Sugai F, Yamamoto Y, Miyaguchi K, Zhou Z, Sumi H, Hamasaki T, et al. Benefit of valproic acid in suppressing disease progression of ALS model mice. *Eur. J. Neurosci.* 2004;20(11):3179-83.

[193] Feng HL, Leng Y, Ma CH, Zhang J, Ren M, Chuang DM. Combined lithium and valproate treatment delays disease onset, reduces neurological deficits and prolongs survival in an amyotrophic lateral sclerosis mouse model. *Neuroscience.* 2008;155(3):567-72.

[194] Aggarwal SP, Zinman L, Simpson E, McKinley J, Jackson KE, Pinto H, et al. Safety and efficacy of lithium in combination with riluzole for treatment of amyotrophic lateral sclerosis: a randomised, double-blind, placebo-controlled trial. *Lancet Neurol.* 2010;9(5):481-8.

[195] Chio A, Borghero G, Calvo A, Capasso M, Caponnetto C, Corbo M, et al. Lithium carbonate in amyotrophic lateral sclerosis: lack of efficacy in a dose-finding trial. *Neurology.* 2010;75(7):619-25.

[196] Crochemore C, Virgili M, Bonamassa B, Canistro D, Pena-Altamira E, Paolini M, et al. Long-term dietary administration of valproic acid does not affect, while retinoic acid decreases, the lifespan of G93A mice, a model for amyotrophic lateral sclerosis. *Muscle Nerve.* 2009;39(4):548-52.

[197] Rouaux C, Panteleeva I, Rene F, Gonzalez de Aguilar JL, Echaniz-Laguna A, Dupuis L, et al. Sodium valproate exerts neuroprotective effects in vivo through CREB-binding protein-dependent mechanisms but does not improve survival in an amyotrophic lateral sclerosis mouse model. *J. Neurosci.* 2007;27(21):5535-45.

[198] Piepers S, Veldink JH, de Jong SW, van der Tweel I, van der Pol WL, Uijtendaal EV, et al. Randomized sequential trial of valproic acid in amyotrophic lateral sclerosis. *Ann. Neurol.* 2009;66(2):227-34.

In: Autophagy: Principles, Regulation and Roles in Disease
Editor: Nikolai Gorbunov
ISBN: 978-1-61942-266-7

Chapter XII

Role of Autophagy in Parkinson's Disease

***Giovanni Stefanoni*[1,2], *Gessica Sala*[1], *Lucio Tremolizzo*[1,2], *Laura Brighina*[2] and *Carlo Ferrarese*[1,2]**
[1] Department of Neuroscience and Biomedical Technologies, University of Milano-Bicocca, Italy
[2] Department of Neurology, San Gerardo Hospital, Monza, Italy

Abstract

Parkinson's disease (PD) is a neurodegenerative movement disorder featured by the degeneration of dopaminergic neurons within the substantia nigra. Although the etiology of sporadic disease still remains unclear, the identification of the gene products involved in hereditary variants has provided critical information about the molecular physiopathology of PD. Thus, it has been shown that morphologically this disorder is characterized by accumulation of misfolded protein aggregates and aberrant organelles in the neuronal structures and that these alterations can be triggered by different factors such as mitochondrial-induced oxidative stress and/or α-synuclein-mediated cytotoxicity. Recently, a growing line of evidence indicates that dysregulation of the two major autophagic pathways, macroautophagy and chaperone mediated autophagy, can be driving mechanisms of neurodegeneration.

Aim of this chapter is to provide in-depth analysis of the experimental evidence connecting malfunction of autophagy with the PD pathogenesis.

1. Introduction

Parkinson's disease (PD) is a disabling neurodegenerative disorder, characterized by progressive loss of dopaminergic neurons within the substantia nigra pars compacta of basal ganglia. Patients are usually diagnosed after the appearance of typical motor manifestations, notoriously representing hallmarks of very advanced stages of the degenerative process. A late diagnosis, together with the multifactorial etiopathogenesis of the disease, explains the fact that clinical and neuropathological deterioration cannot actually benefit from any neuroprotective interventions. In such a context, the investigation of PD molecular pathogenesis is necessary in order to identify novel targets for neuroprotective therapies and early diagnostic biomarkers.

Several mechanisms are implicated in the degeneration of dopaminergic neurons in PD [Dawson 2003]. Among them, a significant role is assigned to the intraneuronal accumulation and aggregation of misfolded α-synuclein. The latest one is the main component of typical protein inclusions, also known as Lewy bodies, that appear in the substantia nigra of PD patients [Spillantini 1997]. The formation of oligomers is likely to be the key step for α-synuclein toxic gain-of-function [Vekrellis 2004]. Hereditary variants of PD linked to missense mutations or multiplications of α-synuclein gene [Polymeropoulos 1997; Singleton 2003] have provided crucial evidence indicating that the accumulation of this protein or specific alterations in its aminoacid sequence can promote self-aggregation and lead to dopaminergic neurodegeneration. These observations were further corroborated by experiments with transgenic cellular and animal models [Dawson 2002, Vekrellis 2004]. In more frequent idiopathic PD the tendency of α-synuclein to aggregate has been associated with post-translational modifications induced by oxidative stress. Indeed, the generation of free radicals, strongly related to the dysfunction of mitochondrial respiratory chain [Mancuso 2006], notoriously represents another pathogenetic mechanism of PD [Zhang 2000]. Since the accumulation and aggregation of misfolded proteins are known to be key pathogenetic mechanisms in the neurodegenerative process, the efficiency of cell repair machinery has been proposed as essential part of neuro- resistance to stress factors and to cytotoxicity of abnormal proteins. It has been proposed recently that degenerative process in the ***substantia nigra*** is associated with the impairment of autophagic-lysosomal pathways [Nixon 2006, Lee 2009, Yang 2010, Xilouri 2011]. At this stage a key-role in the pathogenesis of PD was tentatively assigned to dysregulation of autophagy, including CMA and macroautophagy [Xilouri 2011]. CMA is a selective pathway for degradation of aberrant proteins, which are directly transported into the lysosomal lumen by specific carrier proteins; macroautophagy mediates lysosomal degradation of organelles and proteins after their sequestration within a double-membrane-limited vacuole called autophagosome. Macroautophagy proceeds through four main steps, each requiring the presence of specific autophagy related genes (Atg) [He 2009]: nucleation of the autophagosome; sequestration of a region of cytoplasm containing substrates to be degraded; maturation of the autophagosome; fusion with lysosome and degradation of substrates. In PD brain, histopatological studies have demonstrated an accumulation of autophagosomes within degenerating dopaminergic neurons [Anglade 1997]. Alterations in autophagosome dynamics are responsible for this abnormality: both induction of autophagosome formation [Webb 2003], secondary to a CMA blockage [Cuervo 2004], and reduction of lysosomal pool [Dehay 2010] contribute to autophagosome accumulation. Furthermore, recognition of specific cargoes, such as mitochondria, by the phagophore membrane has been identified as another defective step of macroautophagy in PD [Burbulla 2010]. In this chapter we are presenting a detailed analysis of current data on the role of macroautophagy, chaperone-mediated autophagy (CMA) and protein homeostasis in conjunction with conventional concepts of the PD pathogenesis. Furthermore, we have analyzed a possible sequence of events connecting dysfunction of autophagy with the death of dopaminergic neurons. In conclusion, we will offer a glimpse on the possible use of autophagy parameters as peripheral biomarkers and we will report data regarding the putative neuroprotective effects of pharmacologic induction of autophagy.

2. Autophagy and Neurodegeneration

A large number of data indicate that autophagy impairment exerts a pathogenetic role in several neurodegenerative diseases. Accumulation of aberrant proteins and mitochondria within senescent neurons contributes to neurodegeneration [Mancuso 2006, Soto 2008, Tatsuta 2008]; the progressive dysfunction of different degradative pathways represents a major cause of these alterations. In particular, autophagy plays a key role in neuronal protein homeostasis: CMA is responsible for the degradation of several potentially neurotoxic proteins; macroautophagy represents the only pathway for mitochondrial turnover and an important degradative system for protein aggregates [Brunk 2002]. Hence, autophagy appears as a control system able to rescue neurons from the deleterious mechanisms involved in the degenerative process. According to this hypothesis, the progressive neuronal degeneration starts only when the efficiency of this pathway is compromised enough to allow the development of mechanisms exerting a direct toxic function. Therefore, the poor efficiency of autophagy might be a primum movens in the physiopathology of neurodegenerative diseases. When other pathogenetic mechanisms arise, they determine a compensatory induction of autophagy pathways; if this activation is not appropriate to reestablish neuronal homeostasis and prevent accumulation of neurotoxic products, the neurodegenerative process develops. As a matter of fact, cytoplasmatic protein inclusions and a high amount of mitochondria can be detected in aged post-mitotic cells; these alterations have been demonstrated to correlate with a loss of function of almost every step of CMA and macroautophagy [Martinez-Vicente 2005]. A presence of link between autophagy and neurodegenerative diseases was suggested when proteins with specific neuropathologic relevance were identified as substrates of autophagic pathways. Thus α-synuclein [Webb 2003], APP [Pickford 2008] and huntingtin [Ravikumar 2002] were shown to be processed by macroautophagy and to contain the KFERQ motif, which characterizes all CMA substrates. Furthermore, post-mortem studies revealed an intraneuronal accumulation of autophagosomes in brain areas affected by PD [Anglade 1997], Huntington's disease [Sapp 1997], Alzheimer's disease [Nixon 2005] and spongiform encephalopathies [Sikorska 2004]. This neuropathologic feature is far to be univocally interpreted. Indeed, there are at least three different reasonable explanations for interplay between accumulation of autophagosome and neurodegenerative events.

1) hyper-activation of macroautophagy promotes neuronal death and contributes to brain damage in neurodegenerative diseases.
2) macroautophagy induction mediates a compensatory and protective mechanism activated in neuronal stress in order to delay the degenerative process.
3) accumulation of vacuoles results from a major impairment in maturation and turnover of autophagosomes: this defect further promotes neurodegeneration.

The first hypothesis is supported by the identification of autophagic programmed cell death and pro-apoptotic effect exerted by strong autophagy induction in particular conditions

[Baehrecke 2005]; nevertheless, compelling evidence confirming a deleterious role of autophagy in neurodegenerative diseases does not exist. On the other hand, increasing experimental data suggest that autophagosome accumulation could mirror a selective impairment in the maturation step of macroautophagy pathway. The reduced efficiency of the autophagic process may be an early pathogenetic mechanism possibly contributing to neuronal proteinopathy; anyway, the deleterious processes triggered by autophagy dysfunction and contributing to neurodegeneration, such as the accumulation of aberrant substrates, oxidative stress, ATP and neurotrophic factors defects, might stimulate the residual functional capacity of autophagy. Therefore, the second and third hypotheses cited above are not mutually exclusive [Nixon 2006, Lee 2009]. The assumption that autophagy impairment could contribute to neurodegeneration is firstly based on biological premises: on one hand autophagy is, under stress conditions, a well known cytoprotective mechanism [Boya 2005], on the other this system gradually loses its efficiency during cell aging [Martinez-Vicente 2001]; in particular, autophagosome formation and lysosomal function are impaired in senescent cells, as well as CMA function. These premises, together with the fact that neurodegenerative diseases arise in elderly and show a slow and progressive natural history, makes it improbable an involvement of autophagic cell death, which causes rapid and massive involutional changes in the animal models where it has been described [Baehrecke 2005]. Instead, a causal link between the gradual failure of autophagic pathways and the progressive neurodegeneration, both features typical of advanced age, is more likely. Several experimental data obtained in animal models clearly demonstrate that the loss of autophagy machinery is a sufficient cause of neuronal death: the deletion of genes encoding for key components of macroautophagy, such as Atg5 or Atg7, leads to diffuse neurodegeneration in mice [Hara T 2006, Komatsu 2007]. The molecular mechanisms linking autophagy impairment to neuronal death are still unclear. Intracellular protein aggregates accumulate following macroautophagy inhibition but they do not seem to be directly toxic for neurons [Komatsu 2007]. A key factor may be the reduced turnover of damaged organelles and monomeric or oligomeric misfolded proteins, which are likely to exert a direct neurotoxic effect.

3. Autophagy in Parkinson's Disease

Autophagy impairment may promote processes traditionally recognized as major features characterizing the PD pathogenesis, namely α-synuclein toxicity [Kazanstev 2008], mitochondrial dysfunction [Schapira 1998] and oxidative stress [Zhang 2008]. In turn these events can negatively affect the degradative systems leading to formation of vicious circles and, finally, to death of dopaminergic neurons.

3.1. Chaperone Mediated Autophagy

The hypothesis that a down-regulation of CMA may be involved in PD pathogenesis firstly emerged when this pathway was identified as the major degradative system for α-synuclein [Cuervo 2004]. Studies performed on transgenic cell and animal models of PD have demonstrated that the blockage of CMA-mediated protein degradation is a toxic mechanism shared by different hereditary forms of PD [Cuervo 2004, Kabuta 2008, Xilouri 2009].

The existence of a defect in CMA activity has been confirmed even in sporadic PD by post mortem studies on brain samples from affected patients: in fact, a significant reduction in the levels of lamp2A and hsc70, main components of CMA machinery, has been demonstrated in the brain of patients compared to age-matched controls [Alvarez-Erviti 2010].

3.1.1. Correlations between CMA Dysfunction and A-Synuclein Toxicity

The detection of α-synuclein on the membrane and in the lumen of lysosomes has suggested that this protein is subjected to lysosomal clearance [Lee 2004]. Furthermore, α-synuclein protein contains the KFERQ motif, which is required for CMA-mediated degradation. The inhibition of α-synuclein uptake in isolated lysosomes incubated with high levels of other specific CMA substrates has definitely proved that this protein is degraded by CMA [Cuervo 2004]. Moreover, the blockage of CMA activity through siRNA-mediated suppression of lamp2A expression determines an increase in α-synuclein half-life and a consequent intracellular accumulation of this protein [Alvarez-Erviti 2010].

The A53T and A30P mutant α-synuclein variants, responsible for autosomic dominant hereditary PD, bind hsc70 and lamp2A with a higher affinity than the wild type protein. This prevents the translocation of the mutant protein to lysosomal lumen and subtracts CMA mediators from the chance to bind other substrates. Hence, mutated α-synuclein acts as a blocker of CMA, thus inhibiting both its own degradation and that of other proteins [Cuervo 2004]. The final result is an accumulation of α-synuclein and other potentially neurotoxic proteins. Similarly, the over-expression of wild type α-synuclein, typical of familial forms of PD linked to α-synuclein gene amplification, causes a reduction in lysosomal degradation of proteins through competitive antagonism for the binding to CMA components, as proved by the evidence that an α-synuclein variant lacking the KFERQ motif does not produce any modification in lysosomal protein catabolism. KFERQ deletion partially prevents α-synuclein toxicity, thus demonstrating that a fraction of α-synuclein neurotoxic effect is directly related to the capacity of this protein, when mutated or accumulated, to compromise the CMA-mediated clearance of other proteins [Xilouri 2009]. Therefore, it is conceivable to conclude that α-synuclein intraneuronal status is critical for the modulation of the autophagic-lysosomal pathway and, as a consequence, for neuronal proteostasis. Experiments performed by our group show a significant increase of lamp2A levels in SH-SY5Y over-expressing A53T mutant α-synuclein, as a putative compensatory reaction to the blockage of CMA [manuscript in preparation].

These assumptions highlight the existence of a strong link between CMA efficiency, α-synuclein toxicity and dopaminergic neurodegeneration.

Oxidant-induced PD models show an increase in lamp2A levels [Marin 2011], consistent with the well known upregulation of lamp2A expression under oxidative conditions [Kiffin 2004]. This alteration determines a CMA activation that represents an important bias in the reproduction of PD pathology in toxin-induced cell and animal models. Therefore, creating a model that combines mitochondrial impairment and oxidative stress with inhibition of CMA and other autophagic-lysosomal pathways might represent a promising approach to limit the physiopathologic gap between PD models and human disease.

3.1.2. Mutated UCHL1 Impairs CMA

PARK5 is the gene encoding for the ubiquitin carboxyl-terminal esterase L1 (UCHL1), known to be involved in recycling ubiquitin monomers within the ubiquitin-proteasome system [Gong 2007]. The I93M mutation of PARK5 has been identified in patients affected by a hereditary PD variant with an autosomic dominant pattern of inheritance [Leroy 1998]. However, this mutation has previously been associated with a loss of the protein enzymatic activity [Nishikawa 2003], which is hardly consistent with a dominant transmission. Therefore, a gain of function of the mutated protein has been hypothesized to explain its toxic effect on dopaminergic neurons. Co-immunoprecipitation experiments have demonstrated an interaction of wild type UCHL1 with lamp2A and hsc70 [Kabuta 2008]. The I93M mutation causes a marked increase of the affinity between UCHL1 and the CMA machinery, thus determining the suppression of CMA activity and the consequent intracellular accumulation of α-synuclein and other substrates. The aberrant interaction between mutated UCH-L1 and CMA mediators, leading to the impairment of this autophagic pathway, may be a key mechanism underlying the nigral degeneration related to I93M UCHL1 mutation.

Based on the detrimental effects that mutations linked to hereditary PD exert on CMA activity, the dysfunction of this clearance system appears to be a mechanism able to cause the death of dopaminergic neurons [Bandopadhyay 2010]. This outcome is probably mediated by the accumulation of aberrant proteins with a putative neurotoxic effect [Pan 2008]; among these, α-synuclein is of particular importance, due to its notorious involvement in PD pathogenesis and its capacity to esacerbate CMA impairment.

3.1.3. Role of MEF2D in CMA-Deficient Neurons

Alpha-synuclein is not the only CMA substrate characterized by a specific neuropathologic interest in PD. Indeed, other proteins have been identified as specific CMA substrates and as components of molecular pathways contributing to dopaminergic neurodegeneration; the efficiency of this degradative system might play a central role in modulating the activity of these proteins.

An example is represented by myocyte enhancer factor 2D (MEF2D), a neuronal transcriptional factor with a cytoprotective and anti-apoptotic function [Liu 2003]. This protein is phosphorylated by cyclin-dependent kinase 5 (cdk5), a pro-apoptotic factor that inactivates MEF2D [Gong 2003, Tang 2005]. Cdk5 has been found to favor the death of dopaminergic neurons in cell and animal MPTP-induced models of disease [Lee 2000, Smith 2003]; the cdk5-mediated inactivation of MEF2D represents a key step linking cdk5 activation with neuronal cell death [Smith 2006]. Recently, MEF2D has been identified as a substrate of CMA: this nuclear protein can translocate to the cytosol, interact with hsc70 and lamp2A and undergo lysosomal degradation. Furthermore, CMA efficiency has been demonstrated to modulate not only the intracellular levels of this protein, but also its transcriptional activity. The inhibition of lysosomal degradative function, obtained through treatment of cells with ammonium chloride, leads to an accumulation of the total and cytoplasmatic fraction of MEF2D and to the decrease of nuclear levels and DNA-binding capacity of the protein. Mutated α-synuclein, as well as the over-expressed wild type protein, impairs the CMA-mediated degradation of MEF2D and suppresses its function in vitro [Yang 2009]. Unpublished data obtained by our group confirm an increase of MEF2D levels in SH-SY5Y cells over-expressing A53T mutant α-synuclein [manuscript in submission]. The expression of a MEF2D variant unable to translocate to the cytosol prevents the modifications of protein levels and activity, thus rescuing cell viability from α-synuclein and ammonium chloride toxicity [Yang 2009]. This indicates that MEF2D cytosolic accumulation links CMA impairment to the inhibition of nuclear expression and activity of this transcriptional factor.

Sporadic PD patients and α-synuclein transgenic mice show higher cytosolic [Yang 2009] and decreased nuclear [Chu 2011] levels of MEF2D in their brain, both alterations consistent with a CMA impairment in PD.

This evidence suggests a correlation between CMA failure and the suppression of MEF2D neuroprotective activity. In particular, MEF2D appears as a putative key link between the CMA impairment and dopaminergic neurodegeneration.

3.2. Macroautophagy

As already reminded, the hypothesis that macroautophagy dysfunction might be involved in PD pathogenesis, as well as in other neurodegenerative diseases, has raised from the identification of the key pathologic role of misfolded proteins [Soto 2008] and aberrant mitochondria [Mancuso 2003], together with the autoptic finding of a number of autophagic vacuoles sensibly increased in the substantia nigra of patients compared to age-matched controls [Anglade 1997]. This last feature is likely to be secondary to a selective impairment in autophagosome maturation and fusion with lysosomes, rather than in vacuole formation [Nixon 2006, Dehay 2010]. The impairment in autophagosome turnover occurs physiologically during aging [Brunk 2002, Martinez-Vicente 2005]; in PD and other neurodegenerative diseases this defect may be particularly relevant and constitute an early pathogenetic mechanism.

Macroautophagy is induced by toxic agents inhibiting mitochondrial complex I, commonly used to create experimental models of disease. Such a paradox is easily explained by the fact that these toxins exert an acute deleterious action on proliferating cells that can possess an efficient rescue machinery, including macroautophagy; on the other hand, neurodegeneration in idiopathic PD represents the outcome of chronic processes involving senescent neurons, which are post-mitotic cells whose surveillance mechanisms have lost a great part of their efficiency. This represents an important gap between human disease and toxin-induced models. Furthermore, data on the biological role of macroautophagy induction in these models are contrasting: autophagy inhibition has been reported to rescue non neuronal cells from rotenone toxicity [Chen 2007], whereas it would favor the death of neuronal cells exposed to paraquat [Gonzales-Polo 2007].

Anyway, the correlations existing between proteins linked to hereditary PD and the macroautophagy pathway have provided some elements useful to clarify the role of this degradative system in PD.

3.2.1. Macroautophagy and α-Synuclein Toxicity

Several studies demonstrate that mutated or over-expressed α-synuclein causes an induction of macroautophagy through a mechanism dependent on the α-synuclein-mediated blockage of CMA; indeed, macroautophagy activation does not occur if the protein loses its KFERQ motif [Xilouri 2009]. However, this assumption remains controversial, since other studies showed an inhibition of macroautophagy in α-synuclein transgenic cell and animal models of disease. Macroautophagy suppression would be dependent on the α-synuclein-mediated inhibition of Rab1A, a cytoprotective factor involved in the regulation of intracellular vesicular trafficking [Winslow 2010]. It is likely that the cell type and the expression level of α-synuclein influence the prevalent inhibitory effect of α-synuclein on CMA or Rab1A, thus justifying the different effects observed on macroautophagy activity.

The role of macroautophagy on the survival of α-synuclein over-expressing neurons remains unclear. Indeed, while some reports have shown that suppression of macroautophagy protects against over-expression of α-synuclein [Xilouri 2009], other communications have indicated that upregulation of macroautophagy can diminish cytotoxicity of both mutated and wild-type α-synuclein expressed at particular levels, i.e., close to those observed in human pathology [Sarkar 2007]. The recent demonstration that pro-autophagic activity of Rab1a normalizes physiological functions of the α-synuclein transgenic animals [Cooper 2006] supports hypothesis that macroautophagy is a process which mitigates α-synuclein toxicity.

Moreover, Spencer and co-authors have demonstrated that the co-expression of beclin1, a key component of macroautophagy induction pathway, can stimulate α-synuclein clearance and prevent the neurodegenerative pathology in α-synuclein over-expressing cell and animal models; these effects can be enhanced by treatment with rapamycin, a well-known macroautophagy activator [Spencer 2009]."The degradation of α-synuclein and other intraneuronal proteins probably becomes more dependent on macroautophagy when their aggregation tendency increases. This assumption is corroborated by remarkable ability of macroautophagy to specifically degrade mutated rather than wild type α-synuclein [Sarkar 2007]. Post-translational modifications produced by oxidative stress on wild type α-synuclein are known to favor the tendency of this protein to aggregate [Paxinou 2001, Hodara 2004]; therefore, the role of macroautophagy in the clearance of α-synuclein may be particularly relevant in oxidative conditions characterizing the substantia nigra of patients with sporadic PD [Zang 2008].

3.2.2. LRRK2 Interferes with Macroautophagy

A functional correlation between leucine-rich repeat kinase 2 (LRRK2) (also known as dardarin) and macroautophagy has been recently established. This cytosolic and mitochondrial kinase, linked to autosomal dominant inherited PD [Zimprich 2004], acts as a negative regulator of macroautophagy. In particular, it can prevent the fusion between autophagosomes and multi-vesicular bodies (MVBs), thus interfering with the maturation of autophagic vacuoles. There is a line of evidence indicating that the deletion of LRRK2 gene increases LC3II turnover and protects cells against the toxicity of autophagy inhibitors while the pathogenetic R1441C mutation of LRRK2 results in intraneuronal accumulation of MVBs and autophagosomes [Alegre-Abarrategui 2009]. Therefore, autophagy impairment seems to be a pathogenetic mechanism triggered by the R1441C mutation, which would be responsible for the enhancement of the physiologic suppressive effect of dardarin on macroautophagy.

3.2.3. Role of PINK1 and Parkin in Mitochondria Autophagy

The impairment of protein turnover is not likely a major negative outcome triggered by macroautophagy dysfunction. A major feature linking the loss of autophagy function to neurodegeneration is the intraneuronal accumulation of aberrant mitochondria. The dysregulation of mitochondrial function is known to be a key mechanism in PD pathogenesis [Zhu 2010] that was proven by the selective degeneration of dopaminergic neurons after exposure to toxic agents inhibiting the complex I of the mitochondrial respiratory chain. The neurotoxicity of mitochondrial dysfunction derives from the energetic deficiency and the oxidative stress triggered by the defective oxygen metabolism in aberrant mitochondria. Furthermore, the damaged mitochondria release factors able to activate pro-apoptotic molecular pathways.

The rapid turnover of damaged mitochondria is necessary to prevent the generation of reactive oxygen species and the release of pro-apoptotic agents [Burbulla 2010]. The catabolism of mitochondria is mediated by a specific macroautophagic pathway named mitophagy. Impairment of mitophagy determines the accumulation of aberrant mitochondria within neurons; this mechanism represents a putative triggers for neurodegeneration [Tanaka 2010].

The molecular pathway regulating mitophagy is now under investigation. The assumption that mitophagy impairment could be a pathogenetic mechanism of PD is supported by the identification of a specific role of recessive PD-related genes within mitophagy machinery. Indeed, PTEN induced putative kinase 1 (PINK1) and parkin, encoded by genes linked to autosomic recessive forms of hereditary PD [Kitada 1998, Valente 2004], have been demonstrated as key factors for the autophagic degradation of mitochondria. PINK1 is a mitochondrial kinase, set in the outer mitochondrial membrane and able to phosphorylate cytosolic proteins [Zhou 2008]; parkin is an E3-ligase enzyme involved in protein ubiquitination and degradation through the proteasome [Tanaka 2001]. In intact mitochondria, PINK1 is rapidly cleaved, whereas membrane depolarization of damaged mitochondria causes the blockage of PINK1 degradation and the accumulation of this kinase on mitochondrial membrane [Narendra 2010]; this determines the phosphorylation and the consequent translocation of parkin from the cytosol to the aberrant organelles. The pro-mitophagic functions of PINK1 and parkin are mediated by two proteins, p62, which binds ubiquitinated proteins and targets them to the phagophore, and voltage-dependent anion channel 1 (VDAC1), which localizes in the outer mitochondrial membrane [Geisler 2010]. Parkin, when phosphorylated by PINK1, determines the ubiquitination of VDAC1; this allows p62 to bind aberrant mitochondria and carry them to the phagophore. Pathogenetic mutations of PINK1 and parkin cause the loss of the capacity to ubiquitinate VDAC1 and to induce mitophagy after mitochondrial damage. Based on this evidence we suggest that the dysregulation of mitochondrial turnover can be a mechanism responsible for the neuropathologic and clinical phenotype linked to mutations inactivating PINK1 and parkin.

Mitofusins (Mfn) 1 and 2 are other proteins involved in mitochondrial turnover within dopaminergic neurons. These proteins are localized in the outer mitochondrial membrane, where they mediate mitochondrial fusion [Legros 2002]. They have recently been identified as substrates of parkin, which activates the ubiquitin proteasome system for Mfn degradation [Ziviani 2010]. The loss of parkin or PINK1 causes an increase of Mfn intracellular levels and a consequent elongation of mitochondria [Clark 2006]. Therefore, Mfn ubiquitination by the PINK1/parkin pathway determines Mfn degradation, thus inducing the fission and the consequent shortening of mitochondria; these modifications might be determinant to allow the sequestration of damaged mitochondria within the autophagic vacuole. A more complex mitochondrial morphology has been detected in fibroblasts from patients affected by parkin-related hereditary PD compared to healthy controls, thus confirming the key role of the PINK1/parkin pathway in regulating mitochondrial dynamics in humans [Mortiboys 2008].

Proteasome inhibition completely abrogates parkin-mediated mitophagy, suggesting that remodeling of the mitochondrial outer membrane proteome through proteasome activation is essential for mitophagy [Chan 2011]. Such a synergy between proteasome function and mitophagy suggests the existence of a strong correlation between the failure of these two pathways in PD physiopathology.

Besides the parkin-dependent pro-mitophagic function, PINK1 can also directly modulate key components of the macroautophagy regulatory pathway. In particular, an interaction between PINK1 and beclin1 is responsible for the PINK1-mediated induction of macroautophagy; pathogenetic mutations of PINK1 have been shown to prevent the bound to beclin1 and to impair the pro-autophagic function of the protein [Michiorri 2010].

3.2.4. The ERK2-Mediated Pro-Autophagic Role of DJ-1

DJ-1 is a protein that exerts an antioxidant function [Abou-Sleiman 2006]. Although it is present in various subcellular compartments, it has cytoprotective function in mitochondria. Mutations of the DJ-1 gene are responsible for autosomic recessive PD [Hering 2004]. Recently, DJ-1 has been demonstrated to play a regulatory role on macroautophagy and mitophagy [Krebiehl 2010]. In fact, DJ-1 knockout mice show reduced rate of autophagosome formation and intracellular accumulation of mitochondria linked to a reduction in their lysosomal degradation. Therefore, evidently DJ-1 can induce macroautophagy and mitophagy. Autophagy impairment could represent a key feature in the dopaminergic neurodegeneration linked to DJ-1 mutations.

The molecular mechanisms that mediate the pro-autophagic function of DJ-1 are still unclear. Reduced phosphorylation of mitogen-activated protein kinase 1 (MAPK1/ERK2) has been detected in mitochondria of cells lacking DJ-1 gene. ERK2 is a kinase involved in different molecular pathways favoring cell survival and proliferation [Fukunaga 1998]. It is activated through phosphorylation. The kinase activity of mitochondrial ERK2 has been demonstrated to exert a pro-autophagic and pro-mitophagic function [Dagda 2008]. Since the loss of DJ-1 gene is associated with a reduced activity of mitochondrial ERK2, the effect of DJ-1 on autophagy may be mediated by the activation of this mitochondrial kinase.

3.3. Lysosomal Function

A defect in lysosomal degradative activity can lead to autophagosome accumulation and can impair the efficiency of all autophagic pathways. Since both CMA and macroautophagy can be defective in PD [Xilouri 2011] and an accumulation of autophagosomes is shown in PD brains [Anglade 1997], lysosomal dysfunction has been proposed as a putative pathogenetic mechanism in this syndrome. Toxins able to inhibit lysosomal proteolysis, such as ammonium chloride and tunicamycin, have recently been found to produce a PD-like motor impairment and neuropathologic phenotype when injected in cerebrospinal fluid of medaka fish, thus demonstrating the critical role played by lysosomal function for the survival of dopaminergic neurons (Matsui 2010). A significant reduction of lysosomal content has been shown in autoptic samples of substantia nigra from patients affected by sporadic PD compared to age-matched controls [Dehay 2010]. The same abnormality has been found to precede the accumulation of autophagosomes in cellular and animal MPTP-induced models. Oxidative stress, triggered by MPTP and other inhibitors of mitochondrial complex I [Perier 2005], represents the mechanism responsible for lysosomal depletion in these models; indeed, pre-treatment with antioxidants rescues the MPTP-induced lysosomal loss in neuronal cultures [Dehay 2010]. These data support the existence of a lysosomal dysfunction in PD. The accumulation of autophagosomes [Anglade 1997] and LC3II [Alvarez-Erviti 2010] in the substantia nigra of PD patients may be consequences of this defect. Once again, genetics of PD provides further data supporting the crucial role of lysosomal function in the homeostasis of dopaminergic neurons. Indeed, ATPase type 13A2 (ATP13A2) and glucocerebrosidase (GBA) are proteins that guarantee the correct function of the degradative enzymes within lysosomes and, at the same time, their genotype influence the risk of developing PD. ATP13A2 is indispensable for creating an acid environment within lysosomes [Bandopadhyay 2010]; since acidity contributes to the degradation of the internalized substrates by optimizing the catabolic activity of lysosomal hydrolases, ATP13A2 plays an important role in the degradation of both CMA and macroautophagy substrates. Mutations inactivating ATP13A2 are responsible for Kufor-Rakeb syndrome, an autosomic recessive disease that represents parkinsonism as one of its manifestations [Ramirez 2006]. Studies performed on α-synuclein over-expressing cell lines have demonstrated that loss of ATP13A2 function impairs α-synuclein clearance, whereas ATP13A2 over-expression determines a decrease in the amount of α-synuclein-positive inclusions and rescues cells from mutated α-synuclein toxicity [Gitler 2009]. These data provide evidence that ATP13A2 plays a key role in contrasting α-synuclein accumulation by favoring its lysosomal degradation; therefore, it is conceivable that mutations inactivating ATP13A2 and responsible for Kufor-Rakeb syndrome can compromise neuronal survival by interfering with α-synuclein clearance within lysosomes. GBA is a lysosomal enzyme that catalyzes the hydrolysis of the lipid glucosylceramide into glucose and ceramide. Homozygous mutations of the gene encoding this enzyme are responsible for Gaucher syndrome, a multisystem lysosomal storage disease characterized by neurodegenerative manifestations. Recently, carriers of these mutations have been identified as subjects with higher risk of developing sporadic PD [Sidransky 2009, Neumann 2009, Lesage 2010]. A recent study has shown that the over-expression of GBA gene with pathogenetic mutations leads to α-synuclein accumulation in a dose-dependent manner, whereas the complete loss of GBA enzymatic function, that causes. Gaucher syndrome in animal models, is not sufficient to raise α-synuclein levels [Cullen 2011]. This suggests that, beside the loss of the enzymatic activity, pathogenetic GBA mutations produce a gain of function that interferes with α-synuclein catabolism and may explain the increased PD risk in heterozygous mutation carriers.

4. Diagnostic and Therapeutic Implications

Considering a crucial role of autophagy impairment in PD and in other neurodegenerative diseases, a modulation of autophagy activity can be promising tool for neuroprotective therapies aimed at amelioration of the PD pathogenesis. No drugs are currently available to enhance CMA function, whereas a large amount of macroautophagy enhancers are under investigation. It is likely that different pathological conditions are characterized by malfunction of specific steps of the macroautophagy pathway [Wong 2010]. Analyses of these alterations can provide with essential information for new drug target therapy of diseases associated with autophagy impairment. For example, when lysosomal elimination of autophagosomes is compromised, the most reasonable therapeutic approach would be improvement of lysosomal function, and activation of the transcriptional factor EB (TFEB) may be a putative strategy to achieve this goal [Sardiello 2009].

Preclinical studies on PD models have demonstrated neuroprotective effects of drugs activating macroautophagy. The first tested molecule, rapamycin, is known to activate macroautophagy by the mTOR inhibition; this drug has been shown to rescue dopaminergic neurons in various transgenic and sporadic models of disease, both in vitro and in vivo [Tain 2009, Spencer 2009, Malagelada 2010]. Since the neuroprotective effect of rapamycin may be dependent on mTOR functions not related to autophagy [Malagelada 2010], mTOR-independent autophagy enhancers have been tested in order to confirm the therapeutic potentialities of autophagy induction in PD. Among them, trehalose has shown to produce a protective effect in an animal model of PD and this effect has been demonstrated to be dependent on macroautophagy enhancement [Rodriguez-Navarro 2010]. The induction of lysosomal biogenesis by genetical or pharmacological activation of TFEB has been demonstrated to restore the intracellular lysosomal pool and to rescue from neuronal death in MPTP cell and animal models [Dehay 2010]. Interestingly, both rapamycin and trehalose, known to induce the autophagosome nucleation and the consequent sequestration of substrates, have also been demonstrated to stimulate lysosomal biogenesis through the activation of TFEB. This double mechanism of action, which guarantees the parallel increase of substrate sequestration and degradation, probably represents a basic requirement to obtain neuroprotective effectiveness by the pharmacologic macroautophagy induction, especially in diseases, such as PD, where lysosomal function appears to be one of the most defective steps of autophagy.

These results have encouraged development of compounds with a selective stimulatory activity on autophagic pathways compromised in PD dopaminergic neurons. Indeed, the neuroprotective effectiveness of the most known autophagy inducers, such as rapamycin, is limited by the fact that these drugs, known to intensely stimulate autophagy in non neuronal cells, show poor effect on neuronal autophagy. Furthermore, the systemic activity of rapamycin and other compounds determine a series of side effects that limit their clinical usefulness. Therefore, aiming at future therapeutic applications, it would be useful for identification of macroautophagy enhancers with a relevant and selective activity in neurons. In this regard, N^{10} substituted phenoxazine (10-NCP) has been identified as a molecule able to induce a relevant activation of macroautophagy in neurons; moreover, this drug has been demonstrated to promote the clearence of protein aggregates and to protect neurons in an in vitro model of Huntington's disease [Tsvetkov 2010]. Experiments testing the same molecule in PD models are actually in progress. Several compounds with a 10-NCP-like chemical structure have revealed to enhance neuronal macroautophagy; a great part of these molecules, such as chlorpromazine and promethazine, are currently used in clinical practice, although their anti-dopaminergic properties limit their use in PD patients

The preclinical demonstration of the neuroprotective effectiveness of drugs activating macroautophagy highlights the need for identifying easily accessible peripheral biomarkers related to this process, in order to:

- study drug mechanisms;
- establish the biochemical effectiveness of pharmacologic intervention targeted on autophagy in clinical trials;
- predict patients responsiveness to therapy.

It is reasonable to hypothesize that autophagy dysregulation in the central nervous system of PD patients may be associated with systemic alterations of the same process, which could be secondary to the biochemical imbalance occurring in PD brains or, more probably, might be markers of a genetic or acquired defect involving all tissues; such a systemic imbalance would be responsible for the exclusive death of the most vulnerable cells, such as dopaminergic neurons, thus predisposing to PD. A relevant increase in LC3II levels has been demonstrated in lymphomonocytes from sporadic PD patients compared to age-matched healthy controls [Prigione 2010]. This proves that a systemic dysregulation of macroautophagy exists in idiopathic PD. Hence, biochemical parameters of autophagy appear as promising targets of the research on peripheral biomarkers of disease.

Conclusion

In this chapter we emphasized that the products of several genes specifically linked to hereditary PD have functional correlations with the activity of different autophagic-lysosomal pathways. This reinforces a view that autophagy impairment plays a pathogenetic role in PD. This idea is corroborated by clinical histopathology of idiopathic syndrome and by numerous in vivo and in vitro models. Summary of this idea is schematically presented in figure 1.

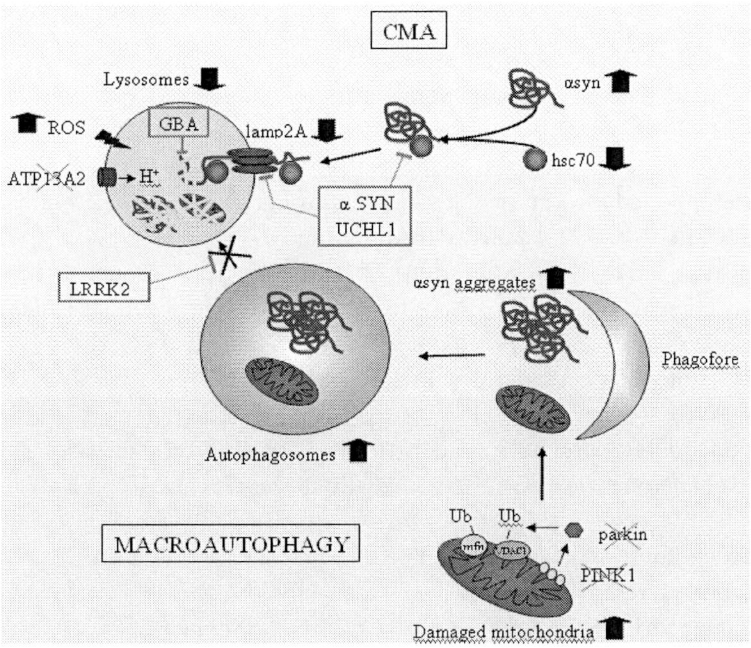

Figure 1. Steps of macroautophagy and chaperone mediated autophagy (CMA) altered in hereditary and sporadic Parkinson's disease (PD). αsyn, α-synuclein; CMA, chaperone mediated autophagy; ROS, reactive oxygen species; Ub, ubiquitin.

=gain-of-function mutations; = loss-of-function mutations;

= increased in sporadic PD; decreased in sporadic PD; = damage.

The dysfunction of the autophagic pathways represents a physiological outcome that is progressively developing during neuronal aging. Rapid decline of autophagy function may constitute an early pathogenetic factor that predisposes to the development of other neurotoxic mechanisms, such as the accumulation of α-synuclein and aberrant mitochondria. However, it is known that these last processes are not only consequences of autophagy failure, but can themselves potentiate autophagy dysfunction. For instance, α-synuclein leads to a blockage of CMA, while oxidative stress triggered by the accumulation of damaged mitochondria, disrupts the lysosomal system, thus impairing both CMA and macroautophagy. This indicates a presence of major metabolic dysregulations that implicate the autophagy-lysosomal pathway and the mechanisms traditionally considered central in nigral degeneration.

Since increasing evidence supports the key pathogenetic role of autophagy impairment in PD, the development of therapies aimed at stimulating this system represents a promising strategy to preserve neuronal viability and interfere with the disease progression.

References

Abou-Sleiman PM, Muqit MM, Wood NW. Expanding insights of mitochondrial dysfunction in Parkinson's disease. *Nat. Rev. Neurosci.* 2006 Mar;7(3):207-19. Review.

Alegre-Abarrategui J, Christian H, Lufino MM, Mutihac R, Venda LL, Ansorge O, Wade-Martins R. LRRK2 regulates autophagic activity and localizes to specific membrane microdomains in a novel human genomic reporter cellular model. *Hum. Mol. Genet.* 2009 Nov 1;18(21):4022-34.

Alvarez-Erviti L, Rodriguez-Oroz MC, Cooper JM, Caballero C, Ferrer I, Obeso JA, Schapira AH. Chaperone-mediated autophagy markers in Parkinson disease brains. *Arch. Neurol.* 2010 Dec;67(12):1464-72. Epub 2010 Aug 9.

Anglade P, Vyas S, Javoy-Agid F, Herrero MT, Michel PP, Marquez J, Mouatt-Prigent A, Ruberg M, Hirsch EC, Agid Y. Apoptosis and autophagy in nigral neurons of patients with Parkinson's disease. *Histol. Histopathol.* 1997 Jan;12(1):25-31.

Baehrecke EH. Autophagy: dual roles in life and death? *Nat. Rev. Mol. Cell Biol.* 2005 Jun;6(6):505-10. Review.

Bandopadhyay R, de Belleroche J. Pathogenesis of Parkinson's disease: emerging role of molecular chaperones. *Trends Mol Med.* 2010 Jan;16(1):27-36. Epub. 2009 Dec 24. Review.

Boya P, González-Polo RA, Casares N, Perfettini JL, Dessen P, Larochette N, Métivier D, Meley D, Souquere S, Yoshimori T, Pierron G, Codogno P, Kroemer G. Inhibition of macroautophagy triggers apoptosis. *Mol. Cell Biol.* 2005 Feb;25(3):1025-40.

Brunk UT, Terman A. The mitochondrial-lysosomal axis theory of aging: accumulation of damaged mitochondria as a result of imperfect autophagocytosis. Eur J Biochem. 2002 Apr;269(8):1996-2002. Review.

Burbulla LF, Krebiehl G, Krüger R. Balance is the challenge--the impact of mitochondrial dynamics in Parkinson's disease. *Eur. J. Clin. Invest.* 2010 Nov;40(11):1048-60. doi: 10.1111/j.1365-2362.2010.02354.x. Review.

Chan NC, Salazar AM, Pham AH, Sweredoski MJ, Kolawa NJ, Graham RL, Hess S, Chan DC. Broad activation of the ubiquitin-proteasome system by Parkin is critical for mitophagy. *Hum. Mol. Genet.* 2011 May 1;20(9):1726-37. Epub 2011 Feb 4.

Chen Y, McMillan-Ward E, Kong J, Israels SJ, Gibson SB. Mitochondrial electron-transport-chain inhibitors of complexes I and II induce autophagic cell death mediated by reactive oxygen species. *J. Cell Sci.* 2007 Dec 1;120(Pt 23):4155-66.

Chu Y, Mickiewicz AL, Kordower JH. α-Synuclein aggregation reduces nigral myocyte enhancer factor-2D in idiopathic and experimental Parkinson's disease. *Neurobiol. Dis.* 2011 Jan;41(1):71-82. Epub 2010 Sep 9.

Clark IE, Dodson MW, Jiang C, Cao JH, Huh JR, Seol JH, Yoo SJ, Hay BA, Guo M. Drosophila pink1 is required for mitochondrial function and interacts genetically with parkin. *Nature.* 2006 Jun 29;441(7097):1162-6. Epub 2006 May 3.

Cooper AA, Gitler AD, Cashikar A, Haynes CM, Hill KJ, Bhullar B, Liu K, Xu K, Strathearn KE, Liu F, Cao S, Caldwell KA, Caldwell GA, Marsischky G, Kolodner RD, Labaer J, Rochet JC, Bonini NM, Lindquist S. Alpha-synuclein blocks ER-Golgi traffic and Rab1 rescues neuron loss in Parkinson's models. *Science.* 2006 Jul 21;313(5785):324-8. Epub 2006 Jun 22.

Cuervo AM, Stefanis L, Fredenburg R, Lansbury PT, Sulzer D. Impaired degradation of mutant alpha-synuclein by chaperone-mediated autophagy. *Science.* 2004 Aug 27;305(5688):1292-5.

Cullen V, Sardi SP, Ng J, Xu YH, Sun Y, Tomlinson JJ, Kolodziej P, Kahn I, Saftig P, Woulfe J, Rochet JC, Glicksman MA, Cheng SH, Grabowski GA, Shihabuddin LS, Schlossmacher MG. Acid β-glucosidase mutants linked to gaucher disease, parkinson disease, and lewy body dementia alter α-synuclein processing. *Ann. Neurol.* 2011 Feb 18. doi: 10.1002/ana.22400.

Dagda RK, Zhu J, Kulich SM, Chu CT. Mitochondrially localized ERK2 regulates mitophagy and autophagic cell stress: implications for Parkinson's disease. *Autophagy.* 2008 Aug 16;4(6):770-82. Epub 2008 Jun 16.

Dawson T, Mandir A, Lee M. Animal models of PD: pieces of the same puzzle? *Neuron.* 2002 Jul 18;35(2):219-22. Review.

Dawson TM, Dawson VL. Molecular pathways of neurodegeneration in Parkinson's disease. *Science.* 2003 Oct 31;302(5646):819-22. Review.

Dehay B, Bové J, Rodríguez-Muela N, Perier C, Recasens A, Boya P, Vila M. Pathogenic lysosomal depletion in Parkinson's disease. *J. Neurosci.* 2010 Sep 15;30(37):12535-44.

Fukunaga K, Miyamoto E. Role of MAP kinase in neurons. *Mol. Neurobiol.* 1998 Feb;16(1):79-95. Review.

Geisler S, Holmström KM, Skujat D, Fiesel FC, Rothfuss OC, Kahle PJ, Springer W. PINK1/Parkin-mediated mitophagy is dependent on VDAC1 and p62/SQSTM1. *Nat. Cell Biol.* 2010 Feb;12(2):119-31. Epub 2010 Jan 24.

Gitler AD, Chesi A, Geddie ML, Strathearn KE, Hamamichi S, Hill KJ, Caldwell KA, Caldwell GA, Cooper AA, Rochet JC, Lindquist S. Alpha-synuclein is part of a diverse and highly conserved interaction network that includes PARK9 and manganese toxicity. *Nat. Genet.* 2009 Mar;41(3):308-15. Epub 2009 Feb 1.

Gong B, Leznik E. The role of ubiquitin C-terminal hydrolase L1 in neurodegenerative disorders. *Drug News Perspect.* 2007 Jul-Aug;20(6):365-70. Review.

Gong X, Tang X, Wiedmann M, Wang X, Peng J, Zheng D, Blair LA, Marshall J, Mao Z. Cdk5-mediated inhibition of the protective effects of transcription factor MEF2 in neurotoxicity-induced apoptosis. *Neuron.* 2003 Apr 10;38(1):33-46.

González-Polo RA, Niso-Santano M, Ortíz-Ortíz MA, Gómez-Martín A, Morán JM, García-Rubio L, Francisco-Morcillo J, Zaragoza C, Soler G, Fuentes JM. Inhibition of paraquat-induced autophagy accelerates the apoptotic cell death in neuroblastoma SH-SY5Y cells. *Toxicol. Sci.* 2007 Jun;97(2):448-58. Epub 2007 Mar 6.

Hara T, Nakamura K, Matsui M, Yamamoto A, Nakahara Y, Suzuki-Migishima R, Yokoyama M, Mishima K, Saito I, Okano H, Mizushima N. Suppression of basal autophagy in neural cells causes neurodegenerative disease in mice. *Nature.* 2006 Jun 15;441(7095):885-9. Epub 2006 Apr 19.

He C, Klionsky DJ. Regulation mechanisms and signaling pathways of autophagy. *Annu. Rev. Genet.* 2009;43:67-93.

Hering R, Strauss KM, Tao X, Bauer A, Woitalla D, Mietz EM, Petrovic S, Bauer P, Schaible W, Müller T, Schöls L, Klein C, Berg D, Meyer PT, Schulz JB, Wollnik B, Tong L, Krüger R, Riess O. Novel homozygous p.E64D mutation in DJ1 in early onset Parkinson disease (PARK7). *Hum. Mutat.* 2004 Oct;24(4):321-9.

Hodara R, Norris EH, Giasson BI, Mishizen-Eberz AJ, Lynch DR, Lee VM, Ischiropoulos H. Functional consequences of alpha-synuclein tyrosine nitration: diminished binding to lipid vesicles and increased fibril formation. *J. Biol. Chem.* 2004 Nov 12;279(46):47746-53. Epub 2004 Sep 9.

Kabuta T, Furuta A, Aoki S, Furuta K, Wada K. Aberrant interaction between Parkinson disease-associated mutant UCH-L1 and the lysosomal receptor for chaperone-mediated autophagy. *J. Biol. Chem.* 2008 Aug 29;283(35):23731-8. Epub 2008 Jun 12.

Kazantsev AG, Kolchinsky AM. Central role of alpha-synuclein oligomers in neurodegeneration in Parkinson disease. *Arch. Neurol.* 2008 Dec;65(12):1577-81. Review.

Kiffin R, Christian C, Knecht E, Cuervo AM. Activation of chaperone-mediated autophagy during oxidative stress. *Mol. Biol. Cell.* 2004 Nov;15(11):4829-40. Epub 2004 Aug 25.

Kitada T, Asakawa S, Hattori N, Matsumine H, Yamamura Y, Minoshima S, Yokochi M, Mizuno Y, Shimizu N. Mutations in the parkin gene cause autosomal recessive juvenile parkinsonism. *Nature.* 1998 Apr 9;392(6676):605-8.

Komatsu M, Waguri S, Koike M, Sou YS, Ueno T, Hara T, Mizushima N, Iwata J, Ezaki J, Murata S, Hamazaki J, Nishito Y, Iemura S, Natsume T, Yanagawa T, Uwayama J, Warabi E, Yoshida H, Ishii T, Kobayashi A, Yamamoto M, Yue Z, Uchiyama Y, Kominami E, Tanaka K. Homeostatic levels of p62 control cytoplasmic inclusion body formation in autophagy-deficient mice. *Cell.* 2007 Dec 14;131(6):1149-63.

Komatsu M, Wang QJ, Holstein GR, Friedrich VL Jr, Iwata J, Kominami E, Chait BT, Tanaka K, Yue Z. Essential role for autophagy protein Atg7 in the maintenance of axonal homeostasis and the prevention of axonal degeneration. *Proc. Natl. Acad. Sci. U. S. A.* 2007 Sep 4;104(36):14489-94. Epub 2007 Aug 28.

Krebiehl G, Ruckerbauer S, Burbulla LF, Kieper N, Maurer B, Waak J, Wolburg H, Gizatullina Z, Gellerich FN, Woitalla D, Riess O, Kahle PJ, Proikas-Cezanne T, Krüger R. Reduced basal autophagy and impaired mitochondrial dynamics due to loss of Parkinson's disease-associated protein DJ-1. PLoS One. 2010 Feb 23;5(2):e9367.

Lee HJ, Khoshaghideh F, Patel S, Lee SJ. Clearance of alpha-synuclein oligomeric intermediates via the lysosomal degradation pathway. *J. Neurosci.* 2004 Feb 25;24(8):1888-96.

Lee JA. Autophagy in neurodegeneration: two sides of the same coin. *BMB Rep.* 2009 Jun 30;42(6):324-30. Review.

Lee MS, Kwon YT, Li M, Peng J, Friedlander RM, Tsai LH. Neurotoxicity induces cleavage of p35 to p25 by calpain. *Nature.* 2000 May 18;405(6784):360-4.

Legros F, Lombès A, Frachon P, Rojo M. Mitochondrial fusion in human cells is efficient, requires the inner membrane potential, and is mediated by mitofusins. *Mol. Biol. Cell.* 2002 Dec;13(12):4343-54.

Leroy E, Boyer R, Auburger G, Leube B, Ulm G, Mezey E, Harta G, Brownstein MJ, Jonnalagada S, Chernova T, Dehejia A, Lavedan C, Gasser T, Steinbach PJ, Wilkinson KD, Polymeropoulos MH. The ubiquitin pathway in Parkinson's disease. *Nature.* 1998 Oct 1;395(6701):451-2.

Lesage S, Anheim M, Condroyer C, Pollak P, Durif F, Dupuits C, Viallet F, Lohmann E, Corvol JC, Honoré A, Rivaud S, Vidailhet M, Dürr A, Brice A; French Parkinson's Disease Genetics Study Group. Large-scale screening of the Gaucher's disease-related glucocerebrosidase gene in Europeans with Parkinson's disease. *Hum. Mol. Genet.* 2011 Jan 1;20(1):202-10. Epub 2010 Oct 14.

Liu L, Cavanaugh JE, Wang Y, Sakagami H, Mao Z, Xia Z. ERK5 activation of MEF2-mediated gene expression plays a critical role in BDNF-promoted survival of developing but not mature cortical neurons. *Proc. Natl. Acad. Sci. U. S. A.* 2003 Jul 8;100(14):8532-7.

Malagelada C, Jin ZH, Jackson-Lewis V, Przedborski S, Greene LA. Rapamycin protects against neuron death in in vitro and in vivo models of Parkinson's disease. *J. Neurosci.* 2010 Jan 20;30(3):1166-75.

Mancuso M, Coppede F, Migliore L, Siciliano G, Murri L. Mitochondrial dysfunction, oxidative stress and neurodegeneration. *J. Alzheimers Dis.* 2006 Sep;10(1):59-73. Review.

Marin C, Aguilar E. In vivo 6-OHDA-induced neurodegeneration and nigral autophagic markers expression. *Neurochem. Int.* 2011 Mar;58(4):521-6. Epub 2011 Jan

Martinez-Vicente M, Sovak G, Cuervo AM. Protein degradation and aging. *Exp. Gerontol.* 2005 Aug-Sep;40(8-9):622-33. Review.

Matsui H, Ito H, Taniguchi Y, Takeda S, Takahashi R. Ammonium chloride and tunicamycin are novel toxins for dopaminergic neurons and induce Parkinson's disease-like phenotypes in medaka fish. *J. Neurochem.* 2010 Dec;115(5):1150-60.

Michiorri S, Gelmetti V, Giarda E, Lombardi F, Romano F, Marongiu R, Nerini-Molteni S, Sale P, Vago R, Arena G, Torosantucci L, Cassina L, Russo MA, Dallapiccola B, Valente EM, Casari G. The Parkinson-associated protein PINK1 interacts with Beclin1 and promotes autophagy. *Cell Death Differ.* 2010 Jun;17(6):962-74.

Mortiboys H, Thomas KJ, Koopman WJ, Klaffke S, Abou-Sleiman P, Olpin S, Wood NW, Willems PH, Smeitink JA, Cookson MR, Bandmann O. Mitochondrial function and morphology are impaired in parkin-mutant fibroblasts. *Ann. Neurol.* 2008 Nov;64(5):555-65.

Narendra DP, Jin SM, Tanaka A, Suen DF, Gautier CA, Shen J, Cookson MR, Youle RJ. PINK1 is selectively stabilized on impaired mitochondria to activate Parkin. *PLoS Biol.* 2010 Jan 26;8(1):e1000298.

Neumann J, Bras J, Deas E, O'Sullivan SS, Parkkinen L, Lachmann RH, Li A, Holton J, Guerreiro R, Paudel R, Segarane B, Singleton A, Lees A, Hardy J, Houlden H, Revesz T, Wood NW. Glucocerebrosidase mutations in clinical and pathologically proven Parkinson's disease. *Brain.* 2009 Jul;132(Pt 7):1783-94. Epub 2009 Mar 13.

Nishikawa K, Li H, Kawamura R, Osaka H, Wang YL, Hara Y, Hirokawa T, Manago Y, Amano T, Noda M, Aoki S, Wada K. Alterations of structure and hydrolase activity of parkinsonism-associated human ubiquitin carboxyl-terminal hydrolase L1 variants. *Biochem. Biophys. Res. Commun.* 2003 Apr 25;304(1):176-83.

Nixon RA, Wegiel J, Kumar A, Yu WH, Peterhoff C, Cataldo A, Cuervo AM. Extensive involvement of autophagy in Alzheimer disease: an immuno-electron microscopy study. *J. Neuropathol. Exp. Neurol.* 2005 Feb;64(2):113-22.

Nixon RA. Autophagy in neurodegenerative disease: friend, foe or turncoat? *Trends Neurosci.* 2006 Sep;29(9):528-35. Epub 2006 Jul 20. Review.

Pan T, Kondo S, Le W, Jankovic J. The role of autophagy-lysosome pathway in neurodegeneration associated with Parkinson's disease. *Brain.* 2008 Aug;131(Pt 8):1969-78. Epub 2008 Jan 10. Review.

Paxinou E, Chen Q, Weisse M, Giasson BI, Norris EH, Rueter SM, Trojanowski JQ, Lee VM, Ischiropoulos H. Induction of alpha-synuclein aggregation by intracellular nitrative insult. *J. Neurosci.* 2001 Oct 15;21(20):8053-61.

Perier C, Tieu K, Guégan C, Caspersen C, Jackson-Lewis V, Carelli V, Martinuzzi A, Hirano M, Przedborski S, Vila M. Complex I deficiency primes Bax-dependent neuronal apoptosis through mitochondrial oxidative damage. *Proc. Natl. Acad. Sci. U. S. A.* 2005 Dec 27;102(52):19126-31. Epub 2005 Dec 19.

Pickford F, Masliah E, Britschgi M, Lucin K, Narasimhan R, Jaeger PA, Small S, Spencer B, Rockenstein E, Levine B, Wyss-Coray T. The autophagy-related protein beclin 1 shows reduced expression in early Alzheimer disease and regulates amyloid beta accumulation in mice. *J. Clin. Invest.* 2008 Jun;118(6):2190-9.

Polymeropoulos MH, Lavedan C, Leroy E, Ide SE, Dehejia A, Dutra A, Pike B, Root H, Rubenstein J, Boyer R, Stenroos ES, Chandrasekharappa S, Athanassiadou A, Papapetropoulos T, Johnson WG, Lazzarini AM, Duvoisin RC, Di Iorio G, Golbe LI, Nussbaum RL. Mutation in the alpha-synuclein gene identified in families with Parkinson's disease. *Science.* 1997 Jun 27;276(5321):2045-7.

Prigione A, Piazza F, Brighina L, Begni B, Galbussera A, Difrancesco JC, Andreoni S, Piolti R, Ferrarese C. Alpha-synuclein nitration and autophagy response are induced in peripheral blood cells from patients with Parkinson disease. *Neurosci. Lett.* 2010 Jun 14;477(1):6-10. Epub 2010 Apr 24.

Ravikumar B, Duden R, Rubinsztein DC. Aggregate-prone proteins with polyglutamine and polyalanine expansions are degraded by autophagy. *Hum. Mol. Genet.* 2002 May 1;11(9):1107-17.

Rodríguez-Navarro JA, Rodríguez L, Casarejos MJ, Solano RM, Gómez A, Perucho J, Cuervo AM, García de Yébenes J, Mena MA. Trehalose ameliorates dopaminergic and tau pathology in parkin deleted/tau overexpressing mice through autophagy activation. *Neurobiol. Dis.* 2010 Sep;39(3):423-38. Epub 2010 May 28.

Sapp E, Schwarz C, Chase K, Bhide PG, Young AB, Penney J, Vonsattel JP, Aronin N, DiFiglia M. Huntingtin localization in brains of normal and Huntington's disease patients. *Ann. Neurol.* 1997 Oct;42(4):604-12.

Sardiello M, Palmieri M, di Ronza A, Medina DL, Valenza M, Gennarino VA, Di Malta C, Donaudy F, Embrione V, Polishchuk RS, Banfi S, Parenti G, Cattaneo E, Ballabio A. A gene network regulating lysosomal biogenesis and function. *Science.* 2009 Jul 24;325(5939):473-7. Epub 2009 Jun 25.

Sarkar S, Davies JE, Huang Z, Tunnacliffe A, Rubinsztein DC. Trehalose, a novel mTOR-independent autophagy enhancer, accelerates the clearance of mutant huntingtin and alpha-synuclein. *J. Biol. Chem.* 2007 Feb 23;282(8):5641-52. Epub 2006 Dec 20.

Schapira AH, Gu M, Taanman JW, Tabrizi SJ, Seaton T, Cleeter M, Cooper JM. Mitochondria in the etiology and pathogenesis of Parkinson's disease. *Ann. Neurol.* 1998 Sep;44(3 Suppl 1):S89-98. Review.

Sidransky E, Nalls MA, Aasly JO, Aharon-Peretz J, Annesi G, Barbosa ER, Bar-Shira A, Berg D, Bras J, Brice A, Chen CM, Clark LN, Condroyer C, De Marco EV, Dürr A, Eblan MJ, Fahn S, Farrer MJ, Fung HC, Gan-Or Z, Gasser T, Gershoni-Baruch R, Giladi N, Griffith A, Gurevich T, Januario C, Kropp P, Lang AE, Lee-Chen GJ, Lesage S, Marder K, Mata IF, Mirelman A, Mitsui J, Mizuta I, Nicoletti G, Oliveira C, Ottman R, Orr-Urtreger A, Pereira LV, Quattrone A, Rogaeva E, Rolfs A, Rosenbaum H, Rozenberg R, Samii A, Samaddar T, Schulte C, Sharma M, Singleton A, Spitz M, Tan EK, Tayebi N, Toda T, Troiano AR, Tsuji S, Wittstock M, Wolfsberg TG, Wu YR, Zabetian CP, Zhao Y, Ziegler SG. Multicenter analysis of glucocerebrosidase mutations in Parkinson's disease. *N. Engl. J. Med.* 2009 Oct 22;361(17):1651-61.

Ramirez A, Heimbach A, Gründemann J, Stiller B, Hampshire D, Cid LP, Goebel I, Mubaidin AF, Wriekat AL, Roeper J, Al-Din A, Hillmer AM, Karsak M, Liss B, Woods CG, Behrens MI, Kubisch C. Hereditary parkinsonism with dementia is caused by mutations in ATP13A2, encoding a lysosomal type 5 P-type ATPase. *Nat. Genet.* 2006 Oct;38(10):1184-91.

Sikorska B, Liberski PP, Giraud P, Kopp N, Brown P. Autophagy is a part of ultrastructural synaptic pathology in Creutzfeldt-Jakob disease: a brain biopsy study. *Int. J. Biochem. Cell Biol.* 2004 Dec;36(12):2563-73.

Singleton AB, Farrer M, Johnson J, Singleton A, Hague S, Kachergus J, Hulihan M, Peuralinna T, Dutra A, Nussbaum R, Lincoln S, Crawley A, Hanson M, Maraganore D, Adler C, Cookson MR, Muenter M, Baptista M, Miller D, Blancato J, Hardy J, Gwinn-Hardy K. alpha-Synuclein locus triplication causes Parkinson's disease. *Science.* 2003 Oct 31;302(5646):841.

Smith PD, Crocker SJ, Jackson-Lewis V, Jordan-Sciutto KL, Hayley S, Mount MP, O'Hare MJ, Callaghan S, Slack RS, Przedborski S, Anisman H, Park DS. Cyclin-dependent kinase 5 is a mediator of dopaminergic neuron loss in a mouse model of Parkinson's disease. *Proc. Natl. Acad. Sci. U. S. A.* 2003 Nov 11;100(23):13650-5. Epub 2003 Oct 31.

Smith PD, Mount MP, Shree R, Callaghan S, Slack RS, Anisman H, Vincent I, Wang X, Mao Z, Park DS. Calpain-regulated p35/cdk5 plays a central role in dopaminergic neuron death through modulation of the transcription factor myocyte enhancer factor 2. *J. Neurosci.* 2006 Jan 11;26(2):440-7.

Soto C, Estrada LD. Protein misfolding and neurodegeneration. *Arch. Neurol.* 2008 Feb;65(2):184-9. Review.

Spencer B, Potkar R, Trejo M, Rockenstein E, Patrick C, Gindi R, Adame A, Wyss-Coray T, Masliah E. Beclin 1 gene transfer activates autophagy and ameliorates the neurodegenerative pathology in alpha-synuclein models of Parkinson's and Lewy body diseases. *J. Neurosci.* 2009 Oct 28;29(43):13578-88.

Spillantini MG, Schmidt ML, Lee VM, Trojanowski JQ, Jakes R & Goedert M. (1997). Alpha-synuclein in Lewy bodies. *Nature,* Vol.388, No.6645, (August 1997), pp. 839-40, ISSN 0028-0836

Tain LS, Mortiboys H, Tao RN, Ziviani E, Bandmann O, Whitworth AJ. Rapamycin activation of 4E-BP prevents parkinsonian dopaminergic neuron loss. *Nat. Neurosci.* 2009 Sep;12(9):1129-35. Epub 2009 Aug 16.

Tanaka A. Parkin-mediated selective mitochondrial autophagy, mitophagy: Parkin purges damaged organelles from the vital mitochondrial network. *FEBS Lett.* 2010 Apr 2;584(7):1386-92.

Tanaka K, Suzuki T, Chiba T, Shimura H, Hattori N, Mizuno Y. Parkin is linked to the ubiquitin pathway. *J. Mol. Med.* 2001 Sep;79(9):482-94. Review.

Tang X, Wang X, Gong X, Tong M, Park D, Xia Z, Mao Z. Cyclin-dependent kinase 5 mediates neurotoxin-induced degradation of the transcription factor myocyte enhancer factor 2. *J. Neurosci.* 2005 May 11;25(19):4823-34.

Tatsuta T, Langer T. Quality control of mitochondria: protection against neurodegeneration and ageing. *EMBO J.* 2008 Jan 23;27(2):306-14. Review.

Tsvetkov AS, Miller J, Arrasate M, Wong JS, Pleiss MA, Finkbeiner S. A small-molecule scaffold induces autophagy in primary neurons and protects against toxicity in a Huntington disease model. *Proc. Natl. Acad. Sci. U. S. A.* 2010 Sep 28;107(39):16982-7. Epub 2010 Sep 10.

Valente EM, Abou-Sleiman PM, Caputo V, Muqit MM, Harvey K, Gispert S, Ali Z, Del Turco D, Bentivoglio AR, Healy DG, Albanese A, Nussbaum R, González-Maldonado R, Deller T, Salvi S, Cortelli P, Gilks WP, Latchman DS, Harvey RJ, Dallapiccola B, Auburger G, Wood NW. Hereditary early-onset Parkinson's disease caused by mutations in PINK1. *Science.* 2004 May 21;304(5674):1158-60

Vekrellis K, Rideout HJ, Stefanis L. Neurobiology of alpha-synuclein. *Mol. Neurobiol.* 2004 Aug;30(1):1-21.

Webb JL, Ravikumar B, Atkins J, Skepper JN, Rubinsztein DC. Alpha-Synuclein is degraded by both autophagy and the proteasome. *J. Biol. Chem.* 2003 Jul 4;278(27):25009-13. Epub 2003 Apr 28.

Winslow AR, Chen CW, Corrochano S, Acevedo-Arozena A, Gordon DE, Peden AA, Lichtenberg M, Menzies FM, Ravikumar B, Imarisio S, Brown S, O'Kane CJ, Rubinsztein DC. α-Synuclein impairs macroautophagy: implications for Parkinson's disease. *J. Cell Biol.* 2010 Sep 20;190(6):1023-37.

Wong E, Cuervo AM. Autophagy gone awry in neurodegenerative diseases. *Nat. Neurosci.* 2010 Jul;13(7):805-11. Review.

Xilouri M, Vogiatzi T, Vekrellis K, Park D, Stefanis L. Abberant alpha-synuclein confers toxicity to neurons in part through inhibition of chaperone-mediated autophagy. *PLoS One.* 2009;4(5):e5515. Epub 2009 May 13.

Xilouri M, Stefanis L. Autophagic pathways in Parkinson disease and related disorders. *Expert Rev. Mol.Med.* 2011 Mar 21;13:e8. Review

Yang Q, She H, Gearing M, Colla E, Lee M, Shacka JJ, Mao Z. Regulation of neuronal survival factor MEF2D by chaperone-mediated autophagy. *Science.* 2009 Jan 2;323(5910):124-7.

Yang Q, Mao Z. Parkinson disease: a role for autophagy? *Neuroscientist.* 2010 Aug;16(4):335-41. Epub 2010 Apr 1. Review.

Zhang Y, Dawson VL, Dawson TM. Oxidative stress and genetics in the pathogenesis of Parkinson's disease. *Neurobiol. Dis.* 2000 Aug;7(4):240-50.

Zhou C, Huang Y, Shao Y, May J, Prou D, Perier C, Dauer W, Schon EA, Przedborski S. The kinase domain of mitochondrial PINK1 faces the cytoplasm. *Proc. Natl. Acad. Sci. U. S. A.* 2008 Aug 19;105(33):12022-7. Epub 2008 Aug 7.

Zhu J, Chu CT. Mitochondrial dysfunction in Parkinson's disease. *J. Alzheimers Dis.* 2010;20 Suppl 2:S325-34. Review.

Zimprich A, Biskup S, Leitner P, Lichtner P, Farrer M, Lincoln S, Kachergus J, Hulihan M, Uitti RJ, Calne DB, Stoessl AJ, Pfeiffer RF, Patenge N, Carbajal IC, Vieregge P, Asmus F, Müller-Myhsok B, Dickson DW, Meitinger T, Strom TM, Wszolek ZK, Gasser T. Mutations in LRRK2 cause autosomal-dominant parkinsonism with pleomorphic pathology. *Neuron.* 2004 Nov 18;44(4):601-7

Ziviani E, Tao RN, Whitworth AJ. Drosophila parkin requires PINK1 for mitochondrial translocation and ubiquitinates mitofusin. *Proc. Natl. Acad. Sci .U. S. A.* 2010 Mar 16;107(11):5018-23. Epub 2010 Mar 1.

In: Autophagy: Principles, Regulation and Roles in Disease
Editor: Nikolai Gorbunov
ISBN: 978-1-61942-266-7

Chapter XIII

Autophagy and Parkinson's Disease: A Love-Hate Relationship

José M. Fuentes, José M. Bravo-San Pedro, Rubén Gómez-Sánchez, Mireia Niso-Santano and Rosa A. González-Polo

Centro de Investigación Biomédica en Red sobre Enfermedades Neurodegenerativas (CIBERNED): Departamento de Bioquímica y Biología Molecular y Genética, E. Enfermería y T.O., Universidad de Extremadura, Cáceres, Spain

Abstract

Parkinson's disease (PD) is a disorder characterized by a progressive degeneration of midbrain neurons in the *substantia nigra* and by the presence of cytoplasmic protein inclusions. Therefore, the pathogenesis of this disease has been linked to dysfunctions of cellular degradation. The etiology of this disease is unknown, although it seems clear that the pathogenesis involves a multifactorial process combining genetic predisposition (mutations, deletions, etc.) and exposition to certain environmental factors.

Autophagic dysfunction is emerging as an important topic in the study of neurodegenerative diseases in which appears aggregated proteins, including PD. Autophagy involves the sequestration of portions of the cytoplasm (cytosol and/or organelles) in structures called autophagosomes and their subsequent lysosomal enzyme degradation. Although, some early studies identified the autophagy as a mechanism of cell death, in recent years it is placing value as a process by which cells can adapt to changes and stress, including nutrient deprivation, hypoxia, DNA damage, mitochondrial alterations or endoplasmic reticulum stress, among others. In this sense, it has recently demonstrated that genetic ablation of autophagy in mice induces neurodegeneration and accumulation of ubiquitinated proteins. In addition, some genetic mutations that cause neurodegenerative diseases can directly affect the proteolytic systems responsible for the degradation of the mutant protein. In PD, the death of dopaminergic neurons in the *substantia nigra* is associated with the accumulation of α-synuclein, which has synaptic functions within inclusions called Lewy bodies. Reflecting a multifactorial disease process, the pattern of cell death in PD is complex, with characteristics of apoptosis and necrosis as well as accumulations autophagosome-like structures. Interestingly, different types of mutant genes that trigger familial PD encode proteins that act in ubiquitin-dependent proteasomal system, underlining the pathogenic importance of failure of proteolysis in PD. Furthermore, various parkinsonian-related toxicants have been shown to impair mitochondrial function, redox balance, and to some extent protein degradation machinery.

The knowledge of the role of autophagy in the pathogenesis of PD and the study of the regulation of the interaction between environmental agents and genes mutations can open new lines of research to improve human health and the clarification of different alternatives therapeutic potential of PD.

1. Introduction

1.1. Parkinson's Disease

1.1.1. Definition and Epidemiology

Parkinson's disease (PD) was first formally described in modern times in "*An Essay on the Shaking Palsy*" published in 1817 by a London physician named James Parkinson (1755-1824) [1]. James Parkinson systematically described the medical history of six individuals who had symptoms of the disease that eventually bore his name. It was not until 1861 and 1862 that Jean-Martin Charcot [2] with Alfred Vulpian added more symptoms to James Parkinson's clinical description and then subsequently confirmed James Parkinson's place in medical history by attaching the name PD to the syndrome. Charcot added to the list of symptoms the mask face, various forms of contractions of hands and feet, akathesia as well as rigidity.

Actually, PD is the most common form of parkinsonism, an general term which is used to describe all movement disorders with parkinsonian features independent of their (unknown) etiology. In any case, PD is the second most frequent neurodegenerative disorder (after Alzheimer's disease) and the most common movement disorder. We can divide its symptoms into two main groups:

1) Symptoms motors as tremor or trembling in hands, arms, legs, jaw, and face; rigidity or stiffness of the limbs and trunk; bradykinesia or slowness of movement; and postural instability or impaired balance and coordination. As these symptoms become more pronounced, patients may have difficulty walking, talking, or completing other simple tasks.
2) Non-motors symptoms as depression, emotional changes, difficulty in swallowing and chewing, speech changes, urinary problems or constipation, sleep problems or hyposmia.

PD is more common in men, with a male/female ratio of 3 to 2. No ethnical association is observed. Its prevalence can vary from 31 to 340 per 100,000-population group. Average is considered to have a prevalence of 160-200 cases per 100,000 citizens and an incidence of 20 cases per 100,000 populations per year. Adult (late)-onset PD is most common, but early-onset PD (onset between 21-40 years), and juvenile-onset PD (onset before age 21) also exist increasing in the last years.

The diagnosis of PD is a clinical one, meaning that it is based on your medical history, observations of your symptoms, and a neurologic examination. There are no confirmatory biological assays and routine genetic testing in PD not being used yet [3]. Other medical assays as Computerized Tomography (CT scan), Magnetic Resonance Imaging (MRI), Positron Emission Tomography (PET) Scan or Single-photon Emission Computerized Tomography (SPECT) usually show normal or only minimal changes. Although, at times, alterations in brain imaging studies may be helpful to suggest alternative diagnoses. Significant improvement with supplementation of L-DOPA (L-3,4-dihydroxyphenylalanine), in the absence of other diagnosis known to cause parkinsonism is also evidence to support the diagnosis.

1.1.2. Origin

Neuropathological changes in PD are on the one hand characterized by preceding depigmentation and progressive death of dopaminergic neurons of the *substantia nigra pars compacta*. Within the *pars compacta*, the neuronal loss tends to be greatest in the ventrolateral layer followed by the medial ventral and the dorsal layer [4]. Clinically a parkinsonian syndrome becomes evident when 50% of the dopaminergic neurons in the *substantia nigra* have died, respectively an abatement of the dopamine concentration in the striatum for about 70-80% [5]. On the other hand, Lewy bodies (LB) intracytoplasmic, eosinophilic inclusion bodies are characteristic depositions in surviving neurons [6]. LB is typically in brains of PD patients but they are not specific for the diagnosis. Within LB there is a dense accumulation of misfolded and aggregated alpha-synuclein, ubiquitin and TorsinA [7-9]. Although the pathogenesis of PD remains elusive, evidence from environmental risk studies and genetic approaches suggest a convergence between energy metabolism and the disposal of damaged proteins in the pathogenesis of PD. The active turnover of damaged proteins has a great importance.

1.1.3. Risk Factors

Over time, different studies have suggested that PD has a multifactorial etiology, which includes both genetic and environmental factors [10]. While recent genetic discoveries have led to a number of different genetic rodent models of PD, none of these recapitulates the typical degeneration of dopaminergic neurons. Moreover, the majority of PD cases are sporadic and do not develop from obvious genetic defects. Therefore, to fully address the questions associated with the underlying mechanisms responsible for the selective neurodegeneration in PD, experimental models using neurotoxins may provide valuable insights. A number of exogenous chemicals and toxins mediate neurotoxicity and have been shown to be involved in PD pathogenesis. In any case, oxidative stress appears to be a key factor to influence the progression of PD development.

In the late nineteenth century, Gowers et al., [11] said of the EP "*... in any occasions the influence of the inheritance could be important ....*" This was the general opinion among researchers until 1980, when there is the discovery of the toxin 1-methyl-4-phenyl-1,2,3,6-tetrahydropyridine (MPTP) [12], which produces a selective death of neurons in the *substantia nigra* in humans and experimental models [13], and induces motor symptoms similar to those observed in PD. This discovery opened the door to further investigation of possible environmental factors that may be related to that disease. Today, there are a large number of compounds such as pesticides, herbicides or industrial chemicals involved in the development of PD. Among the most studied compounds are rotenone, a plant substance used as an insecticide and acts as a potent inhibitor of mitochondrial respiratory chain complex I, 6-hydroxydopamine (6-OHDA) neurotoxin-inducing oxidative stress [14], bipyridine compounds such as MPTP or the herbicide paraquat (PQ), which has been widely studied its involvement in neurodegenerative diseases[14-17]. On the other hand, in 1988 Gowers [11] introduced the possibility that it is an inherited disease due to the considerable number of sick patients with a family history. Thus, knowledge of genetic factors involved in this disease is essential when elucidating the possible causes and mechanisms that occur during its development.

Today, because epidemiological studies, it is known that most cases of individuals with the disease are sporadic and only 5-10% have a familial inheritance pattern, highlighting the importance of environmental factors in origin of the disease. Therefore, it is postulated that the cause of the disease may be due to an interaction between hereditary and environmental factors, where the genetic factor predisposes but does not determine the development of the disease. In 1880 Leroux and Lhirondel, two of Charcot's students at the "Hôpital de la Salpêtrière" in Paris, first described a familial component to PD, "a true cause of paralysis agitans, and may be the only true cause, is heredity" [18]. Since this publication enormous reports dealing with hereditary parkinsonism appeared in the literature.

The typical PD phenotype is produced by mutations in several single genes indicating those gene products pivotal to PD pathogenesis. The functional loss of normal gene product resulting from haplotype insufficiency and/or the toxic gain-function of mutant gene product is considered as the pathogenesis of familial PD. In addition, gene multiplications and promoter polymorphisms could influence the expression levels of these pivotal gene products and consequently alter important interactions at gene–gene, gene–protein, and protein–protein levels, and further cellular interactions leading to typical neuronal degeneration in PD.

There are studies that corroborate this interaction, an example may occur in studies that demonstrate how individuals with mutations in the gene *PARK9* increased susceptibility to excess manganese poisoning, and subsequent neuronal death, so that these individuals if they are in environments with high concentrations of lead, present a predisposition to PD [19]. On the other hand, in some animal PD models, α-synuclein-positive aggregation in nigral cells has been observed following MPTP exposure, which is again suggestive of an interaction between genetic and environmental factors [20, 21].

1.1.3.1. Environmental

There are many epidemiological studies found an association between PD and certain environmental factors.

A shared etiological component in PD and psychiatric diseases has been hypothesized because psychiatric symptoms and diseases, primarily depression and anxiety disorders, are common in PD patients. Epidemiological evidence indicates an association between depression and PD that may be explained by shared etiological factors; alternatively, depression may be an early manifestation of PD. Anxiety disorders also seem to be related to PD, but such studies are few [22].

In boxers, brain damage due to repeated head trauma leading to progressive cognitive dysfunction, psychiatric, as well as parkinsonian symptoms, has been recognized since the 1960s [23]. This syndrome is distinct from PD but a role of head trauma in PD still has been investigated chiefly in hospital-based case-control studies, in which recall bias is an obvious concern. Further, head injury was most commonly assessed by self-report. In this sense, although some retrospective studies reported an association between head trauma and PD, these studies are prone to recall bias. Because the association was not confirmed in prospective studies, there is no compelling evidence for a role of head trauma in PD.

A possible infectious cause of PD has been suggested because patients suffering from encephalitis lethargica in the 1920s sometimes developed parkinsonism. Because influenza A virus may affect brain regions implicated in PD (including the *substantia nigra*) [24], it has been hypothesized to play a role in PD [25]. Parkinsonism, however, is also observed (either in the acute phase or as a long-term complication) in viral encephalitis due to viruses other than influenza [26]. Overall, however, there is no clear evidence for involvement of an infectious agent in PD. It has been suggested that an intrauterine infection affected development of dopaminergic neurons resulting in lower reserves, which might be associated with increased vulnerability to PD in adulthood [27]. Thus, conclusive evidence regarding the possible relation between viral infections and PD is lacking.

Living in certain rural environments in which there is sewage, stagnant water consumption or contact with certain pesticides could be a risk factor for PD [28], however it is not easy to say such risk factors as agent's triggers of the disease. Toxicological studies have showed that sub-acute exposure to the common pesticides rotenone and PQ induces behavioral and pathological changes characteristic of PD in animal models [14, 29]. The positive association between environmental neurotoxicant exposure and PD indicates the important role of environmental factors in the process of nigrostriatal degeneration in PD.

1.1.3.1.1. MPTP

The first time that was clearly associated the development of PD with the exposition to an environmental factor was in the early 1980s. In this year, several drug abusers from Northern California developed an acute state of akinesia following the intravenous injection of a street preparation of 1-methyl-4-phenylpyridinium (MPP^+), an analog of the narcotic meperidine [12]. After investigation, J.W. Langston [30] was found that MPTP, which was unintentionally produced during the illicit synthesis of MPP^+, was the causative agent of this dramatic clinical outcome.

In addition, post mortem studies confirmed the presence of LB in the substantia nigra of these patients [31]. The MPTP was thereafter regarded as a powerful inducer of nigral degeneration in animals and causes symptoms similar to Parkinson's in some species such as rats, mice, dogs, cats and monkeys [32].

This discovery opened the door to further investigation of other environmental factors, mainly pesticides, which may be related to the origin of the disease. Nowadays, it known a large number of compounds that may be involved in disease development. Among the most studied compounds are rotenone, a plant substance used as an insecticide and acts as a potent inhibitor of mitochondrial respiratory chain complex I [33], 6-OHDA, pesticides such as dieldrin and maneb, which selectively inhibit complex III of the mitochondrial transport chain [34, 35] or herbicides such as PQ which has studied its involvement in neurodegenerative diseases [15-17, 36, 37].

Since MPTP is a lipophilic compound it can easily crosses the blood-brain barrier. In brain it is converted to more reactive molecule, i.e., the ion MPP^+, by the glial monoamine oxidase-B (MAO-B). For this reason, inhibitors of MAO-B exert protective effects against toxicity induced by MPTP [38]. There are several hypotheses on the mechanisms involved in the transmembrane transport of MPP^+; among them the dopamine transporter (DAT) [39], the amino acid transporter cationic [40], or glutamate transporter [41].

Inside the cell, MPP^+ is transported into the mitochondria acts as a potent inhibitor of α-ketoglutarate dehydrogenase and mitochondrial complex I [42]. This inhibition of electron transport chain results in decreased intracellular ATP levels, thereby decreasing cellular energy [43, 44]. In addition to the inhibition of mitochondrial electron transport chain, MPP^+ causes loss of membrane potential and promotes oxidative stress and homeostatic imbalance [44]. Besides that, exposure to MPP^+ results in a decrease in levels of dopamine and the appearance of protein aggregates.

1.1.3.1.2. 6- Hydroxydopamine

6-OHDA is a neurotoxin which uses the same catecholamine transport system as dopamine. It shares some structural similarities with dopamine and norepinephrine, exhibiting a high affinity for several catecholaminergic plasma membrane transporters such as dopamine and norepinephrine transporters. The cell death due accumulation of 6-OHDA is mediated by the reactive oxygen species produced via Haber-Weiss or Fenton type reactions in the presence of 6-OHDA [45].

1.1.3.1.3. Paraquat

PQ (*N,N'*-dimethyl-4,4′-bipyridinium dichloride) is one of the most widely used herbicides in the world. While the United States and the European Union have banned its use, PQ is still widely used in developed countries [46]. PQ is an herbicide that has a structure very similar to the metabolite MPP^+. Because of this structural homology and epidemiological studies showing a correlation between pesticide exposure and the development of PD, the studies focused on studying the relationship between the PQ and the development of PD. The PQ is a widely used herbicide for weed control. Not at high risk of environmental contamination, however, has proven to be a highly toxic compound for humans and animals, and in recent decades there have been many cases of poisoning and death by this herbicide [47, 48].

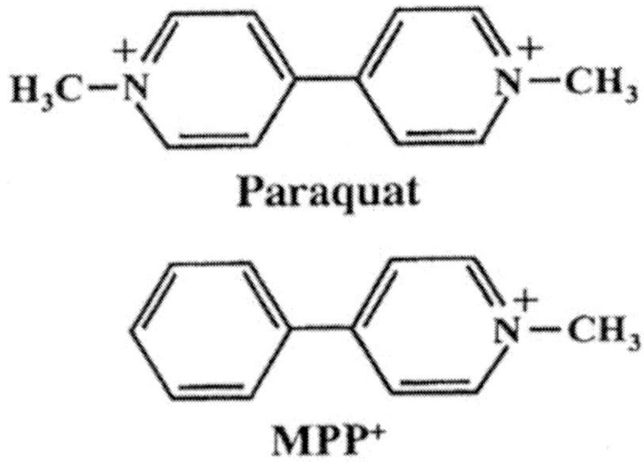

Figure 1. Chemical structure of PQ and MPP^+.

The PQ exposure pathways in humans and animals are ingestion or skin contact. The most common route of poisoning is oral, although there have been reports of poisoning by intravenous and subcutaneous. Although theoretically the PQ is not absorbed through the skin, there are cases of poisoning by the pesticide dermal [49] and this is the main route of exposure of workers from prolonged contact. It was observed that acute exposure to PQ causes lung injury, whereas continued exposure to the pesticide induced parkinsonism [29]. The ability of PQ to cross the blood brain barrier is limited because it is a charged molecule with a hydrophilic structure. Therefore, it is unlikely the passive entrance of PQ through the blood brain barrier. Thus, it has suggested the involvement of an active transport system, the neutral amino acid transporter at the entrance of PQ to the nervous system, and that pretreatment with L-valine and L-phenylalanine prevents PQ-mediated neurodegeneration in mice [50, 51]. Once inside the central nervous system, the toxicity of PQ is largely due to its redox cycle. PQ is capable of accepting electrons from a suitable donor [52]. Many diaphorase enzymes that transfer an electron from NAD(P)H, are capable of reducing PQ to PQ cation radical. This radical is readily oxidized in the presence of oxygen generating the superoxide radical. Subsequently produced a series of chain reactions that lead to the generation of reactive oxygen species (ROS), mainly hydrogen peroxide and hydroxyl radical [29]. Hydroxyl radicals are involved in the damage of cell membranes by lipid peroxidation. Besides the generation of ROS, PQ produces depletion of reducing equivalents, due to increased oxidation, resulting in the disruption of important biochemical processes that require NADPH [48, 52, 53]. It has been studied the implication of mitochondria in ROS production induced by PQ. In this regard, it has been identified PQ-induced toxic effects at the level of the different complexes of the mitochondrial respiratory chain [54] especially at complex I.

1.1.3.1.4. Rotenone

Rotenone is another environmental toxin that has been associated with complex I dysfunction in PD. It is a naturally occurring compound commonly used as an organic insecticide. It is highly lipophilic and thus readily gains access to all organs including the brain. It can also freely cross the plasma membrane and be accumulated in subcellular organelles such as mitochondria. In the mitochondria, it has been described as a potent inhibitor of complex I through impairing oxidative phosphorylation by inhibiting reduced NADH-ubiquinone reductase activity [55]. Interestingly, despite its predicted inhibition of complex I in all the regions of the brain, rotenone caused selective degeneration of the nigrostriatal dopaminergic neurons indicated by selective striatal oxidative damage, and formation of LB-like ubiquitin- and α-synuclein-positive inclusions in nigral cells.

1.1.3.2. Genetics

The existence of a family history that has had the PD is a risk factor in developing PD [56]. The observation by scientists working cases of parkinsonism that there were these familial cases led to increased research on possible genetic predisposition to the development of PD.

Thus, in 1997, in Italian and Greek families who had an inherited form of PD, it was identified an autosomal dominant mutation of the gene *PARK1,* encoding the protein α-synuclein [57]. This fact, together with the discovery that the α-synuclein is a major component of LB [58], enabling a greater interest in research on genetic aspects of PD. In subsequent years yet genes were discovered that are involved in PD (Table 1). In 1998 it was identified the *PARK2* gene, which encodes the protein parkin [59], which appears a variant mutated in inherited juvenile PD. Recent studies have identified new key mutations in PD as in the case of a mutation in in *PARK7* gene, which encodes protein DJ-1 in Dutch and Italian families [60, 61], responsible for autosomal recessive variant of the PD. Besides that it was identified a mutation in the *PARK6* gene, encoding the PINK1 protein. This mutation was tentatively associated with metabolic failure and neuronal death in the *substantia nigra* [62]. The role of LRRK2 protein in the development of PD is also under current investigation. LRRK2 (encoded by the gene *PARK8)* is considered to be a one of key elements in the "molecular puzzle" of the etiology and pathogenesis of PD.

In 2004, *PARK8* gene mutations were discovered as a major genetic causes associated with hereditary parkinsonism [63-65]. *PARK8* gene was first studied in the Japanese Sagamihara family, which responded positively to treatment with L-DOPA and parkinsonism presenting late with unknown etiology of the disease [66].

1.1.3.2.1. α-Synuclein (PARK1, PARK4)

PD is characterized by the presence of intracellular inclusions called LB, which are mainly composed of α-synuclein. Missense mutations (A53T, A30P and E46K) or multiplications (duplications and triplications) in the *α-synuclein* gene (*PARK1/4*) cause autosomal-dominant parkinsonism [57, 67-70], but it is still unclear whether fibrils of aggregated α-synuclein, as seen in LB, have a critical causative role in the more common forms of PD or they are simply markers of the underlying pathogenetic process. α-synuclein appears in three common forms, monomers, dimers, and protofibrils. It was shown that excess of the protofibril form can inhibit ubiquitin-proteasome system (UPS) *in vitro* [71-73], although these effect has not been confirmed in *in vivo* experiments [74]. Different reports have documented that a small fraction of α-synuclein from LB is monoubiquitinated [75, 76], and it may be developed by SIAH [77, 78], but it is still unknown if this is or not a protective mechanism, depending on the stage of LB maduration.

1.1.3.2.2. PINK1/PARKIN (PARK6/PARK2)

Another typical PD characteristic is the mitochondrial dysfunction. It has been reported that *post-mortem* analysis in the *substantia nigra* of some patients with PD showed complex I deficiency [79]. Also, the presence of the mitochondrial DNA ($_{mt}$DNA) deletions [80] and oxidative stress [81] are higher in patients with parkinsonism. In this sense, there are 2 genes related with PD, *PINK1* (*PARK6*) and *Parkin* (*PARK2*) that may be involved in the regulation of mitochondrial homeostasis.

Parkin mutations have been first linked to a rare form of autosomal resessive juvenile-onset form of PD in Japanese families [59, 82]. Parkin-mutations are common in families with early onset of PD and are found in up to 50% of early-onset individuals with a positive family history of PD. Since then a wide variety of parkin mutations have been described including deletions, multiplications and missense mutations [64, 83, 84] parkin protein acts as an E3 ubiquitin protein ligase [85] in the UPS. Ubiquitination of proteins leads to proteasomal protein degradation. Consequently, parkin mutations should lead to an incorrect ubiquitination and an invalid targeting to the proteasome leading to protein accumulation.

Table 1. Family of genes identified in the PD

LOCI	CROMOSOME	EXPRESSED PROTEIN	YEAR OF IDENTIFICATION
park1	4q21-q23	α-synuclein	1996
park2	6q25.2-q27	Parkin	1997
park3	2p13	¿?	1998
park4	4p15	α-synuclein (triplication)	1999
park5	4p14	UCHL1	1998
park6	1p35-36	PINK1	2001
park7	1p36.33-p36.12	DJ-1	2001
park8	12p11.23-q13.11	LRRK2	2002
park9	1p36	ATP13A2	2001
park10	1p32	¿?	2002
park11	2q36-q37	GIFYF2	2003
park12	Xq21-q25	¿?	2003
park13	2q12	HtrA2/Omi	2005
park14	22q13.1	PLA2G6	2006
park15	22q12-q13	FBXO7	2008
park16	1q32	RAB7L1?, NUCKS1?	2009
park17	4p	GAK/DGKQ	2010
park18	6p	HLA-DRA	2010

PINK1 (PTEN-induced putative kinase-1) is a serine/threonine kinase capable of autophosphorilation [86], with an N-terminal mitochondrial targeting signal (MTS) and is synthesized as a full-length version (FL) and is processed into at least two cleaved forms (Δ1 and Δ2) [87]. Mutations in *PINK1* are a cause of autosomal recessive PD [88], and *PINK1* knockout (KO) flies have a motor deficit and disorganized mitochondrial morphology [89]. On the other hand, parkin is an E3 ubiquitin ligase [85, 90], and contains two ring-finger domains separated by an in-between RING-finger domain, but its function is unknown. PINK1 is considered to be a mitochondrial protein with a role in protecting against oxidative stress and apoptosis in *in vitro* models [91]. Very recently PINK1 has been related with the regulation of autophagy and more concretely mitophagy [92-94].

1.1.3.2.3. DJ-1 (PARK7)

DJ-1 is a small protein that belongs to the ThiJ/PfpI protein superfamily and has homologs in all biological kingdoms [95]. DJ1 was initially identified as an oncogene that transformed cell in cooperation with H-Ras [96] and was found to be a potential serum biomarker secreted from breast cancer and uveal malignant melanoma [97, 98].

The involvement of DJ-1 with neurodegeneration became apparent when it was discovered as the DJ-1 gene (*PARK7*) causing autosomal recessive PD in a Dutch family with multiple consanguinity loops [60, 99]. *PARK7* gene is localized to chromosome 1p36 and spans 24 kb at genomic level. It contains eight exons, the first two of which are non-coding and alternatively spliced, while exons 2-7 comprise the open reading frame (ORF) which encodes a 189 amino acid protein of approximately 20 kDa [61, 100].

The expression of the DJ-1 transcripts is ubiquitous [96]. In brain, DJ-1 is moderately expressed in neurons and in astrocytes throughout the central nervous system [95, 101, 102].

The DJ-1 gene is highly conserved in evolution. Mouse and human DJ-1 is 83% and 90% identical at the DNA and protein level, respectively [100].

Different pathogenic mutations have been identified in the *PARK7* gene, including truncation, exonic deletions and homozygous and heterozygous missense mutations [99, 103]. The most dramatic point mutation, L166P, causes protein instability, whereas other mutations such as A104T and M26I have a weaker destabilizing effect on the protein structure. However, the frequency of DJ-1 mutations seems rather low, being estimated at about 1-2 % in early-onset PD.

Structurally, the DJ-1 protein appears to exist in a homodimeric form. Monomer of DJ-1 adopts a structure with six-strand parallel sheet sandwiched by α-helical arrangements (flavodoxin fold) [104, 105]. DJ-1 posses a flavodoxin-like core fold and is homodimer both in the crystal and in the solution. In the dimer, two DJ-1 monomers make extensive contacts mainly through interactions among α-helices 1, 7 and 8 from each monomer. Intringuingly, the parkinsonism phenotype L166P mutant exists only in monomeric form but not as a dimer, whereas the sumoylation deficient Lys130Arg (K130R) mutant exists normally as a dimer [106]. The L166P mutation is located in the center of α-helix 7, which is a major part of the hydrophobic patch. This mutation appears to destabilize the dimeric structure of DJ-1 by promoting the unfolding of its C-terminal region resulting in rapid degradation [107-109]. Hence, stabilization of DJ-1 may serve to prevent PD-associated degradation. X-ray crystallographic analysis indicates a loss of stability for M26I, A104T, and E163K [110], but they appear to retain dimeric structure in solution. M26I disrupts the hydrophobic core; A104T causes packing defects in a critical β-sheet near the C-terminus, and E163K disrupts a structurally critical salt bridge.

Structural alterations of DJ-1 are closely associated with its functions. Thus far, three structural modifications of the DJ-1 dimeric structure have been reported. First, DJ-1 is sumoylated at lysine residue 130 that is thought to be essential in *ras*-dependent cellular transformation, cell growth promotion and anti-UV- induced apoptosis [111]. Second, a cysteine residue (Cys^{106}) of DJ-1 is oxidized to sulfinic acid under oxidative conditions [104, 112]. Although DJ-1 is present in various subcellular compartiments including cytoplasm and nucleus, targeting of protein to mitochondria contributes to its physiological cytoprotective role [112, 113]. Under standard conditions, DJ-1 is localized to the cytosol, but with oxidative stress DJ-1 itself becomes oxidized at Cys^{106} and translocates to the mitochondria. Mitochondrial localization and protection against oxidative stress both depend on oxidation of these cysteine residues. In this sense, a C106A mutant fails to prevent cell death against oxidative stress, indication the importance of oxidation at Cys^{106} by DJ-1 function [100, 112]. Third, insoluble aggregates of DJ-1 are dramatically increased in brains of sporadic PD patients [114]. DJ-1 colocalizes with tau-positive inclusions over a range of neurodegenerative tautopathies, including Alzheimer disease, and with α-synuclein-positive glial inclusions in multiple system atrophy [101, 102, 114]. The physiological function of DJ-1 is unclear but it may have a role in protecting against mitochondrial damage in response to oxidative stress [112].

1.1.3.2.4. LRRK2 (PARK8)

The up-to-date findings suggest that mutations in the *leucine-rich repeat kinase 2* (*LRRK2*) gene are the most common genetic cause of late-onset PD. In 2002, *PARK8* gene mutations were discovered as a major genetic causes associated with hereditary parkinsonism [63-65]. *PARK8* gene was first studied in the japanese Sagamihara family, which responded positively to treatment with L-DOPA and parkinsonism presenting late with unknown etiology of the disease [66]. This protein was associated later with the EP by other studies in two other families (German and Canadian) who also had an autosomal dominant parkinsonism, late-onset [115]. However, little is known about the involvement of LRRK2 in the molecular pathogenesis of PD

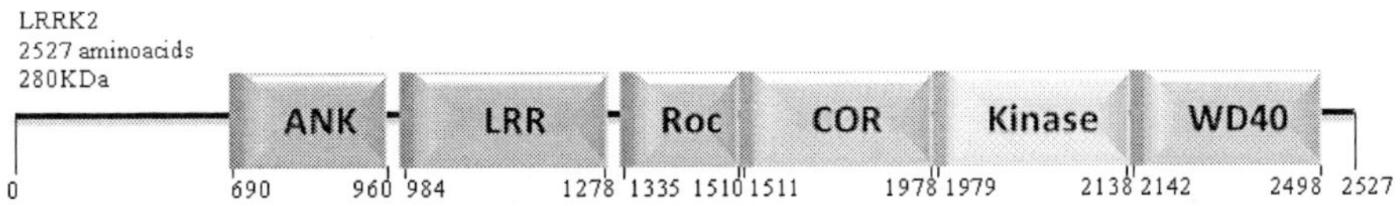

Figure 2. Schematic representation of LRRK2 domain structure: The ideogram in the figure shows a LRRK2 molecule with each of the proposed domains labelled, from N- to C-termini; ANK, ankyrin repeats region; LRR, leucine-rich repeats region; ROC, Ras of complex proteins, GTPase domain; COR, C-terminal of ROC domain; kinase; WD40 region.

In the LRRK2 structure, there exist two well-known functional domains, kinase and GTPase domains. The LRRK2 kinase domain has high homology with other MAPKKK group of tyrosine kinases [116]. In this domain important mutations are related to PD (G2019S, I2020T or K1906M). The G2019S mutation is present in a conserved region of this segment (ADYGIAQYCC), the binding site for Mg^{2+} of the kinase domain [117]. The change of glycine to a serine facilitates access of the kinase domain to its substrates, increasing by 2.5 times the capacity of autophosphorylation and 3 times the capacity of phosphorylation of other substrates like myelin basic protein (MBP) for the LRRK2 phosphorylation capacity without the presence of this mutation [118, 119], responsible for an increased effect of the toxicity of this molecule [120]. Patients with this mutation have an early onset of disease of PD. Mutation is more prevalent in Parkinson patients and is present in 6-7% of cases of familial parkinsonism, and up to 1-2% in sporadic cases [121, 122]. The frequency of this mutation is even higher (13-40%) in certain populations such as North African Arab-Berbers and Ashkenazi Jews [123, 124]. The I2020T mutation is adjacent to the residue in 2019, therefore, also influences the activation site of the kinase domain, increasing 40% capacity of LRRK2 autophosphorylation. It could also modify the substrate specificity and like the G2019S mutation leads to increased toxicity [125, 126]. However, K1906M mutation has the opposite effect in the above study, resulting in elimination of kinase activity [127].

As for the R1441C mutation, there is also controversy regarding the kinase activity. Thus, while some studies have found that this mutation up-regulates kinase activity [128], other ones do not confirm this effect [119]. However, there is a clear relationship between the two domains of LRRK2 in terms of the ability of dimerization of the molecule, because kinase activity is necessary for homodimerization of the molecule [129], while the binding capacity of LRRK2 to GTP or GDP is also involved in homodimerization ability of the molecule [130], so there is a clear cooperation between the two domains.

LRRK2 is a protein that is expressed in the central nervous system, as in other locations outside of the system, such as kidneys, lungs, liver, heart, or leukocytes [131]. LRRK2 is expressed in different brain regions, i.e., in cortex, basal ganglia, cerebellum and hippocampus [65, 132, 133]. As to cellular localization, LRRK2 is mainly presented in cytosolic protein pool; although 10% of LRRK2 pool is located in the mitochondrial outer membrane [65]. LRRK2 is also associated with the plasma membrane, Golgi apparatus, microtubules [134], synaptic vesicles [135], and lipids rafts (lipid rafts) [136].

Currently there is very limited information on the role of LRRK2 in the cell. LRRK2 function is tentatively assigned to control of the cytoskeleton [119, 133], and maintenance of cell morphology [127]. It also participates in protein transport across the synaptic vesicles [135], and the process of ubiquitination [137]. There is a line of evidence indicating a connection of LRRK2 with apoptosis via FADD/caspase 8 mechanism [138]. Recent studies have also shown a relationship between LRRK2 and chaperone proteins. Thus, when we inhibited the interaction between Hsp90 protein (Heat shock protein 90) and LRRK2 we observed latter degradation by the proteasome. Thus Hsp90 may be responsible for maintaining the stability of the molecule and when that stability is altered may result in deletion of LRRK2. [139]. aCHIP (Hsp70-interaction protein) is another protein that has been studied by the interaction of LRRK2 [137], and would intervene in the regulation of molecular stability of LRRK2. There are also a number of studies linking LRRK2 with other proteins involved in PD, such as parkin protein [140, 141], PINK-1 and DJ-1 [142], as well as with α-synuclein. These studies show that an increase in LRRK2 protein regulates the progression of neuropathology induced Parkinson's disease, related to the mutant alpha-synuclein [143].

Proteins phosphorylation is essential in MAPK-linked signal transduction pathways in pathologic states [144]. Two important pathways like JNK and p38 MAPK signaling cascade are implicated in modulating of stress and inflammatory processes and are activated during age-associated neurodegeneration [145]. In 2007, it was showed that activation of Src, Hsp27, and JNK was reduced in patients with G2019S-associated PD, nonsymptomatic G2019S carriers, and patients with sporadic PD compared to healthy controls [146]. Studies indicate relationship between LRRK2 and another MAPK signaling protein like ERK, so as in response to U0126 (ERK inhibition) is induced changes in neuritic autophagy and neurite shortening in dopaminergic neuroblastoma cell line [127]. These results suggest that changes in MAPK signaling may be common in the PD pathophysiology.

1.2. Autophagy

1.2.1. Definition

A correct balance between synthesis and degradation of the cell constituents it is essential for the normal cell function. There are two main mechanisms of degradation of cellular components, namely the UPS [147], and autophagy [148].

The term autophagy comes from two Greek words "auto" meaning self and "phagy" indicating an action of eating, which literally means, "eat yourself". Autophagy is an intracellular catabolic mechanism, highly conserved through evolution, through which the cell recycles or degrade damaged proteins or cytoplasmic organelles [149]. It was first described by Christian de Duve in the 60's, however it was not until the 90's when identified in yeast genes involved in this process. These genes were named thereafter as *ATGs* (Autophagy-Related-Genes) genes, and their discovery was a breakthrough in understanding the autophagy mechanism [150]. At present the number of articles published in relation to autophagy grows exponentially. This demonstrates the growing importance the study of this mechanism in understanding the development of living beings and of many diseases.

Autophagy has been linked to many vital cellular processes such as immune response [151] development [152], aging [153] and neonatal development [154]. It has also been associated with many diseases including cancer [155], various cardiomyopathies [156], musculoskeletal problems [157], or alterations in the metabolism of adipose tissue [158]. Many studies establish relationship between autophagy and neurodegenerative diseases [17, 159, 160]. In this sense, different studies have shown that in the substantia nigra of PD patients there is increased formation of autophagic vacuoles [161]. Paradoxically, although autophagy is an initial process of protection for the cells involved, for example, the recycling of aggregates of proteins that could be toxic to the cell [162] may also play a key role in triggering cell death by execution of an irreversible self-destruction program [163]. In addition, autophagy is associated with apoptosis [164]. This relationship between the two phenomena is very important in the neurodegenerative process that Parkinson's patients suffer.

Autophagy can be classified by the mechanism of delivery of constituents to the autolysosomes and by the selected autophagy target [165]. Thus, there are several distinguishable types of autophagy:

1. Macroautophagy (hereafter called autophagy). In this process, the material degrading is sequestered in double-membrane vesicles forming a typical structure called autophagosome. Depending on the material to degrade within autophagosomes has developed a classification of different types of macroautophagy:
   a) Mitophagy: abduction and selective degradation of mitochondria.
   b) Xenophagy: selective degradation of microbes (bacteria, fungi, parasites or viruses).
   c) Ribophagy: abduction and selective degradation of ribosomes.
   d) Agrephagy: selective degradation of protein aggregates.
   e) Pexophagy: abduction and selective degradation of peroxisomes (microautophagy can also occur.).
   f) Crinophagy: merger direct secretory vesicles to lysosomes.
   g) Reticulophagy: abduction and selective degradation of endoplasmic reticulum.
2. Microautophagy: In this process the material to degrade is sequestered by lysosomes by invagination of its membrane. Once introduced into the interior of the lysosome, the material is degraded by the action of lysosomal enzymes identical to macroautophagy.
3. Chaperone-mediated autophagy (CMA): In this case, material to degrade damaged or misfolded proteins are translocated into the lumen of the lysosome through the membrane. This translocation is mediated by cytosolic and lysosomal chaperones, involving conveyor LAMP-2A (lysosome-associated membrane protein 2A) and is a highly selective degradation.

1.2.2. Regulation

Autophagy pathway has a very complex mechanism of regulation. Despite recent extensive research of this field, there is still considerable lack of understanding these regulatory mechanisms mainly due to complicity of the metabolic and signaling pathways involved into autophagy-mediated constituent recycling and remodeling.

Currently there are several known negative regulators of autophagy. The mammalian target of rapamycin (mTOR) protein is one of the most studied repressor of autophagy. mTOR is a protein kinase, which under favorable conditions for the cell is active, suppressing autophagy by interacting with the phosphatase PP2A [166]. The route fosfoinositido3-kinase (PI3K) class I is also involved in the negative regulation of autophagy through direct interaction with mTOR [167]. NF-kβ, as well as the PI3K class I, also exerts its negative regulation of mTOR activation [168]. Another molecule that has a negative regulation of autophagy is Bcl-2 because it has the ability to inhibit the activation pathway via PI3K class III (by interaction with beclin-1) and the protective Bcl-2 exerts on the membrane mitochondrial [169].

Besides, there are also many key signaling pathways for up-regulation of autophagy. A one of the best investigated cascades is the path-dependent class III PI3K beclin-1, involved in the activation of the early stages of formation of autophagosomes [170]. It is also known that there is a stimulation of autophagy via the extracellular signal-regulated kinases (ERK) [171]. Several recent models implicate the production of reactive oxygen species (ROS) and associated redox signaling mechanisms in the regulation of autophagy [17, 172].

2. Autophagy Deregulation in Parkinson's Disease

In recent years, converging evidence suggest that impairment of homeostasis mechanisms processing unwanted and misfolded proteins plays a central role in the pathogenesis in PD [173]. A proper balance between the formation and degradation of cellular proteins is critical for neuronal survival, and alterations in degradation may lead to aberrant accumulation and aggregation of proteins, resulting in cellular toxicity and neurodegeneration, as seen in PD [174, 175] Emerging evidence supports a critical role for autophagy in the pathogenic process of neurodegeneration [160]. Accumulating biochemical, toxicological and genetic evidence indicate that dysregulation of the autophagic process may be part of the etiological process of PD.

Impairment of the autophagy-lysosomal pathway was related to the development of PD [176]. Activation of autophagy was also identified within peripheral blood mononuclear cells from PD patients [177].

Several lines of evidence suggest that once the UPS is inhibited, autophagy is upregulated and the remaining aggregated proteins are degraded [178]. This mechanism may be one of the explanations of the increased number of autophagic vacuoles in the brains of PD patients, in the early stage of the (MPTP)-injected mouse model of PD and in a progressive mouse model of PD with lysosomal malfunction accompanied by α-synuclein aggregation [21, 161, 179]. However, with pathogenic deterioration, this compensatory mechanism is ultimately unable to maintain the cellular balance and eventually results in neuronal death [180]. Altered autophagy is associated with a variety of neuronal diseases characterized by neurodegeneration and movement disorders [159]. Thus, the upregulation of autophagic markers raises the argument whether autophagy is a cause or a protective factor of neuronal death.

It is encouraging to consider enhancing autophagic capacity as a therapeutic strategy in the prevention of neurodegeneration. Besides, emerging evidence shows that abnormal regulation of autophagic pathways may lead to apoptosis and cell death [181].

2.1. By Toxic Agents Exposure

The MPP^+ is related to the deregulation of the autophagy – mediated recycling mechanism [182, 183]. In the case of toxins that are weak bases, such as MPP^+, it is not clear whether increase in the amount of autophagosomes reflect up-regulation of autophagy or impairment of autophagosomes degradation due to increase in lysosomal pH. Moreover, the role of autophagy in cell survival and cell death has been model-dependent. In some cases the "autophagy inhibitor" 3-Methyladenine, (3-MA) exacerbated cell death [184] and autophagy stimulation conferred protection [185], but in other cases 3-MA ameliorated cell death [186, 187]. Increased amount of autophagosomes have been described in acute injury models involving MPP^+ [182]. While neuroprotective effects of autophagy in toxin models have been correlated with α-synuclein sequestration [184], noncanonical beclin-1-independent autophagy/mitophagy contributes to MPP^+ toxicity, as shown by RNA interference knockdown of Atg7 and LC3/Atg8 [182]. This implicates excessive activation of autophagy, since the participation of beclin 1 in autophagy is downregulated by binding to Bcl-2 [169] or rubicon [188, 189]. Blunting the autophagy response has also been shown using regulator Vps34 to prevent hydrogen peroxide-mediated lysosomal leakage and caspase activation [184]. Likewise, a compound that activates mTOR to suppress autophagy confers protection from oxidative stress in neurons [187] while rapamycin exacerbates toxicity in this system and in primary neurons treated with MPP^+ [190].

It has been demonstrated that oxidative stress has an aggravating effect on the failure of the degradation systems during the late stages of neurodegeneration [191]. Persistent chronic oxidative stress interferes with different aspects of the lysosomal function [191, 192]. The incomplete degradation of lysosomal cargo during oxidative stress could result in the formation of products further impairing lysosomal degradation[192]. 6-OHDA has been shown in neuronal cells to trigger activity-related localization of ERK1/2 in mitochondria and induction of autophagy (mitophagy) [171]. Furthermore, infusion of 6-OHDA in rats was shown to increase the expression of LC3 in the nigral DA neurons [193]. Accumulation of autophagic vacuoles and activation of lysosomes were observed in 6-OHDA injured rat nigral neurons. Coincidently, nuclear lever of cathepsin L was also downregulated by 3-MA. These studies provide the evidence of simultaneous activation of autophagy and nuclear translocation of cathepsin L in PD patients and rodent PD models. The association between oxidative stress and activation of LC3 has been reported after dopamine oxidization in SH-SY5Y neuroblastoma cells [194].

Levels of α-synuclein were reported as elevated in both the frontal cortex and ventral midbrain as well as α-synuclein positive inclusions in substantia nigra neurons of mice treated with PQ [195]. The association of dopaminergic neuronal death with α-synuclein up-regulation and aggregation makes such elevation of α-synuclein observed after PQ toxication particularly relevant as a PD model. PQ seems to be able to trigger the sequential activation of c-Jun N-terminal kinase (JNK), c-Jun, and caspase-3, indicating that JNK signaling pathways could be involved in PQ-induced neurodegeneration [16, 37, 196]. Moreover, our group showed that PQ-exposure induces an early reticulum stress response that is correlated with the adaptive activation of autophagy [17, 36, 197]. This response is accelerated in cells that overexpress wild type of ASK1 (apoptosis signal kinase 1) protein. However, when the toxic stimuli remain, the cell succumbs to an apoptotic cell death. These results suggest that autophagy has an important role in the cell death/survival events produced by PQ and ASK1 that contribute to neuronal degeneration.

Clearance of α-synuclein through macroautophagy depends on the metabolic state of cells. Thus, brief exposure to rotenone, which is not compromised metabolism, cells were able to remove oligomeric α-synuclein, however, with chronic exposure to rotenone, the ability of autophagy to degrade α-synuclein is significantly lower. Rotenone-induced α-synuclein aggregates can be cleared following rapamycin stimulation of autophagy. Chronic rotenone exposure and commensurate reduction of metabolic activity can interfere with the efficacy of rapamycin to promote autophagy, suggesting that cellular metabolism is critical for determining autophagic activity [198].

2.2. By Genetic Factors

Numerous connections exist between proteins genetically linked to autosomal dominant PD, in particular α-synuclein and LRRK2, and autophagic pathways. However, proteins involved in recessive PD, such as PINK1 and Parkin, function in the process of mitophagy, whereby damaged mitochondria are selectively eliminated by macroautophagy.

One proposed mechanism to how α-synuclein exerts its neurotoxic effect is through direct impairment of protein degradation over the ubiquitin-proteasome system (UPS). Other mechanism have been described as proteasomal inhibition [199], and inhibition of protein degradation over the lysosome/autophagy pathway [200].

This protein exhibits consensus motifs for lysosomal degradation via CMA. Several studies have been published focusing in the relation between α-synuclein and autophagy. In 2004, Cuervo et al. [201] reported that wild-type (WT) α-synuclein is properly degraded in lysosomes by CMA, meanwhile the pathogenic mutations A30P and A53T are poorly degraded by CMA despite a high affinity for the CMA receptor, which may promote toxic gain-of-functions by contributing to its aggregation or modifications. Vogiatzi et al. [202] have found how the degradation of monomeric WT is mediated by macroautophagy and CMA, and LAMP2a and α-synuclein interact in the rat and mouse cortex and midbrain, regions which are affected in PD. This research group, in 2009, published that the expression of A53T α-synuclein leads to CMA dysfunction in neuronal cells, and this in turn leads to compensatory induction of macroautophagy [179]. In 2009, Yu et al. [198], showed that stimulation of macroautophagy leads to a partial elimination of the aggregates of α-synuclein in neurones. It was also noted in A53T transgenic mice, increased levels of LC3-II and beclin-1 is associated with activation of autophagy. The examination of mutant A53T transgenic animals reveals that A53T localizes to the mitochondria membranes in dopaminergic neurons in monomeric or oligomeric forms. These events coincide with an age-related mitochondrial complex I inhibition and decreased substrate-specific respiration along with increases in mitochondrial autophagy (mitophagy) [203]. In 2010, Winslow et al. [204] reported a contradictory relationship between α-synuclein and autophagy. Their findings appeared specific to wild-type α-synuclein, as the PD-associated point mutants A53T and

A30P had no effect on LC3-II levels and demonstrated that α-synuclein overexpression impairs macroautophagy in mammalian cells and in transgenic mice. Attractively, in *C. elegans*, low-dose bafilomycin A1 is cytoprotective in part through its maintenance of the autophagy-lysosome pathway, and can protect against α-synuclein accumulation [205]. A recent report by Choubey, V., et al. (2011) suggests that the activation of macroautophagy in primary cortical neurons that overexpress mutant A53T α-synuclein leads to mitochondrial destruction and loss [206]. Loss-of-function mutations of *Parkin* gene are a known to cause early-PD. It may also indicate that the involvement of the UPS in the clearance of misfolded or aggregated proteins is crucial in the pathogenesis of PD. Several studies in *Drosophila melanogaster* showed that loss of Parkin results in the impairment of mitochondria and dopaminergic neurons loss [207, 208] and this fact shows that the phenotype of *PARK6* and *PARK2* KO flies are similar. Interestingly, the overexpression of Parkin could rescue the mitochondrial dysfunction caused by the deficiency of PINK1, but not vice versa [89, 209]. This fact suggests that PINK1 and Parkin participate in the same pathway, and PINK1 is upstream of Parkin.

Recently, it was found that both proteins are involved in the selective degradation of damaged mitochondria by autophagy (mitophagy). A key finding for the function was observed PINK1/Parkin after treatment of cells with the mitochondrial depolarizing agent carbonyl cyanide m-chlorophenylhydrazone (CCCP) or carbonyl cyanide p-trifluoromethoxyphenylhydrazone (FCCP). Narendra et al. [210] reported that in HEK293 cells Parkin is recruited to dysfunctional mitochondria where it mediates the engulfment of mitochondria by autophagosomes and their subsequent degradation. Another study has shown that translocation of Parkin is PINK1-dependent and the both proteins promote mitochondrial aggregates and perinuclear clusters [211]. However, Parkin also directly binds to Bcl-2 and mediates the Bcl-2 monoubiquitination to stabilized it, leading an increase in its binding to Beclin 1, and ultimately repressing autophagy [212]. Thus, depending on target accessibility and/or subcellular localization, Parkin can operate as either promoter or suppressor of macroautophagy. Michiorri et al [213] have found that FL-PINK1 interacts with Beclin 1, and may promote basal and starvation-induced autophagy. Returning to mitophagy, some findings show that PINK1 mutants are diffusely distributed and not necessarily present in mitochondria. There are also obsercations that PINK1 can interact with LC3-II, mediating the mitochondrial elimination by the PINK1/Parkin pathway [214]. Studies in *D. melanogaster* have shown that PINK1 and Parkin promote the ubiquitination of the profusion factor Mfn, suggesting a mechanism to recognize the damaged mitochondria for degradation by autophagy [215]. In this perspective, Gegg et al. [94] have reported that mitofusin 1 and 2 are ubiquitinated in a PINK1/parkin-dependent manner after CCCP treatment in neuroblastoma cells. Recent studies suggest that PINK1 is constitutively synthesized and imported into mitochondria, but cleaved if mitochondria is healthy and maintained if it is damaged, leading its accumulation on the outer mitochondrial membrane, which is sufficient for Parkin recruitment and mitophagy. Also, they have shown that patient with mutations in *PINK1* and *Parkin* interfere this pathway in different steps [216-218]. Geisler et al. [219] have found that, in depolarized mitochondria, Parkin is translocated and mediates the Lys 27 polyubiquitination of VDAC1 (voltage-dependent anion channel 1). In addition, p62 is also recruited to mitochondrial clusters and is basic for the mitophagy.

The physiological role of DJ-1 implicates broad biological functions including modulation of transcription, chaperone-like functions and antioxidant properties [220]; [221].

Furthermore it may protect against endoplasmic reticulum stress, and proteasomal inhibition [222]. It is suggested that DJ-1 may act as a component of the UPS acting as a chaperone or protease to refold or promote the degradation of misfolded proteins [114]. In this respect DJ-1 has been also identified as a modulator of autophagy [223, 224]

Serving as a transcriptional co-activator, DJ-1 stabilizes the activity of a master regulator of response to oxidative stress, Nrf2 [225, 226], sequesters cell death protein Daxx to the nucleus, thus preventing apoptosis induces after oxidative stress [227] and can upregulate glutathione synthesis [228, 229]. Recently, DJ-1 has been detected a direct physical interaction between DJ-1 and ASK1 protein [230, 231] demonstrating that DJ-1 is recruited to the ASK1 signalosome under oxidative conditions where it inhibits ASK1 activity. Moreover, it has been reported to be a redox-dependent chaperone [221]. Its chaperone action may contribute to inhibit the aggregation of α-synuclein [221, 232]. Overexpression of *WT DJ-1* inhibits A53T human α-synuclein protein aggregation and reduces neural toxicity in N27 cells [229].

Diverse functions in stress response have been already described for DJ-1. Loss of function of DJ-1 leads to increased stress levels in cellular models [112, 222, 233, 234]. DJ-1 null mutants in mice and *Drosophila* also show exarcebated sensitivity to agents that increase oxidative stress [235-239] and the same is true for *Caenorhabitis* worms in which DJ-1 is downregulated using RNA interference [228, 240]. Recent reports show that DJ-1 can work in parallel with the PINK1/parkin pathway to maintain mitochondrial function in the presence of an oxidative environment [224]. Loss of DJ-1 function causes a prominent disturbance on mitochondrial function and morphology that is linked to decreased basal autophagy and lysosomal degradation [241]. Mitochondrial dysfunction observed in DJ-1 null cells is accompanied by decrease in the activated LC3-II band [241, 242]. Moreover, DJ-1 siRNA has also reported to inhibit PQ-induced autophagy [223].

On the other hand, the association of DJ-1 with autophagy is varied. DJ-1 function may either contribte to the regulation of autophagy or mitigate the downstream effects of ROS, as studies have shown that ROS can upregulate autophagy [243-245]. It has been suggested that ERK1/2 regulates lysosomal degradation and more specifically, ERK2 kinase activation is necessary for autophagy [171]. Recently, decreased ERK2 phosphorylation was noted in the mitochondrial fraction of DJ-1-deficient cells [241] and it was inferred that DJ-1-mediated ERK2 phosphorylation could putatively control autophagic and lysosomal functions.

The involvement of parkinsonian LRRK2 protein in the regulation of autophagy seems clear, however there is still much to study. It seems that changes in GTPase activity and kinase activity of LRRK2 is involved in modulating autophagy. In yeasts it has been observed that alterations in GTPase activity is closely associated with defects in endocytic vesicular trafficking and autophagy [246]. Vesicular trafficking and autophagic defects closely correlate with the level of toxicity induced by each truncated GTPase variant suggesting that defects in trafficking may underlie LRRK2-induced toxicity in this model. On the other hand, it is known from previous works that the R1441G mutation causes an increase in autophagosomes but not produce an efficient degradation of their contents [247]. However, a mutation that only affects the kinase domain of LRRK2 such as G2019S, induces an increase in autophagic flux, showing an increased number of autophagosomes and increased degradation of the content in retinoic acid-differentiated SH-SY5Y cells [127]. The same results of alterations in autophagy levels by *LRRK2* mutations are corroborated in a work with fibroblasts from PD patients [248]. This information indicates that LRRK2 is implicated in regulation of autophagy, but the complex mechanism comprising the different steps of autophagy and high complexity of LRRK2 protein make it difficult to understand this relationship.

Conclusion

Autophagy has emerged as a central response observed in multiple models of parkinsonian neurodegeneration. Only, in the first six months of 2011, it has been published 977 articles related to autophagy, of which, 41 are related to autophagy and neurodegeneration, and 21 related to autophagy and PD.

In several chronic models, autophagy induction plays beneficial roles in clearing protein aggregates or damaged mitochondria. In this sense, a proper balance in catabolic and anabolic activities is critical for survival of neurons. In some instances, autophagy plays beneficial, homeostatic roles that include clearance of protein aggregates, and maintenance of mitochondria and axonal-synaptic morphology.

On the other hand, excessive or imbalanced autophagy creates a condition of "autophagic stress" that contributes to neurodegeneration and cell death. In particular, excessive mitochondrial degradation may be detrimental to neurons due to their high dependence on oxidative phosphorylation.

Cell type specific mechanisms must also be considered, as parkin translocation is not observed in CCCP-treated cortical and striatal/midbrain neurons. Redundancy in mechanisms that underlie the removal of damaged mitochondria may account for the relatively minor symptoms observed in knockout mice and form the basis for future therapies to heighten neuroprotective responses in PD patients.

Neurons are particularly vulnerable to autophagic stress, and factors that promote autophagic stress lead to the aberrant accumulation of autophagic vacuoles and lysosomes. As mitochondrial dysfunction plays a prominent role in PD pathogenesis, ROS mediated mitophagy may initially represent an adaptive response to "sequester" damaged mitochondria. However, excessive mitochondrial autophagy may commit neurons to cell death. Autophagy-induced neuritic remodeling would also be expected to be crucial in normal physiology and pathophysiology, as observed in a mutant LRRK2 model of PD.

While stimulation of autophagy holds promise for therapies to clear aggregated proteins implicated in neurodegenerative diseases, the prospects for future therapies may depend upon our ability to correct age- or disease-related factors that promote autophagic stress. Thus, relevant questions extend beyond debates over whether autophagy is "good" or "bad", towards a greater emphasis on factors that contribute to pathological imbalances in the system and how this relates to disease.

References

[1] Parkinson, J., An essay on the shaking palsy. *Medical Classics.* 1817. 10: p. 963-997.

[2] Charcot, P., Lecture V. on paralysis agitans. *Lectures on diseases of the nervous system, ed. Sigerson G.* 1879, Philadelphia: HC Lea

[3] Janko, K.A. and K.M. Shannon, Parkinsonism and Parkinson's disease. Nurse Pract. *Forum.* 1996. 7(4): p. 174-8.
[4] Fearnley, J.M. and A.J. Lees, Ageing and Parkinson's disease: substantia nigra regional selectivity. *Brain.* 1991. 114 (Pt 5): p. 2283-301.
[5] Bernheimer, H., et al., Brain dopamine and the syndromes of Parkinson and Huntington. Clinical, morphological and neurochemical correlations. *J. Neurol. Sci.* 1973. 20(4): p. 415-55.
[6] Kish, S.J., K. Shannak, and O. Hornykiewicz, Uneven pattern of dopamine loss in the striatum of patients with idiopathic Parkinson's disease. Pathophysiologic and clinical implications. *N. Engl. J. Med.* 1988. 318(14): p. 876-80.
[7] Gibb, W.R. and A.J. Lees, The significance of the Lewy body in the diagnosis of idiopathic Parkinson's disease. *Neuropathol. Appl. Neurobiol.* 1989. 15(1): p. 27-44.
[8] Lucking, C.B. and A. Brice, Alpha-synuclein and Parkinson's disease. *Cell Mol. Life Sci.* 2000. 57(13-14): p. 1894-908.
[9] Cole, N.B. and D.D. Murphy, The cell biology of alpha-synuclein: a sticky problem? *Neuromolecular Med.* 2002. 1(2): p. 95-109.
[10] Benmoyal-Segal, L. and H. Soreq, Gene-environment interactions in sporadic Parkinson's disease. *J. Neurochem.* 2006. 97(6): p. 1740-55.
[11] Gowers, W.R. and A.E. Barker, On a Case of Abscess of the Temporo-Sphenoidal Lobe of the Brain, Due to Otitis Media, Successfully Treated by Trephining and Drainage. *Br. Med. J.* 1886. 2(1354): p. 1154-6.
[12] Langston, J.W., et al., Chronic Parkinsonism in humans due to a product of meperidine-analog synthesis. *Science.* 1983. 219(4587): p. 979-80.
[13] Gerlach, M., et al., Acute MPTP treatment produces no changes in mitochondrial complex activities and indices of oxidative damage in the common marmoset ex vivo one week after exposure to the toxin. *Neurochem. Int.* 1996. 28(1): p. 41-9.
[14] Bove, J., et al., Toxin-induced models of Parkinson's disease. NeuroRx, 2005. 2(3): p. 484-94.
[15] Gonzalez-Polo, R.A., et al., Paraquat-induced apoptotic cell death in cerebellar granule cells. *Brain Res.* 2004. 1011(2): p. 170-6.
[16] Niso-Santano, M., et al., Low concentrations of paraquat induces early activation of extracellular signal-regulated kinase 1/2, protein kinase B, and c-Jun N-terminal kinase 1/2 pathways: role of c-Jun N-terminal kinase in paraquat-induced cell death. *Toxicol. Sci.* 2006. 92(2): p. 507-15.
[17] Gonzalez-Polo, R.A., et al., Inhibition of paraquat-induced autophagy accelerates the apoptotic cell death in neuroblastoma SH-SY5Y cells. *Toxicol. Sci.* 2007. 97(2): p. 448-58.
[18] Leroux, P., Contribution à l'Étude des Causes de la Paralysie Agitante. 1880, Paris: PARIS.
[19] Gitler, A.D., et al., Alpha-synuclein is part of a diverse and highly conserved interaction network that includes PARK9 and manganese toxicity. *Nat. Genet.* 2009. 41(3): p. 308-15.
[20] Kowall, N.W., et al., MPTP induces alpha-synuclein aggregation in the substantia nigra of baboons. *Neuroreport.* 2000. 11(1): p. 211-3.

[21] Meredith, G.E., et al., Lysosomal malfunction accompanies alpha-synuclein aggregation in a progressive mouse model of Parkinson's disease. *Brain Res.* 2002. 956(1): p. 156-65.
[22] Wirdefeldt, K., et al., Epidemiology and etiology of Parkinson's disease: a review of the evidence. *Eur. J. Epidemiol.* 2011. 26 Suppl 1: p. S1-58.
[23] Lacava, G., Boxer's Encephalopathy. *J. Sports Med. Phys. Fitness.* 1963. 168: p. 87-92.
[24] Takahashi, M. and T. Yamada, Viral etiology for Parkinson's disease--a possible role of influenza A virus infection. *Jpn J. Infect. Dis.* 1999. 52(3): p. 89-98.
[25] Takahashi, M. and T. Yamada, A possible role of influenza A virus infection for Parkinson's disease. *Adv. Neurol.* 2001. 86: p. 91-104.
[26] Lai, B.C., et al., Occupational and environmental risk factors for Parkinson's disease. *Parkinsonism Relat. Disord.* 2002. 8(5): p. 297-309.
[27] Logroscino, G., et al., Physical activity and risk of Parkinson's disease: a prospective cohort study. *J. Neurol. Neurosurg. Psychiatry.* 2006. 77(12): p. 1318-22.
[28] Di Monte, D.A., The environment and Parkinson's disease: is the nigrostriatal system preferentially targeted by neurotoxins? *Lancet Neurol.* 2003. 2(9): p. 531-8.
[29] Dinis-Oliveira, R.J., et al., Paraquat exposure as an etiological factor of Parkinson's disease. *Neurotoxicology.* 2006. 27(6): p. 1110-22.
[30] Langston, J.W., et al., Evidence of active nerve cell degeneration in the substantia nigra of humans years after 1-methyl-4-phenyl-1,2,3,6-tetrahydropyridine exposure. *Ann. Neurol.* 1999. 46(4): p. 598-605.
[31] Davis, G.C., et al., Chronic Parkinsonism secondary to intravenous injection of meperidine analogues. *Psychiatry Res.* 1979. 1(3): p. 249-54.
[32] Hantraye, P., et al., Stable parkinsonian syndrome and uneven loss of striatal dopamine fibres following chronic MPTP administration in baboons. *Neuroscience.* 1993. 53(1): p. 169-78.
[33] Sherer, T.B., et al., Mechanism of toxicity in rotenone models of Parkinson's disease. *J. Neurosci.* 2003. 23(34): p. 10756-64.
[34] Bergen, W.G., The in vitro effect of dieldrin on respiration of rat liver mitochondria. *Proc. Soc. Exp. Biol. Med.* 1971. 136(3): p. 732-5.
[35] Zhang, Z., et al., [Contents of organochlorine pesticides in water, pore water and sediment in Minjiang River Estuary of China]. *Huan Jing Ke Xue.* 2003. 24(1): p. 117-20.
[36] Gonzalez-Polo, R.A., et al., Relationship between autophagy and apoptotic cell death in human neuroblastoma cells treated with paraquat: could autophagy be a "brake" in paraquat-induced apoptotic death? *Autophagy.* 2007. 3(4): p. 366-7.
[37] Peng, J., et al., The herbicide paraquat induces dopaminergic nigral apoptosis through sustained activation of the JNK pathway. *J. Biol. Chem.* 2004. 279(31): p. 32626-32.
[38] Chiba, K., A. Trevor, and N. Castagnoli, Jr., Metabolism of the neurotoxic tertiary amine, MPTP, by brain monoamine oxidase. *Biochem. Biophys. Res. Commun.* 1984. 120(2): p. 574-8.
[39] Bezard, E., et al., Absence of MPTP-induced neuronal death in mice lacking the dopamine transporter. Exp Neurol, 1999. 155(2): p. 268-73.
[40] Gonzalez-Polo, R.A., et al., Mechanisms of MPP(+) incorporation into cerebellar granule cells. *Brain Res. Bull.* 2001. 56(2): p. 119-23.

[41] Sheldon, A.L. and M.B. Robinson, The role of glutamate transporters in neurodegenerative diseases and potential opportunities for intervention. *Neurochem. Int.* 2007. 51(6-7): p. 333-55.
[42] Nicklas, W.J., I. Vyas, and R.E. Heikkila, Inhibition of NADH-linked oxidation in brain mitochondria by 1-methyl-4-phenyl-pyridine, a metabolite of the neurotoxin, 1-methyl-4-phenyl-1,2,5,6-tetrahydropyridine. *Life Sci.* 1985. 36(26): p. 2503-8.
[43] Gonzalez-Polo, R.A., et al., MPP(+) causes inhibition of cellular energy supply in cerebellar granule cells. *Neurotoxicolog.* 2003. 24(2): p. 219-25.
[44] Gonzalez-Polo, R.A., et al., Protection against MPP+ neurotoxicity in cerebellar granule cells by antioxidants. *Cell Biol. Int.* 2004. 28(5): p. 373-80.
[45] Sachs, C. and G. Jonsson, Mechanisms of action of 6-hydroxydopamine. *Biochem. Pharmacol.* 1975. 24(1): p. 1-8.
[46] Drechsel, D.A. and M. Patel, Role of reactive oxygen species in the neurotoxicity of environmental agents implicated in Parkinson's disease. *Free Radic. Biol. Med.* 2008. 44(11): p. 1873-86.
[47] Bismuth, C., et al., Paraquat poisoning. An overview of the current status. *Drug Saf.* 1990. 5(4): p. 243-51.
[48] Gram, T.E., Chemically reactive intermediates and pulmonary xenobiotic toxicity. *Pharmacol. Rev.* 1997. 49(4): p. 297-341.
[49] Garnier, R., et al., Paraquat poisoning by skin absorption: report of two cases. *Vet. Hum. Toxicol.* 1994. 36(4): p. 313-5.
[50] Shimizu, K., et al., Carrier-mediated processes in blood--brain barrier penetration and neural uptake of paraquat. *Brain Res.* 2001. 906(1-2): p. 135-42.
[51] McCormack, A.L. and D.A. Di Monte, Effects of L-dopa and other amino acids against paraquat-induced nigrostriatal degeneration. *J. Neurochem.* 2003. 85(1): p. 82-6.
[52] Bus, J.S. and J.E. Gibson, Paraquat: model for oxidant-initiated toxicity. *Environ. Health Perspect.* 1984. 55: p. 37-46.
[53] Heath, D. and P. Smith, Pulmonary vascular disease. *Med. Clin. North Am.* 1977. 61(6): p. 1279-307.
[54] Richardson, J.R., et al., Paraquat neurotoxicity is distinct from that of MPTP and rotenone. *Toxicol. Sci.* 2005. 88(1): p. 193-201.
[55] Betarbet, R., et al., Chronic systemic pesticide exposure reproduces features of Parkinson's disease. *Nat. Neurosci.* 2000. 3(12): p. 1301-6.
[56] De Michele, G., et al., Environmental and genetic risk factors in Parkinson's disease: a case-control study in southern Italy. *Mov. Disord.* 1996. 11(1): p. 17-23.
[57] Polymeropoulos, M.H., et al., Mutation in the alpha-synuclein gene identified in families with Parkinson's disease. *Science.* 1997. 276(5321): p. 2045-7.
[58] Spillantini, M.G., et al., Filamentous alpha-synuclein inclusions link multiple system atrophy with Parkinson's disease and dementia with Lewy bodies. *Neurosci. Lett.* 1998. 251(3): p. 205-8.
[59] Kitada, T., et al., Mutations in the parkin gene cause autosomal recessive juvenile parkinsonism. *Nature.* 1998. 392(6676): p. 605-8.
[60] van Duijn, C.M., et al., Park7, a novel locus for autosomal recessive early-onset parkinsonism, on chromosome 1p36. *Am. J. Hum. Genet.* 2001. 69(3): p. 629-34.
[61] Bonifati, V., et al., DJ-1(PARK7), a novel gene for autosomal recessive, early onset parkinsonism. *Neurol. Sci.* 2003. 24(3): p. 159-60.

[62] Morais, V.A., et al., Parkinson's disease mutations in PINK1 result in decreased Complex I activity and deficient synaptic function. *EMBO Mol. Med.* 2009. 1(2): p. 99-111.
[63] Paisan-Ruiz, C., et al., Cloning of the gene containing mutations that cause PARK8-linked Parkinson's disease. *Neuron.* 2004. 44(4): p. 595-600.
[64] Zimprich, A., et al., The PARK8 locus in autosomal dominant parkinsonism: confirmation of linkage and further delineation of the disease-containing interval. *Am. J. Hum. Genet.* 2004. 74(1): p. 11-9.
[65] Biskup, S., et al., Localization of LRRK2 to membranous and vesicular structures in mammalian brain. *Ann. Neurol.* 2006. 60(5): p. 557-69.
[66] Funayama, M., et al., A new locus for Parkinson's disease (PARK8) maps to chromosome 12p11.2-q13.1. *Ann. Neurol.* 2002. 51(3): p. 296-301.
[67] Kruger, R., et al., Ala30Pro mutation in the gene encoding alpha-synuclein in Parkinson's disease. *Nat. Genet.* 1998. 18(2): p. 106-8.
[68] Zarranz, J.J., et al., The new mutation, E46K, of alpha-synuclein causes Parkinson and Lewy body dementia. *Ann. Neurol.* 2004. 55(2): p. 164-73.
[69] Ibanez, P., et al., Causal relation between alpha-synuclein gene duplication and familial Parkinson's disease. *Lancet.* 2004. 364(9440): p. 1169-71.
[70] Singleton, A.B., et al., alpha-Synuclein locus triplication causes Parkinson's disease. *Science.* 2003. 302(5646): p. 841.
[71] McNaught, K.S., et al., Failure of the ubiquitin-proteasome system in Parkinson's disease. *Nat. Rev. Neurosci.* 2001. 2(8): p. 589-94.
[72] Snyder, H., et al., Aggregated and monomeric alpha-synuclein bind to the S6' proteasomal protein and inhibit proteasomal function. *J. Biol. Chem.* 2003. 278(14): p. 11753-9.
[73] Zhang, N.Y., Z. Tang, and C.W. Liu, alpha-Synuclein protofibrils inhibit 26 S proteasome-mediated protein degradation: understanding the cytotoxicity of protein protofibrils in neurodegenerative disease pathogenesis. *J. Biol. Chem.* 2008. 283(29): p. 20288-98.
[74] Dyllick-Brenzinger, M., et al., Reciprocal effects of alpha-synuclein overexpression and proteasome inhibition in neuronal cells and tissue. *Neurotox. Res.* 2010. 17(3): p. 215-27.
[75] Hasegawa, M., et al., Phosphorylated alpha-synuclein is ubiquitinated in alpha-synucleinopathy lesions. J Biol Chem, 2002. 277(50): p. 49071-6.
[76] Tofaris, G.K., et al., Ubiquitination of alpha-synuclein in Lewy bodies is a pathological event not associated with impairment of proteasome function. *J. Biol. Chem.* 2003. 278(45): p. 44405-11.
[77] Liani, E., et al., Ubiquitylation of synphilin-1 and alpha-synuclein by SIAH and its presence in cellular inclusions and Lewy bodies imply a role in Parkinson's disease. *Proc. Natl. Acad. Sci. U. S. A.* 2004. 101(15): p. 5500-5.
[78] Rott, R., et al., Monoubiquitylation of alpha-synuclein by seven in absentia homolog (SIAH) promotes its aggregation in dopaminergic cells. *J. Biol. Chem.* 2008. 283(6): p. 3316-28.
[79] Schapira, A.H., et al., Mitochondrial complex I deficiency in Parkinson's disease. *Lancet.* 1989. 1(8649): p. 1269.

[80] Bender, A., et al., High levels of mitochondrial DNA deletions in substantia nigra neurons in aging and Parkinson disease. *Nat. Genet.* 2006. 38(5): p. 515-7.

[81] Jenner, P., Oxidative stress in Parkinson's disease. *Ann. Neurol.* 2003. 53 Suppl 3: p. S26-36; discussion S36-8.

[82] Matsumine, H., et al., Localization of a gene for an autosomal recessive form of juvenile Parkinsonism to chromosome 6q25.2-27. *Am. J. Hum. Genet. 1*997. 60(3): p. 588-96.

[83] West, A.B. and N.T. Maidment, Genetics of parkin-linked disease. *Hum. Genet.* 2004. 114(4): p. 327-36.

[84] Hattori, N. and Y. Mizuno, Pathogenetic mechanisms of parkin in Parkinson's disease. *Lancet.* 2004. 364(9435): p. 722-4.

[85] Shimura, H., et al., Familial Parkinson disease gene product, parkin, is a ubiquitin-protein ligase. *Nat. Genet.* 2000. 25(3): p. 302-5.

[86] Nakajima, A., et al., BRPK, a novel protein kinase showing increased expression in mouse cancer cell lines with higher metastatic potential. *Cancer Lett.* 2003. 201(2): p. 195-201.

[87] Lin, W. and U.J. Kang, Characterization of PINK1 processing, stability, and subcellular localization. *J. Neurochem.* 2008. 106(1): p. 464-74.

[88] Valente, E.M., et al., Hereditary early-onset Parkinson's disease caused by mutations in PINK1. *Science.* 2004. 304(5674): p. 1158-60.

[89] Clark, I.E., et al., Drosophila pink1 is required for mitochondrial function and interacts genetically with parkin. *Nature.* 2006. 441(7097): p. 1162-6.

[90] Zhang, Y., et al., Parkin functions as an E2-dependent ubiquitin- protein ligase and promotes the degradation of the synaptic vesicle-associated protein, CDCrel-1. Proc Natl Acad Sci U S A, 2000. 97(24): p. 13354-9.

[91] Valente, E.M., et al., PINK1 mutations are associated with sporadic early-onset parkinsonism. *Ann. Neurol.* 2004. 56(3): p. 336-41.

[92] Rakovic, A., et al., Mutations in PINK1 and Parkin impair ubiquitination of Mitofusins in human fibroblasts. *PLoS One.* 2011. 6(3): p. e16746.

[93] Rakovic, A., et al., Effect of endogenous mutant and wild-type PINK1 on Parkin in fibroblasts from Parkinson disease patients. *Hum. Mol. Genet.* 2010. 19(16): p. 3124-37.

[94] Gegg, M.E., et al., Mitofusin 1 and mitofusin 2 are ubiquitinated in a PINK1/parkin-dependent manner upon induction of mitophagy. *Hum. Mol. Genet.* 2010. 19(24): p. 4861-70.

[95] Bandyopadhyay, S. and M.R. Cookson, Evolutionary and functional relationships within the DJ1 superfamily. *BMC Evol. Biol.* 2004. 4: p. 6.

[96] Nagakubo, D., et al., DJ-1, a novel oncogene which transforms mouse NIH3T3 cells in cooperation with ras. *Biochem. Biophys. Res. Commun.* 1997. 231(2): p. 509-13.

[97] Le Naour, F., et al., Proteomics-based identification of RS/DJ-1 as a novel circulating tumor antigen in breast cancer. *Clin. Cancer Res.* 2001. 7(11): p. 3328-35.

[98] Pardo, M., et al., The characterization of the invasion phenotype of uveal melanoma tumour cells shows the presence of MUC18 and HMG-1 metastasis markers and leads to the identification of DJ-1 as a potential serum biomarker. *Int. J. Cancer.* 2006. 119(5): p. 1014-22.

[99] Bonifati, V., et al., Mutations in the DJ-1 gene associated with autosomal recessive early-onset parkinsonism. *Science.* 2003. 299(5604): p. 256-9.
[100] Taira, T., et al., Molecular cloning of human and mouse DJ-1 genes and identification of Sp1-dependent activation of the human DJ-1 promoter. *Gene.* 2001. 263(1-2): p. 285-92.
[101] Rizzu, P., et al., DJ-1 colocalizes with tau inclusions: a link between parkinsonism and dementia. *Ann. Neurol.* 2004. 55(1): p. 113-8.
[102] Neumann, M., et al., Pathological properties of the Parkinson's disease-associated protein DJ-1 in alpha-synucleinopathies and tauopathies: relevance for multiple system atrophy and Pick's disease. *Acta Neuropathol.* 2004. 107(6): p. 489-96.
[103] Hague, S., et al., Early-onset Parkinson's disease caused by a compound heterozygous DJ-1 mutation. *Ann. Neurol.* 2003. 54(2): p. 271-4.
[104] Lee, S.J., et al., Crystal structures of human DJ-1 and Escherichia coli Hsp31, which share an evolutionarily conserved domain. *J. Biol. Chem.* 2003. 278(45): p. 44552-9.
[105] Wei, Y., et al., Identification of functional subclasses in the DJ-1 superfamily proteins. *PLoS Comput. Biol.* 2007. 3(1): p. e10.
[106] Tao, X. and L. Tong, Crystal structure of human DJ-1, a protein associated with early onset Parkinson's disease. *J. Biol. Chem.* 2003. 278(33): p. 31372-9.
[107] Miller, D.W., et al., L166P mutant DJ-1, causative for recessive Parkinson's disease, is degraded through the ubiquitin-proteasome system. *J. Biol. Chem.* 2003. 278(38): p. 36588-95.
[108] Moore, D.J., et al., A missense mutation (L166P) in DJ-1, linked to familial Parkinson's disease, confers reduced protein stability and impairs homo-oligomerization. *J. Neurochem.* 2003. 87(6): p. 1558-67.
[109] Olzmann, J.A., et al., Familial Parkinson's disease-associated L166P mutation disrupts DJ-1 protein folding and function. *J. Biol. Chem.* 2004. 279(9): p. 8506-15.
[110] Lakshminarasimhan, M., et al., Structural impact of three Parkinsonism-associated missense mutations on human DJ-1. *Biochemistry*. 2008. 47(5): p. 1381-92.
[111] Shinbo, Y., et al., Proper SUMO-1 conjugation is essential to DJ-1 to exert its full activities. *Cell Death Differ.* 2006. 13(1): p. 96-108.
[112] Canet-Aviles, R.M., et al., The Parkinson's disease protein DJ-1 is neuroprotective due to cysteine-sulfinic acid-driven mitochondrial localization. *Proc. Natl. Acad. Sci. U. S. A.* 2004. 101(24): p. 9103-8.
[113] Blackinton, J., et al., Formation of a stabilized cysteine sulfinic acid is critical for the mitochondrial function of the parkinsonism protein DJ-1. *J. Biol. Chem.* 2009. 284(10): p. 6476-85.
[114] Moore, D.J., et al., Association of DJ-1 and parkin mediated by pathogenic DJ-1 mutations and oxidative stress. *Hum. Mol. Genet.* 2005. 14(1): p. 71-84.
[115] Zimprich, A., et al., Mutations in LRRK2 cause autosomal-dominant parkinsonism with pleomorphic pathology. *Neuron*. 2004. 44(4): p. 601-7.
[116] Guo, L., W. Wang, and S.G. Chen, Leucine-rich repeat kinase 2: relevance to Parkinson's disease. *Int. J. Biochem. Cell Biol.* 2006. 38(9): p. 1469-75.
[117] Kachergus, J., et al., Identification of a novel LRRK2 mutation linked to autosomal dominant parkinsonism: evidence of a common founder across European populations. *Am. J. Hum. Genet.* 2005. 76(4): p. 672-80.

[118] West, A.B., et al., Parkinson's disease-associated mutations in leucine-rich repeat kinase 2 augment kinase activity. *Proc. Natl. Acad. Sci. U. S. A.* 2005. 102(46): p. 16842-7.

[119] Jaleel, M., et al., LRRK2 phosphorylates moesin at threonine-558: characterization of how Parkinson's disease mutants affect kinase activity. *Biochem. J.* 2007. 405(2): p. 307-17.

[120] Greggio, E., et al., Kinase activity is required for the toxic effects of mutant LRRK2/dardarin. *Neurobiol. Dis.* 2006. 23(2): p. 329-41.

[121] Paisan-Ruiz, C., et al., LRRK2 gene in Parkinson disease: mutation analysis and case control association study. *Neurology.* 2005. 65(5): p. 696-700.

[122] Giasson, B.I. and V.M. Van Deerlin, Mutations in LRRK2 as a cause of Parkinson's disease. *Neurosignals.* 2008. 16(1): p. 99-105.

[123] Ozelius, L.J., et al., LRRK2 G2019S as a cause of Parkinson's disease in Ashkenazi Jews. *N. Engl. J. Med.* 2006. 354(4): p. 424-5.

[124] Lesage, S., et al., LRRK2 G2019S as a cause of Parkinson's disease in North African Arabs. *N. Engl. J. Med.* 2006. 354(4): p. 422-3.

[125] Gloeckner, C.J., et al., The Parkinson disease causing LRRK2 mutation I2020T is associated with increased kinase activity. *Hum. Mol. Genet.* 2006. 15(2): p. 223-32.

[126] Smith, W.W., et al., Kinase activity of mutant LRRK2 mediates neuronal toxicity. *Nat. Neurosci.* 2006. 9(10): p. 1231-3.

[127] Plowey, E.D., et al., Role of autophagy in G2019S-LRRK2-associated neurite shortening in differentiated SH-SY5Y cells. *J. Neurochem.* 2008. 105(3): p. 1048-56.

[128] Guo, L., et al., The Parkinson's disease-associated protein, leucine-rich repeat kinase 2 (LRRK2), is an authentic GTPase that stimulates kinase activity. *Exp. Cell Res.* 2007. 313(16): p. 3658-70.

[129] Sen, S., P.J. Webber, and A.B. West, Dependence of leucine-rich repeat kinase 2 (LRRK2) kinase activity on dimerization. *J. Biol. Chem.* 2009. 284(52): p. 36346-56.

[130] Deng, J., et al., Structure of the ROC domain from the Parkinson's disease-associated leucine-rich repeat kinase 2 reveals a dimeric GTPase. *Proc. Natl. Acad. Sci. U. S. A.* 2008. 105(5): p. 1499-504.

[131] Miklossy, J., et al., LRRK2 expression in normal and pathologic human brain and in human cell lines. *J. Neuropathol. Exp. Neurol.* 2006. 65(10): p. 953-63.

[132] Melrose, H., et al., Anatomical localization of leucine-rich repeat kinase 2 in mouse brain. *Neuroscience.* 2006. 139(3): p. 791-4.

[133] Higashi, S., et al., Expression and localization of Parkinson's disease-associated leucine-rich repeat kinase 2 in the mouse brain. *J. Neurochem.* 2007. 100(2): p. 368-81.

[134] Gandhi, P.N., et al., The Roc domain of leucine-rich repeat kinase 2 is sufficient for interaction with microtubules. *J. Neurosci. Res.* 2008. 86(8): p. 1711-20.

[135] 1hin, N., et al., LRRK2 regulates synaptic vesicle endocytosis. *Exp. Cell Res.* 2008. 314(10): p. 2055-65.

[136] Hatano, T., et al., Leucine-rich repeat kinase 2 associates with lipid rafts. *Hum. Mol. Genet.* 2007. 16(6): p. 678-90.

[137] Ko, H.S., et al., CHIP regulates leucine-rich repeat kinase-2 ubiquitination, degradation, and toxicity. *Proc. Natl. Acad. Sci. U. S. A.* 2009. 106(8): p. 2897-902.

[138] Ho, C.C., et al., The Parkinson disease protein leucine-rich repeat kinase 2 transduces death signals via Fas-associated protein with death domain and caspase-8 in a cellular model of neurodegeneration. *J. Neurosci.* 2009. 29(4): p. 1011-6.

[139] Wang, Y., et al., Modulation of gold(III) porphyrin 1a-induced apoptosis by mitogen-activated protein kinase signaling pathways. *Biochem. Pharmacol.* 2008. 75(6): p. 1282-91.

[140] Smith, W.W., et al., Leucine-rich repeat kinase 2 (LRRK2) interacts with parkin, and mutant LRRK2 induces neuronal degeneration. *Proc. Natl. Acad. Sci. U. S. A.* 2005. 102(51): p. 18676-81.

[141] Ng, C.H., et al., Parkin protects against LRRK2 G2019S mutant-induced dopaminergic neurodegeneration in Drosophila. *J. Neurosci.* 2009. 29(36): p. 11257-62.

[142] Venderova, K., et al., Leucine-Rich Repeat Kinase 2 interacts with Parkin, DJ-1 and PINK-1 in a Drosophila melanogaster model of Parkinson's disease. *Hum. Mol. Genet.* 2009. 18(22): p. 4390-404.

[143] Lin, X., et al., Leucine-rich repeat kinase 2 regulates the progression of neuropathology induced by Parkinson's-disease-related mutant alpha-synuclein. *Neuron.* 2009. 64(6): p. 807-27.

[144] Kyosseva, S.V., The role of the extracellular signal-regulated kinase pathway in cerebellar abnormalities in schizophrenia. *Cerebellum.* 2004. 3(2): p. 94-9.

[145] Ho, R.C., et al., Dissociation of AMP-activated protein kinase and p38 mitogen-activated protein kinase signaling in skeletal muscle. *Biochem. Biophys. Res. Commun.* 2007. 362(2): p. 354-9.

[146] White, L.R., et al., MAPK-pathway activity, Lrrk2 G2019S, and Parkinson's disease. *J. Neurosci. Res.* 2007. 85(6): p. 1288-94.

[147] Leroy, E., et al., The ubiquitin pathway in Parkinson's disease. *Nature.* 1998. 395(6701): p. 451-2.

[148] Ciechanover, A., Intracellular protein degradation: From a vague idea thru the lysosome and the ubiquitin-proteasome system and onto human diseases and drug targeting. *Biochim. Biophys. Acta.* 2005.

[149] Levine, B. and D.J. Klionsky, Development by self-digestion: molecular mechanisms and biological functions of autophagy. *Dev. Cell.* 2004. 6(4): p. 463-77.

[150] Suzuki, K. and Y. Ohsumi, Molecular machinery of autophagosome formation in yeast, Saccharomyces cerevisiae. *FEBS Lett.* 2007. 581(11): p. 2156-61.

[151] Singh, S.B., et al., Human IRGM induces autophagy to eliminate intracellular mycobacteria. *Science.* 2006. 313(5792): p. 1438-41.

[152] Juhasz, G., et al., The Drosophila homolog of Aut1 is essential for autophagy and development. *FEBS Lett.* 2003. 543(1-3): p. 154-8.

[153] Colman, R.J., et al., Caloric restriction delays disease onset and mortality in rhesus monkeys. *Science.* 2009. 325(5937): p. 201-4.

[154] Komatsu, M., et al., Loss of autophagy in the central nervous system causes neurodegeneration in mice. *Nature.* 2006. 441(7095): p. 880-4.

[155] Kondo, Y. and S. Kondo, Autophagy and cancer therapy. *Autophagy.* 2006. 2(2): p. 85-90.

[156] Nakai, A., et al., The role of autophagy in cardiomyocytes in the basal state and in response to hemodynamic stress. *Nat. Med.* 2007. 13(5): p. 619-24.

[157] Masiero, E., et al., Autophagy is required to maintain muscle mass. *Cell Metab.* 2009. 10(6): p. 507-15.

[158] Singh, R., et al., Autophagy regulates adipose mass and differentiation in mice. *J. Clin. Invest.* 2009. 119(11): p. 3329-39.

[159] Ravikumar, B. and D.C. Rubinsztein, Can autophagy protect against neurodegeneration caused by aggregate-prone proteins? *Neuroreport.* 2004. 15(16): p. 2443-5.
[160] Martinez-Vicente, M. and A.M. Cuervo, Autophagy and neurodegeneration: when the cleaning crew goes on strike. *Lancet.* Neurol, 2007. 6(4): p. 352-61.
[161] Anglade, P., et al., Apoptosis and autophagy in nigral neurons of patients with Parkinson's disease. *Histol. Histopathol.* 1997. 12(1): p. 25-31.
[162] Baehrecke, E.H., Autophagy: dual roles in life and death? *Nat. Rev. Mol. Cell Biol.* 2005. 6(6): p. 505-10.
[163] Bursch, W., The autophagosomal-lysosomal compartment in programmed cell death. *Cell Death Differ.* 2001. 8(6): p. 569-81.
[164] Bursch, W., et al., Programmed cell death (PCD). Apoptosis, autophagic PCD, or others? *Ann. N. Y. Acad. Sci.* 2000. 926: p. 1-12.
[165] Klionsky, D.J., et al., How shall I eat thee? *Autophagy.* 2007. 3(5): p. 413-6.
[166] Reggiori, F. and D.J. Klionsky, Autophagy in the eukaryotic cell. *Eukaryot. Cell.* 2002. 1(1): p. 11-21.
[167] Meijer, A.J. and P. Codogno, Regulation and role of autophagy in mammalian cells. *Int. J. Biochem. Cell Biol.* 2004. 36(12): p. 2445-62.
[168] Djavaheri-Mergny, M., et al., NF-kappaB activation represses tumor necrosis factor-alpha-induced autophagy. *J. Biol. Chem.* 2006. 281(41): p. 30373-82.
[169] Pattingre, S., et al., Bcl-2 antiapoptotic proteins inhibit Beclin 1-dependent autophagy. *Cell.* 2005. 122(6): p. 927-39.
[170] Petiot, A., et al., Distinct classes of phosphatidylinositol 3'-kinases are involved in signaling pathways that control macroautophagy in HT-29 cells. *J. Biol. Chem.* 2000. 275(2): p. 992-8.
[171] Dagda, R.K., et al., Mitochondrially localized ERK2 regulates mitophagy and autophagic cell stress: implications for Parkinson's disease. *Autophagy.* 2008. 4(6): p. 770-82.
[172] Kulich, S.M. and C.T. Chu, Role of reactive oxygen species in extracellular signal-regulated protein kinase phosphorylation and 6-hydroxydopamine cytotoxicity. *J. Biosci.* 2003. 28(1): p. 83-9.
[173] Olanow, C.W., The pathogenesis of cell death in Parkinson's disease--2007. *Mov. Disord.* 2007. 22 Suppl 17: p. S335-42.
[174] Dawson, T.M. and V.L. Dawson, Rare genetic mutations shed light on the pathogenesis of Parkinson disease. *J. Clin. Invest.* 2003. 111(2): p. 145-51.
[175] Ventruti, A. and A.M. Cuervo, Autophagy and neurodegeneration. *Curr. Neurol. Neurosci. Rep.* 2007. 7(5): p. 443-51.
[176] Pan, T., et al., The role of autophagy-lysosome pathway in neurodegeneration associated with Parkinson's disease. *Brain.* 2008. 131(Pt 8): p. 1969-78.
[177] Prigione, A., et al., Alpha-synuclein nitration and autophagy response are induced in peripheral blood cells from patients with Parkinson disease. *Neurosci. Lett.* 2010. 477(1): p. 6-10.
[178] Massey, A.C., et al., Consequences of the selective blockage of chaperone-mediated autophagy. *Proc. Natl. Acad. Sci. U. S. A.* 2006. 103(15): p. 5805-10.
[179] Xilouri, M., et al., Abberant alpha-synuclein confers toxicity to neurons in part through inhibition of chaperone-mediated autophagy. *PLoS One.* 2009. 4(5): p. e5515.

[180] Trojanowski, J.Q. and V.M. Lee, "Fatal attractions" of proteins. A comprehensive hypothetical mechanism underlying Alzheimer's disease and other neurodegenerative disorders. *Ann. N. Y. Acad. Sci.* 2000. 924: p. 62-7.

[181] Walls, K.C., et al., Lysosome dysfunction triggers Atg7-dependent neural apoptosis. *J. Biol. Chem.* 2010. 285(14): p. 10497-507.

[182] Zhu, J.H., et al., Regulation of autophagy by extracellular signal-regulated protein kinases during 1-methyl-4-phenylpyridinium-induced cell death. *Am. J. Pathol.* 2007. 170(1): p. 75-86.

[183] Cai, Z.L., et al., MPP+ impairs autophagic clearance of alpha-synuclein by impairing the activity of dynein. *Neuroreport.* 2009. 20(6): p. 569-73.

[184] Castino, R., et al., Suppression of autophagy precipitates neuronal cell death following low doses of methamphetamine. *J. Neurochem.* 2008. 106(3): p. 1426-39.

[185] Dadakhujaev, S., et al., Autophagy protects the rotenone-induced cell death in alpha-synuclein overexpressing SH-SY5Y cells. *Neurosci. Lett.* 2010. 472(1): p. 47-52.

[186] Gomez-Santos, C., et al., Dopamine induces autophagic cell death and alpha-synuclein increase in human neuroblastoma SH-SY5Y cells. *J. Neurosci. Res.* 2003. 73(3): p. 341-50.

[187] Choi, K.C., et al., A novel mTOR activating protein protects dopamine neurons against oxidative stress by repressing autophagy related cell death. *J. Neurochem.* 2010. 112(2): p. 366-76.

[188] Zhong, Y., et al., Distinct regulation of autophagic activity by Atg14L and Rubicon associated with Beclin 1-phosphatidylinositol-3-kinase complex. *Nat. Cell Biol.* 2009. 11(4): p. 468-76.

[189] Matsunaga, K., et al., Two Beclin 1-binding proteins, Atg14L and Rubicon, reciprocally regulate autophagy at different stages. *Nat. Cell Biol.* 2009. 11(4): p. 385-96.

[190] Chu, C.T., J. Zhu, and R. Dagda, Beclin 1-independent pathway of damage-induced mitophagy and autophagic stress: implications for neurodegeneration and cell death. *Autophagy.* 2007. 3(6): p. 663-6.

[191] Kiffin, R., et al., Activation of chaperone-mediated autophagy during oxidative stress. *Mol. Biol. Cell.* 2004. 15(11): p. 4829-40.

[192] Kiffin, R., U. Bandyopadhyay, and A.M. Cuervo, Oxidative stress and autophagy. *Antioxid. Redox Signal.* 2006. 8(1-2): p. 152-62.

[193] Fei, X.F., et al., Olomoucine inhibits cathepsin L nuclear translocation, activates autophagy and attenuates toxicity of 6-hydroxydopamine. *Brain Res.* 2009. 1264: p. 85-97.

[194] Gimenez-Xavier, P., et al., Effects of dopamine on LC3-II activation as a marker of autophagy in a neuroblastoma cell model. *Neurotoxicology.* 2009. 30(4): p. 658-65.

[195] Manning-Bog, A.B., et al., The herbicide paraquat causes up-regulation and aggregation of alpha-synuclein in mice: paraquat and alpha-synuclein. *J. Biol. Chem.* 2002. 277(3): p. 1641-4.

[196] Niso-Santano, M., et al., Activation of apoptosis signal-regulating kinase 1 is a key factor in paraquat-induced cell death: modulation by the Nrf2/Trx axis. *Free Radic. Biol. Med.* 2010. 48(10): p. 1370-81.

[197] Niso-Santano, M., et al., ASK1 overexpression accelerates paraquat-induced autophagy via endoplasmic reticulum stress. *Toxicol. Sci.* 2011. 119(1): p. 156-68.

[198] Yu, W.H., et al., Metabolic activity determines efficacy of macroautophagic clearance of pathological oligomeric alpha-synuclein. *Am. J. Pathol.* 2009. 175(2): p. 736-47.

[199] Pals, P., et al., alpha-Synuclein promoter confers susceptibility to Parkinson's disease. *Ann. Neurol.* 2004. 56(4): p. 591-5.

[200] Webb, J.L., et al., Alpha-Synuclein is degraded by both autophagy and the proteasome. *J. Biol. Chem.* 2003. 278(27): p. 25009-13.

[201] Cuervo, A.M., et al., Impaired degradation of mutant alpha-synuclein by chaperone-mediated autophagy. *Science.* 2004. 305(5688): p. 1292-5.

[202] Vogiatzi, T., et al., Wild type alpha-synuclein is degraded by chaperone-mediated autophagy and macroautophagy in neuronal cells. *J. Biol. Chem.* 2008. 283(35): p. 23542-56.

[203] Chinta, S.J., et al., Mitochondrial alpha-synuclein accumulation impairs complex I function in dopaminergic neurons and results in increased mitophagy in vivo. *Neurosci. Lett.* 2010. 486(3): p. 235-9.

[204] Winslow, A.R., et al., alpha-Synuclein impairs macroautophagy: implications for Parkinson's disease. *J. Cell Biol.* 2010. 190(6): p. 1023-37.

[205] Pivtoraiko, V.N., et al., Low-dose bafilomycin attenuates neuronal cell death associated with autophagy-lysosome pathway dysfunction. *J. Neurochem.* 2010. 114(4): p. 1193-204.

[206] Choubey, V., et al., Mutant A53T alpha-synuclein induces neuronal death by increasing mitochondrial autophagy. *J. Biol. Chem.* 2011. 286(12): p. 10814-24.

[207] Greene, J.C., et al., Mitochondrial pathology and apoptotic muscle degeneration in Drosophila parkin mutants. *Proc. Natl. Acad. Sci. U. S. A.* 2003. 100(7): p. 4078-83.

[208] Whitworth, A.J., et al., Increased glutathione S-transferase activity rescues dopaminergic neuron loss in a Drosophila model of Parkinson's disease. *Proc. Natl. Acad. Sci. U. S. A.* 2005. 102(22): p. 8024-9.

[209] Park, Y.S., et al., Activity-dependent potentiation of large dense-core vesicle release modulated by mitogen-activated protein kinase/extracellularly regulated kinase signaling. *Endocrinology.* 2006. 147(3): p. 1349-56.

[210] Narendra, D., et al., Parkin is recruited selectively to impaired mitochondria and promotes their autophagy. *J. Cell Biol.* 2008. 183(5): p. 795-803.

[211] Vives-Bauza, C., et al., PINK1-dependent recruitment of Parkin to mitochondria in mitophagy. *Proc. Natl. Acad. Sci. U. S. A.* 2010. 107(1): p. 378-83.

[212] Chen, D., et al., Parkin mono-ubiquitinates Bcl-2 and regulates autophagy. *J. Biol. Chem.* 2010. 285(49): p. 38214-23.

[213] Michiorri, S., et al., The Parkinson-associated protein PINK1 interacts with Beclin1 and promotes autophagy. *Cell Death Differ.* 2010. 17(6): p. 962-74.

[214] Kawajiri, S., et al., PINK1 is recruited to mitochondria with parkin and associates with LC3 in mitophagy. *FEBS Lett.* 2010. 584(6): p. 1073-9.

[215] Ziviani, E., R.N. Tao, and A.J. Whitworth, Drosophila parkin requires PINK1 for mitochondrial translocation and ubiquitinates mitofusin. *Proc. Natl. Acad. Sci. U. S. A.* 2010. 107(11): p. 5018-23.

[216] Narendra, D.P., et al., PINK1 is selectively stabilized on impaired mitochondria to activate Parkin. *PLoS Biol.* 2010. 8(1): p. e1000298.

[217] Matsuda, N., et al., PINK1 stabilized by mitochondrial depolarization recruits Parkin to damaged mitochondria and activates latent Parkin for mitophagy. *J. Cell Biol.* 2010. 189(2): p. 211-21.

[218] Geisler, S., et al., The PINK1/Parkin-mediated mitophagy is compromised by PD-associated mutations. *Autophagy*. 2010. 6(7): p. 871-8.

[219] Geisler, S., et al., PINK1/Parkin-mediated mitophagy is dependent on VDAC1 and p62/SQSTM1. *Nat. Cell Biol.* 2010. 12(2): p. 119-31.

[220] Cookson, M.R., Pathways to Parkinsonism. *Neuron.* 2003. 37(1): p. 7-10.

[221] Shendelman, S., et al., DJ-1 is a redox-dependent molecular chaperone that inhibits alpha-synuclein aggregate formation. *PLoS Biol.* 2004. 2(11): p. e362.

[222] Yokota, T., et al., Down regulation of DJ-1 enhances cell death by oxidative stress, ER stress, and proteasome inhibition. *Biochem. Biophys. Res. Commun.* 2003. 312(4): p. 1342-8.

[223] Gonzalez-Polo, R., et al., Silencing DJ-1 reveals its contribution in paraquat-induced autophagy. *J. Neurochem.* 2009. 109(3): p. 889-98.

[224] Thomas, K.J., et al., DJ-1 acts in parallel to the PINK1/parkin pathway to control mitochondrial function and autophagy. *Hum. Mol. Genet.* 2011. 20(1): p. 40-50.

[225] Clements, C.M., et al., DJ-1, a cancer- and Parkinson's disease-associated protein, stabilizes the antioxidant transcriptional master regulator Nrf2. *Proc. Natl. Acad. Sci. U. S. A.* 2006. 103(41): p. 15091-6.

[226] Malhotra, D., et al., Decline in NRF2-regulated antioxidants in chronic obstructive pulmonary disease lungs due to loss of its positive regulator, DJ-1. *Am. J. Respir. Crit. Care Med.* 2008. 178(6): p. 592-604.

[227] Junn, E., et al., Interaction of DJ-1 with Daxx inhibits apoptosis signal-regulating kinase 1 activity and cell death. *Proc. Natl. Acad. Sci. U. S. A.* 2005. 102(27): p. 9691-6.

[228] van der Brug, M.P., et al., RNA binding activity of the recessive parkinsonism protein DJ-1 supports involvement in multiple cellular pathways. *Proc. Natl. Acad. Sci. U. S. A.* 2008. 105(29): p. 10244-9.

[229] Zhou, W., et al., Phenylbutyrate Up-regulates the DJ-1 Protein and Protects Neurons in Cell Culture and in Animal Models of Parkinson Disease. *J. Biol. Chem.* 2011. 286(17): p. 14941-51.

[230] Mo, J.S., et al., DJ-1 modulates the p38 mitogen-activated protein kinase pathway through physical interaction with apoptosis signal-regulating kinase 1. *J. Cell Biochem.* 2010. 110(1): p. 229-37.

[231] Waak, J., et al., Oxidizable residues mediating protein stability and cytoprotective interaction of DJ-1 with apoptosis signal-regulating kinase 1. *J. Biol. Chem.* 2009. 284(21): p. 14245-57.

[232] Zhou, W., et al., The oxidation state of DJ-1 regulates its chaperone activity toward alpha-synuclein. *J. Mol. Biol.* 2006. 356(4): p. 1036-48.

[233] Martinat, C., et al., Sensitivity to oxidative stress in DJ-1-deficient dopamine neurons: an ES- derived cell model of primary Parkinsonism. *PLoS Biol.* 2004. 2(11): p. e327.

[234] Xu, J., et al., The Parkinson's disease-associated DJ-1 protein is a transcriptional co-activator that protects against neuronal apoptosis. *Hum. Mol. Genet.* 2005. 14(9): p. 1231-41.

[235] Goldberg, M.S., et al., Nigrostriatal dopaminergic deficits and hypokinesia caused by inactivation of the familial Parkinsonism-linked gene DJ-1. *Neuron.* 2005. 45(4): p. 489-96.

[236] Kim, R.H., et al., Hypersensitivity of DJ-1-deficient mice to 1-methyl-4-phenyl-1,2,3,6-tetrahydropyrindine (MPTP) and oxidative stress. *Proc. Natl. Acad. Sci. U. S. A.* 2005. 102(14): p. 5215-20.

[237] Menzies, F.M., S.C. Yenisetti, and K.T. Min, Roles of Drosophila DJ-1 in survival of dopaminergic neurons and oxidative stress. *Curr. Biol.* 2005. 15(17): p. 1578-82.

[238] Meulener, M., et al., Drosophila DJ-1 mutants are selectively sensitive to environmental toxins associated with Parkinson's disease. *Curr. Biol.* 2005. 15(17): p. 1572-7.

[239] Park, J., et al., Drosophila DJ-1 mutants show oxidative stress-sensitive locomotive dysfunction. *Gene.* 2005. 361: p. 133-9.

[240] Ved, R., et al., Similar patterns of mitochondrial vulnerability and rescue induced by genetic modification of alpha-synuclein, parkin, and DJ-1 in Caenorhabditis elegans. *J. Biol. Chem.* 2005. 280(52): p. 42655-68.

[241] Krebiehl, G., et al., Reduced basal autophagy and impaired mitochondrial dynamics due to loss of Parkinson's disease-associated protein DJ-1. *PLoS One.* 2010. 5(2): p. e9367.

[242] Irrcher, I., et al., Loss of the Parkinson's disease-linked gene DJ-1 perturbs mitochondrial dynamics. *Hum. Mol. Genet.* 2010. 19(19): p. 3734-46.

[243] Azad, M.B., Y. Chen, and S.B. Gibson, Regulation of autophagy by reactive oxygen species (ROS): implications for cancer progression and treatment. *Antioxid. Redox Signal.* 2009. 11(4): p. 777-90.

[244] Chen, Y. and S.B. Gibson, Is mitochondrial generation of reactive oxygen species a trigger for autophagy? *Autophagy.* 2008. 4(2): p. 246-8.

[245] Filomeni, G., et al., Under the ROS...thiol network is the principal suspect for autophagy commitment. *Autophagy.* 2010. 6(7): p. 999-1005.

[246] Xiong, Y., et al., GTPase activity plays a key role in the pathobiology of LRRK2. *PLoS Genet.* 2010. 6(4): p. e1000902.

[247] Alegre-Abarrategui, J., et al., LRRK2 regulates autophagic activity and localizes to specific membrane microdomains in a novel human genomic reporter cellular model. *Hum. Mol. Genet.* 2009. 18(21): p. 4022-34.

[248] Bravo San-Pedro JM, Gómez Sánchez R, Niso Santano M, Fuentes JM, González Polo RA, LRRK2 mutant G2019S exacerbates basal autophagy and sensitizes cells against neurotoxins related to Parkinson's disease. *Cell Death and Differentiation.* Under review

In: Autophagy: Principles, Regulation and Roles in Disease ISBN: 978-1-61942-266-7
Editor: Nikolai Gorbunov

Chapter XIV

Autophagy and Liver Ischemia-Reperfusion Injury

Raffaele Cursio*[1,2*]*, Pascal Colosetti*[2]*,
***Patrick Auberger*[2] *and Jean Gugenheim*[1,2]**
[1] Service de Chirurgie et Transplantation Hépatique, Hôpital l'Archet 2, Université de Nice Sophia-Antipolis, Nice, France
[2] INSERM U526, Equipe 2: Cell death, Differentiation and Cancer, Université de Nice Sophia Antipolis, Nice, France

Abstract

Liver ischemia-reperfusion (I-R) injury occurs during hemorrhagic shock and liver surgery, particularly during liver transplantation. After liver transplantation, I-R injury causes up to 10% of early transplant failures and can lead to acute and chronic rejection. Originally, the mode of cell death after I-R was considered to be necrotic, then it was shown that apoptosis too plays an important role, and now there is increasing evidence that autophagy is also involved. Autophagy is a conserved cellular process that controls protein and organelle degradation and recycling, and has essential roles in survival, development and homeostasis. Thus autophagy is involved in physiology, lifespan and a wide range of diseases, including cancer, neurodegeneration and microbial infection. In the normal liver, autophagy is enhanced in starved animals or following glucagon infusion. During Hepatitis B and C infections, acute liver injury and alpha1-antitrypsin deficit autophagy is increased; whereas in the alcoholic or steatotic liver and in hepatocellular carcinoma it is decreased. Modulation of autophagy by its inhibitors and/or stimulators may reduce liver I-R injury during liver surgery.

The objectives of this chapter are to provide a review of the mechanisms mediating I-R liver injury in experimental and clinical settings with particular focus on the role of autophagy, its interconnection with apoptosis and necrosis and, finally to summarize the

---

* Corresponding author: Raffaele Cursio, MD, PhD, Service de Chirurgie et Transplantation Hépatique, Hôpital l'Archet 2, 151 Route Saint Antoine de Ginestière, B.P. 3079, 06202 Nice, France. Phone: +33-4-92-03-64-76/4-92-03-63-81; Fax: +33-4-92-03-65-62; e-mail: cursio@unice.fr.

most promising surgical and pharmacological autophagy target therapies ameliorating sequelae of the I-R injury.

Keywords: Autophagy; Apoptosis; Necrosis; Ischemia; Reperfusion injury; Liver; Ischemic preconditioning

Introduction

Liver ischemia-reperfusion (I-R) injury is characterized by partial or complete interruption of the liver blood flow and with this, its oxygen supply, followed by reperfusion, which means the reestablishment of blood flow and oxygen supply. Reperfusion itself aggravates the cellular injuries caused by the ischemic period [1-2]. Originally, the mode of cell death after I-R was considered to be necrotic [3], then it was shown that apoptosis too plays an important role [1-4-5-6], and now there is increasing evidence that autophagy is also involved [7-8-9-10-11-12-13-14-15]. Autophagy, from the Greek word "autophagus" (αυτοφαγια), *"self eating"* is a conserved cellular process that controls protein and organelle degradation and recycling, and has essential roles in survival, development and homeostasis [16-17]. The term "autophagy" first appeared in 1965-66 [18-19]; two years later a link between lysosomes, degradative process and autophagy was found [16-17]. The main functional role of autophagy is to recycle cellular components, to sustain metabolism during nutrient deprivation and to prevent the accumulation of damaged, toxic proteins and organelles during stress [20]. In most situations, autophagy promotes survival to stress and starvation, but it seems that if left unabated or if over stimulated, autophagy can progress to cell death. Autophagy is involved in physiology, lifespan and a wide range of diseases, including cancer, neurodegeneration and microbial infection. In the normal liver, autophagy is enhanced in starved animals or following glucagon infusion [17]. During Hepatitis B and C infections, acute liver injury and alpha1-antitrypsin deficit autophagy is increased; whereas in the alcoholic or steatotic liver and in hepatocellular carcinoma it is decreased [21].

The autophagy mechanism of liver cell remodeling following I-R injury is very complex and remain poorly understood. Nevertheless, the current evidence from literature indicates that the autophagy target therapy is a promising therapeutic approach to control pathogenesis of I-R injury.

Clinical Relevance of Ischemia-Reperfusion in Liver Injury

Liver I-R injury is commonplace not only in liver surgery [in particular, in liver transplantation (LT)], but occurs also during hemorrhagic shock and other clinical conditions that cause low liver blood flow resulting in insufficient perfusion of the organ. Depending on the clinical setting, the liver can be subjected to different types of ischemia, cold, warm and rewarming ischemia [22]. Whereas cold ischemia occurs typically in LT following harvesting and static cold preservation of the graft, warm ischemia is characteristic for liver trauma, hypovolemic shock, or for blood inflow occlusion realized during liver surgery in order to achieve a vascular control. Rewarming ischemia occurs in LT after anastomosis of the liver vessels. Pathomechanisms associated with cold ischemia is different from those of warm ischemia, although all types of ischemia share common mechanisms [22]. Cold ischemia of the liver is characterized mainly by injury to sinusoidal lining cells, whereas warm ischemia results primarily in damage of hepatocytes [22-23].

A) Liver Transplantation

LT represents the primary therapy for end–stage liver disease and acute liver failure. During transplantation the liver is subjected to I-R injury at the following three distinct stages [5]. The first stage, when the liver is explanted from the donor and stored on ice at 0° to 4°C, is a variable but generally long period of cold ischemia. The time of vascular anastomosis, when the liver is removed from ice until its implantation in the recipient, represents the second, relatively shorter stage of I-R injury. In this period of ischemia the liver warms slowly up towards body temperature at a rate of 0.5°C per minute. Normothermic reperfusion of the implanted liver with the recipient's blood at 37°C delineates the third stage.

Liver I-R injury causes up to 10% of early transplant failures and may be responsible for primary graft non-function (< 5 % of grafts) and initial poor graft function (10-30% of grafts). This leads to a higher incidence of both, acute and chronic rejection increasing the need for re-transplantation [22-24-25-26]. Poor initial liver graft function secondary to I-R also correlates with higher rates of early rejection [27]. Liver I-R may also be responsible for late allograft failure [28-29].

Due to organ shortage the patient waiting lists for LT are growing with an increased gap between the demand for and the number of available organs. This has led to a strong incentive for using extended criteria donor organs, which include organs distinguished by hepatic steatosis, old donor age, prolonged cold ischemia, or donation after cardiac death [30]. Extended criteria donor livers are more susceptible to I-R injury and present more graft dysfunction events as mentioned above [22-24-25].

B) Liver Surgery for Tumours

Liver resection remains the only curative treatment for most patients with primary or secondary malignant liver tumours. The Pringle manoeuvre, clamping of the portal venous and hepatic arterial inflow, is used to minimize intra-operative blood loss during parenchymal resection for liver tumours or trauma [31], but it may result in liver I-R injury [32-33]. Extended resections, especially in case of liver steatosis, may be responsible for liver dysfunction [34]. Steatosis or accumulation of lipids in the liver, is considered pathologic when the hepatic fat content, consisting mainly of triglycerides, exceeds 5% of the actual wet weight of the liver [35]. Steatosis is the most common chronic liver disease in the world with a prevalence of 10 to 20% in lean population, 60 to 74% among the obese and over 90% in the morbidly obese [36-37], and it is a major risk factor in patient outcome after liver resection. The increased vulnerability of the steatotic liver to I-R injury is due to a combination of increased hepatic reactive oxygen species (ROS) production and cellular pH changes, increased inflammatory responses and reduced blood microcirculation by sinusoidal vasoconstriction [38-39-40-41]. Lipid accumulation perturbs hepatic energy homeostasis and induces hepatocellular damage which subsequently affects hepatocellular regeneration and recovery [42].

C) Liver Surgery for Trauma

The liver is the most commonly injured intra-abdominal organ in the setting of both, blunt and penetrating abdominal trauma [43]. If the majority of liver injuries can be managed non-operatively [44], the remnant injuries require surgical repair as well as parenchymal resection [45-46-47]. In order to achieve a vascular control of the injured liver, portal triad occlusion (Pringle manoeuvre) may by performed, but it results in liver I-R injury [48-49]. Major hepatic trauma is associated with a high mortality of patients [50-51].

In rare cases, such as uncontrollable hemorrhage, irreversible liver failure, or life-threatening post-reperfusion injury, total hepatectomy and LT may be performed as the only life-saving measure [52-53].

During liver resection for trauma, pre-existing liver disease, as steatosis and cirrhosis, may increase the vulnerability of the liver to I-R injury [54-55].

D) Hemorrhagic Shock and Resuscitation

Hemorrhagic shock remains a major cause of morbidity and mortality following trauma [56]. Liver dysfunction occurs in approximately 20% of trauma patients admitted in hemorrhagic shock [57]. During hemorrhagic shock there is a regional redistribution of the hepatic blood flow within the liver and low-flow ischemia followed by restoration of normal flow results in ischemia in the pericentral regions with sinusoidal perfusion failure in the periportal region of the liver [58].

Despite the fact that hepatic blood flow remains unchanged or drops modestly following hemorrhagic shock, oxygen requirements may increase and supercede eventual compensatory increases in the splanchnic vessels, producing regional imbalances of oxygen supply and demand. Reduced or absent oxygen delivery causes primary (hypoxic) or secondary (ischemia and reperfusion) hepatocellular injury, which may result in mild to moderate increase of aminotransferases and bilirubin blood levels [59]. In fact hepatocellular necrosis is not a prominent feature of the postshock liver whereas apoptosis might even be accentuated by the type of resuscitation fluid [60-61].

The duration and severity of hemorrhagic shock may result in expression of inducible nitric oxyde synthase as a response to inflammatory cytokines, which may enhance microcirculatory damage on reperfusion [62]. Finally production and release of various cytokines by prolonged hemorrhagic shock induce additional inflammatory response mediated, principally, by neutrophils [63].

Adhesion of neutrophils to the endothelium disturbs hepatic microcirculation and contributes to the flow-no flow phenomenon of reperfusion that adds to regional cellular hypoxia [64-65-66-67].

E) Other Clinical Conditions Causing Insufficient Liver Blood Perfusion

Other conditions that cause low blood flow to the liver resulting in insufficient perfusion are cardiogenic or septic shock with subsequent fluid resuscitation [68], extracorporeal circulation in cardiovascular surgery [69], laparoscopic surgery [70-71] and, abdominal compartment syndromes [72].

Pathophysiology of Liver Ischemia-Reperfusion Injury

The lack of oxygen and nutrient supply to the liver during the ischemic period induces several cellular functional changes promoting liver cell injury [24]. Liver I-R rapidly leads to an acute inflammatory response, causing significant hepatocellular damage, followed by organ dysfunction [24-38-73]. In animal and human the severity of injury ranges from mildly raised levels of serum aminotransferases to primary non-function of liver allograft or post-hepatectomy liver insufficiency. Under extreme circumstances multiple organ failure and death may occur.

A) Functional Liver Unit and I-R Injury

The liver is highly perfused, receiving approximately one quarter of total blood output from the heart. Liver blood flow derives from both, the hepatic artery (20%) and the portal vein (80%). The terminal branches of these vessels surround the hexagonal-shaped hepatic lobule. The smallest microcirculatory and functional unit of the liver is the acinus, a parenchymal cell mass arranged around the terminal vessels consisting of the classical portal triad: bile duct, hepatic artery and portal vein [74] (Fig. 1A). The blood stream is unidirectional, and starts from the terminal portal and terminal hepatic arteriole which deliver their blood into the sinusoids. The central vein collects the blood from the sinusoids and flows into the hepatic vein [74]. The hepatocytes are polarized parenchymal cells that form a trabecular network within the acinus. The basolateral membranes of the hepatocytes form the limits of the sinusoidal space. Hepatic sinusoids consist of fenestrated endothelial cells which are surrounded by perisinusoidal (Ito or fat storing) cells. Kupffer cells and pit cells (large granular lymphocytes) bulge into the sinusoidal lumen [75]. Thus hepatocytes are directly exposed to plasma in the perisinusoidal space of Disse, which is the space between the basal microvilli-rich surfaces of hepatocytes and the sinusoidal lining cells.

Because of liver metabolism, the composition of blood changes during the passage through the sinusoids. As blood flows through the liver sinusoids, its partial oxygen pressure (free concentration) drops by about 50%, from about 60-65 mm Hg (84-91 μmol/L) in the periportal blood to 30-35 mm Hg (42-49 μmol/L) in the perivenous blood [75]. Following the changes of blood composition there are three zones in the acinus (Fig. 1B). Zone 1 is the periportal (afferent) zone in which concentration of oxygen, hormones, substrates and nutrients is high. Zone 2, the intermediate zone, is intermediate in oxygen and nutrients. In cells of zone 3 (pericentral, centrolobular, efferent), zone close the central vein, the concentration of oxygen, hormones, and substrates is low, but rich in carbon dioxide and metabolic products [76]. Whereas periportal cells of zone 1 are subjected to oxidant injury produced by reperfusion, cells of zone 3 are most sensitive to ischemia, hypoxemia [76] or hypoxia [77]. In case of reduced blood supply to the liver, the oxygen delivery is reduced and leads to perivenous hypoxia, which may result in perivenous cellular damage [77]. Following hemorrhagic shock depletion of adenosine triphosphate (ATP) as oxygen delivery progressively diminishes in transit across the acinus. Reperfusion of the intermediate zone cell 2, results in formation of ROS, producing cell injury and death [78]. Under physiological conditions in the liver, glycogen degradation starts periportally and ends perivenously and, glycogen stores are refilled in the perivenous and ending in the periportal zone [79]. However, in perfused livers carbohydrate metabolism is significantly modified by the oxygen supply [80]. So in human liver transplantation, during ischemia, the mean loss of glycogen was 48% in periportal zones and 74% in pericentrolobular zones [81]. As result of glycogen consumption and lack of oxygen supply, Kupffer cells, sinusoidal endothelial cells and hepatocytes suffer from the interruption of ATP production [82]. Kupffer and sinusoidal endothelial cell swelling associated with the increase of vasoconstrictor factors lead to sinusoidal narrowing, resulting in reduction of microcirculatory blood flow in the liver [82].

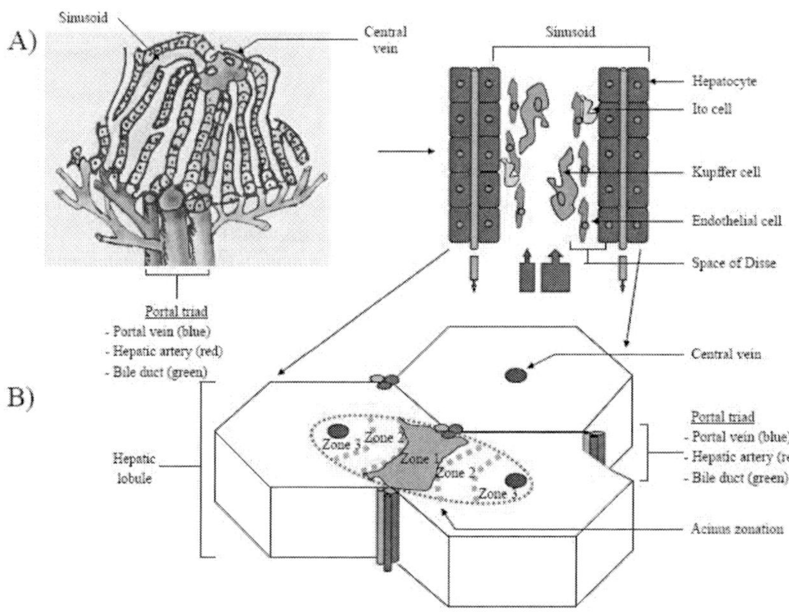

Figure 1. Liver architecture, blood flow and functional liver unit.

A) The terminal branches of the hepatic artery and the portal vein surround the hexagonal-shaped hepatic lobule. B) The liver acinus superimposed on the traditional lobular architecture. The acinus, a parenchymal cell mass arranged around the terminal vessels consisting of the classical portal triad, bile duct, hepatic artery and portal vein, is the smallest microcirculatory and functional unit of the liver. Following the changes of blood composition there are three zones of the acinus. Zone 1 is the periportal (afferent), zone 2, the intermediate zone, and zone 3 (pericentral, centrolobular, efferent), zone close the central vein.

B) Cellular and Molecular Mechanisms of Liver I-R Injury

In the normal liver ischemia becomes critical with a period of 15-20 min in animals [58] and of above 60 min in humans [31-32-33]. Reperfusion of the liver after a longer period than the critical ischemia time point results in endothelial and parenchymal injury. Reperfusion injury consists of initial phase injury (within 2 h after reperfusion) and late phase injury (6-48 h after reperfusion) [58].

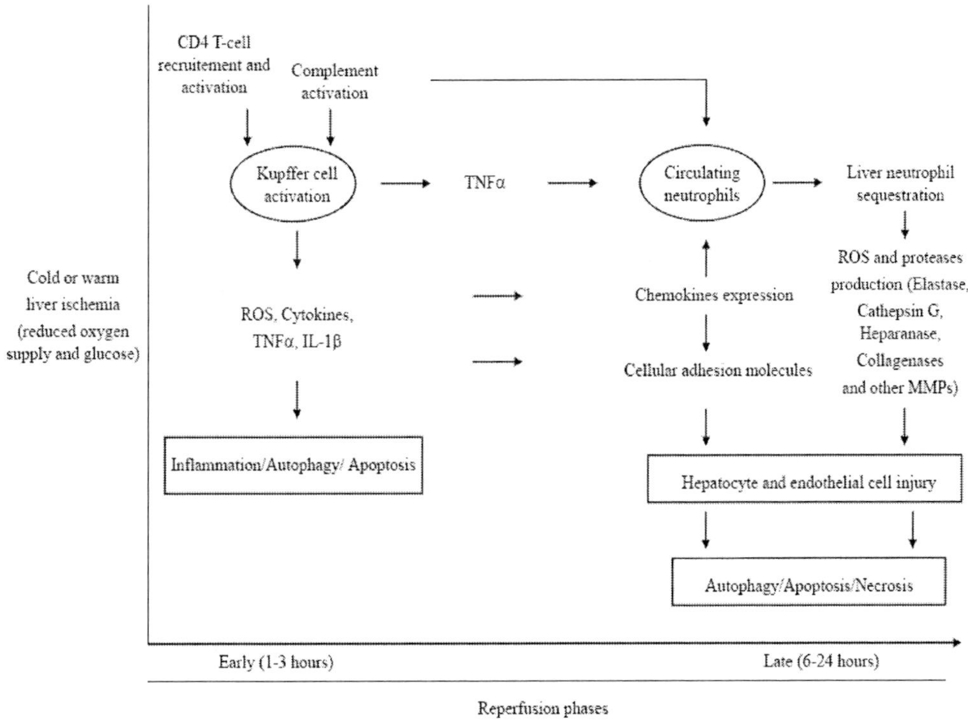

Figure 2. Schematic presentation of mechanisms involved in the pathophysiology of liver I-R injury. Reperfusion injury consists of initial phase injury (within 3 h after reperfusion) and late phase injury (6-24 h after reperfusion). The main event in the initial phase of reperfusion injury is activation of Kupffer cells, whereas neutrophils are the principal cells involved in the late phase of reperfusion injury.

As shown in Fig. 2, the main event in the early phase of reperfusion injury is the activation of Kupffer cells, which are located in the peri-portal regions of the normal liver. Generation of ROS by these activated Kupffer cells initiates reperfusion injury, with endothelial lesions (through protein oxidation and degradation, lipid peroxidation, and DNA damage) and a further release of pro-inflammatory cytokines, complement factors and neutrophils, which are the principal cells involved in the late phase of reperfusion injury [83]. Neutrophils damage hepatocytes through ROS and proteases release [73]. In addition to the generation of oxidants, activated neutrophils release a number of mediators by granule exocytosis. The contents of neutrophil granules include large amounts of proteases (i.e., elastase, cathepsin G, heparanase, collagenase) and hydrolytic enzymes that may be directly cytotoxic to hepatocytes [84-85]. Serine proteases, such as elastase and cathepsin G, may directly damage membrane components of hepatocytes, whereas metalloproteinases primarily degrade basement membrane and matrix components [86]. Loss of the equilibrium between nitric oxide and Endothelin-1 (ET-1) induces vasoconstriction, increases platelet aggregation and narrowing of the sinusoidal lumen, compromises leukocyte flow, resulting in microcirculatory disturbances, and consequently aggravating liver damage [39-87].

C) Autophagy and Liver I-R Injury

Necroapoptosis is a major feature of I-R indicating that both cell death mechanisms, necrosis and apoptosis, occur simultaneously during the ischemic phase, and that they even imbricate during reperfusion phase [1-2-3-4-5-6-88]. However, autophagy as mode of cell death after I-R injury plays an important role too [7-8-9-10-11-12-13-14-15].

Autophagy is an active process at a basal level in most of the cells in the organism, although individual autophagic capacity may vary [89]. The mitochondria organelles occupying 20% of the cytoplasmic volume of hepatocytes, are a major target of autophagy [90]. When autophagy is impaired, deformed mitochondria accumulate in the liver [91]. Autophagy is a prominent survival mechanism under nutrient deprivation in the liver [92]. Under stress conditions, such as starvation or transient hypoxia, autophagy can have protective functions degrading cytoplasmic materials to produce amino acids and fatty acids that can be used to synthesize new proteins or are oxidized by mitochondria to produce ATP for cell survival [93-94]. Also the autophagic removal of damaged mitochondria which might otherwise release pro-apoptotic factors and trigger the JNK/c-Jun mitogen-activated protein kinases (MAPKs) signaling pathway, mediating apoptosis, underline the protective effects of induced autophagy [95]. However, when autophagy is excessively induced, it can result in autophagic cell death, so called type II programmed cell death (PDC), which is distinct from type I PCD (apoptosis) and from necrosis [93-96-97-98].

Whether autophagy protects from or promotes liver injury following I-R remains to be elucidated (Tables 1 and 2).

In LT, cold protects tissues by reducing the rate of cell death, and the use of preservation solutions helps to prolong the cold storage time [99]. However, the introduction of the host blood supply during the warm reperfusion phase, induces critical changes in the mitochondria with production of ROS and cell death [100]. ROS can regulate autophagy [101-102-103]. They stimulate the activation of AMP-activated kinase (AMPK) [104], a kinase involved in activating energy-generating pathways to maintain or restore ATP levels, and the activation of Snf1 protein kinase [105], which both induce autophagy. In animal studies, autophagosomes/autolysosomes in graft hepatocytes were observed not only during the cold preservation phase but also during the reperfusion phase [13]. In particular, abundant autophagosomes/autolysosomes were associated with dying hepatocytes at the beginning of the warm reperfusion phase after LT in rats [13] and after reperfusion following 120 min normothermic liver ischemia (Fig. 3) [7]. A decrease in autophagic responsiveness may induce tissue injury earlier, due to increased accumulation of damaged protein aggregates and large dysfunctional mitochondria, leading to increased ROS, defective ATP generation, less efficient metabolic pathways and a decrease in lysosomal stability [106].

Table 1. Autophagy and liver I-R injury: animal studies

Species	Type of I-R	Autophagy function	Autophagy inhibition	Autophagy stimulation	Effect on I-R injury	References
Rat	Partial warm 120 min	Increased ?	No	No	Protective ?	Cursio et al. [7]
Mice	Partial warm 60 min	Decreased	No	Cisplatin	Protective	Cardinal et al. [8]
Rat	Cold/warm liver transplantation	Increased	Proteinase inhib. Pepstatin Leupeptin	No	Protective	Gotoh et al. [9]
Rat	Cold storage 20 hours	Decreased	No	Hypothermic reconditioning	Protective	Minor et al. [10]
Old mice	Partial warm 30, 60 and 90 min	Decreased	No	Rosiglitazone	Protective	Shin et al. [11]
Rat	Total liver warm 45 min Anoxia/reoxygenation of hepatocytes	Decreased	No	Calpain2 inhibitor Adenoviral overexpression of Atg7 and Beclin-1 Nutrient depletion	Protective	Kim et al. [12]
Rat	Cold/warm liver transplantation	Increased	No	No	Deletere	Lu et al. [13]

Table 2. Autophagy and liver I-R injury: human studies

Liver disease	Type of I-R	Type of liver surgery	Surgical strategy	Autophagy function	Ischemic preconditioning	Effect on I-R injury	References
Steatosis	Cold/warm	Transplantation	Total liver vascular exclusion	Decreased	Yes	Protective	Degli Esposti et al. [14]
Steatosis post-chemotherapy	Warm	Resection	Pringle's manoeuvre	Decreased	Yes	Protective	Domart et al. [15]

In animal studies, cold ischemia during liver transplantation reduces the graft metabolism and ATP production [107], and therefore prolongs the time period during which anoxic cells can retain essential metabolic functions. Warm reperfusion may facilitate autophagosome formation in hepatocytes under ATP exhaustion as a stress response [13]. Induction of excessive autophagy in hepatocytes in the early stage of the reperfusion period, 30 to 60 min, may constitute an additional stress stimulus to the liver graft [13]. Hepatocytes with numerous vacuolar/lysosomal structures often undergo degeneration and are phagocytosed by Kupffer cells late in the reperfusion phase [13]. When the autophagic process is deregulated or excessive, it may result in excessive catabolism, "misrecognition of cargo, and/or hijac of the apoptosis machinery" [108]. Between autophagic genes, ATGs (for AuTophagy Genes), which are conserved in higher organisms [109], over-expressed Atg5 can promote apoptosis via its interaction with FADD [110] or Bcl-XL, a member of B-cell CLL/lymphoma 2 (Bcl-2) family protein [111].

Anoxia/reperfusion of hepatocytes causes loss of both, Atg7 and Beclin-1 by Calpain 2 activation, which in turn limits mitochondrial autophagy, mitophagy, a particular form of autophagy requiring fragmentation of the mitochondrial network in order to segregate the dysfunctional units to be removed [12]. Defective mitophagy after reperfusion and reoxygenation subsequently culminates in onset of the mitochondrial permeability transition (MPT) leading to uncoupling of oxidative phosphorylation and necrotic cell death of hepatocytes [12]. The MPT onset induces large-amplitude swelling of mitochondria and apoptosis by releasing proapoptotic factors that are normally sequestered in the mitochondrial intermembrane space [112]. Thus, MPT is the common mechanism that culminates either in liver necrosis or in apoptosis following I-R [113]. Stimulation of autophagy through nutrient depletion and overexpression of Atg7 or beclin-1 inhibits "MIT-dependent-hepatocyte-necrosis" and apoptosis and enhances ATP recovery after reperfusion [12]. Although not constitutively expressed in normal livers [114] Bcl-2, seems to play a role in I-R [15]. In fact Bcl-2 can have a dual function, by turning autophagy or apoptosis on or off when required [115]. Autophagic cell death was observed under conditions in which apoptosis was blocked [97].

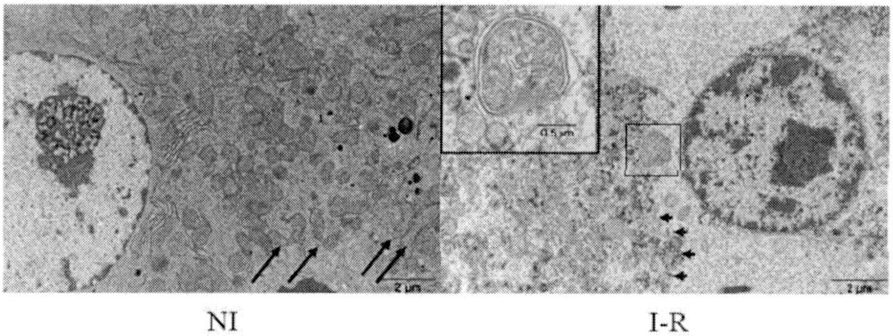

Figure 3. Autophagy and liver I-R injury. Liver autophagy in rats subjected to (I-R) or not (NI) for 120 min normothermic I-R, evaluated by transmission electron microscopy 6 hours after reperfusion. On the I-R picture, a square is surrounding an autophagolysosome with a double plasma membrane. Fusion of mitochondria (black arrows) was observed in non-ischemic liver lobes (NI), while in ischemic liver lobes (I) mitochondria were subjected to the fission and fragmentation process (arrowheads).

During the lifetime of the cell, the mitochondrial network, which facilitates transfer of the mitochondrial membrane potential from oxygen-rich to oxygen-poor cellular regions, is continuously shaped by fusion and fission events [116]. Mitochondrial fusion processes, which serve to maintain a tubular mitochondrial network and optimal mitochondrial function, seem to be associated with cell differentiation and senescence whereas fission processes, which includes a profound mitochondrial remodeling and changes of its membrane curvature, are mandatory during cell proliferation [117]. Unbalanced mitochondrial fusion and fission may result in breakdown of the mitochondrial network, loss of the mitochondrial DNA, and respiratory defects and in increase of ROS [118-119]. Mitochondria undergo rapid and excessive fission evoked by several stimuli in the early apoptotic pathway [120]. The fission process, which precedes the release of apoptogenic factors by mitochondria [121-122], is associated with progression of the apoptotic execution pathway [123] and followed by mitochondrial fragmentation [124]. As shown in Fig. 3, following 120 min normothemic liver I-R in rats, fusion of mitochondria was observed in non-ischemic liver lobes, while in ischemic liver lobes mitochondria were subjected to the fission and fragmentation process. There was an increase of Fas ligand (FasL) mRNA and protein expression in these ischemic liver lobes [125]. This increase was not modified by caspase inhibition [125]. The mitochondrial fission and fragmentation observed may be the result of Fas ligation. In fact it has been reported that mitochondrial fission and fragmentation in apoptotic lymphoid cells was associated to Fas ligation [126]. Furthermore the mitochondrial fission and fragmentation seem to be independent of effector caspases [127]. While mitochondria may be essential to provide membranes for the formation of the autophagosome [128], a part of it should be excluded from immediate fragmentation, to ensure the progression of the autophagic process [129-130].

Mitochondrial morphology determines the cellular response to autophagy [130]. When autophagy is triggered, mitochondria elongate [130]. These elongated mitochondria are spared from autophagic degradation, possess more cristae, increased levels of dimerization and activity of ATP synthase, and maintain ATP production [130]. Thus mitochondrial elongation may be an initial response to a low level of stress, which provides a protective response against a future insult [131].

Liver I-R injury seems to be age-dependent [132]. Indeed there has been a distinct difference in the response to liver I-R injury in mice of different ages [132]. The activation of Peroxisome Proliferators-Activated Receptor gamma (PPARγ), which belongs to the hormone nuclear receptor superfamily [133], seems to decrease liver I-R injury by modulating inflammatory response and apoptosis [134], and has been associated to autophagy [11]. During liver ischemia its activation is suppressed [135]. In mice pretreatment with roziglitazone, an autophagy inducer, activates PPRAγ and significantly reduces liver I-R injury [135]. The PPARγ activation is significantly prolonged in young mice compared to adult and old mice [11].

In summary liver I-R leads to induction of autophagy, which can be protective or detrimental to hepatocytes. Low levels of autophagy can promote hepatocyte survival by degradation of intracellular contents to maintain ATP production and removal of damaged organelles, and protein aggregates. Excessive and long-term upregulation of autophagy, as it occurs during severe ischemic insult, eventually leads to destruction of essential proteins and organelles resulting in hepatocellular apoptosis and necrosis (Fig. 4).

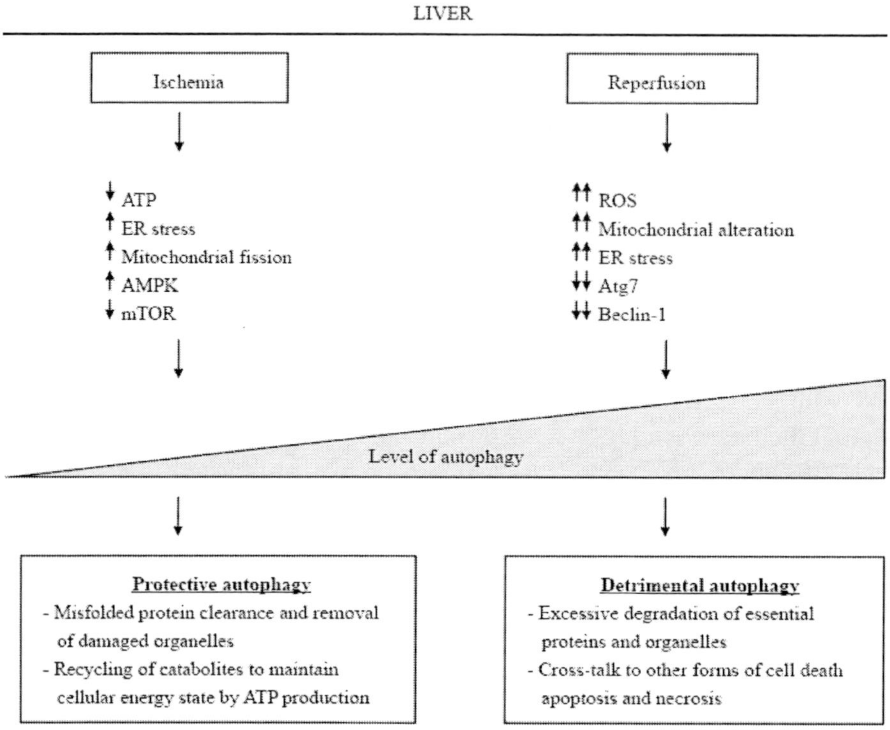

Figure 4. Autophagy and liver I-R injury. Liver I-R leads to induction of autophagy, which can be protective or detrimental to hepatocytes. Low levels of autophagy can promote hepatocyte survival by degradation of intracellular contents to maintain ATP production and removal of damaged organelles, and protein aggregates. Excessive and long-term upregulation of autophagy, i.e. during severe ischemic insult, eventually leads to destruction of essential proteins and organelles resulting in hepatocellular apoptosis and necrosis.

D) Autophagy and Apoptosis During Liver I-R Injury

In liver transplantation, the initial liver graft damage occurs during cold preservation, but at the time of reperfusion, the rapid metabolism of accumulated metabolic products leads to the production of toxic molecules, for example free radicals, and to the activation of apoptosis. In transplanted livers of patients with graft dysfunction, such as IPGF, a high expression of some genes involved in the regulation of apoptosis and an increase of caspase-3 activity was observed [136]. Caspases are cysteine proteinases specifically involved in the initiation and execution of apoptosis. As shown in Fig. 5, an increased *in situ* caspase-3 and -7 activity, located mainly in the perivascular areas of the reperfused liver followed by a marked increase of apoptotic cells was observed in warm ischemic/ reperfused liver lobes [137]. Thus the extent of liver injury following I-R seems to be correlated to the increased caspase activity [1]. Ischemia by itself can trigger apoptosis, but reperfusion accelerates the process. Caspases and autophagy can be simultaneously activated within a dying cell [138]. Autophagy which not only limits metabolic stress but also minimizes genomic instability [139], may be the first line of defence against a cellular insult [97]. Autophagy helps cells to avoid a metabolic crisis, modifies the downstream cell death pathways, and protects the cell itself from dying.

Thus autophagic activity seems to have an impact on the onset of other death modes, apoptosis and necrosis, through intricating the cell's metabolic balance sheet [140]. This dynamic cellular response provides cellular adaptability to a given stressful condition, for example liver I-R injury, which can, context specific, extend into the process and endpoint of death [140]. Autophagy may end with apoptosis and apoptosis may begin with autophagy [106].

NI I-R

Figure 5. Apoptosis and liver I-R injury. Liver apoptosis evaluated by fluorescent green probe FLIVO (Fluorescence in vIVO) associated to in situ caspase -3 and -7 activity (green) in rats submitted (I-R) or not (NI) to 120 min normothermic I-R. Rats were injected with FLIVO via portal vein, 10 min before the end of 6 hours of reperfusion. Nuclei are DAPI stained (blue).

E) Autophagy and Necrosis During Liver I-R Injury

Necrosis is a degenerative process with the complete loss of cellular integrity leading to cell death characterized by cytoplasm swelling, irreversible plasma membrane damage and irreversible changes in the nucleus such as pyknosis, kariorhexis and karyolysis [140]. Normothermic liver I-R results in increased necrosis in the reperfused liver lobes (Fig. 6) [1].

Autophagy seems to be induced as a stress response to increase likelihood of survival [141] and can be activated temporarily to avoid cell death by necrosis and its subsequent inflammatory process. In case of insufficient autophagic activity to maintain a favorable cellular environment, necrosis is induced [142]. As shown in Fig. 7, hepatocytes may respond to I-R gradually depending on the severity of the ischemic insult.

As described above, apoptosis and autophagy can simultaneously be activated within a dying cell during liver I-R and autophagosomes and lysosomes may synergize as proteolitic system to facilitate necrosis [142-143-144]. Accumulation of apoptotic hepatocytes obstructs the sinusoidal stream during reperfusion after prolonged cold liver preservation, and results in secondary necrosis of liver grafts and liver dysfunction [145]. In case of profound depletion of ATP because of uncoupling of oxidative phosphorylation and accelerated ATP hydrolysis by the mitochondrial ATPase, the autophagic process may be bypassed and cells quickly die by necrosis [146].

Finally autophagy may induce indirectly necrosis by delaying apoptosis [106].

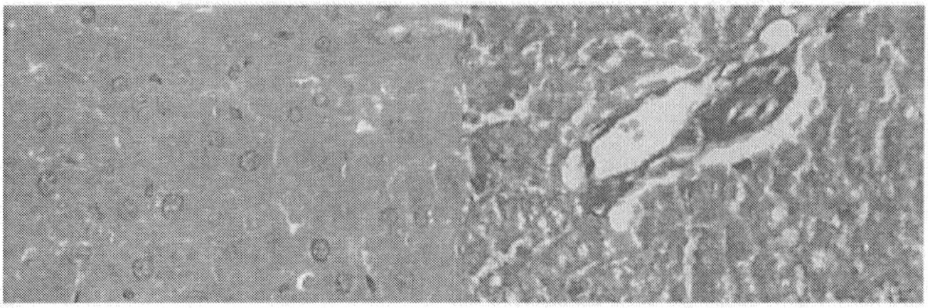

Figure 6. Necrosis and liver I-R injury. Liver necrosis evaluated by hematoxylin and eosin stained tissue sections at 6 hours of reperfusion in rats submitted (I-R) or not (NI) to 120 min normothermic I-R. X200.

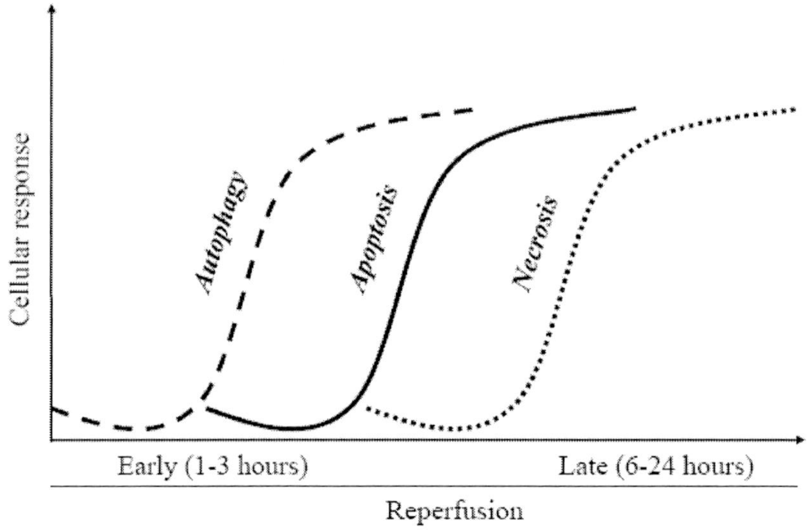

Figure 7. Mode of hepatocyte death following warm I-R. Hepatocytes may respond to I-R in a graded fashion depending on the severity of the ischemic insult. The hepatocyte's response to warm I-R is highly dynamic and morphological features between the death modes can overlap.

F) Modulation of Autophagy by Inhibitor and Inducer Molecules

Autophagy is linked to the organelle lysosome, which is a unique supplier of hydrolases/proteases. When the lysosomal membrane is destabilized, cathepsins are released, which trigger MPT. This causes autophagy as well as caspase activation and apoptosis [147-148].

Cathepsins, from the Greek word "catepso" (καθέψω) "*to digest*" or "*to boil down*", are proteases among more than 50 acid hydrolases (phosphatases, nucleases, glycosidases, proteases, peptidases, sulphatases and lipases) present within the lysosomes [19]. They are playing an important role during cell signalling, death and autophagy. Lysosomal proteases are expressed both, ubiquitously and in a tissue- or cell-type-specific manner. Cathepsins can be divided into several groups based on the catalytic residue, namely, cysteine, serine and aspartic cathepsins [149].

In lysosomes mainly cathepsins B, D, and L, are found and, with exception of cathepsin D, which is an aspartic protease, they are cysteine proteases. Cathepsins catalyse the hydrolysis of proteins within the endocytic pathway and sometimes also extracellularly with an optimal activity at acidic pH, although cysteine cathepsins are also active at neutral pH [149]. Only a few of these enzymes act as aminopeptidases or carboxypeptidases. Most of them are endopeptidases, preferentially cleaving peptide bonds within a polypeptide chain [150]. As cathepsins catalyse the cleavage of peptide bonds of autophagy substrates, they dispose, together with other factors, of the autophagic flux [150]. Each cathepsin allows induction of different physiological or pathological conditions.

Regulators of apoptosis seem to determine also autophagic activation [140], so that the inhibition of autophagy, either at an early or a late stage of the liver I-R process, may lead to apoptosis of liver cells.

So using cathepsin inhibitors in order to modulate autophagy may reduce apoptosis and consequently liver I-R injury.

Between specific inhibitors for individual cathepsins, CA-074 and CA-030 are commonly used as cathepsin B inhibitors [151]. The cell-permeable form of CA-074, CA-074Me inhibitor, represents its lower specificity form. Similarly to the CA-074 inhibitor, the cathepsin L specific inhibitor CAA0225, inhibits the degradation of long-lived proteins, but its involvement in the autophagosomal membrane degradation is more specific than that of CA-074 inhibitor [151]. Other specific inhibitors of cathepsin L were recently synthetized, CLIK-148 and 195 [152]. These inhibitors reduce body weight gain and serum insulin levels, and increase glucose tolerance due to increased levels of muscle IRB subunit, fibronectin, and Glut-4 in obese mice [152].

E-64 and its cell-permeable form, E-64d, also inhibitor of calpains, and, Z-Phe-Ala-fmk, represent the broad spectrum family of cathepsin inhibitors, [151-153].

As apoptosis and autophagy can be simultaneously activated within a dying cell [138], a close relationship between caspases and the fundamental lysosomal proteases, the cathepsins, can be assumed [154].

Indeed, procaspase-7 was activated by cathepsin G [155] and lysosomes have been identified as a protease source that could proteolytically activate the proinflammatory caspase-11 in liver cells [154]. Caspase 11 is crucial for the activation of caspase-1 and for proIL-1beta maturation [154].

Furthermore, caspase 1 was demonstrated to be cathepsin B processed, while procaspase-2, -6, -7, -14 are weak substrates, procaspase-3 a very poor substrate, and procaspase-12 is not at all a substrate for cathepsin B [156].

Cathepsin B seems also to contribute to bile salt-induced apoptosis and appears to be a downstream effector protease dependent on caspase activity for activation. [157]. Indeed, pharmacological and genetic inactivation of cathepsin B in steatotic livers led to a decrease in hepatocyte apoptosis after cold ischemia-warm reperfusion injury [158].

Cathepsin B is the most stable and abundant cysteine protease which has its greatest destructive potential in cellular compartment outside the lysosome and, contributes the most to the permeabilization of the lysosomal membrane [159]. So cathepsin B may be the best pharmacological target to prevent or reduce liver I-R injury. However, further development of this strategy requires effective specific inhibitors of cathepsin B that pharmacologically are not available at present.

At the Prestwick Chemical Library there are 1,120 US Food and drug (FDA)-approved compounds classed as modulators of autophagy [160].

Molecules currently studied, as inhibitors or inducers of autophagy, in order to elucidate autophagy mechanisms are shown in tables 3 and 4.

Table 3. Inhibitors of autophagy. 3-MA, 3-methylanine; PI3K, phosphatidylinositol-3-kinase; AICAR, 5-amino-4-imidazolecarboxamide ribonucleoside; AMPK, AMP-activated kinase; cAMP, adenosine 3′,5′-cyclic monophosphate; mTOR, mammalian target of rapamycin

Inhibitor	Target
3-MA3MA	Blocks the class III PI3K Blocks the class III PI3K
AICAR	AMPK activator
Amrinone	Increases cAMP levels
Asparagine Bafilomycin A1 Chloroquine Hydroxychloroquine Insulin	Impairs autophagosome/lysosome functions Inhibits vacuolar ATPase Alkalinizes lysosomal pH Alkalinizes lysosomal pH Activates mTOR
Isopreterenol Lucanthone Norclomipramine	PI3K Disrupts lysosomal function Alkalinizes lysosomal pH

Between the most studied autophagy modulators, Rapamycin, FK228, Cucurmin, suberoylanilide hydroxamic *acid* (SAHA), Temozolomide Imiquimod, Timosaponin A-III and Autophagonizer are inducers, whereas 3-methylanine (3-MA), Bafilomycin A1, Wortmannin (WA), PD98059, Chloroquine/hydroxychloroquine and 2-phenylethynesulfonamide (PES) are inhibitors [160].

To our knowledge, at the moment, there are 20 listed protocols using autophagy modulators all of them associated with cancer chemotherapy [161]. The phoshodiesterase III inhibitor, amrinone, a vasodilatator inhibiting autophagy by elevating intracellular cAMP and glucose levels, has been used successfully to reduce liver I-R injury in cirrhotic patients who underwent hepatectomy [162].

Furthermore the development of molecules, which interfere specifically with autophagy signalling or action, would be useful for the treatment of liver I-R injury. Small synthetic peptides, able to interfere with protein function *in vivo*, and peptidomimetic compounds, comprising at least one portion that has the same 3D conformation as the synthetic peptides, can be applied in order to modulate many cell processes, including autophagy [163]. These peptidomimetic compounds are assembled with modified amino acids and /or organic molecules that have the same 3D conformation as L-amino acid but are not recognized by cellular and extra-cellular proteases [163]. In this manner an original peptide screening for *in vivo* cardiac ischemic tissue targeting led to an elective nonamer, a potent new delivering tool that might be also useful for imaging [164]. However, in many cases, the binding between the peptide and the target molecule has the potential to dysregulate alternative functions of the target and to induce adverse phenotype(s) [163].

Finally compounds modulating autophagy may have opposite effects on both basal and activated autophagy [165]. For example, compounds activating autophagy via the AMPK/mammalian target of rapamycin (mTOR) pathway, as resveratrol and rapamycin, exert protective effects in models of cardiovascular disease, but cytotoxic or cytostatic effects in cancer models [166-167-168]. In the same way, inhibition of phosphoinositide 3-kinase (PI3K) by WA or 3-MA may have either inhibitor or activating effects, depending on action on class III PIK3, required for the induction of autophagy, or class I PI3K, which negatively regulates autophagy [169].

Table 4. Inducers of autophagy. CAPS, chloramphenicol succinate; IP3, inositol trisphosphate; cAMP, adenosine 3′,5′-cyclic monophosphate; AKT, protein kinase B; mTOR, mammalian target of rapamycin; HDAC, hystone deacetylase; IMP ase, inositol monophosphatase; SAHA, suberoylanilide hydroxamic acid; 3-MA, 3-methyladenine; PPAR gamma, peroxisome proliferator-activated receptor gamma; PCK, protein kinase C

Inducer	Target
Autophagonizer	New mechanism, autophagic cell death
CAPS	Increases Beclin-1 and LC3 levels
Carmazepine	Decrease IP3 levels
Clonidine	Imidazoline receptor agonist: decreases AMP levels
Cucurmin	Aminopeptidases: inhibits AKT/mTOR pathway
Dexamethasone	Upregulates IP3 receptor
FK228	Inhibits HDAC
Glucagon	Liver specific
Imiquimod	Toll-like receptor 7 ligand
Lithium	Inhibits IMPase and reduce inositol and IP3 levels
Loperamide	Ca2+ channel blocker (reduces intracytosolic Ca2+ levels)
MG-132	Proteasome inhibitor
Minoxidil	K+ and ATP channel opener
PP242	ATP-competitive inhibitor of mTORp
Propanolol	Antagonizes cAMP
Rapamycin	mTOR inhibitor
Resveratol	mTOR and S6kinase inhibitor; activates AMPK
Rosiglitazone	PPAR gamma activator
SAHA	HDAC inhibitor
Sulphaphenazole	PKC-dependent
Tamoxifen	Increases ceramide levels; abolishes the inhibitory effects of PI3K
Trehalose	mTOR independent pathway; increases Beclin-1, Atg12, Atg7, Atg5 levels
Troglitazone	PPAR gamma activator
Valproate	Decreases IP3 levels

Protective Strategies to Reduce Liver I-R Injury by Modulation of Autophagy

Modulation of I-R injury in liver surgery, including LT, can be achieved in various ways including hypothermia, ischemic preconditioning (IP), ischemic postconditioning (Ipost-C), machine perfusion, genetic modulation of IR injury and pharmacological agents to reduce IR injury [170]. Several strategies to protect the liver from I-R injury developed in animal models were subsequently utilized efficiently in clinical studies [171-172]. Treatment should start before ischemia by surgical preconditioning procedures and/or by adding specific drugs to the standard preservation solution, in order to "prepare" the liver to the I-R insult [5].

Surgical procedures such as IP and intermittent clamping of the liver vessels have found their way into clinical practice. In contrary pharmacological strategies, which seem ideal because the time of drug exposure is limited and the substance is likewise washed out prior to revascularization [5] remain limited until now [171].

A) IP of the Liver Induces Autophagy

IP consists of brief periods of vascular occlusion followed by short periods of blood reperfusion of organs [173]. This mechanism results in the ability of the organ to withstand a subsequent prolonged period of ischemia [174]. Ischemic tolerance is induced by regulation of endothelial function and blood flow, and by decreased macrophage as well as neutrophil activity [174] resulting in decreased endothelial injury and eventually decreased parenchymal injury. IP can be achieved by a local preconditioning stimulus (direct IP [175] or by remote stimulus (remote IP) [176]. Remote IPC is the mechanism by which IPC of one vascular area (supplied by one artery) protects another vascular area (supplied by another artery) from IR injury [177].

In the human liver, IP decreased aminotransferase levels and endothelial cell injury after liver resection [41]. These beneficial effects on liver I-R injuries have been linked to various mechanisms, such as decreased apoptosis [3], preservation of the ATP content in liver tissue [41-178], or overproduction of pro–survival and anti-inflammatory proteins [179]. IP seems to induce a major expression of Bcl-2, maintaining higher intracellular ATP levels, which are usually decreased in steatotic livers [180], and it may also trigger authophagic pathways by the activation of Beclin-1 expression in cells with nonactivated caspase-3 [181]. Therefore, autophagy can be activated to temporarily avoid cell death by necrosis and its subsequent inflammatory process [15]. In steatotic human livers preconditioning resulted in limited hepatocyte necrosis following I-R injury [15]. Because chemotherapy causes steatosis and vascular injuries that reduce tolerance to reperfusion injury and increase the risk of subsequent liver failure, IP was realized in patients undergoing liver surgery after receiving courses of chemotherapy. Before prolonged ischemia, by occlusion of the liver vessels with preservation of the cava vena flow, IP consisting of 10 min of portal triad clamping (Pringle manoevre) followed by 10 min of reperfusion [15] was performed. IP induced autophagy in these steatotic and peliotic human livers formerly treated by chemotherapy [15] hypothesizing that IP may start up autophagy to keep ATP levels, and avoid hepatocellular necrosis by delaying pro-apoptotic effects [15]. In liver transplantation IP of the graft (preconditioning in the donor before graft removal) [182] induced autophagy, limited necrosis in human recipients of fatty liver grafts and reduced the incidence of rejection episodes [14]. In fact a cellular increase of Beclin-1 and microtubule-associated protein light chain 3 (LC3), two proteins which represent respectively an early and a late marker of the autophagic process [97], was observed after IP of steatotic liver grafts compared with non-IP steatotic liver grafts [14]. Furthermore an inverse correlation between the number of LC3-positive cells and the necrotic index in IP steatotic liver grafts was found, whereas no correlation was observed in non-IP steatotic liver grafts [14]. Interestingly, IP decreased the incidence of both, acute and chronic liver rejection in recipients of steatotic grafts compared to recipients of non-IP steatotic grafts [14].

Impaired autophagic clearance was observed after cold preservation of steatotic livers and correlated with increased necrotic cell death after reperfusion [10]. In fact, the use of hypothermic reconditioning, which means a temporary hypothermic oxygenation of the grafts by using an oxygenated machine perfusion [183-184] or gaseous oxygen persufflation [185-186] may maintain cellular autophagy at a normal rate. This can improve ulterior recovery of cold-stored steatotic liver graft and may effectively limit mitochondrial dysfunction following reperfusion with limited activation of the mitochondrial apoptosis pathway [10].

B) Preservation Solution for the Liver Induces Autophagy

Deceased donor organ preservation involves the rapid cooling of the *ex vivo* donor organ to 0°C. This organ preservation process for transplantation represents both, a cell stress and nutrient deprivation, that may increase the autophagic flux during cold ischemia. Alternatively, it is possible that cold ischemia, by virtue of limiting enzymatic reaction, may reduce the autophagic flux. During LT in rats, 15 minutes after the start of warm reperfusion, grafts showed hepatocytes containing numerous autophagosomes and autolysosomes, frequently dissociated from the hepatic cords and obstructing sinusoids which causes massive necrosis of hepatocytes within 2 hours after warm reperfusion and led to liver dysfunction [9]. These injuries were reduced by treatment with inhibitors of autophagy [9]. In fact autophagy suppression with WM or LY294002, which prevents autophagosome formation by inhibiting the activity of PI3K, reduced both, liver damage and mortality rate of recipient rats [9]. Furthermore suppression of lysosomal proteinases with finally, a reduction of the mortality rate was achieved by treating liver grafts with pepstatin A and leupeptin, aspartic and cysteine proteinase inhibitors, before LT in rats [9].

A considerable part of primary graft dysfunction is due to inadequate cold storage, including an imperfect storage solution [187]. In LT, ATP loss during prolonged cold storage is one of the critical events of cold I-R [187]. ATP depletion results in impairment of mitochondrial respiratory function [188] and acidosis from glycolysis [189]. Cold preservation of the liver followed by reperfusion results in loss of viability of sinusoidal endothelial cells [190] and apoptosis seems to play an important role in this type of injury [191]. Depending on liver ATP concentration, cells may switch from an apoptotic form of cell death to necrosis [187-192]. Repletion of the ATP pool by supplying glucose prevented necrosis and restored the ability of cells to undergo apoptosis [193]. Thus an appropriate energy store for full graft recovery can be either maintained during cold storage or effectively repleted before or during organ reperfusion [187]. Autophagy may be induced or depressed depending on the type of cold preservation solution used [9-10-13]. Amino acid depletion for example rapidly induces autophagy [194]. Liver graft preservation in University of Wisconsin (UW) cold storage solution, which does not contain amino acids, resulted in an increase of autophagy in animal studies [9-13-21]. UW cold storage solution consists of lactobionate as major anion, raffinose as waterproofing agent, adenosine as energetic pool, glutathione and allopurinol as free radical trap, hydroxyethyl starch as osmotic colloid, dexamethasone as anti-inflammatory and immunosuppressant, and insulin as glycolysis stimulator [195]. By the way it should be noted that dexamethasone is an autophagy inducer [196], whereas insulin is an inhibitor of autophagy [197] and furthermore insulin may exacerbate liver ischemic injury by energy depletion [198]. However it seems that the lack of the amino acids in UW cold storage solution is the most important factor for the above mentioned increase of autophagy. When liver graft was conserved in cold-storage solution containing histidine-tryptophan-ketoglutarate amino acids, autophagy was decreased [10]. So modifying the composition of cold-storage solution may provide a therapeutic rationale for the decrease of liver I-R injury.

Conclusion

The pathogenesis of liver I-R injury represents a complex interplay between the different types of cell death including autophagy, apoptosis and necrosis (Fig. 7). The functional role of autophagy in the liver following I-R is complex. Activation of autophagy during ischemia is essential for cell survival and maintenance of liver functions, activation of autophagy during reperfusion could be either protective or detrimental. The extent and the pathophysiological consequences of autophagy depend on the severity and duration of the ischemic insult, its level of induction, and may result from an upstream signaling mechanism. Anyway, the autophagic cell response to liver I-R injury may provide an additional time to the cell death processes, delaying apoptosis and necrosis. Considering that autophagy represents the first line of cellular defense against I-R injury, its pharmacologic modulation may integrate therapeutic strategies, in order to improve clinical outcomes in patients undergoing liver surgery and to reduce graft dysfunction in liver transplantation.

References

[1] Cursio, R; Gugenheim, J; Ricci, JE; Crenesse, D; Rostagno, P; Maulon, L; Saint-Paul, M-C; Ferrua, B; Gugenheim, J. A caspase inhibitor fully protects rats against lethal normothermic liver ischemia by inhibition of liver apoptosis. *FASEB J.* 1999;13:253-261.

[2] Cursio, R. Liver ischemia-reperfusion injury in the rat: role of nonparenchymal liver cells. *Postdoctoral thesis on liver transplantation. University of Paris XI,* 14 Nov 2002, Paris, France.

[3] Jaeschke, H; Lemasters, JJ. Apoptosis versus oncotic necrosis in hepatic ischemia/reperfusion injury. *Gastroenterology.* 2003;125:1246-57.

[4] Cursio, R; Filippa, N; Miele, C; Van Obberghen, E; Gugenheim, J. Involvement of protein kinase B nad mitogen-activated protein kinases in experimental normothermic liver ischaemia-reperfusion injury. *Br. J. Surg.* 2006;93:752-61.

[5] Cursio, R. Caspase inhibition in liver transplantation: from basic research to clinical studies. *HPB.* 2010;12:1-2.

[6] Gao, W; Bentley, RC; Madden, JF; Clavien, PA. Apoptosis of sinusoidal endothelial cells is a critical mechanism of preservation injury in rat liver transplantation. *Hepatology.* 1998;27:1652-60.

[7] Cursio, R; Colosetti, P; Saint Paul M-C; Pagnotta, S; Gounon, P; Iannelli, A; Auberger, P; Gugenheim, J. Induction of different types of cell death after normothermic liver ischemia-reperfusion. *Transplant. Proc.* 2010;42:3977-80.

[8] Cardinal, J; Pan, P; Tsung, A. Protective role of cisplatin in ischemic liver injury through induction of autophagy. *Autophagy.* 2009;8:1211-2.

[9] Gotoh, K; Lu, Z; Morita, M; Shibata, M; Koike, M; Waguri, S; Dono, K; Doki,Y; Kominami, E; Sugioka, A; Monden, M; Uchiyama, Y. Participation of autophagy in the initiation of graft dysfunction after rat liver transplantation. *Autophagy.* 2009;5:351-60.

[10] Minor, T; Stegemann, J; Hirner, A; Koetting, M. Impaired autophagy clearance after cold preservation of fatty livers correlates with tissue necrosis upon reperfusion and is reversed by hypothermic reconditioning. *Liver Transplant.* 2009;15:798-805.

[11] Shin, T; Kuboki, S; Huber, N; Eismann, T; Galloway, E; Schuster, R; Blanchard, J; Pritts, TA; Lentsch, AB. Activation of peroxisome proliferator-activated receptor-gamma during hepatic ischemia is age-dependent. *J. Surg. Res.* 2008;147:200-5.

[12] Kim, JS; Nitta, T; Mohuczy, D; O'Malley, KA; Moldawer, LL; Dunn, WA; Behrns, KE. Impaired autophagy: A mechanism of mitochondrial dysfunction in anoxic rat hepatocytes. *Hepatology.* 2008;47:1725-36.

[13] Lu, Z; Dono, K; Gotoh, H; Shibata, M; Koike, M; Marubashi, S; Miyamoto, A; Takeda, Y; Nagano, H; Umeshita, K, Uchiyama, Y; Monden, M. Participation of autophagy in the degeneration process of rat hepatocytes after transplantation following prolonged cold preservation. *Arch. Histol. Cytol.* 2005;68:71-80.

[14] Degli Esposti, D; Sebagh, M; Pham, P; Reffas, M; Poüs, C; Brenner, C; Azoulay, D; Lemoine, A. Ischemic preconditioning induces autophagy and limits necrosis in human recipients of fatty liver grafts, decreasing the incidence of rejection episodes. *Cell Death Dis.* 2011 Jan 13;2:e111.

[15] Domart, MC; Esposti, DD; Sebagh, M; Olaya, N; Harper, F; Pierron, G; Franc, B; Tanabe, KK; Debuire, B; Azoulay, D; Brenner, C; Lemoine, A. Concurrent induction of necrosis, apoptosis, and autophagy in ischemic preconditioned human livers formerly treated by chemotherapy. *J. Hepatol.* 2009;51:881-9.

[16] Deter, RL; De Duve, C. Influence of glucagon, an inducer of cellular autophagy, on some physical properties of rat liver lysosomes. *J. Cell Biol.* 1967;33:437-449.

[17] Deter, RL; Baudhuin, P; De Duve, C. Participation of lysosomes in cellular autophagy induced in rat liver by glucagons. *J. Cell Biol.* 1967;35:C11-16.

[18] Omodei Zorini, A. Considerations on primary carcinomatous caverns of the lung. Possibility of the intervention of a phenomenon of "autophagy of the neoplastic cells", *Lotta Tuberc.* 1965;35:946-ç68.

[19] Baudhuin, P. Lysosomes and cellular autophagy, *Brux Med.* 1966;46:1059-1070.

[20] White, E. Autophagic cell death unraveled: Pharmacological inhibition of apoptosis and autophagy enables necrosis. *Autophagy.* 2008;4:399-401.

[21] Ratou, PE; Mansouri, A; Lebrec, D; Durand, F; Valla, D; Moreau, R. Autophagy in liver diseases. *J. Hepatol.* 2010;53:1123-34.

[22] Clavien, PA; Harvey, PR; Strasberg, SM. Preservation and reperfusion injuries in liver allografts. An overview and synthesis of current studies. *Transplantation.* 1992;53:957-78.

[23] McKeown, CM; Edwards, V; Phillips, MJ; Harvey, PR; Petrunka, CN; Strasberg, SM. Sinusoidal lining cell damage: the critical injury in cold preservation of liver allografts in the rat. *Transplantation.* 1988;46:178-91.

[24] Farmer, DG; Amersi, F; Kupiec-Weglinski, JW. Current status of ischemia and reperfusion injury in the liver. *Transplant. Rev.* 2000;14:106-126.

[25] Busuttil, R. Liver ischemia and reperfusion injury. *Br. J. Surg.* 2007;94:787-8.

[26] Glanemann, M; Langrehr, JM; Stange, BJ; Neumann, U; Settmacher, U; Steinmüller, T; Neuhaus, P. Clinical implications of hepatic preservation injury after adult liver transplantation. *Am. J. Transplant.* 2003;3:1003-9.

[27] Mueller, AR; Platz, KP; Haak, M; Undi, H; Müller, C; Köttgen, E; Weidemann, H; Neuhaus, P. The release of cytokines, adhesion molecules, and extracellular matrix parameters during and after reperfusion in human liver transplantation. *Transplantation.* 1996;62:1118-26.

[28] Halloran, PF. Call for revolution: a new approach to describing allograft deterioration. *Am. J. Transplant.* 2002;2:195-200.

[29] Gourishankar, S, Halloran, PF. Late deterioration of organ transplants: a problem in injury and homeostasis. *Curr. Opin. Immunol.* 2002;14:576-83.

[30] de Rougemont, O; Lehmann, K; Clavien, PA. Preconditioning, organ preservation, and postconditioning to prevent ischemia-reperfusion injury to the liver. *Liver Transpl.* 2009;15:1172-82.

[31] Huguet, C; Nordlinger, B; Bloch, P; Conard, J. Tolerance of the human liver to prolonged normothermic ischemia. A biological study of 20 patients submitted to extensive hepatectomy. *Arch. Surg.* 1978;113:1448-51.

[32] Huguet, C; Gavelli, A; Chieco, PA; Bona, S; Harb, J; Joseph, JM; Jobard, J; Gramaglia, M; Lasserre, M. Liver ischemia for hepatic resection: where is the limit? *Surgery.* 1992;111:251-9.

[33] Huguet, C; Gavelli, A; Bona, S. Hepatic resection with ischemia of the liver exceeding one hour. *J. Am. Coll. Surg.* 1994;178:454-8.

[34] Belghiti, J; Hiramatsu, K; Benoist, S; Massault, P; Sauvanet, A; Farges, O. Seven hundred forty-seven hepatectomies in the 1990s: an update to evaluate the actual risk of liver resection. *J. Am. Coll. Surg.* 2000;191:38-46.

[35] Sherlock, S. Acute fatty liver of pregnancy and the microvesicular fat diseases. *Gut.* 1983;24:265-9.

[36] Bellentani, S; Saccoccio, G; Masutti, F; Crocè, LS; Brandi, G; Sasso, F; Cristanini, G; Tiribelli, C. Prevalence of and risk factors for hepatic steatosis in Northern Italy. *Ann. Intern. Med.* 2000;132:112-7.

[37] Silverman, JF; O'Brien, KF; Long, S; Leggett, N; Khazanie, PG; Pories, WJ; Norris, HT; Caro, JF. Liver pathology in morbidly obese patients with and without diabetes. *Am. J. Gastroenterol.* 1990;85:1349-55.

[38] Jaeschke, H. Mechanisms of reperfusion injury after warm ischemia of the liver. J Hepatobiliary *Pancreat. Surg.* 1998;5:402-8.

[39] Pannen, BH; Al-Adili, F; Bauer, M; Clemens, MG; Geiger, KK. Role of endothelins and nitric oxide in hepatic reperfusion injury in the rat. *Hepatology.* 1998;27:755-64.

[40] Sun, CK; Zhang, XY; Zimmermann, A; Davis, G; Wheatley, AM. Effect of ischemia-reperfusion injury on the microcirculation of the steatotic liver of the Zucker rat. *Transplantation.* 2001;72:1625-31.

[41] Clavien, PA; Selzner, M; Rüdiger, HA; Graf, R; Kadry, Z; Rousson, V; Jochum, W. A prospective randomized study in 100 consecutive patients undergoing major liver resection with versus without ischemic preconditioning. *Ann. Surg.* 2003;238:843-50.

[42] Bernau, J; Rueff, B; Benhamou, J-P. Fulminant and subfulminant liver failure: definitions and causes. *Semin. Liv. Dis.* 1986;6:97-106.

[43] Velmahos, GC; Toutouzas, K; Radin, R; Chan, L; Rhee, P; Tillou, A; Demetriades, D. High success with nonoperative management of blunt hepatic trauma: the liver is a sturdy organ. *Arch. Surg.* 2003;138:475-80.

[44] Saltzherr, TP; van der Vlies, CH; van Lienden, KP; Beenen, LF; Ponsen, KJ; van Gulik, TM; Goslings, JC. Improved outcomes in the non-operative management of liver injuries. *HPB.* 2011;13:350-5.

[45] Kozar, RA; McNutt, MK. Management of adult blunt hepatic trauma. *Curr. Opin. Crit. Care.* 2010 Sep 9. [Epub ahead of print]

[46] Piper, GL; Peitzman, AB. Current management of hepatic trauma. *Surg. Clin. North Am.* 2010;90:775-85.

[47] Polanco; P; Leon, S; Pineda, J; Puyana, JC; Ochoa, JB; Alarcon, L; Harbrecht, BG; Geller, D; Peitzman, AB. Hepatic resection in the management of complex injury to the liver. *J. Trauma.* 2008;65:1264-9.

[48] Patcher, HL; Spencer, FC; Hofstetter, SR; Liang, HG; Coppa, GF. Significant trends in the treatment of hepatic trauma. Experience with 411 injuries. *Ann. Surg.* 1992;5:492-500.

[49] Bismuth, H; Castaing, D; Garden, OJ. Major hepatic resection under total vascular exclusion. *Ann. Surg.* 1989;210:13-9.

[50] Asensio, JA; Roldán, G; Petrone, P; Rojo, E; Tillou, A; Kuncir, E; Demetriades, D; Velmahos, G; Murray, J; Shoemaker, WC; Berne, TV; Chan, L. Operative management and outcomes in 103 AAST-OIS grades IV and V complex hepatic injuries: trauma surgeons still need to operate, but angioembolization helps. *J. Trauma.* 2003;54:647-53.

[51] Caruso, DM; Battistella, FD; Owings, JT; Lee, SL; Samaco, RC. Perihepatic packing of major liver injuries: complications and mortality. *Arch. Surg.* 1999;134:958-62.

[52] Ringe, B; Pichlmayr, R; Ziegler, H; Grosse; H; Kuse, E; Oldhafer, K; Bornscheuer, A; Gubernatis, G. Management of severe hepatic trauma by two-stage total hepatectomy and subsequent liver transplantation. *Surgery.* 1991;109:792-5.

[53] Tucker, ON; Marriott, P; Rela, M; Heaton, N. Emergency liver transplantation following severe liver trauma. *Liver Transpl.* 2008;14:1204-10.

[54] Patel, MS; Malinoski, DJ; Nguyen, XM; Hoyt, DB. The impact of select chronic diseases on outcomes after trauma: a study from the National Trauma Data Bank. *J. Am. Coll. Surg.* 2011;212:96-104.

[55] Lin; BC; Fang, JF; Wong, YC; Hwang, TL; Hsu, YP. Management of cirrhotic patients with blunt abdominal trauma: Analysis of risk factor of postoperative death with the Model for End-Stage Liver Disease score. *Injury.* 2011 Apr 19. [Epub ahead of print]

[56] Hassoun, HT; Kone, BC; Mercer, DW; Moody, FG; Weisbrodt, NW; Moore, FA. Post-injury multiple organ failure: the role of the gut. *Shock.* 2001;15:1-10.

[57] Harbrecht, BG; Doyle, HR; Clancy, KD; Townsend, RN; Billiar, TR; Peitzman, AB. The impact of liver dysfunction on outcome in patients with multiple injuries. *Am. Surg.* 2001;67:122-6.

[58] Jaeschke, H; Farhood, A. Kupffer cell activation after no-flow ischemia versus hemorrhagic shock. *Free Radic Biol. Med.* 2002;33:210-9.

[59] Champion, HR; Jones, RT; Trump, BF; Decker, R; Wilson, S; Stega, M; Nolan, J; Crowley, RA; Gill, W. Post-traumatic hepatic dysfunction as a major etiology in post-traumatic jaundice. *J. Trauma.* 1976;16:650-7.

[60] Chang, CG; Van Way, CW; Dhar, A; Helling, T; Hahn, Y. The use of insulin and glucose during resuscitation from hemorrhagic shock increases hepatic ATP. *J. Surg. Res.* 2000;92:171-6.

[61] Shires, GT; Browder, LK; Steljes, TP; Williams, SJ; Browder, TD; Barber, AE. The effect of shock resuscitation fluids on apoptosis. *Am. J. Surg*. 2005;189:85-91.

[62] Szabó, C; Billiar, TR. Novel roles of nitric oxide in hemorrhagic shock. *Shock.* 1999;12:1-9.

[63] Shimizu, T; Tani, T; Hanasawa, K; Endo, Y; Kodama, M. The role of bacterial translocation on neutrophil activation during hemorrhagic shock in rats. *Shock.* 2001;16:59-63.

[64] El-Wahsh, M; Seifalian, AM; Fuller, BJ; Rolles, K; Davidson, BR. Induction of adhesion molecule expression in liver ischaemia-reperfusion injury is associated with impaired hepatic parenchymal microcirculation. *Br. J. Surg.* 2004;91:1034-9.

[65] Brundage, SI; Schreiber, MA; Holcomb, JB; Zautke, N; Mastrangelo, MA; Xq, X; Macaitis, J; Tweardy, DJ. Amplification of the proinflammatory transcription factor cascade increases with severity of uncontrolled hemorrhage in swine. *J. Surg. Res.* 2003;113:74-80.

[66] Wang, P; Ayala, A; Ba, ZF; Zhou, M; Perrin, MM; Chaudry, IH. Tumor necrosis factor-alpha produces hepatocellular dysfunction despite normal cardiac output and hepatic microcirculation. *Am. J. Physiol.* 1993;265:G126-32.

[67] Wang, P; Ba, ZF; Chaudry, IH. Endothelial cell dysfunction occurs very early following trauma-hemorrhage and persists despite fluid resuscitation. *Am. J. Physiol.* 1993;265:H973-9.

[68] Yamakawa, Y; Takano, M; Patel, M; Tien, N; Takada, T; Bulkley, GB. Interaction of platelet activating factor, reactive oxygen species generated by xanthine oxidase, and leukocytes in the generation of hepatic injury after shock/resuscitation. *Ann. Surg.* 2000;231:387-98.

[69] Okano, N; Miyoshi, S; Owada, R; Fujita, N; Kadoi, Y; Saito, S; Goto, F; Morita, T. Impairment of hepatosplanchnic oxygenation and increase of serum hyaluronate during normothermic and mild hypothermic cardiopulmonary bypass. *Anesth. Analg.* 2002;95:278-86.

[70] Glantzounis, GK; Tselepis, AD; Tambaki, AP; Trikalinos, TA; Manataki, AD; Galaris, DA; Tsimoyiannis, EC; Kappas, AM. Laparoscopic surgery-induced changes in oxidative stress markers in human plasma. *Surg. Endosc.* 2001;15:1315-9.

[71] Glantzounis, GK; Tsimaris, I; Tselepis, AD; Thomas, C; Galaris, DA; Tsimoyiannis, EC. Alterations in plasma oxidative stress markers after laparoscopic operations of the upper and lower abdomen. *Angiology.* 2005;56:459-65.

[72] Rezende-Neto, JB; Moore, EE; Masuno, T; Moore, PK; Johnson, JL; Sheppard, FR; Cunha-Melo, JR; Silliman, CC. The abdominal compartment syndrome as a second insult during systemic neutrophil priming provokes multiple organ injury. *Shock.* 2003;20:303-8.

[73] Jaeschke, H. Molecular mechanisms of hepatic ischemia-reperfusion injury and preconditioning. *Am. J. Physiol. Gastrointest. Liver Physiol.* 2003;284:G15-26.
[74] Rappaport, AM; Hiraki, GY. The anatomical pattern of lesions in the liver. *Acta Anat.* 1958;32:126-40.
[75] Jungermann, K; Kietzmann, T. Zonation of parenchymal and nonparenchymal metabolism in liver. *Annu. Rev. Nutr.* 1996;16:179-203.
[76] Trump, BF; Goldblatt, PJ; Stowell, RE. Studies of necrosis in vitro of mouse hepatic parenchymal cells. Ultrastructural alterations in endoplasmic reticulum, Golgi apparatus, plasma membrane, and lipid droplets. *Lab. Invest.* 1965;14:2000-28.
[77] Dogan, S, Aslan, M. Hepatic ischemia-reperfusion injury and therapeutic strategies to alleviate cellular damage. *Hepatol. Res.* 2011;41:103-17.
[78] Marotto, ME; Thurman, RG; Lemasters, JJ. Early midzonal cell death during low-flow hypoxia in the isolated, perfused rat liver: protection by allopurinol. *Hepatology.* 1988;8:585-90.
[79] Sasse, D; Spornitz, UM; Maly, IP. Liver architecture. *Enzyme.* 1992;46:8-32.
[80] Jungermann, K; Kietzmann, T. Oxygen: modulator of metabolic zonation and disease of the liver. *Hepatology.* 2000;31:255-60.
[81] Cherid, A; Cherid, N; Chamlian, V; Hardwigsen, J; Nouhou, H; Dodero, F; Benkoel, L; Le Treut, YP; Chamlian, A. Evaluation of glycogen loss in human liver transplants. Histochemical zonation of glycogen loss in cold ischemia and reperfusion. *Cell Mol. Biol.* 2003;49:50914.
[82] Abu-Amara, M; Yang, SY; Quaglia, A; Rowley, P; Fuller, B; Seifalian, A; Davidson, B. Role of endothelial nitric oxide synthase in remote ischemic preconditioning of the mouse liver. *Liver Transpl.* 2011;17:610-9.
[83] Jaeschke, H. Redox considerations in hepatic injury and inflammation. *Antioxid. Redox Signal.* 2002;4:699-700.
[84] Mavier, P; Guigui, B; Preaux, AM; Rosenbaum, J; Lescs, MC; Zafrani, ES; Dhumeaux, D. In vitro toxicity of hydrogen peroxide against normal vs. tumor rat hepatocytes: role of catalase and of the glutathione redox cycle. *Hepatology.* 1988;8:1673-8.
[85] Li, XK; Matin, AF; Suzuki, H; Uno, T; Yamaguchi, T; Harada, Y. Effect of protease inhibitor on ischemia/reperfusion injury of the rat liver. *Transplantation.* 1993;56:1331-6.
[86] Cursio, R; Mari, B; Louis, K; Rostagno, P; Saint-Paul, M-C; Giudicelli, J; Bottero, V; Anglard, P; Yiotakis, A; Dive, V; Gugenheim, J; Auberger, P. Rat liver injury after normothermic ischemia is prevented by a phosphinic matrix metalloproteinase inhibitor. *FASEB J.* (November 14, 2001) 10.1096/fj.01-0279fje
[87] Vollmar, B; Glasz, J; Leiderer, R; Post, S; Menger, MD. Hepatic microcirculatory perfusion failure is a determinant of liver dysfunction in warm ischemia-reperfusion. *Am. J. Pathol.* 1994;145:1421-31.
[88] Lemasters, JJ. V. Necrapoptosis and the mitochondrial permeability transition: shared pathways to necrosis and apoptosis. *Am. J. Physiol.* 1999;276:G1-6.
[89] Codogno, P; Meijer, AJ. Autophagy and signaling: their role in cell survival and cell death. *Cell Death Differ.* 2005;12:1509-18.
[90] Loud, AV. A quantitative stereological description of the ultrastructure of normal rat liver parenchymal cells. *J. Cell Biol.* 1968;37:27-46.

[91] Komatsu, M; Waguri, S; Ueno, T; Iwata, J; Murata, S; Tanida, I; Ezaki, J; Mizushima, N; Ohsumi, Y; Uchiyama, Y; Kominami, E; Tanaka, K; Chiba, T. Impairment of starvation-induced and constitutive autophagy in Atg7-deficient mice. *J. Cell Biol.* 2005;169:425-34.

[92] Yin, XM; Ding, WX; Gao, W. Autophagy in the liver. *Hepatology.* 2008;47:1773-85.

[93] Levine, B; Yuan, J. Autophagy in cell death: an innocent convict? *J. Clin. Invest.* 2005;115:2679-88.

[94] Huang, J; Klionsky, DJ. Autophagy and human disease. *Cell Cycle.* 2007;6:1837-49.

[95] Czaja, MJ. Two types of autophagy are better than one during hepatocyte oxidative stress. *Autophagy.* 2011;7:96-7.

[96] Chen, Y; Azad, MB; Gibson, SB. Methods for detecting autophagy and determining autophagy-induced cell death. *Can. J. Physiol. Pharmacol.* 2010;88:285-95.

[97] Maiuri, MC; Zalckvar, E; Kimchi, A; Kroemer, G. Self-eating and self-killing: crosstalk between autophagy and apoptosis. *Nat. Rev. Mol. Cell Biol.* 2007;8:741-52.

[98] Platini, F; Pérez-Tomás, R; Ambrosio, S; Tessitore, L. Understanding autophagy in cell death control. *Curr. Pharm. Des.* 2010;16:101-13.

[99] Belzer, FO, Southard JH. Principles of solid-organ preservation by cold storage. *Transplantation.* 1988;45:673-6.

[100] Jassem, W; Fuggle, SV; Rela, M; Koo, DD; Heaton, ND. The role of mitochondria in ischemia/reperfusion injury. *Transplantation.* 2002;73:493-9.

[101] Chen, Y; Azad, M; Gibson, SB. Superoxide is the major reactive oxygen species regulating autophagy. *Cell Death Differ.* 2009;16:1040-52.

[102] Huang, J; Canadien, V; Lam, GY; Steinberg, BE; Dinauer, MC; Magalhaes, MA; Glogauer, M; Grinstein, S; Brumell, JH. Activation of antibacterial autophagy by NADPH oxidases. *Proc. Natl. Acad. Sci. U. S. A.* 2009;106:6226-31.

[103] Scherz-Shouval, R; Elazar, Z. ROS, mitochondria and the regulation of autophagy. *Trends Cell Biol.* 2007;17:422-7.

[104] Jung, SN; Yang, WK; Kim, J; Kim, HS; Kim, EJ; Yun, H; Park, H; Kim, SS; Choe, W; Kang, I; Ha, J. Reactive oxygen species stabilize hypoxia-inducible factor-1 alpha protein and stimulate transcriptional activity via AMP-activated protein kinase in DU145 human prostate cancer cells. *Carcinogenesis.* 2008;29:71321.

[105] Hong, SP; Carlson, M. Regulation of snf1 protein kinase in response to environmental stress. *J. Biol. Chem.* 2007;282:16838-45.

[106] Loos, B; Engelbrecht, AM. Cell death: a dynamic response concept. *Autophagy.* 2009;5:590-603.

[107] Lanir, A; Clouse, ME; Lee, RG. Liver preservation for transplant. Evaluation of hepatic energy metabolism by 31P NMR. *Transplantation.* 1987;43:786-90.

[108] Shimizu, S; Kanaseki, T; Mizushima, N; Mizuta, T; Arakawa-Kobayashi, S; Thompson, CB; Tsujimoto, Y. Role of Bcl-2 family proteins in a non-apoptotic programmed cell death dependent on autophagy genes. *Nat. Cell Biol.* 2004;6:1221-8.

[109] Klionsky, DJ; Cregg, JM; Dunn, WA; Emr, SD; Sakai, Y; Sandoval, IV; Sibirny, A; Subramani, S; Thumm, M; Veenhuis, M; Ohsum,i Y. A unified nomenclature for yeast autophagy-related genes. *Dev. Cell.* 2003;5:539-45.

[110] Pyo, JO; Jang, MH; Kwon, YK; Lee, HJ; Jun, JI; Woo, HN; Cho, DH; Choi, B; Lee, H; Kim, JH; Mizushima, N; Oshumi, Y; Jung, YK. Essential roles of Atg5 and FADD in

autophagic cell death: dissection of autophagic cell death into vacuole formation and cell death. *J. Biol. Chem.* 2005;280:20722-9.

[111] Yousefi, S; Perozzo, R; Schmid, I; Ziemiecki, A; Schaffner, T; Scapozza, L; Brunner, T; Simon, HU. Calpain-mediated cleavage of Atg5 switches autophagy to apoptosis. *Nat. Cell Biol.* 2006;8:1124-32.

[112] Du, C; Fang, M; Li, Y; Li, L; Wang, X. Smac, a mitochondrial protein that promotes cytochrome c-dependent caspase activation by eliminating IAP inhibition. *Cell.* 2000;102:33-42.

[113] Kim, JS; Qian, T; Lemasters, JJ. Mitochondrial permeability transition in the switch from necrotic to apoptotic cell death in ischemic rat hepatocytes. *Gastroenterology.* 2003;124:494-503.

[114] Charlotte, F; L'Herminé, A; Martin, N; Geleyn, Y; Nollet, M; Gaulard, P; Zafrani, ES. Immunohistochemical detection of bcl-2 protein in normal and pathological human liver. *Am. J. Pathol.* 1994;144:460-5.

[115] Shimizu, S ; Kanaseki, T ; Mizushima, N ; Mizuta, T; Arakawa-Kobayashi, S ; Thompson, CBT; Sujimoto, Y. Role of Bcl-2 family proteins in a non-apoptotic programmed cell death dependent on autophagy genes. *Nat. Cell Biol.* 2004;6:1221-8.

[116] Bereiter-Hahn, J; Voth, M. Dynamics of mitochondria in living cells: shape changes, dislocations, fusion, and fission of mitochondria. *Microsc. Res. Techn.* 1994;27:198-219.

[117] Nicholls, DG. Mitochondrial function and dysfunction in the cell: its relevance to aging and aging-related disease. *Int. J. Biochem. Cell Biol.* 2002;34:1372-1381.

[118] Zanzami, N; Marchetti, P; Castedo, M; Zanin,C; Vayssiere, JL; Petit, PX; Kroemer, G.B. Reduction in mitochondrial potential constitutes an early irreversible step of programmed lymphocyte death in vivo. *J. Exp. Med.* 1995;181:1661-1672.

[119] Parone, PA; Da Cruz, S; Tondera, D; Mattenberger, Y; James, DI; Maechler, P; Barja, F; Martinou, JC. Preventing mitochondrial fission impairs mitochondrial function and leads to loss of mitochondrial DNA. *PLoS ONE.* 2008;3:e3257.

[120] Scorrano, L; Ashiya, M; Buttle, K; Weller, S; Oakes, SA; Mannella, CA; Korsmeyer, SJ. A distinct pathway remodels mitochondrial cristae and mobilizes cytochrome c during apoptosis. *Dev. Cell.* 2002;2:55-67.

[121] Suen, DF; Norris, KL; Youle, RJ. Mitochondrial dynamics and apoptosis. *Genes Dev.* 2008;22:1577-1590.

[122] Frank, S; Gaume, B; Bergmann-Leitner, ES; Leitner, WW; Robert, EG; Catez F; Smith, CL; Youle, RJ. The role of dynamin-related protein 1, a mediator of mitochondrial fission, in apoptosis. *Dev Cell.* 2001;1:515-525.

[123] Bossy-Wetzel, E; Barsoum, MJ; Godzik, A; Schwarzenbacher, R; Lipton, SA. Mitochondrial fission in apoptosis, neurodegenerating and aging. *Curr. Opin. Cell Biol.* 2003;15:706-716]

[124] Nakamura, N; Kimura, Y; Tokuda, M; Honda, S; Hirose, S. MARCH-V is a novel mitofusin 2- and Drp1-binding protein able to change mitochondrial morphology. *EMBO Rep.* 2006;7:1019-1022.

[125] Cursio, R; Filippa, N; Miele, C; Colosetti, P; Auberger, P; Van Obberghen, E; Gugenheim, J. *Fas ligand expression following normothermic liver ischemia-reperfusion.*

[126] Ciarlo, L; Manganelli, T; Garofalo, T; Matarrese, P; Tinari, A; Misasi, R; Malorni, W; Sorice, M. Association of fission proteins with mitochondrial raft-like domains. *Cell Death Differ.* 2010;17:1047-1058.

[127] Brooks, C; Wei, Q; Feng, L; Dong, G; Tao, Y; Mei, L; Xie, ZJ; Dong, Z. Bak regulates mitochondrial morphology and pathology during apoptosis by interacting with mitofusins. *Proc. Natl. Acad. Sci. U.S.A.* 2007;104:11649-11654.

[128] Hailey, DW; Rambold, AS; Satpute-Krishnan, P; Mitra, K; Sougrat, R; Kim, PK; lippincott-Schwartz, J. Mitochondria supply membranes for autophagosome biogenesis during starvation. *Cell.* 2010;141:656-667.

[129] Gomes, LC; Di Benedetto, G; Scorrano, L. During autophagy mitochondria elongate, are spared from degradation and sustain cell viability. *Nat. Cell Biol.* 2011;13:589-598.

[130] Gomes, LC; Scorrano, L. High levels of Fis1, a pro-fission mitochondrial protein, trigger autophagy. *Biochim. Biophys. Acta.* 2008;1777:860-866.

[131] Tondera, D; grandemange, S; Jourdain, A; Karbowsky, M; Mattenberger, Y; Herzig, S; Da Cruz, S; Clerc, P; Raschke, I; Merkwirth, C; Ehses, S; Krause, F; Chan, DC; Alexander, C; Bauer, C; Youle, R; Langer, T; Martinou, JC. SLP-2 is required for stress-induced mitochondrial hyperfusion. *EMBO J.* 2009;28:1589-1600.

[132] Okaya, T; Blanchard, J; Schuster, R; Kuboki, S; Husted, T; Caldwell, CC; Zingarelli, B; Wong, H; Solomkin, JS; Lentsch, AB. Age-dependent responses to hepatic ischemia/reperfusion injury. *Shock.* 2005;24:421-7.

[133] Green, S. PPAR: a mediator of peroxisome proliferator action. *Mutat. Res.* 1995;333:101-9.

[134] Akahori, T; Sho, M; Hamada, K; Suzaki, Y; Kuzumoto, Y; Nomi, T; Nakamura, S; Enomoto, K; Kanehiro, H; Nakajima, Y. Importance of peroxisome proliferator-activated receptor-gamma in hepatic ischemia/reperfusion injury in mice. *J. Hepatol.* 2007;47:784-92.

[135] Kuboki, S; Shin, T; Huber, N; Eismann, T; Galloway, E; Schuster, R; Blanchard, J; Zingarelli, B; Lentsch, AB. Peroxisome proliferator-activated receptor-gamma protects against hepatic ischemia/reperfusion injury in mice. *Hepatology.* 2008;47:215-24.

[136] Defamie, V; Cursio, R; Lebrigand, K; Moreilhon, C; Saint-Paul, M-C; Laurens, M; Crenesse, D; Cardinaud, B; Auberger, P; Gugenheim, J; Barbry, P; Mari, B. Gene expression profiling of human liver transplants identifies an early transcriptional signature associated with initial poor graft function. *Am. J. Transpl.* 2008;8:1221-1236.

[137] Cursio, R; Colosetti, P; Auberger, P; Gugenheim, J. Liver apoptosis following normothermic ischemia-reperfusion: in vivo evaluation of caspase activity by FLIVO assay in rats. *Transplant. Proc.* 2008;40:2038-2041.

[138] Yu, L; Lenardo, MJ; Baehrecke, EH. Autophagy and caspases: a new cell death program. *Cell Cycle.* 2004;3:1124-6.

[139] Karantza-Wadsworth, V; Patel, S; Kravchuk, O; Chen, G; Mathew, R; jin, S; White, E. Autophagy mitigates metabolic stress and genome damage in mammary tumorigenesis. *Genes Dev.* 2007;21:1621-35.

[140] Kroemer, G; Galluzzi, L; Vandenabeele, P; Abrams, J; Alnemri, ES; Baehrecke, EH; Blagosklonny, MV; El-Deiry, WS; Golstein, P; Green, DR; Hengartner, M; Knight, RA; Kumar, S; Lipton, SA; Malorni, W; Nuñez, G; Peter, ME; Tschopp, J; Yuan, J; Piacentini, M; Zhivotovsky, B; Melino, G; Nomenclature Committee on Cell Death

2009. Classification of cell death: recommendations of the Nomenclature Committee on Cell Death 2009. *Cell Death Differ.* 2009;16:3-11.

[141] Maruyama, R; Goto, K; Takemura, G; Ono, K; Nagao, K; Horie, T; Tsujimoto, A; Kanamori, H; Miyata, S; Ushikoshi, H; Nagashima, K; Minatoguchi, S; Fujiwara, T; Fujiwara, H. Morphological and biochemical characterization of basal and starvation-induced autophagy in isolated adult rat cardiomyocytes. *Am. J. Physiol. Heart Circ. Physiol.* 2008;295:H1599-607.

[142] Samara, C; Syntichaki, P; Tavernarakis, N. Autophagy is required for necrotic cell death in Caenorhabditis elegans. *Cell Death Differ.* 2008;15:105-12.

[143] Artal-Sanz, M; Tavernarakis, N. Proteolytic mechanisms in necrotic cell death and neurodegeneration. *FEBS Lett.* 2005;579:3287-96.

[144] Sybers, HD; Ingwall, J; DeLuca, M. Autophagy in cardiac myocytes. *Recent Adv. Stud. Cardiac. Struct. Metab.* 1976;12:453-63.

[145] Huet, PM; Nagaoka, MR; Desbiens, G; Tarrab, E; Brault, A; Bralet, MP; Bilodeau, M. Sinusoidal endothelial cell and hepatocyte death following cold ischemia-warm reperfusion of the rat liver. *Hepatology.* 2004;39:1110-9.

[146] Lemasters, JJ; Nieminen, AL; Qian, T; Trost, LC; Elmore, SP; Nishimura, Y; Crowe, RA; Cascio, WE; Bradham, CA; Brenner, DA; Herman, B. The mitochondrial permeability transition in cell death: a common mechanism in necrosis, apoptosis and autophagy. *Biochim. Biophys. Acta.* 1998;1366:177-96.

[147] Ferri, KF; Kroemer, G. Organelle-specific initiation of cell death pathways. *Nat. Cell Biol.* 2001;3: E255-63.

[148] Gump, JM; Thorburn, A. Autophagy and apoptosis: what is the connection? *Trends Cell Biol.* 2011;21:387-92. May 9.

[149] Kaminskyy, V; Zhivotovsky, B. Proteases in autophagy. *Biochim. Biophys. Acta.* 2011 May 24. [Epub ahead of print]

[150] Saftig, P; Chapter 3 Physiology of the lysosome, Fabry Disease: Perspectives from 5 Years of FOS, Mehta A, Beck M, Sunder-Plassmann G, editors, Oxford: Oxford PharmaGenesis; 2006.

[151] Takahashi, K ; Ueno, T ; Tanida, I ; Minematsu-Ikeguchi, N ; Murata, M ; Kominami, E. Characterization of CAA0225, a novel inhibitor specific for cathepsin L, as a probe for autophagic proteolysis, *Biol. Pharm. Bull.* 2009;32:475-9.

[152] Yang, M; Zhang, Y; Pan, J; Sun, J; Liu, J; Libby, P, Sukhova, GK; Doria, A; Katunuma, N; Peroni, OD; Guerre-Millo, M; Kahan, BB; Clement, K; Shi, GP. Cathepsin L activity controls adipogenesis and glucose tolerance. *Nat. Cell Biol.* 2007;9:970-7.

[153] Hanada, K; Tamai, M; Yamagishi, M; Ohmura, S; Sawada, J; Tanaka, I. Isolation and characterization of E-64, a new thiol protease inhibitor. *Agric. Biol. Chem.* 1978;42:523-8.

[154] Schotte, P ; Van Criekinge, W ; Van de Craen, M ; Van Loo, G ; Desmedt, M ; Grooten, J ; Cornelissen, M ; De Ridder, L ; Vandekerckhove, J ; Fiers, W ; Vandenabeele, P ; Beyaert, R. Cathepsin B-mediated activation of the proinflammatory caspase-11, *Biochem. Biophys. Res. Commun.* 1998;251:379-87.

[155] Zhou, Q; Salvesen, GS. Activation of pro-caspase-7 by serine proteases includes a non-canonical specificity. *Biochem. J.* 1997;324:361-4.

[156] Vancompernolle, K ; Van Herreweghe, F ; Pynaert, G ; Van de Craen, M ; De Vos, K ; Totty, N ; Sterling, A ; Fiers, W; Vandenabeele, P; Grooten, J. Atractyloside-induced release of cathepsin B, a protease with caspase-processing activity. *FEBS Lett.* 1998;438:150-8.

[157] Jones, B; Roberts, PJ; Faubion, WA; Kominami, E; Gores, GJ. Cystatin A expression reduces bile salt-induced apoptosis in a rat hepatoma cell line. *Am. J. Physiol.* 1998;275:G723-30.

[158] Baskin, ES ; Canbay, A ; Bronk, SF ; Werneburg, N; Guicciardi, ME; Nyberg, SL; Gores, GJ. Cathepsin B inactivation attenuates hepatocyte apoptosis and liver damage in steatotic livers after cold ischemia-warm reperfusion injury. *Am. J. Physiol. Gastrointest. Liver Physiol.* 2005;288:G396-G402.

[159] Guicciardi, ME; Leist, M; Gores, GJ. Lysosomes in cell death. *Oncogene.* 2004;23:2881-90.

[160] Prestwick Chemical Library. www.prestwickchemical.com

[161] Clinical trials . www.clinicaltrials.gov

[162] Orii, R; Sugawara, Y; Hayashida, M; Yamada, Y; Chang, K; Takayama, T; et al. Effects of amrinone on ischaemia-reperfuison injury in cirrhotic patients undergoing hepatectomy: a comparative study with prostaglandin E1. *Br. J. Anaesth.* 2000;85:389-395.

[163] Cecconi, F. Peptides and peptidomimetics: their relevance in autophagy, *Gastroenterol. Hepatol.* 2009;2:S27-33.

[164] Kanki, S; Jaalouk, DE; Lee, S; Yu, AYC; Gannon, J; Lee, RT. Identification of targeting peptides for ischemic myocardium by *in vivo* phage display. *J. Mol. Cell Cardiology.* 2011;50:841-8.

[165] Hundeshagen, P; Hamacher-Brady, A; Eils, R; Brady, NR. Concurrent detection of autolysosome formation and lysosomal degradation by flow cytometry in a high-content screen for inducers of autophagy. *BMC Biol.* 2011;9:38.

[166] Buss, SJ; Muenz, S; Riffel, JH; Malekar, P; Hagenmueller, M; Weiss, CS; Bea, F; Bekeredjian, R; Schinke-Braun, M; Izumo, S; Katus, HA; Hardt, SE. Beneficial effects of Mammalian target of rapmycin inhibition on left ventricular remodelling after myocardial infarction. *J. Am. Coll Cardiol.* 2009;54:2435-46.

[167] Baur, JA; Sinclair, DA. Therapeutic potential of resveratrol: the in vivo evidence. *Nat. Rev. Drug Discov.* 2006;5:493-506]

[168] Easton, JB; Houghton PJ. mTOR and cancer therapy. *Oncogene.* 2006;25:6436-46.].

[169] Wu, Y-T; Tan, HL; Shui, G; Bauvy, C; Huang, Q; Wenk, MR; Ong, C-N; Codogno, P, Shen, H-M. Dual role of 3-methylanine in modulation of autophagy via different temporal patterns of inhibition on class I and class III phosphoinositide 3-kinase. *J. Biol. Chem.* 2010;285:10850-61.

[170] Selzner, N; Rudiger, H; Graf, R; Clavien, PA. Protectives strategies against ischemic injury of the liver. *Gastroenterology.* 2003;125:17-36.

[171] Bahde, R; Spiegel, HU. Hepatic ischaemia-reperfusion injury from bench to bedside. *Br. J. Surg.* 2010;97:1461-75.

[172] Abu-Amara, M; Gurusamy, K; Hori, S; Glantzounis, G; Fuller, B; Davidson, B. Systematic review of randomized controlled trials of pharmacological interventions to reduce ischaemia-reperfusion injury in elective liver resection with vascular occlusion. *HPB.* 2010;12:4-14.

[173] Murry, CE; Jennings, RB; Reimer, KA. Preconditioning with ischemia: a delay of lethal cell injury in ischemic myocardium. *Circulation.* 1986;74:1124-36.

[174] Gurusamy, KS; Gonzalez, HD; Davidson, BR. Current protective strategies in liver surgery. *World J. Gastroenterol.* 2010;16:6098-103.

[175] Gurusamy, KS; Kumar, Y; Sharma, D; Davidson, BR. Ischaemic preconditioning for liver transplantation. *Cochrane Database Syst. Rev.* 2008 Jan 23;(1):CD006315.

[176] Tapuria, N; Kumar, Y; Habib, MM; Abu Amara, M; Seifalian, AM; Davidson, BR. Remote ischemic preconditioning: a novel protective method from ischemia reperfusion injury--a review. *J. Surg. Res.* 2008;150:304-30.

[177] Hausenloy, DJ; Yellon, DM. Remote ischaemic preconditioning: underlying mechanisms and clinical application. *Cardiovasc. Res.* 2008;79:377-86.

[178] Koti, RS; Seifalian, AM; Davidson, BR. Protection of the liver by ischemic preconditioning: a review of mechanisms and clinical applications. *Dig. Surg.* 2003;20:383-96.

[179] Barrier, A; Olaya, N; Chiappini, F; Roser, F; Scatton, O; Artus, C; Franc, B; Dudoit, S; Flahault, A; Debuire, B; Azoulay, D; Lemoine, A. Ischemic preconditioning modulates the expression of several genes, leading to the overproduction of IL-1Ra, iNOS, and Bcl-2 in a human model of liver ischemia-reperfusion. *FASEB J.* 2005;19:1617-26.

[180] Yamabe, K; Shimizu, S; Kamiike, W; Waguri, S; Eguchi, Y; Hasegawa, J; Okuno, S; Yoshioka, Y; Ito, T; Sawa, Y; Uchiyama, Y; Tsujimoto, Y; Matsuda, H. Prevention of hypoxic liver cell necrosis by in vivo human bcl-2 gene transfection. *Biochem. Biophys. Res. Commun.* 1998;243:217-23.

[181] Fazi, B; Bursch, W; Fimia, GM; Nardacci, R; Piacentini, M; Di Sano, F; Piredda, L. Fenretinide induces autophagic cell death in caspase-defective breast cancer cells. *Autophagy.* 2008;4:435-41.

[182] Azoulay, D; Del Gaudio, M; Andreani, P; Ichai, P; Sebag, M; Adam, R; Scatton, O; Min, BY; Delvard, V; Lemoine, A; Bismuth, H; Castaing, D. Effects of 10 minutes of ischemic preconditioning of the cadaveric liver on the graft's preservation and function: the ying and the yang. *Ann. Surg.* 2005;242:133-9.

[183] Stegemann, J; Hirner, A; Rauen, U; Minor, T. Gaseous oxygen persufflation or oxygenated machine perfusion with Custodiol-N for long-term preservation of ischemic rat livers? *Cryobiology.* 2009;58:45-51.

[184] Minor, T; Lüer, B; Efferz, P. Dopamine improves hypothermic machine preservation of the liver. *Cryobiology.* 2011 May 30. [Epub ahead of print]

[185] Koetting, M; Lüer, B; Efferz, P; Paul, A; Minor, T. Optimal time for hypothermic reconditioning of liver grafts by venous systemic oxygen persufflation in a large animal model. *Transplantation.* 2011;91:42-7.

[186] Fuller, BJ; Busza, AL; Proctor, E. Possible resuscitation of liver function by hypothermic reperfusion in vitro after prolonged (24-hour) cold preservation--a 31P NMR study. *Transplantation.* 1990;50:511-3.

[187] Vajdová, K; Graf, R; Clavien, PA. ATP-supplies in the cold-preserved liver: A long-neglected factor of organ viability. *Hepatology.* 2002;36:1543-52.

[188] Leducq, N; Delmas-Beauvieux, MC; Bourdel-Marchasson, I; Dufour, S; Gallis, JL; Canioni, P; Diolez, P. Mitochondrial permeability transition during hypothermic to normothermic reperfusion in rat liver demonstrated by the protective effect of cyclosporin A. *Biochem. J.* 1998;336:501-6.

[189] Delmas-Beauvieux, MC; Gallis, JL; Clerc, M; Canioni, P. 31P NMR studies of rat liver cold preservation with histidine-buffered lactobionate solution. *Cryobiology.* 1993;30:551-61.

[190] Holloway, CM; Harvey, PR; Strasberg, SM. Viability of sinusoidal lining cells in cold-preserved rat liver allografts. *Transplantation.* 1990;49:225-9.

[191] Sindram, D; Porte, RJ; Hoffman, MR; Bentley, RC; Clavien, PA. Synergism between platelets and leukocytes in inducing endothelial cell apoptosis in the cold ischemic rat liver: a Kupffer cell-mediated injury. *FASEB J.* 2001;15:1230-2.

[192] Lelli, JL; Becks, LL; Dabrowska, MI; Hinshaw, DB. ATP converts necrosis to apoptosis in oxidant-injured endothelial cells. *Free Radic. Biol. Med.* 1998;25:694-702.

[193] Leist, M; Single, B; Castoldi, AF; Kühnle, S; Nicotera, P. Intracellular adenosine triphosphate (ATP) concentration: a switch in the decision between apoptosis and necrosis. *J. Exp. Med.* 1997;185:1481-6.

[194] Mortimore, GE; Schworer, CM. Induction of autophagy by amino-acid deprivation in perfused rat liver. *Nature.* 1977;270:174-6.

[195] Cywes, R; Greig, PD; Sanabria, JR; Clavien, PA; Levy, GA; Harvey, PR; Strasberg, SM. Effect of intraportal glucose infusion on hepatic glycogen content and degradation, and outcome of liver transplantation. *Ann. Surg.* 1992;216:235-46.

[196] Xia, X; Kar, R; Gluhak-Heinrich, J; Yao, W; Lane, NE; Bonewald, LF; Biswas, SK; Lo, WK; Jiang, JX. Glucocorticoid-induced autophagy in osteocytes. *J. Bone Miner. Res.* 2010;25:2479-88.

[197] Ezaki, J; Matsumoto, N; Takeda-Ezaki, M; Komatsu, M; Takahashi, K; Hiraoka, Y; Taka, H; Fujimura, T; Takehana, K; Yoshida, M; Iwata, J; Tanida, I; Furuya, N; Zheng, D; Tada, N; Tanaka, K; Kominami, E; Ueno, T. Liver autophagy contributes to the maintenance of blood glucose and amino acid levels. *Autophagy.* 2011;7:727-36.

[198] Li, XL; Man, K; Ng, KT; Lee, TK; Lo, CM; Fan, ST. Insulin in UW solution exacerbates hepatic ischemia/reperfusion injury by energy depletion through the IRS-2/SREBP-1c pathway. *Liver Transpl.* 2004;10:1173-82.

In: Autophagy: Principles, Regulation and Roles in Disease
Editor: Nikolai Gorbunov
ISBN: 978-1-61942-266-7

Chapter XV

Implication of Autophagy in Response to Tubular Ageing and Injuries

Yoshitaka Isaka**, Tomonori Kimura, Yoshitsugu Takabatake and Hiromi Rakugi
Department of Geriatric Medicine and Nephrology,
Osaka University Graduate School of Medicine,
Box B6, 2-2 Yamada-oka, Suita, Osaka, Japan

Abstract

In kidney, proximal tubules consume a large amount of energy to support electrolyte homeostasis. These tubules contain large quantities of mitochondria, which provide the energy for the electrolyte reabsorption. Proximal tubules are susceptible to many kinds of insults such as ischemia-reperfusion injury and nephrotoxic substrates, but little is known about the factors that counteract the cellular stress signaling pathways in tubular epithelial cells. Autophagy mediates bulk degradation and recycling of cytoplasmic constituents to maintain cellular homeostasis. We demonstrated the critical role of autophagy in normal proximal tubule function and protection against acute tubular injury.

1. Introduction

Autophagy was first described in the 1960s as a bulk degradation system [2], but now is found to be a cellular pathway involved in protein and organelle degradation with multiple connections between cellular homeostasis and human disease [3]. While ubiquitin-proteasome system selectively degrades shortlived-proteins by ubiquitination [4], macroautophagy (here called autophagy) was initially believed to be the "non-selective" degradation system for

* Corresponding author; Yoshitaka Isaka; Address: Department of Geriatric Medicine and Nephrology, Osaka University Graduate School of Medicine, Box B6, 2-2 Yamada-oka, Suita, Osaka, Japan 585-0871; Phone: +81-6-6879-3857 ; FAX: +81-6-6879-3857 ; E-mail: isaka@kid.med.osaka-u.ac.jp.

long-lived proteins and damaged organelles [5]. Recent reports showed that autophagy selectively target organelles (mitochondria, endoplasmic reticulum, and so on), protein aggregates, and lipid droplets. Autophagy is now defined as an important pathway under normal and pathological conditions. Studies in animal models suggest the association of autophagy with numerous diseases, including kidney. In this review, we would like to focus on the physiological function of autophagy in cellar homeostasis and protective effect of autophagy under pathological condition in kidney.

2. Assessment of Autophagy Activity

Autophagy consist of at least five steps [5, 6]: (1) the initiation of a double-membrane structure, called isolation membrane (autophagophore), (2) the elongation of this autophagophore, (3) the maturation into the autophagosome (double-membraned vesicles engulfing a portion of cytoplasm), (4) the fusion of autophagosome with lysosome (autophagolysosome), and (5) the degradation of the engulfed biological material. In the step of autophagosome elongation/closure, two dependent ubiquitin-like conjugation systems are involved. Protein Atg12 conjugates with Atg5 in an Atg7- and Atg10-dependent manner, and the resulting conjugate is stabilized by Atg16. This complex is important in the stimulation and localization of the LC3 conjugation reaction. LC3-I is subsequently conjugated with phosphatidylethanolamine (PE) to become LC3-II (LC3-PE) by an ubiquitination-like enzymatic reaction. Thus, conversion of LC3-I to LC3-II is a good marker of autophagosomes. In contrast to the cytoplasmic localization of LC3-I, LC3-II associates with both the outer and inner membranes of the autophagosome. The formation of LC3-PE is dependent on Atg7 and Atg3. Atg4 primes LC3 for conjugation by exposing a glycine residue at the COOH terminus of LC3-I. After closure of the autophagosome, the LC3-II located on the external surface is released from PE by Atg4. On the contrary, LC3 on the inner membrane is degraded by lysosomal enzymes, resulting in decreased LC3 content. It seems that visualization of LC3 by labeling with GFP in transgenic mice is a powerful tool to assess the autophagic activity *in vivo* [7], however, we should not simply regard the GFP-LC3 intensity as an autophagic activity as discussed below.

In the basal states in kidney, GFP-LC3 transgenic mice exhibit the prominent fluorescence in podocytes, but rarely in the tubular epithelial cells [7]. However, this may be a pitfall of understanding the autophagic activity in this mouse [8]. The GFP-LC3 expression of this mouse is driven by the cytomegalovirus enhancer and β-actin (CAG) promoter [9]. We have previously demonstrated that transgenic mice carrying CAG promoter-driven non-tagged GFP display high GFP signal in podocytes, but weak in tubules [10]. Although the number of punctuate LC3 structures per cell is usually an accurate measure of autophagosome number, the above observation imply that the strength of green fluorescence in GFP-LC3 transgenic mice does not represent the actual basal autophagic activity, but the CAG promoter activity in individual cells.

In addition, the increased level of autophagosomes is not always indicative of autophagy induction; it may be the result of either increased generation of autophagosomes (autophagic flux) or suppression of autophagy at its maturation stage (fusion with lysosome and lysosomal degradation). After induction of autophagy, the amount of LC3-II increases, and thereafter degraded by autophagy. Therefore, the disappearance of total LC3 may be paradoxically a good marker of autophagic activity.

3. Basal Autophagic Activity in Kidney

Autophagy occurs at a basal rate in most cells, where it acts as a cytoplasmic quality control mechanism to eliminate long-lived or unfolded protein, damaged organelles, and other nonactive structures [11]. The physiological importance of basal autophagy in maintaining tissue homeostasis has been demonstrated in several organs including brain, liver, heart, etc. using murine *Atg* conditional knockout models [11-14]. In kidney, podocyte-specific [15] and proximal tubules-specific [16] *Atg-5* conditional knockout mice were investigated. Both conditional knockout mice is not lethal, but exhibits gradual accumulation of damaged organelle and unfolded proteins with increasing loss of function. As podocyte-specific knockout mice unexpectedly showed slower progression, we would also like to discuss in this point.

The kidney consists of more than 20 kinds of cells. Among them, podocyte functions in glomerular filtration barrier, deterioration of which leads to glomerulosclerosis. As discussed above, GFP-LC3 transgenic mice exhibit the prominent fluorescence in podocytes [7]. As podocytes are postmitotic, highly differentiated cells, the observation from GFP-LC3 transgenic mice suggests that podocytes possess extremely high basal autophagic activity for intracellular quality control through the proper turnover of cellular constituents. Recently, podocyte-specific deletion of the *Atg5* gene has been shown to result in the accumulation of damaged mitochondria and ubiquitinated protein aggregates. In addition, the accumulation of damaged mitochondria and unfolded proteins in autophagy-deficient podocyte contributed to intracellular buildup of oxidized proteins [15]. Highly-differentiated cells, such as neurons [12, 13] and cardiomyocytes [14], are considered to be ones in which autophagy plays an important role in quality control of proteins and organelles. At first glance, these findings may indicate that, in the kidney, autophagy plays a more important role in podocytes (long-lived cells with high GFP signals) than in tubular cells (highly-proliferative, short-lived cell with low GFP signals). This report actually shows that podocyte-specific *Atg5*-deficient mice develop glomerulosclerosis with an accumulation of ubiquitinated proteins under basal conditions [15]. This phenotype however, was not observed until the mice reached a very old age (22 to 24 months).

The proximal tubular cell is one of the most energy-consuming cells that participate in the absorption of electrolytes and proteins in order to refine primitive urine just after the filtration action of the glomerulus. Proximal tubules also possess abundant organelles that have the capacity to degrade reabsorbed proteins. Therefore, proximal tubules are susceptible to renal damage after ischemic injury or exposure to toxic substrates [17]. Proximal tubule-specific *Atg5*-deficient mice are born normal, but exhibit gradual and massive accumulation of amorphous substrates, deformed mitochondria, and inclusions in the cytosol, thereby resulting in hypertrophy of kidney proximal tubules and eventual degeneration within 6 months of age at the latest, which is not observed in wild-type controls [16]. Thus, our observation suggests that autophagy contributes to cellular homeostasis by eliminating damaged proteins and organelle in the process of aging in kidney proximal tubular cells.

These observations suggest that the basal level of autophagy in proximal tubules has been underestimated compared to podocytes. We suggest that autophagy is more important for maintaining homeostasis in proximal tubules compare to podocytes.

4. Autophagy and Mitochondria

We demonstrated that autophagy-deficient tubular cells accumulate crescent membranous structure-like isolation membrane, possibly intermediate structures of the autophagic process, adjacent to the mitochondria (Fig.1). This observation was consistent with recent data indicating involvement of outer membrane of mitochondria in formation of autophagosomes [18]. As discussed above, quality control of mitochondria is an important role of autophagy, and we suggest that autophagy deficiency results in the accumulation of unfinished isolation membrane-like vacuoles along the damaged mitochondria. Although the damaged mitochondria can produce signals [including reactive oxygen species (ROS) (1)] initiating the mitochondria autophagic degradation (i.e., mitophagy), under the above conditions autophagosomal formation could not be completed.

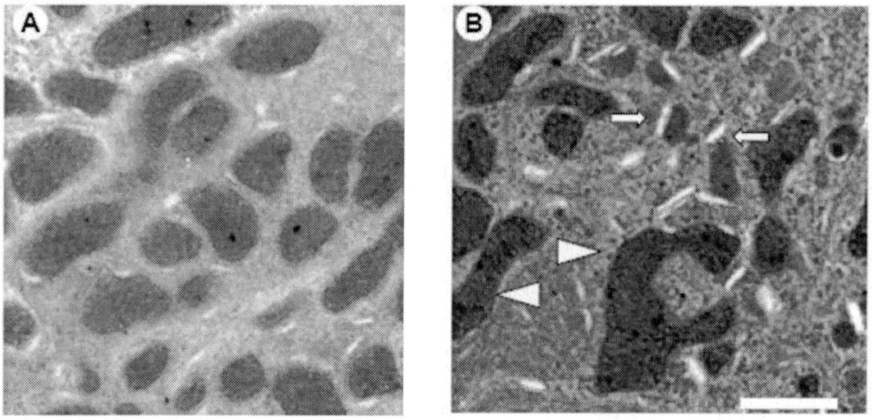

Figure 1. Electron-microscopy of proximal tubular epithelial cell from autophagy deficient mice.

Panel A is TEM image of normal proximal tubular cells. Panel B is TEM image of proximal tubular cells in autophagy-deficient mice at the 9th month. Deformed mitochondria are indicated with arrowhead). Appearance of crescent membranous structures adjacent to mitochondria is shown with arrows. Scale bar 2mm.

Mitophagy can occur under nutrient-rich conditions to remove damaged mitochondria or to mediate mitochondrial remodeling [19]. Mitophagy was first described in mammalian cells by early electron microscopy studies, identifying increased mitochondrial sequestration in lysosomes after stimulation of hepatocyte catabolism with glucagon [20]. Mitophagy is preceded by mitochondrial fission [21], which splits elongated mitochondria into pieces of controllable size for encapsulation and also quality control segregation of damaged mitochondrial material for selective removal by mitophagy. Thus, one of the mechanisms of mitochondrial turnover is autophagic sequestration and delivery to lysosomes for hydrolytic degradation.

Recently, excessive mitochondrial damage has been reported associated to Parkinson's disease [22]. Two of the autosomal recessive genes, PTEN-induced putative kinase protein 1 (*PINK1*) and *PARK2* (which encodes parkin), are implicated in elimination of damaged mitochondria [23]. When mitochondria are damaged, PINK1 accumulates around the mitochondria, and recruits the E3 ubiquitin ligase parkin from the cytosol specifically to the damaged mitochondria. Thereafter, parkin ubiquitylates mitochondrial proteins, and causes mitochondria degraded by autophagy. This may mediate mitochondrial quality control. Note it has been proposed recently that suppression of autophagic degradation of damaged mitochondria can cause neurodegeneration in pathogenesis of Parkinson's disease.

Contrary to above observation, another group of investigators have shown that abnormal mitochondrial autophagy can promote cellular injury in nephropathic cystinosis [24]. Sansanwal et al observed abnormal patterns of mitophagy with a high number of autophagic vacuoles and fewer mitochondria in nephropathic cystinosis. In addition, they showed that 3-methyl adenine, an autophagy inhibitor, rescued cell death in cystinotic cells. In case of nephropathic cystinosis, excessive elimination of mitochondria by autophagy may result in a significant decrease in ATP generation, which leads to cell death [24]. The above discrepancies between autophagic functions in cell survival and cell death may have a quantitative relationship, e.g., a low or moderate autophagy activity is protective, whereas a high level of autophagy activity promotes cell death via excessive degradation of essential proteins or organelles. It has to be noted that the interpretation of the above results is limited by reliability of the applied techniques, i.e., assessment of autophagy activity by conversion LC3-I to LC3-II using an autophagy inhibitor. In this respect more reliable results could be obtained using depletion of *Atg* genes in animal models.

As a major source of reactive oxygen species (ROS), mitochondria are especially responsible for ROS-associated injuries. Oxidative stress and various disease processes cause mitochondrial damage and dysfunction. Appropriate elimination of aged and dysfunctional mitochondria is critical to protect cells from ROS-induced apoptosis. Moreover, apoptotic and autophagic pathways are not mutually exclusive; they may be both synergistic and antagonistic under different conditions. In the turning point of these pathways, mitochondria and Bcl-2-related proteins play important roles. Thus, we previously demonstrated that overexpression of the anti-apoptotic factor, Bcl-2, inhibited tubular apoptosis and subsequent tubulointerstitial damage after ischemia-reperfusion (I/R) injury [25]. I/R injury promoted the formation of autophagosomes containing mitochondria in tubular epithelial cells. On the other hand, Bcl-2 augmentation suppressed the formation of autophagosomes, and subsequently ameliorated tubular damage [26]. Thus, we suggest that Bcl-2 augmentation protected renal tubular epithelial cells from I/R injury.

5. Autophagy and Aging

It has been reported that aging leads to a reduction in autophagosome formation and autophagosome-lysosome fusion [27, 28]. In the conditional autophagy-deficient brain, polyubiquitinated proteins accumulate in neurons [12, 13], resulting in neurological abnormalities and neuronal death. These observations suggest that basal autophagy plays a crucial role in maintaining neural function and preventing neuronal aging. Autophagy functions as anti-aging quality control of cellular components by eliminating aggregation-prone proteins and damaged mitochondria, thereby suppresses the production of ROS [27]. This may be also important for cells with a low turnover rate, such as podocytes. If autophagic activity is reduced with increasing age, this may lead to the accumulation of damaged organelle and unfolded proteins, which are 'wastes' in cells. Therefore, the maintenance of autophagic activity by calorie restriction may help extend life span [27, 29]. It was shown that several autophagy genes are necessary for life span extension in *Caenorhabditis elegans* (30). Recently, it was demonstrated that mitochondrial damage in aged kidney is associated with Sirt1 heterogeneous deficiency, and that Sirt1 may promote cell adaptation to hypoxia through autophagy in aged kidney [31].

The autophagic insufficiency results in the age-related 'waste' accumulation within long-lived post-mitotic cells. We demonstrated that proximal tubule-specific autophagy-deficient mice exhibited the accumulation of numerous crescent membranous structures in the tubular epithelial cells with age, mainly adjacent to the mitochondria. Furthermore, accumulation of deformed mitochondria was exclusively observed in autophagy-deficient mice [16]. Many mitochondria in aged post-mitotic cells are enlarged and structurally deteriorated, showing swelling and disintegration of cristae, often resulting in usually called 'giant' mitochondria [32] with the formation of amorphous material [33]. Senescent mitochondria are defective in ATP production [34] and are reported to produce increased amounts of ROS [35], which are harmful for cells and nevertheless cannot be eliminated. The mechanisms underlying age-related mitochondrial changes remain obscure, however, our observation [16] suggests that damaged mitochondria should be autophagocytosed and degraded, but accumulated according to the decreased autophagic activity with age.

6. Autophagy against Ischemia-Reperfusion Injury

Ischemia-reperfusion (I/R) injury is the most important cause of acute kidney injury, and proximal tubular cells are the main lesion of this injury [36]. GFP-LC3 transgenic mice exhibited massive accumulation of LC3-positive autophagosomes in the proximal tubules after I/R injury, suggesting that autophagy is induced during I/R in proximal tubular epithelial cells. Previous *in vitro* and *in vivo* studies indicate that autophagy responds to hypoxia, oxidative stress and/or defective energy production [37-39]. However, the precise role of autophagic flux on I/R injury remained debating whether autophagy functions in cell death or survival. Indeed it has been shown that autophagy can, on the one hand, have a protective role for several diseases, whereas, on the other hand, it can also play a role in cell death [40].

It was shown that the expression of Beclin-1 and LC3 was increased in renal tubules during I/R injury in kidney, and that expression of Bcl-xL and Bcl-2 could ameliorate both autophagy and apoptosis accompanied by the amelioration of ischemic kidney injury [41] as we previously reported [26]. Although this observation suggests that autophagy might contribute to tubular cell injury or death, these results might indicate that autophagy is active when the cell damage is severe. On the contrary, Jiang et al demonstrated the renoprotective function of autophagy on I/R injury [42]. They demonstrated that hypoxia induced autophagy in cultured renal proximal tubular cells, and that blockade of autophagy by 3-methyladenine

(3-MA) or siRNA for *beclin-1* or *Atg-5* enhanced hypoxia-induced apoptosis. In addition, they showed *in vivo* that inhibition of autophagy by chloroquine or 3-MA worsened renal ischemia/reperfusion injury. One of the most serious problems in the research of autophagy is lack of specific autophagic inhibitors or activators, although several modulators as above (3-MA, chloroquine or siRNA) are now available [8]. In addition, autophagy chemical inhibitors can reduce or prevent cell death. Therefore, we employed proximal tubule-specific *Atg5*-deficient mice to determine the role of autophagy in proximal tubule cells against I/R injury [16].

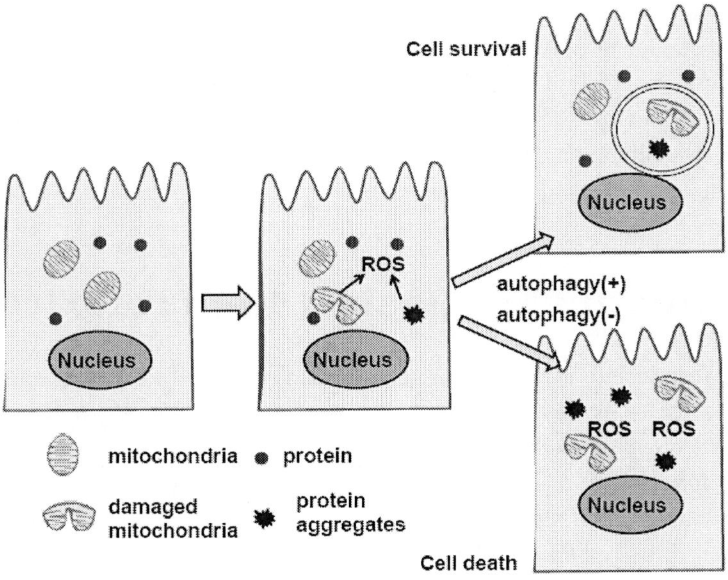

Figure 2. Autophagy acts under pathological stress.

Autophagy guard against the cellular stress by eliminating damaged proteins and organelle, but deficiency of autophagy accelerates cell death with accumulated damaged mitochondria and aggregated proteins, which stimulates ROS production and so on.

We found that I/R injury was much more severe in the proximal tubule-specific autophagy-deficient mice than in controls. This result clearly suggests that autophagy plays a protective role in kidney ischemic injury. One of the most efficient responses to ischemia is a reduction in metabolic activity [43], autophagy may improve the metabolic status in proximal tubules, where high energy consumption is required. Moreover, I/R injury induces fragmentation of the mitochondria, which leads to mitochondrial outer membrane permeabilization, the release of apoptogenic factors, and consequent apoptosis [44]. In addition, damaged mitochondrial could also become a source of intrinsic ROS. Thereby, it is possible that failure to eliminate damaged mitochondria due to defective autophagy might augment cell injury (Fig. 2).

Another possibility is that autophagy protects tubular cells by avoiding accumulation of aggregate-prone proteins, which are generated in response to I/R injury. The proximal tubule-specific autophagy-deficient mice showed rapid accumulation of p62- and ubiquitin-positive inclusion bodies in response to I/R injury [16]. These inclusion bodies markedly increased in response to I/R injury, while only a small number of p62- and ubiquitin-positive inclusion bodies were observed in sham-operated autophagy-deficient mice. These observations indicate that massive amount of abnormal proteins generated in response to ischemic injury is degraded by autophagy, thus, corroborating a recent report by Kraft et al. 2010 [45]. Recently, Komatsu et al. demonstrated that overproduction of p62 or deficiency in autophagy affects the interaction between Nrf2 (nuclear factor erythroid 2-related factor 2) and Keap1 (Kelch-like ECH-associated protein 1), resulting in induction of cellular stress genes that in turn can lead to a disease [46]. Degradation systems are functionally coupled, and it is quite possible that, under cellular stress, ubiquitinated protein accumulation might exceed capacity of proteasomal degradation, resulting in a buildup of protein aggregates. These protein aggregates also induce ROS production, and in turn, ROS alone have been suggested to induce autophagy *in vitro* [47], indicating that autophagy may be a cellular response to oxidative stress. Although the precise underlying mechanism remains unknown, autophagy plays a critical in cellular protection by reduction of oxidative stress *in vivo*, and decreased function of autophagy, at least, cannot guard proximal tubules against I/R injury.

7. Autophagy against Nephrotoxicity

Autophagy response to injury is non-specific and along with I/R can be induced in kidney tubular epithelial cells by several drugs, such as cisplatin [48, 49] and cyclosporine [50]. Immediately after cisplatin treatment, cisplatin accumulates in proximal tubular cells resulting in autophagy in tubular epithelial cells.

The mechanism of cisplatin-induced autophagy is not clear; however, it has been recently shown that it can directly damage DNA and activate oxidative stress that trigger a variety of signaling cascades ultimately leading to apoptosis and/or autophagy.

There is evidence that inhibition of autophagy by 3-MA or siRNA for *beclin-1* enhances cisplatin- stimulated caspase activation and apoptosis in cultured tubular epithelial cells [48]. Similar to cisplatin, cyclosporine induces autophagy via endoplasmic reticulum (ER) stress, and inhibition of autophagy by *beclin-1* siRNA increased tubular cell death [50]. Although the precise mechanism to determine the balance between apoptosis and autophagy remains obscure, these observations suggest that the agent to stimulate autophagy might promote cellular survival and delays the onset of an apoptotic response in cisplatin injury in tubular cells. To support this hypothesis, the absence of heme oxygenase-1 (HO-1) was shown to result in impaired autophagy and increased apoptosis in tubular epithelial cells under cisplatin [49]. Restoring HO-1 expression reversed the impaired autophagic response and decreased susceptibility to cisplatin-induced apoptosis [49]. On the contrary, the other study reported that suppression of autophagy by similar reagents resulted in reduction of apoptosis *in vitro*, and thereby, concluded that autophagy may contribute to tubular cell death. These controversial observations could be explained by their experimental design, because most chemical inhibitors of autophagy are not entirely specific. Further study using *Atg* gene depletion is mandatory in this field.

Conclusion

Autophagy-mediated cellular remodeling is a recent emerging area in kidney diseases. In this chapter we highlight a new perspective on role of autophagy in function of kidney tubules, in particular on tubular aging. Recent study conducted using autophagy-deficient conditional knockout mice have shown that suppression of autophagic function results in the gradual accumulated damaged organelle and aggregated proteins, leading to so-called "aged cells". Vice versa, maintaining of autophagic activity may prevent cellular senescence." Although autophagy supports tubular cells homeostasis by eliminating damaged organelle and misfolded protein aggregates [15], excessive autophagy may promote cell death (24). Autophagy can ether protect [15] or kill cells [24] under certain conditions. Currently, there are no techniques available for assessment of autophagic activity under stress conditions.

The important conclusion from the available data is that autophagy is primarily a survival mechanism in response to cellular stress. Removal of damaged mitochondria and misfolded proteins in proximal tubules under aging or ischemia-reperfusion injury supports this idea. How is the damage of organelle or misfolded protein transmitted to initiate autophagy? How is the autophagic activity controlled at proper level? What are the molecular targets of autophagy in the cell stress response? These questions are the goals of future studies. However, we can say now that the answer to the above questions can crucial steps toward discovery of novel molecular targets for precise pharmacological correction of autophagy mechanism in order to improve the survival of tubular cells in kidney diseases.

References

[1] Scherz-Shouval R, Elazar Z. Regulation of autophagy by ROS: physiology and pathology. *Trends Biochem. Sci.* 2011; 36: 30-38.

[2] Deter RL, Baudhuin P, De Duve C. Participation of lysosomes in cellular autophagy induced in rat liver by glucagon. *J. Cell Biol.* 1967; 35: C11-16.

[3] Mizushima N, Levine B, Cuervo AM, Klionsky DJ. Autophagy fights disease through cellular self-digestion. *Nature.* 2008; 451: 1069-1075.

[4] Rubinsztein DC. The roles of intracellular protein-degradation pathways in neurodegeneration. *Nature.* 2006; 443: 780-786.

[5] Mizushima N, Levine B. Autophagy in mammalian development and differentiation. *Nat. Cell Biol.* 2010; 12: 823-830.

[6] Yang Z, Klionsky DJ. Eaten alive: a history of macroautophagy. *Nat. Cell Biol.* 2010; 12: 814-822.

[7] Mizushima N, Yamamoto A, Matsui M, Yoshimori T, Ohsumi Y.. In vivo analysis of autophagy in response to nutrient starvation using transgenic mice expressing a fluorescent autophagosome marker. *Mol. Biol. Cell.* 2004; 15: 1101-1111.

[8] Mizushima N, Yoshimori T, Levine B. Methods in mammalian autophagy research. *Cell.* 2010; 140: 313-326.

[9] Niwa H, Yamamura K, Miyazaki J. Efficient selection for high-expression transfectants with a novel eukaryotic vector. *Gene.* 1991; 108: 193-199.

[10] Akagi Y, Isaka Y, Akagi A, Ikawa M, Takenaka M, Moriyama T, Yamauchi A, Horio M, Ueda N, Okabe M, Imai E. Transcriptional activation of a hybrid promoter composed of cytomegalovirus enhancer and beta-actin/beta-globin gene in glomerular epithelial cells in vivo. *Kidney Int.* 1997; 51: 1265-1269.

[11] Komatsu M, Ueno T, Waguri S, Uchiyama Y, Kominami E, Tanaka K.. Constitutive autophagy: vital role in clearance of unfavorable proteins in neurons. *Cell Death Differ.* 2007; 14: 887-894.

[12] Hara T, Nakamura K, Matsui M, Yamamoto A, Nakahara Y, Suzuki-Migishima R, Yokoyama M, Mishima K, Saito I, Okano H, Mizushima N.Suppression of basal autophagy in neural cells causes neurodegenerative disease in mice. *Nature.* 2006; 441: 885-889.

[13] Komatsu M, Waguri S, Chiba T, Murata S, Iwata J, Tanida I, Ueno T, Koike M, Uchiyama Y, Kominami E, Tanaka K. Loss of autophagy in the central nervous system causes neurodegeneration in mice. *Nature.* 2006; 441: 880-884.

[14] Nakai A, Yamaguchi O, Takeda T, Higuchi Y, Hikoso S, Taniike M, Omiya S, Mizote I, Matsumura Y, Asahi M, Nishida K, Hori M, Mizushima N, Otsu K.. The role of autophagy in cardiomyocytes in the basal state and in response to hemodynamic stress. *Nat. Med.* 2007; 13: 619-624.

[15] Hartleben B, Godel M, Meyer-Schwesinger C, Liu S, Ulrich T, Kobler S, Wiech T, Grahammer F, Arnold SJ, Lindenmeyer MT, Cohen CD, Pavenstadt H, Kerjaschki D, Mizushima N, Shaw AS, Walz G, Huber TB. Autophagy influences glomerular disease susceptibility and maintains podocyte homeostasis in aging mice. *J. Clin. Invest.* 2010; 120: 1084-1096.

[16] Kimura T, Takabatake Y, Takahashi A, Kaimori JY, Matsui I, Namba T, Kitamura H, Niimura F, Matsusaka T, Soga T, Rakugi H, Isaka Y. Autophagy protects the proximal tubule from degeneration and acute ischemic injury. *J. Am. Soc. Nephrol.* 2011; 22: 902-913.

[17] Jones DP. Renal metabolism during normoxia, hypoxia, and ischemic injury. *Annu. Rev. Physiol.* 1986; 48: 33-50.

[18] Okamoto K, Kondo-Okamoto N, Ohsumi Y. Mitochondria-anchored receptor Atg32 mediates degradation of mitochondria via selective autophagy. *Dev. Cell.* 2009; 17: 87-97.

[19] Youle RJ, Narendra DP. Mechanisms of mitophagy. *Nat. Rev. Mol. Cell. Biol* 2011; 12: 9-14.

[20] De Duve C, Wattiaux R. Functions of lysosomes. *Annu. Rev. Physiol.* 1966; 28: 435-492.

[21] Twig G, Elorza A, Molina AJ, Mohamed H, Wikstrom JD, Walzer G, Stiles L, Haigh SE, Katz S, Las G, Alroy J, Wu M, Py BF, Yuan J, Deeney JT, Corkey BE, Shirihai OS. Fission and selective fusion govern mitochondrial segregation and elimination by autophagy. *Embo. J.* 2008; 27: 433-446.

[22] Schapira AH. Mitochondria in the aetiology and pathogenesis of Parkinson's disease. *Lancet Neurol.* 2008; 7: 97-109.

[23] Geisler S, Holmstrom KM, Skujat D, Fiesel FC, Rothfuss OC, Kahle PJ, Springer W. PINK1/Parkin-mediated mitophagy is dependent on VDAC1 and p62/SQSTM1. *Nat. Cell Biol.* 2010; 12: 119-131.

[24] Sansanwal P, Yen B, Gahl WA, Ma Y, Ying L, Wong LJ, Sarwal MM. Mitochondrial autophagy promotes cellular injury in nephropathic cystinosis. *J. Am. Soc. Nephrol.* 2010; 21: 272-283.

[25] Suzuki C, Isaka Y, Shimizu S, Tsujimoto Y, Takabatake Y, Ito T, Takahara S, Imai E. Bcl-2 protects tubular epithelial cells from ischemia reperfusion injury by inhibiting apoptosis. *Cell Transplant.* 2008; 17: 223-229.

[26] Isaka Y, Suzuki C, Abe T, Okumi M, Ichimaru N, Imamura R, Kakuta Y, Matsui I, Takabatakc Y, Rakugi H, Shimizu S, Takahara S. Bcl-2 protects tubular epithelial cells from ischemia/reperfusion injury by dual mechanisms. *Transplant. Proc.* 2009; 41: 52-54.

[27] Cuervo AM, Bergamini E, Brunk UT, Droge W, Ffrench M, Terman A. Autophagy and aging: the importance of maintaining "clean" cells. *Autophagy.* 2005; 1: 131-140.

[28] McCray BA, Taylor JP. The role of autophagy in age-related neurodegeneration. *Neurosignals.* 2008; 16: 75-84.

[29] Bergamini E, Cavallini G, Donati A, Gori Z. The role of macroautophagy in the ageing process, anti-ageing intervention and age-associated diseases. *Int. J. Biochem. Cell Biol.* 2004; 36: 2392-2404.

[30] Melendez A, Talloczy Z, Seaman M, Eskelinen EL, Hall DH, Levine B. Autophagy genes are essential for dauer development and life-span extension in C. elegans. *Science.* 2003; 301: 1387-1391.

[31] Kume S, Uzu T, Horiike K, Chin-Kanasaki M, Isshiki K, Araki S, Sugimoto T, Haneda M, Kashiwagi A, Koya D. Calorie restriction enhances cell adaptation to hypoxia through Sirt1-dependent mitochondrial autophagy in mouse aged kidney. *J. Clin. Invest.* 2010; 120: 1043-1055.

[32] Coleman R, Silbermann M, Gershon D, Reznick AZ. Giant mitochondria in the myocardium of aging and endurance-trained mice. *Gerontology.* 1987; 33: 34-39.

[33] Terman A, Dalen H, Eaton JW, Neuzil J, Brunk UT. Mitochondrial recycling and aging of cardiac myocytes: the role of autophagocytosis. *Exp. Gerontol.* 2003; 38: 863-876.

[34] Ermini M. Ageing changes in mammalian skeletal muscle: biochemical studies. *Gerontology.* 1976; 22: 301-316.

[35] Sohal RS, Sohal BH. Hydrogen peroxide release by mitochondria increases during aging. *Mech. Ageing Dev.* 1991; 57: 187-202.

[36] Lameire N, Van Biesen W, Vanholder R. Acute kidney injury. *Lancet.* 2008; 372: 1863-1865.

[37] Adhami F, Liao G, Morozov YM, Schloemer A, Schmithorst VJ, Lorenz JN, Dunn RS, Vorhees CV, Wills-Karp M, Degen JL, Davis RJ, Mizushima N, Rakic P, Dardzinski BJ, Holland SK, Sharp FR, Kuan CY. Cerebral ischemia-hypoxia induces intravascular coagulation and autophagy. *Am. J. Pathol.* 2006; 169: 566-583.

[38] Matsui Y, Takagi H, Qu X, Abdellatif M, Sakoda H, Asano T, Levine B, Sadoshima J. Distinct roles of autophagy in the heart during ischemia and reperfusion: roles of AMP-activated protein kinase and Beclin 1 in mediating autophagy. *Circ. Res.* 2007; 100: 914-922.

[39] Yan L, Vatner DE, Kim SJ, Ge H, Masurekar M, Massover WH, Yang G, Matsui Y, Sadoshima J, Vatner SF. Autophagy in chronically ischemic myocardium. *Proc. Natl. Acad. Sci. U. S. A.* 2005; 102: 13807-13812.

[40] Shintani T, Klionsky DJ. Autophagy in health and disease: a double-edged sword. *Science.* 2004; 306: 990-995.

[41] Chien CT, Shyue SK, Lai MK. Bcl-xL augmentation potentially reduces ischemia/reperfusion induced proximal and distal tubular apoptosis and autophagy. *Transplantation.* 2007; 84: 1183-1190.

[42] Jiang M, Liu K, Luo J, Dong Z. Autophagy is a renoprotective mechanism during in vitro hypoxia and in vivo ischemia-reperfusion injury. *Am. J. Pathol.* 2010; 176: 1181-1192.

[43] Epstein FH. Oxygen and renal metabolism. *Kidney Int.* 1997; 51: 381-385.

[44] Brooks C, Wei Q, Cho SG, Dong Z. Regulation of mitochondrial dynamics in acute kidney injury in cell culture and rodent models. *J. Clin. Invest.* 2009; 119: 1275-1285.

[45] Kraft C, Peter M, Hofmann K. Selective autophagy: ubiquitin-mediated recognition and beyond. *Nat. Cell Biol.* 2010; 12: 836-841.

[46] Komatsu M, Kurokawa H, Waguri S, Taguchi K, Kobayashi A, Ichimura Y, Sou YS, Ueno I, Sakamoto A, Tong KI, Kim M, Nishito Y, Iemura S, Natsume T, Ueno T, Kominami E, Motohashi H, Tanaka K, Yamamoto M. The selective autophagy substrate p62 activates the stress responsive transcription factor Nrf2 through inactivation of Keap1. *Nat. Cell Biol.* 2010; 12: 213-223.

[47] Scherz-Shouval R, Shvets E, Fass E, Shorer H, Gil L, Elazar Z. Reactive oxygen species are essential for autophagy and specifically regulate the activity of Atg4. *Embo J.* 2007; 26: 1749-1760.

[48] Periyasamy-Thandavan S, Jiang M, Wei Q, Smith R, Yin XM, Dong Z. Autophagy is cytoprotective during cisplatin injury of renal proximal tubular cells. *Kidney Int.* 2008; 74: 631-640.

[49] Bolisetty S, Traylor AM, Kim J, Joseph R, Ricart K, Landar A, Heme oxygenase-1 inhibits renal tubular macroautophagy in acute kidney injury. *J. Am. Soc. Nephrol.* 2010; 21: 1702-1712.

[50] Pallet N, Bouvier N, Legendre C, Gilleron J, Codogno P, Beaune P, Thervet E, Anglicheau D. Autophagy protects renal tubular cells against cyclosporine toxicity. *Autophagy.* 2008; 4: 783-791.

In: Autophagy: Principles, Regulation and Roles in Disease ISBN: 978-1-61942-266-7
Editor: Nikolai Gorbunov

Chapter XVI

Autophagy Induced by Nanomaterials

***Jasmine Jia'en Li*[1,2], *Olivia Scully*[1], *Karthik Subramhanya Harve*[1], *Lin-Yue Lanry Yung*[2] *and Boon-Huat Bay*[1]**

[1] Department of Anatomy, Yong Loo Lin School of Medicine, National University of Singapore, Singapore 117597, Singapore
[2] Department of Chemical & Biomolecular Engineering, Faculty of Engineering, National University of Singapore, Singapore 117576, Singapore

Abstract

In recent paradigm, autophagy is considered as a fundamental cellular process which helps to maintain homeostasis under a variety stress-conditions that include exposure to toxic agents, injury, inflammation, and neoplastic transformations. Interestingly, autophagy can result in cell death under certain conditions and yet in other circumstances, the process can promote cell survival. Although there are relatively fewer studies on nanomaterial induced autophagy, the autophagic process is known to be induced by carbon nanomaterials, protein conjugated nanoparticles, gold nanoparticles, fullerene-derived nanomaterials, cadmium selenide and indium gallium phosphide quantum dots, nano neodymium oxide and rare earth oxide nanocrystals, among others. On the other hand, cerium oxide nanoparticles have been reported to inhibit autophagy by impeding reactive oxygen species formation and endoplasmic reticulum stress. There is a need for an in-depth understanding of the mechanisms. such as the role of ROS in mediating the cross-talk of endocytosed nanoparticles and autophagy, and the consequences of nanomaterial-induced autophagy requires further investigation.

1. Introduction

The word "autophagy" coined by Christian de Duve in 1963, is derived from a Greek word meaning "self" (*auto*) and "eating" (*phagy*). Autophagy may be described as a process of degradation of cellular components mediated via the lysosomal pathway (de Duve and Wattiaux, 1966). Although there are still aspects of autophagy which are not clearly understood, it is generally thought to be a fundamental process important for maintaining cellular homeostasis (Yang and Klionsky, 2010). At basal level, autophagy helps to maintain cell homeostasis by the recycling and renewal of old cell organelles (Ebato et al., 2008). Under acute stress, autophagy protects and promotes cell survival by removing damaged organelles and providing much needed nutrients such as during starvation (Lum et al., 2005). However if the cell is subjected to prolonged stress, autophagy may be overexpressed resulting in cell death (Schweichel and Merker, 1973; Levine and Yuan, 2005). The dual roles of autophagy in cell survival and cell death are closely related to each other (Baehrecke, 2005; Maiuri et al., 2007), as both pathways share some common cell-signaling proteins (for example the ATG6 protein, Beclin-1) which play prominent roles in driving the cell towards either fate, cell survival or cell death (Scarlatti et al., 2009). Autophagy has been implicated in several diseases and cellular stresses such as starvation, development, innate and adaptive immunity, aging, tumorigenesis and stress-induced differentiation (Martinez-Vicente and Cuervo 2007; Galluzzi et al., 2008; Beau et al., 2011). Recent concept of autophagy defines two distinct pathways for delivery of cargo materials to phagophore/autophagosome constructions; they are specific and non-specific autophagy (Nair and Klionsky, 2005). The latter is the predominant description of autophagy, where bulk materials in the cellular cytoplasm are indiscriminately degraded in the lysosomal-autophagic pathway. Specific autophagy in contrast, is the degradation of selected proteins or protein aggregates, which are "marked" by polyubiquitination where P62 and NBR1 are two of the ubiquitin- and LC3-binding proteins that recognize the marked proteins and tie them to the autophagic processes (Kirkin et al., 2009; Filimonenko et al., 2010); or specific targeting of distinct cellular organelles including damaged mitochondria ("mitophagy") (Green et al., 2011). This chapter gives an overview of autophagy induced by nanomaterials.

2. The Autophagic Cascade

Conventionally autophagy is classified as macroautophagy, microautophagy and chaperone-mediated autophagy based on the processes by which the autophagic contents are delivered to the lysosomes (Li et al., 2011). Macroautophagy is the most prevalent form of autophagy in cells and will be referred to hereafter as autophagy (Klionsky, 2005). In macroautophagy, parts of the cytoplasm and organelles targeted for degradation are sequestered into a double membrane vesicle forming an autophagosome which then fuses with a lysosome to form an autophagolysosome (Figure 1). This eventually becomes an autolysosome with digestion of the contents taking place. The products of lysosomal degradation are subsequently recycled for use in the cell or exocytosed (Noda et al., 2009; Mehrpour et al., 2010).

The critical events in macroautophagy occur in five identifiable stages. First, is initiation and nucleation, where the phagophore is assembled. Second is expansion, where extension of the phagophore and elongation of the autophagosome membrane takes place. Two enzymatic cascades are required for this process which involves two ubiquitin-like proteins *viz.*, Atg8 (LC3 in mammals) and Atg12 proteins, respectively (Liu et al., 2010; Roy and Debnath,

2010). Third is completion, where formation of the autophagosome (of size 300 to 900 nm) is completed and the characteristic double membrane is clearly seen (Tanida, 2010). Fourth is fusion, in which the autophagosome is fused with a lysosome via the actions of lysosomal proteins LAMP1 and LAMP2 (Eskelinen, 2006), as well as Rab7 (Jäger et al., 2004) and UVRAG (Liang et al., 2008) resulting in the formation of autolysosome which is a single membrane-bound acidic vesicle where the delivered cytoplasmic cargoes are digested and subsequently recycled by lysosomal hydrolases (Liu et al., 2010; Roy and Debnath, 2010). Fifth is degradation, when the autolysosomal contents are degraded by the lysosomal hydrolases, and later recycled in the cytoplasm or targeted for exocytosis.

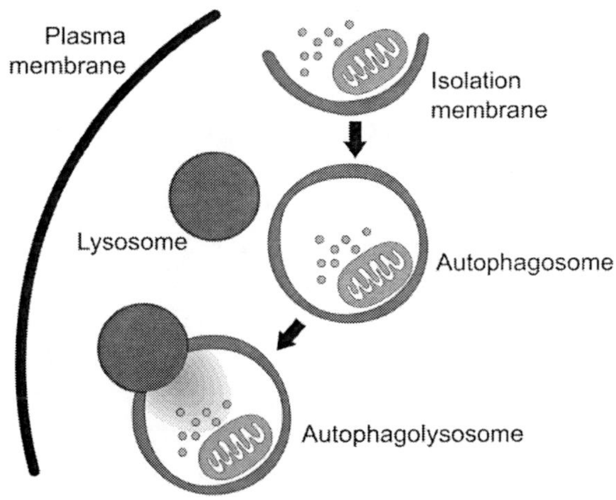

Figure 1. Schematic diagram showing formation of the autophagosome and autophagolysosome.

The whole process of autophagy is tightly regulated by a host of autophagy proteins and their complexes with other related molecules. The consensus nomenclature for autophagy proteins are ATG (AuTophaGy-related) and genes are denoted as Atg. Thus far, 3 or 5 subgroups of the 31 ATG proteins identified (Klionsky et al., 2003; Yang and Klionsky, 2009), are involved in the formation of the autophagosome that make up the core machinery of the autophagic process (Yang and Klionsky, 2009; Tanida, 2010).

3. Nanomaterials-Induced Autophagy

A variety of signals which could trigger and regulate autophagy include nutrient deprivation, hypoxia, hormones, temperature, cytokines and DNA damage (Levine and Kroemer, 2008, Liu et al., 2010). Kim et al. (2009) have also shown that depletion of a single type of amino acid arginine by arginine deiminase (ADI) in castration-resistant prostate cancer CWR22Rν1 cell line, induced autophagy 30 minutes after ADI treatment. The data presented in several recent reports suggest that nanomaterials can be considered as a new class of autophagy inducer (Zabirnyk et al., 2007).

Stern and co-workers observed the presence of autophagy in porcine renal cells treated with cadmium-selenium and indium gallium phosphide quantum dots (Stern et al., 2008). Other nanomaterials inducing autophagy include protein-conjugated quantum dots in human mesenchymal stem cells (Seleverstov et al., 2006) and fullerenes in endothelial cells (Yamawaki and Iwai, 2006) and cancer cells (Wei et al., 2010). Nano neodymium oxide has been reported to induce autophagic cell death in non-small cell lung cancer cells (Chen et al., 2005). Afeseh Ngwa et al. (2011) have observed that manganese NPs exert toxic effects on dopaminergic neuronal cells inducing cell death via apoptosis and autophagy. Dose-dependent and time-dependent induction of the autophagic process have also been observed in HeLa cells exposed to rare earth oxide nanocrystals (Yu et al., 2009).

Li et al. (2010) showed that AuNPs were able to induce the formation of autophagosomes in human lung fibroblast cells. The authors demonstrated a significantly higher expression of ATG 7 and MAP-LC3-II in treated samples. ATG 7 is an essential enzyme which is known to participate in the two major autophagy conjugation systems (Tanida et al., 2001). This leads to activation of ATG 12 with downstream conjugation to ATG 5 resulting in the conversion of MAP-LC3-I to MAP-LC3-II (which is widely used as an autophagosomal marker). Autophagosome membrane formation requires the participation of both ATG 5 and ATG 12 proteins (Geng and Klionsky, 2008; He and Klionsky, 2009). Interestingly, when Atg 7 was deleted or knocked down, there was impairment of cell survival and increased fatality in irradiated mice, demonstrating the importance of this gene in the protection of cell viability against environmental stresses (Mortensen et al., 2011). A very recent study by Tian et al. (2011) showed that a panel of gold(I)–phosphine complexes (Au-4a, Au-4b and Au-4c) known to display anticancer activities, also exhibited significant autophagy-inducing properties in HeLa cervical cancer cells.

Since autophagy is observed across different cell types and with different kinds of NPs, Zabirnyk et al. (2007) posited that NP-induced autophagy could be a common cellular response. Formation of the autophagosome is also dependent on the size and speed of metabolic clearance of the NPs. NPs with smaller diameters and longer presence inside the cell, seem to favour formation of the autophagosome.

4. Oxidative Stress Mediates Nanoparticle-Induced Autophagy

Oxidative stress is known to be a leading cause of NP-induced autophagy (Chen et al., 2008). Reactive oxygen species (ROS) are the main players in the autophagic response to oxidative stress (Kaushik and Cuervo, 2006; Scherz-Shouval and Elazar, 2011). This group of small, highly reactive molecules (which comprise superoxide and hydroxyl radicals arising from oxygen, amongst others), are capable of initiating free radical cascades producing damage to biomolecules (Cerutti, 1994). Under normal circumstances, ROS are produced as by-products of oxidative metabolism and maintained at a low level useful for cell signaling purposes (Droge, 2002; Morgan and Liu, 2011). Extensive stress conditions can lead to dysregulation of ROS homeostasis resulting in wide-spread damage to cell constituents, with prolonged exposure in such situations often leading to cell death or neoplastic transformation (Chua et al., 2009). A large scale accumulation of ROS products can promote tissue degenerative diseases or cancer (Azad et al., 2009).

It is likely that the oxidative environment in cells arising from exposure to NPs could trigger the autophagic process (Moore, 2008; Huang and Shen, 2009), rather than the presence of NPs inciting a direct cellular response. Other investigators including Funnell and Maysinger (2006) have suggested that clusters of intracellular metal-based NPs may only induce localized areas of oxidative stress around the NPs, thereby limiting the extent of cell damage. Li et al. (2010) noted that AuNPs generated significantly more lipid hydroperoxides and malondialdehyde modified protein adducts (a by-product of lipid oxidation), concurrent with ultrastructural and biochemical evidence of autophagy in lung fibroblasts. The authors concluded that AuNPs-induced autophagy is mediated by ROS (Figure 2).

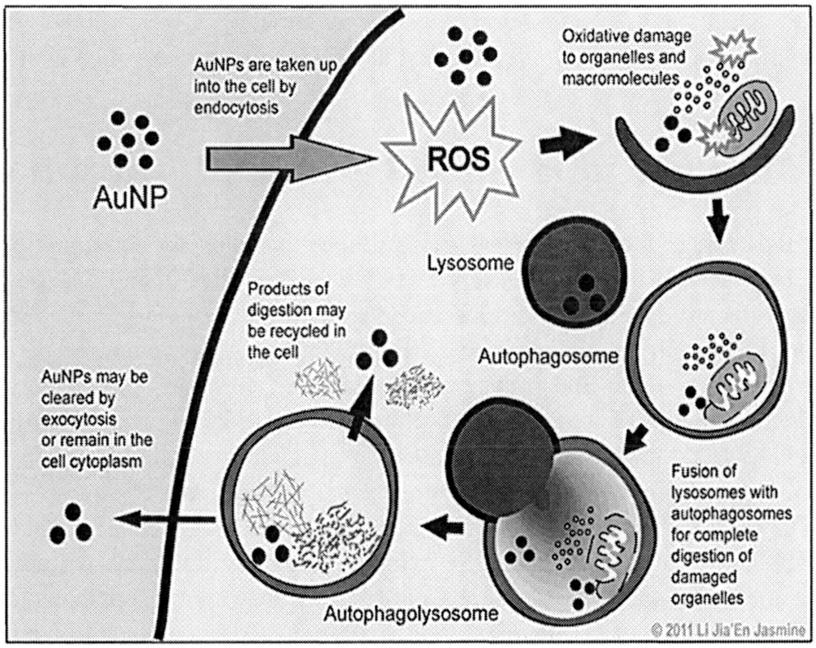

Figure 2. Induction of autophagy after AuNPs treatment in lung fibroblasts. Presence of AuNPs in the cytoplasm generate ROS which in turn causes oxidative damage to cell constituents, inducing the autophagic process (see Li et al., 2010). AuNPs may be subsequently engulfed by autophagosomes and along with the products of lysosomal degradation are either excreted or retained in the cell.

The lung is especially susceptible to oxygen toxicity as there is contact with oxygen in atmospheric air all the time. Several groups of researchers have reported increased expression of cytokines and other pro-inflammatory responses in the lung upon exposure to NPs. For instance, exposure to TiO_2NPs increases production of neutrophil, attracting the chemokines CXCL1, tumor necrosis factor-α (TNF-α) and IL-8, thereby causing pulmonary neutrophilia (Gonclaves et al., 2009; Rossi et al., 2009) as well as recruitment of eosinophils and lymphocytes (Larsen et al., 2009). Studies in rodents exposed to multi-walled carbon nanotubes (MWCNT) by inhalation, have shown localization of MWCNT in the lung macrophages and the subpleural surface of the lung (Ryman-Rasmussen et al., 2009) while interstitial injection of single-walled carbon nanotubes (SWCNT) is able to induce formation of granulomas (Lam et al., 2003). Inflammatory phagocytes, neutrophils and macrophages are ROS generators which can activate a series of cytokine cascades, including up-regulation of interleukins, kinases and tumor necrosis factor (TNF-α) pro-inflammatory signaling processes as a reaction to oxidative stress (Fujii et al., 2001). Although the lung possesses endogenous anti-oxidant defenses (Ryter and Choi, 2010), there are times when the first tier of defenses are unable to restore cell homeostasis. Hence, it is possible that when lung cells enter into a stress situation, there would be induction of autophagy to arrest and contain the damage inflicted by ROS (Moore, 2008). On another note, NPs may also be able to induce autophagy through ROS-independent pathways, suggesting that not all NP-induced autophagy processes are mediated by oxidative stress. Johnson-Lyles et al. (2010) observed the presence of cytoskeleton disruption concomitant with autophagic vacuolation, but no evidence of appreciable oxidative stress when renal proximal tubule cells were exposed to fullerenol. Since actin is known to be involved in mammalian autophagy, it was proposed that cytoskeleton disruption can lead to initiation of the autophagic process (Lee et al., 2010).

5. Implications for NP Induced Autophagy

As a fundamental process, autophagy has specific roles in determining the normal functioning and survival in a wide variety of tissues/cells in health and disease (Beau et al., 2011). When cells are exposed to various stress conditions, autophagy is also induced as an adaptive response which supply required nutrients, through degradation of parts of the cell components to meet cellular and organismal demands (Takeshige et al., 1992; Jentsch and Ulrich, 1998, Levine and Kroemer, 2008; Rabinowitz et al., 2010) such as during starvation (Kuma et al., 2004; Lum et al., 2005). Besides cell survival, autophagy has also been recognized as type II programmed cell death mechanism in times of stress (Tsujimoto and Shimizu, 2005; Kiffin et al., 2006). The mushrooming of the nanotechnology industry has led to the widespread use of NPs such as zinc oxide and titanium dioxide in common consumer products that include cosmetics, sunscreens and coatings for vitamin supplements (Yu and Li, 2011). However, exposure to nanomaterials may induce oxidative stress culminating in autophagic cell death and injury to tissues. Starburst polyamidoamine dendrimer, a class of nanomaterials has been observed to induce autophagic cell death and lung injury *in vivo* by deregulating the Akt-TSC2-mTOR signaling pathway (Li et al., 2009). Yet, cellular organelles, proteins or lipids damaged by ROS may be efficiently cleared by internalization and subsequent digestion in autophagosomes (Kiffin et al., 2006), thereby rescuing the cell from apoptosis. Stern and Johnson (2008) have also hypothesized that pollution particulates such as nanoscale carbon black and fullerene could enter the brain via the olfactory epithelium and affect susceptibility to neuropathological disorders, such as Parkinson's Disease and Alzheimer's Disease which are known to be linked to autophagy (Anglade et al., 1997; Nixon et al., 2005). The observation of nanomaterial-induced autophagy has provided new insights into the design of nanomaterial-based biotherapeutics. Wei et al. (2010) reported that autophagy induced by fullerene C60 Nano nC60/nC60(Nd) sensitized doxorubicin-killing of HeLa cells and MCF-7 doxorubicin resistant breast cancer cells. They speculated that the C60 nano nC60/nC60(Nd) engulfed by the autophagosomes in the cells could not be degraded, leading to blockage of the autophagic pathway with resultant accumulation of autolysosomes. Their work suggests that modulation of autophagy could potentiate chemosensitivity and reduce chemoresistance, thereby, providing novel strategies in cancer therapy (Fleming et al., 2011). Very recently, Wu et al., (2011) have demonstrated that NPs with an iron core and gold shell exerted differential cytotoxicity mediated via mitochondria-mediated autophagy in oral cancerous cells *in vitro*, but sparing normal oral human keratinocytes.

Conclusion

In summary, nanomaterial-induced autophagy is a complex process which involves regulation and coordination with other pathways such as apoptosis, necrosis and the mTOR pathway. It is also evident that autophagy plays a crucial role in cell homeostasis and remodelling. However, there is evidence that NPs may also inhibit rather than promote autophagy. Cerium oxide NPs have been observed to attenuate ER stress and consequently glucose-induced autophagy in cardiomyocytes (Younce et al., 2010). The subject of autophagic cell death and survival induced by nanomaterials is still an under-developed field where more work needs to be done to unravel the significance of this phenomenon. The burgeoning field of nanotechnology has given rise to the enhanced use of innovative devices and novel applications (Thomas et al., 2009). This in turn has lead to the scrutiny of possible cellular toxicity and genotoxicity associated with the use of nanomaterials which may have potential negative and harmful effects on human health (Krug and Wick, 2011). It is increasingly recognized that there should be a balance between the further development of nanotechnology and the need to conduct research to identify potential hazards and perform risk assessments (Oberdörster et al., 2005). Krug and Wick (2011) has very recently advocated the necessity to improve the quality of nanotoxicity studies by conducting experiments under realistic test conditions and using reliable experimental protocols. For the risks and hazard assessments of nanomaterials to be meaningfully assessed, it would be essential to engage the expertise of the exposure scientists, material scientists, biomolecular engineers, toxicologists and cell biologists with input from relevant health professionals in an integrated approach (Thomas et al., 2009).

Acknowledgments

The authors thank Ms Song-Lin Bay for technical assistance. Research funding was from the Singapore Ministry of Education Academic Research Fund Tier 2 Grant MOE2008-T2-1-046, and the National University of Singapore (NUS) Environmental Research Institute (NERI) Grant R706-000-002-646.

References

Afeseh Ngwa H, Kanthasamy A, Gu Y, Fang N, Anantharam V, Kanthasamy AG. Manganese nanoparticle activates mitochondrial dependent apoptotic signaling and autophagy in dopaminergic neuronal cells. *Toxicol Appl Pharmacol.* 2011;256(3):227-40.

Anglade P, Vyas S, Javoy-Agid F, Herrero MT, Michel PP, Marquez J, Mouatt-Prigent A, Ruberg M, Hirsch EC, Agid Y. Apoptosis and autophagy in nigral neurons of patients with Parkinson's disease. *Histol Histopathol.* 1997;12(3):25-31.

Azad MB, Chen Y, Gibson SB. Regulation of autophagy by reactive oxygen species (ROS): implications for cancer progression and treatment. *Antioxid Redox Signal.* 2009;11(4):777-90.

Baehrecke EH (2005). Autophagy: dual roles in life and death? *Nat Rev Mol Cell Biol.* 2005;6(6):505-10.

Beau I, Mehrpour M, Codogno P. Autophagosomes and human diseases. *Int J Biochem Cell Biol.* 2011;43(4):460-4.

Cerutti PA. Oxy-radicals and cancer. *Lancet.* 1994;344(8926):862-3

Chen Y, Yang L, Feng C, Wen LP. Nano neodymium oxide induces massive vacuolization and autophagic cell death in non-small cell lung cancer NCI-H460 cells. 2005;337(1):52-60.

Chen Y, McMillan-Ward E, Kong J, Israels SJ, Gibson SB. Oxidative stress induces autophagic cell death independent of apoptosis in transformed and cancer cells. *Cell Death Differ.* 2008;15(1):171-82.

Chua PJ, Yip GW, Bay BH. Cell cycle arrest induced by hydrogen peroxide is associated with modulation of oxidative stress related genes in breast cancer cells. *Exp Biol Med* (Maywood). 2009; 234(9):1086-94.

de Duve C, Wattiaux R. Functions of lysosomes. *Annu Rev Physiol.* 1966; 28: 435-92.

Dröge W. Free radicals in the physiological control of cell function. *Physiol Rev.* 2002; 82(1):47-95.

Ebato C, Uchida T, Arakawa M, Komatsu M, Ueno T, Komiya K, Azuma K, Hirose T, Tanaka K, Kominami E, Kawamori R, Fujitani Y, Watada H. Autophagy is important in islet homeostasis and compensatory increase of beta cell mass in response to high-fat diet. *Cell Metab.* 2008;8(4):325-32.

Eskelinen EL. Roles of LAMP-1 and LAMP-2 in lysosome biogenesis and autophagy. *Mol Aspects Med.* 2006;27(5-6):495-502.

Filimonenko M, Isakson P, Finley KD, Anderson M, Jeong H, Melia TJ, Bartlett BJ, Myers KM, Birkeland HC, Lamark T, Krainc D, Brech A, Stenmark H, Simonsen A, Yamamoto A. The selective macroautophagic degradation of aggregated proteins requires the PI3P-binding protein Alfy. *Mol Cell.* 2010;38(2):265-79.

Fleming A, Noda T, Yoshimori T, Rubinsztein DC. Chemical modulators of autophagy as biological probes and potential therapeutics. *Nat Chem Biol.* 2011;7(1):9-17.

Fujii T, Hayashi S, Hogg JC, Vincent R, Van Eeden SF. Particulate matter induces cytokine expression in human bronchial epithelial cells. *Am J Respir Cell Mol Biol.* 2001;25(3):265-71.

Funnell WR, Maysinger D. Three-dimensional reconstruction of cell nuclei, internalized quantum dots and sites of lipid peroxidation. *J Nanobiotechnology.* 2006;4:10.

Galluzzi L, Morselli E, Vicencio JM, Kepp O, Joza N, Tajeddine N, Kroemer G. Life, death and burial: multifaceted impact of autophagy. *Biochem Soc Trans.* 2008;36(Pt 5):786-90.

Geng J, Klionsky DJ. The Atg8 and Atg12 ubiquitin-like conjugation systems in macroautophagy. 'Protein modifications: beyond the usual suspects' review series. *EMBO Rep.* 2008;9(9):859-64.

Gonçalves DM, Chiasson S, Girard D. Activation of human neutrophils by titanium dioxide (TiO2) nanoparticles. *Toxicol In Vitro.* 2010;24(3):1002-8.

Green DR, Galluzzi L, Kroemer G. Mitochondria and the autophagy-inflammation-cell death axis in organismal aging. *Science.* 2011;333(6046):1109-12.

He C, Klionsky DJ. Regulation mechanisms and signaling pathways of autophagy. *Annu Rev Genet.* 2009;43:67-93.

Huang Q, Shen HM. To die or to live: the dual role of poly(ADP-ribose) polymerase-1 in autophagy and necrosis under oxidative stress and DNA damage. *Autophagy.* 2009;5(2):273-6.

Jäger S, Bucci C, Tanida I, Ueno T, Kominami E, Saftig P, Eskelinen EL. Role for Rab7 in maturation of late autophagic vacuoles. *J Cell Sci.* 2004;117(Pt 20):4837-48.

Jentsch S, Ulrich HD. Protein breakdown. Ubiquitous déjà vu. *Nature.* 1998;395(6700):321-23.

Johnson-Lyles DN, Peifley K, Lockett S, Neun BW, Hansen M, Clogston J, Stern ST, McNeil SE. Fullerenol cytotoxicity in kidney cells is associated with cytoskeleton disruption, autophagic vacuole accumulation, and mitochondrial dysfunction. *Toxicol Appl Pharmacol.* 2010;248(3):249-58.

Kaushik S, Cuervo AM. Autophagy as a cell-repair mechanism: activation of chaperone-mediated autophagy during oxidative stress. *Mol Aspects Med.* 2006;27(5-6):444-54.

Kiffin R, Bandyopadhyay U, Cuervo AM. Oxidative stress and autophagy. *Antioxid Redox Signal.* 2006;8(1-2):152-62.

Kim RH, Bold RJ, Kung HJ. ADI, autophagy and apoptosis: metabolic stress as a therapeutic option for prostate cancer. *Autophagy.* 2009;5(4):567-8.

Kirkin V, Lamark T, Sou YS, Bjørkøy G, Nunn JL, Bruun JA, Shvets E, McEwan DG, Clausen TH, Wild P, Bilusic I, Theurillat JP, Øvervatn A, Ishii T, Elazar Z, Komatsu M, Dikic I, Johansen T. A role for NBR1 in autophagosomal degradation of ubiquitinated substrates. *Mol Cell.* 2009;33(4):505-16.

Klionsky DJ, Cregg JM, Dunn WA Jr, Emr SD, Sakai Y, Sandoval IV, Sibirny A, Subramani S, Thumm M, Veenhuis M, Ohsumi Y. A unified nomenclature for yeast autophagy-related genes. *Dev Cell.* 2003;5(4):539-45.

Klionsky DJ. The molecular machinery of autophagy: unanswered questions. *J Cell Sci.* 2005;118(Pt 1):7-18.

Krug HF, Wick P. Nanotoxicology: an interdisciplinary challenge. *Angew Chem Int Ed Engl.* 2011;50(6):1260-78.

Kuma A, Hatano M, Matsui M, Yamamoto A, Nakaya H, Yoshimori T, Ohsumi Y, Tokuhisa T, Mizushima N. The role of autophagy during the early neonatal starvation period. *Nature.* 2004;432(7020):1032-6.

Lam CW, James JT, McCluskey R, Hunter RL. Pulmonary toxicity of single-wall carbon nanotubes in mice 7 and 90 days after intratracheal instillation. *Toxicol Sci.* 2004;77(1):126-34.

Larsen ST, Roursgaard M, Jensen KA, Nielsen GD. Nano titanium dioxide particles promote allergic sensitization and lung inflammation in mice. *Basic Clin Pharmacol Toxicol.* 2010;106(2):114-7.

Lee JY, Koga H, Kawaguchi Y, Tang W, Wong E, Gao YS, Pandey UB, Kaushik S, Tresse E, Lu J, Taylor JP, Cuervo AM, Yao TP. HDAC6 controls autophagosome maturation essential for ubiquitin-selective quality-control autophagy. *EMBO J.* 2010;29(5):969-80.

Levine B, Yuan J. Autophagy in cell death: an innocent convict? *J Clin Invest.* 2005;115(10):2679-88.

Levine B, Kroemer G. Autophagy in the pathogenesis of disease. *Cell.* 2008;132(1):27-42.

Li C, Liu H, Sun Y, Wang H, Guo F, Rao S, Deng J, Zhang Y, Miao Y, Guo C, Meng J, Chen X, Li L, Li D, Xu H, Wang H, Li B, Jiang C. PAMAM nanoparticles promote acute lung injury by inducing autophagic cell death through the Akt-TSC2-mTOR signaling pathway. *J Mol Cell Biol.* 2009;1(1):37-45.

Li JJ, Hartono D, Ong CN, Bay BH, Yung LY. Autophagy and oxidative stress associated with gold nanoparticles. *Biomaterials.* 2010;31(23):5996-6003.

Li W, Yang Q, Mao Z. Chaperone-mediated autophagy: machinery, regulation and biological consequences. *Cell Mol Life Sci.* 2011;68(5):749-63.

Liang C, Lee JS, Inn KS, Gack MU, Li Q, Roberts EA, Vergne I, Deretic V, Feng P, Akazawa C, Jung JU. Beclin1-binding UVRAG targets the class C Vps complex to coordinate autophagosome maturation and endocytic trafficking. *Nat Cell Biol.* 2008;10(7):776-87.

Liu B, Cheng Y, Liu Q, Bao JK, Yang JM. Autophagic pathways as new targets for cancer drug development. *Acta Pharmacol Sin.* 2010;31(9):1154-64.

Lum JJ, DeBerardinis RJ, Thompson CB. Autophagy in metazoans: cell survival in the land of plenty. *Nat Rev Mol Cell Biol.* 2005;6(6):439-48.

Maiuri MC, Zalckvar E, Kimchi A, Kroemer G. Self-eating and self-killing: crosstalk between autophagy and apoptosis. *Nat Rev Mol Cell Biol.* 2007;8(9):741-52.

Martinez-Vicente M, Cuervo AM. Autophagy and neurodegeneration: when the cleaning crew goes on strike. *Lancet Neurol.* 2007;6(4):352-61.

Mehrpour M, Esclatine A, Beau I, Codogno P. Overview of macroautophagy regulation in mammalian cells. *Cell Res.* 2010;20(7):748-62.

Moore MN. Autophagy as a second level protective process in conferring resistance to environmentally-induced oxidative stress. *Autophagy.* 2008;4(2):254-6.

Morgan MJ, Liu ZG. Crosstalk of reactive oxygen species and NF-κB signaling. *Cell Res.* 2011;21(1):103-15.

Mortensen M, Soilleux EJ, Djordjevic G, Tripp R, Lutteropp M, Sadighi-Akha E, Stranks AJ, Glanville J, Knight S, Jacobsen SE, Kranc KR, Simon AK. The autophagy protein Atg7 is essential for hematopoietic stem cell maintenance. *J Exp Med.* 2011;208(3):455-67.

Nair U, Klionsky DJ. Molecular mechanisms and regulation of specific and nonspecific autophagy pathways in yeast. *J Biol Chem.* 2005;280(51):41785-8.

Nixon RA, Wegiel J, Kumar A, Yu WH, Peterhoff C, Cataldo A, Cuervo AM. Extensive involvement of autophagy in Alzheimer disease: an immuno-electron microscopy study. *J Neuropathol Exp Neurol.* 2005;64(2):113-22.

Noda T, Fujita N, Yoshimori T. The late stages of autophagy: how does the end begin? *Cell Death Differ.* 2009;16(7):984-90.

Oberdörster G, Oberdörster E, Oberdörster J. Nanotoxicology: an emerging discipline evolving from studies of ultrafine particles. *Environ Health Perspect.* 2005;113(7):823-39.

Rabinowitz JD, White E. Autophagy and metabolism. *Science.* 2010;330(6009):1344-8.

Rossi EM, Pylkkänen L, Koivisto AJ, Vippola M, Jensen KA, Miettinen M, Sirola K, Nykäsenoja H, Karisola P, Stjernvall T, Vanhala E, Kiilunen M, Pasanen P, Mäkinen M, Hämeri K, Joutsensaari J, Tuomi T, Jokiniemi J, Wolff H, Savolainen K, Matikainen S, Alenius H. Airway exposure to silica-coated TiO2 nanoparticles induces pulmonary neutrophilia in mice. *Toxicol Sci.* 2010;113(2):422-33.

Roy S, Debnath J. Autophagy and tumorigenesis. Semin Immunopathol. 2010;32(4):383-96.

Ryman-Rasmussen JP, Cesta MF, Brody AR, Shipley-Phillips JK, Everitt JI, Tewksbury EW, Moss OR, Wong BA, Dodd DE, Andersen ME, Bonner JC. Inhaled carbon nanotubes reach the subpleural tissue in mice. *Nat Nanotechnol.* 2009;4(11):747-51.

Ryter SW, Choi AM. Autophagy in the lung. *Proc Am Thorac Soc.* 2010;7(1):13-21.

Scarlatti F, Granata R, Meijer AJ, Codogno P. Does autophagy have a license to kill mammalian cells? *Cell Death Differ.* 2009;16(1):12-20.

Scherz-Shouval R, Elazar Z. Regulation of autophagy by ROS: physiology and pathology. *Trends Biochem Sci.* 2011;36(1):30-8.

Schweichel JU, Merker HJ. The morphology of various types of cell death in prenatal tissues. *Teratology.* 1973;7(3):253-66.

Seleverstov O, Zabirnyk O, Zscharnack M, Bulavina L, Nowicki M, Heinrich JM, Yezhelyev M, Emmrich F, O'Regan R, Bader A. Quantum dots for human mesenchymal stem cells labeling. A size-dependent autophagy activation. *Nano Lett.* 2006;6(12):2826-32.

Stern ST, Johnson DN. Role for nanomaterial-autophagy interaction in neurodegenerative disease. *Autophagy.* 2008;4(8):1097-100.

Stern ST, Zolnik BS, McLeland CB, Clogston J, Zheng J, McNeil SE. Induction of autophagy in porcine kidney cells by quantum dots: a common cellular response to nanomaterials? *Toxicol Sci.* 2008;106(1):140-52.

Takeshige K, Baba M, Tsuboi S, Noda T, Ohsumi Y. Autophagy in yeast demonstrated with proteinase-deficient mutants and conditions for its induction. *J Cell Biol.* 1992;119(2):301-11.

Tanida I. Autophagosome formation and molecular mechanism of autophagy. *Antioxid Redox Signal.* 2011;14(11):2201-14.

Tanida I, Tanida-Miyake E, Ueno T, Kominami E. The human homolog of Saccharomyces cerevisiae Apg7p is a Protein-activating enzyme for multiple substrates including human Apg12p, GATE-16, GABARAP, and MAP-LC3. *J Biol Chem.* 2001;276(3):1701-6.

Tian S, Siu FM, Kui SC, Lok CN, Che CM. Anticancer gold(I)-phosphine complexes as potent autophagy-inducing agents. *Chem Commun* (Camb). 2011;47(33):9318-20.

Thomas T, Bahadori T, Savage N, Thomas K. Moving toward exposure and risk evaluation of nanomaterials: challenges and future directions. *Wiley Interdiscip Rev Nanomed Nanobiotechnol.* 2009;1(4):426-33.

Tsujimoto Y, Shimizu S. Another way to die: autophagic programmed cell death. *Cell Death Differ.* 2005;12 Suppl 2:1528-34.

Wei P, Zhang L, Lu Y, Man N, Wen L. C60(Nd) nanoparticles enhance chemotherapeutic susceptibility of cancer cells by modulation of autophagy. *Nanotechnology.* 2010;21(49):495101.

Wu YN, Yang LX, Shi XY, Li IC, Biazik JM, Ratinac KR, Chen DH, Thordarson P, Shieh DB, Braet F. The selective growth inhibition of oral cancer by iron core-gold shell nanoparticles through mitochondria-mediated autophagy. *Biomaterials.* 2011; 32 (20): 4565-73.

Yamawaki H, Iwai N. Cytotoxicity of water-soluble fullerene in vascular endothelial cells. *Am J Physiol Cell Physiol.* 2006;290(6):C1495-502.

Yang Z, Klionsky DJ. An overview of the molecular mechanism of autophagy. *Curr Top Microbiol Immunol.* 2009;335:1-32.

Yang Z, Klionsky DJ. Eaten alive: a history of macroautophagy. *Nat Cell Biol.* 2010;12(9):814-22.

Younce CW, Wang K, Kolattukudy PE. Hyperglycaemia-induced cardiomyocyte death is mediated via MCP-1 production and induction of a novel zinc-finger protein MCPIP. *Cardiovasc Res.* 2010;87(4):665-74.

Yu JX, Li TH. Distinct biological effects of different nanoparticles commonly used in cosmetics and medicine coatings. *Cell Biosci.* 2011;1(1):19.

Yu L, Lu Y, Man N, Yu SH, Wen LP. Rare earth oxide nanocrystals induce autophagy in *HeLa cells. Small.* 2009;5(24):2784-7.

Zabirnyk O, Yezhelyev M, Seleverstov O. Nanoparticles as a novel class of autophagy activators. *Autophagy.* 2007;3(3):278-81.

Index

A

B

C

D

E

F

G

I

M

N

O

P

Q

R

S

T

U

V

W

X

Y

Z